Mathematical Physics Studies

More information about this series at http://www.springer.com/series/6316

Daisuke Fujiwara

Rigorous Time Slicing Approach to Feynman Path Integrals

Springer

Daisuke Fujiwara
Department of Mathematics
Gakushuin University
Tokyo
Japan

ISSN 0921-3767 ISSN 2352-3905 (electronic)
Mathematical Physics Studies
ISBN 978-4-431-56551-2 ISBN 978-4-431-56553-6 (eBook)
DOI 10.1007/978-4-431-56553-6

Library of Congress Control Number: 2017939312

Printed on acid-free paper

This Springer imprint is published by Springer Nature
The registered company is Springer Japan KK
The registered company address is: Chiyoda First Bldg. East, 3-8-1 Nishi-Kanda, Chiyoda-ku, Tokyo 101-0065, Japan

Preface

The Feynman path integral is a method of quantization using the Lagrangian function, while Schrödinger's quantization uses the Hamiltonian function. Since it provides a different view point from Schrödinger's, it is a very useful basic tool in quantum physics. These two methods are believed to be equivalent. But equivalence is not fully proved mathematically, because, compared with Schrödinger's method, there is still much to be done concerning rigorous mathematical treatment of Feynman's method. The difficulty lies in the fact that the Feynman path integral is not an integral by means of a countably additive measure.

Feynman himself defined a path integral as the limit of a sequence of integrals over finite-dimensional spaces. To construct this approximating sequence he divided the time interval into small pieces. This method is called the time slicing approximation method or sequential method.

The aim of this book is to carry out rigorously the time slicing approximation method under reasonable assumptions and to prove that Feynman's original definition of a path integral actually gives the fundamental solution of the Schrödinger equation. By the way, it is proved that the obtained fundamental solution has a number of interesting properties which are hard to prove via Schrödinger's method. The semiclassical asymptotic behavior of the fundamental solution is one of them.

A prerequisite of this book is standard knowledge of functional analysis. Mathematical techniques required here are explained from scratch because they are considerably different from techniques usually used in treating the Schrödinger equation. These tools used in this book are collected in Part II, which occupies a large part of the book.

Although there are a large number of publications on the Feynman path integrals, only some of approaches other than the time slicing approximation method are mentioned in Sect. 1.3. It does not give thorough knowledge on up to date state of researches of the Feynman path integral. The author invites readers to check other literature too in this respect, see for instance [9, 48, 50] and references therein.

The book was first published in Japanese in 1999. The present book keeps the main part of it but includes some results obtained after the Japanese edition was published.

The Japanese edition was based on lectures given at Gakushuin University, the University of Tokyo, Hokkaido University and the University of Toronto. The author would like to thank the students in these lectures. He is especially grateful to his colleague Kenji Yajima who encouraged him to write the Japanese edition of the book, which makes the basis of the present book.

Tokyo, Japan Daisuke Fujiwara

Contents

Part I
Convergence of Time Slicing Approximation of Feynman Path Integrals

Chapter 1
Feynman's Idea

Abstract Before going to mathematical discussions, we rapidly explain, for convenience of readers, the notion of Feynman path integrals following Feynman without mathematical rigor. Afterward, some examples are given.

1.1 Feynman Path Integrals

1.1.1 Two Methods of Quantization

In quantum mechanics, the state of a particle in the Euclidean space $\mathbf{R}^d$ is described by an element φ of the complex Hilbert space $L^2(\mathbf{R}^d)$ with unit norm, cf. [16] or [64]. The vector φ is represented by a complex-valued function $\varphi(x)$, called a wave function, with the property

$$\|\varphi\|^2 = \int_{\mathbf{R}^d} |\varphi(x)|^2 dx = 1.$$

The integral of $|\varphi(x)|^2$ over a domain G in $\mathbf{R}^d$

$$\int_G |\varphi(x)|^2 dx$$

gives the probability for the particle to be found in G.

If a particle moves, then the state φ_t of the particle changes as time t changes, i.e. motion of the particle is a one-parameter family $\{\varphi_t\}_{t \in \mathbf{R}}$ of vectors parameterized by time t. The family is represented by a function $\varphi(t, x)$ of $(t, x) \in \mathbf{R}^{1+d}$ with the property

$$\int_{\mathbf{R}^d} |\varphi(t, x)|^2 dx = 1.$$

If the motion occurs under the influence of a given force, e.g. force with the potential $V(x)$, then for any $t, s \in \mathbf{R}$ with $s < t$ the vector φ_t at t is determined by

D. Fujiwara, *Rigorous Time Slicing Approach to Feynman Path Integrals*,
Mathematical Physics Studies, DOI 10.1007/978-4-431-56553-6_1

the vector φ_s of previous time s. The map $U(t,s) : L^2(\mathbf{R}^d) \ni \varphi_s \to \varphi_t \in L^2(\mathbf{R}^d)$ is a linear unitary operator, which is called the evolution operator. Since $U(t,s)$ is a linear operator, it is represented, at least formally, as an integral transformation:

$$\varphi(t,x) = \int_{\mathbf{R}^d} K(t,s,x,y)\varphi(s,y)dy. \tag{1.1.1}$$

The function $K(t,s,x,y)$ is called the propagator or the fundamental solution of the Schrödinger equation.

Quantization is the process to determine the evolution operator $U(t,s)$ or equivalently the propagator from the potential $V(x)$. There exist two ways of quantization. One is Schrödinger's method and the other is Feynman's method. Schrödinger's method is to obtain $U(t,s)$ and Feynman's method is to obtain the propagator. The latter is based on a notion of a new integral, which is called the Feynman path integral.

We will explain Feynman's idea in Sect. 1.1.4 after some preparations in Sects. 1.1.2 and 1.1.3.

From now on we restrict ourselves to the case that the dimension of the configuration space is one, i.e. $d = 1$, for the sake of simplicity. Generalization to the case of several dimensions is obvious.

1.1.2 Classical Mechanics and Action Functional

We recall some concepts in classical mechanics which are used in the later sections. For simplicity we suppose the configuration space is the 1-dimensional euclidean space $\mathbf{R}$. We discuss the case that the motion of a particle of mass m occurs in the potential field $V(x)$.

Hamiltonian formalism

In classical mechanics the motion of a particle of mass m in the Euclidean space $\mathbf{R}$ is described by a curve $(p(t), q(t))$ in the phase space $\mathbf{R}^2$, where $q(t)$ is the position of the particle at time t and $p(t)$ is the momentum at t. $(p(t), q(t))$ is the solution of the Hamiltonian form of the equation of motion, cf. for example [65]:

$$\begin{aligned} \frac{d}{dt}q(t) &= \partial_p H(p,q), \\ \frac{d}{dt}p(t) &= -\partial_q H(p,q), \end{aligned} \tag{1.1.2}$$

where ∂_p and ∂_q are abbreviations of $\dfrac{\partial}{\partial p}$ and $\dfrac{\partial}{\partial q}$ respectively, and $H(p,q)$ is the Hamiltonian function

$$H(p,q) = \frac{1}{2m}p^2 + V(q). \tag{1.1.3}$$

Lagrangian formalism

In classical mechanics the motion of a particle is also described by the Lagrangian function

$$L(\dot{x}, x) = \frac{m}{2}\dot{x}^2 - V(x). \tag{1.1.4}$$

Here x is the position of the particle and $\dot{x}$ is the velocity. The Lagrangian function $L(\dot{x}, x)$ is a function of $(\dot{x}, x) \in \mathbf{R}^2$.

Let $[a, b]$ be a time interval and γ be a path $\gamma : [a, b] \ni t \to \gamma(t) \in \mathbf{R}$. To any path γ we define its action $S(\gamma)$ by

$$S(\gamma) = \int_a^b L(\dot{\gamma}(t), \gamma(t))dt. \tag{1.1.5}$$

$S(\gamma)$ is a functional of γ. Let x, y be arbitrary points of $\mathbf{R}^d$. Let Ω be the set of all paths $\gamma : [a, b] \to \mathbf{R}^d$ such that

$$\gamma(a) = y, \quad \gamma(b) = x.$$

Although Ω contains a huge number of paths, Hamilton's principle of classical mechanics (cf. for example, [65]) states that the actual trajectory γ_0 that is realized under Newton's law of motion is the solution of the variational problem:

$$\delta S(\gamma_0) = 0, \quad \gamma_0(a) = y, \quad \gamma_0(b) = x. \tag{1.1.6}$$

We call a solution of the variational problem (1.1.6) a classical path. In other words, a classical path γ_0 is the critical or stationary point of $S(\gamma)$. We define the classical action by

$$S(b, a, x, y) = \min_{\gamma \in \Omega_{x,y}} S(\gamma) \tag{1.1.7}$$

if the right-hand side exists. The path that attains the minimum is a classical path.

If a path $\gamma(t)$ is a solution of Euler's equation and boundary conditions:

$$m\frac{d^2}{dt^2}\gamma(t) = -\partial_x V(\gamma), \qquad \gamma(a) = y \quad \gamma(b) = x, \tag{1.1.8}$$

then γ is a classical path.

1.1.3 Schrödinger's Quantization—the Schrödinger Equation

Now we sketch Schrödinger's quantization, cf. for eg. [64] and [16].

Let $H(p, q)$ be the Hamiltonian function (1.1.3). Then we make an operator H in $L^2(\mathbf{R})$ by replacing the variable q with the multiplication operator $x\cdot$ and p with the partial differential operator $i^{-1}\hbar\partial_x$ in $H(p, q)$. Here $i = \sqrt{-1}$ and $\hbar$ is a very small positive constant, which plays an important role in quantum mechanics.[1] The operator H is called the Hamiltonian operator, or Hamiltonian for short, which is the partial differential operator:

$$H = \frac{1}{2m}\Big(\frac{\hbar}{i}\partial_x\Big)^2 + V(x). \tag{1.1.9}$$

The one parameter family of vectors φ_t satisfies:

$$-\frac{\hbar}{i}\frac{d}{dt}\varphi_t = H\varphi_t.$$

This means that the wave function $\varphi(t, x)$ is the solution of the partial differential equation, the Schrödinger equation,

$$-\frac{\hbar}{i}\partial_t\varphi(t, x) = \frac{1}{2m}\Big(\frac{\hbar}{i}\partial_x\Big)^2\varphi(t, x) + V(x)\varphi(t, x). \tag{1.1.10}$$

If initial condition $\varphi(s, x)$ is given, $\varphi(t, x)$ is determined uniquely. The correspondence $\varphi(s, x) \to \varphi(t, x)$ is the evolution operator $U(t, s)$. If the potential $V(x)$ is good, then H is a self-adjoint operator and

$$U(t, s) = \mathrm{e}^{-\frac{i}{\hbar}(t-s)H} = \mathrm{e}^{-i\nu(t-s)H}. \tag{1.1.11}$$

Here and hereafter we write ν for $\hbar^{-1}$, cf. for example [43, 56] or [55].

1.1.4 Feynman's Quantization—the Feynman Path Integral

While Schrödinger's quantization uses the Hamiltonian function to construct the evolution operator $U(t, s)$, Feynman's quantization introduced by [21] is a method to construct the propagator $K(t, s, x, y)$ using the Lagrangian function of classical mechanics.

Let $[a, b]$ be an interval of time. Let $\Omega_{x,y}$ be the space of all paths starting from the point y at time a and arriving at point x at time b. Owing to Feynman [21, 22], the propagator $K(b, a, x, y)$ is expressed by the following formula:

$$K(b, a, x, y) = \frac{1}{N}\sum_{\gamma\in\Omega_{x,y}} \exp(i\nu S(\gamma)). \tag{1.1.12}$$

[1] $\hbar = h/2\pi$, where h is Planck's constant.

Here $\nu = \hbar^{-1}$, the symbol $\sum_{\gamma \in \Omega_{x,y}}$ means to take the sum over all paths $\gamma \in \Omega_{x,y}$ and N is the normalizing factor. Since $\Omega_{x,y}$ is a continuum, it is not strange to replace the symbol $\sum_{\gamma \in \Omega_{x,y}}$ by symbol of integration $\int_{\Omega_{x,y}} \mathscr{D}\gamma$, i.e.

$$K(b, a, x, y) = \int_{\Omega_{x,y}} \exp i\nu S(\gamma)\mathscr{D}. \tag{1.1.13}$$

The right-hand side is called the Feynman path integral.

The formula (1.1.12) or (1.1.13) is quite formal. Feynman himself gave a rigorous definition of it in [21] as follows: Let $[a, b]$ be an interval of time and Δ be an arbitrary division of $[a, b]$,

$$\Delta : a = T_0 < T_1 < \cdots < T_J < T_{J+1} = b. \tag{1.1.14}$$

Let $\tau_j = T_j - T_{j-1}, (j = 1, 2, \ldots, J + 1)$ and $|\Delta| = max_{1 \le j \le J+1} \tau_j$. We call $|\Delta|$ the size of the division. For $j = 1, 2, \ldots, J$, choose an arbitrary point $x_j \in \mathbf{R}$ and set $x_0 = y$ and $x_{J+1} = x$. Then we have thus $J + 2$ points $\{(T_j, x_j)\}$ in time-space $\mathbf{R} \times \mathbf{R}$. For each pair (T_{j-1}, x_{j-1}) and (T_j, x_j) in time-space $\mathbf{R}^2$, we write $S(T_j, T_{j-1}, x_j, x_{j-1})$ for the classical action function in (1.1.7), i.e.

$$S(T_j, T_{j-1}, x_j, x_{j-1}) = \min_{\gamma \in \Omega_{x_j, x_{j-1}}} S(\gamma), \tag{1.1.15}$$

where $\min_{\gamma \in \Omega_{x_j, x_{j-1}}}$ means the minimum over all paths γ starting from the point x_{j-1} at time T_{j-1} and reaching the point x_j at time T_j. We write γ_j for the classical path that attains the minimum.

Connecting all of $J + 1$ classical paths γ_j $(j = 1, 2, \ldots, J + 1)$, we obtain a long path connecting (T_0, x_0) and (T_{J+1}, x_{J+1}) in time-space. We write $\gamma_\Delta(x_{J+1}, \ldots, x_1, x_0)$ for this path, because it depends on the division Δ and points $(x_0, x_1, \ldots, x_{J+1})$. Although this is a continuous curve, it is not, in general, a smooth one. It may have edges at (T_j, x_j), $j = 1, 2, \ldots, J$. We call such a path a piecewise classical path. Some times we write γ_Δ as an abbreviation of $\gamma_\Delta(x_{J+1}, \ldots, x_1, x_0)$. If a piecewise classical path is of class C^1, then it is a classical path.

Figure 1.1 is an example of a piecewise classical path with $J = 3$.

For fixed Δ, the action of γ_Δ is

$$S(\gamma_\Delta)(x_{J+1}, \ldots, x_1, x_0) = \sum_{j=1}^{J+1} S(T_j, T_{j-1}, x_j, x_{j-1}). \tag{1.1.16}$$

Let $\gamma \in \Omega_{x,y}$ be an arbitrary path. It is approximated as close as one like by a piecewise classical path γ_Δ if $|\Delta|$ and $\{x_j\}$ are suitably chosen. Taking this fact in mind, Feynman defined:

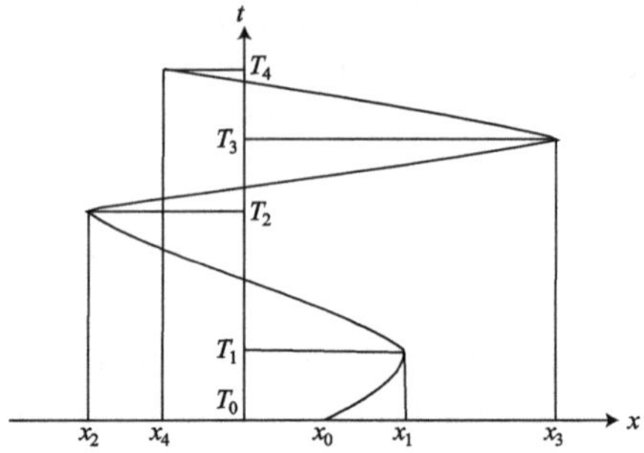

Fig. 1.1 Example of piecewise classical path (the case $J = 3$)

$$\frac{1}{N}\sum_{\gamma\in\Omega}\exp i\nu S(\gamma) = \lim_{|\Delta|\to 0}\prod_{j=1}^{J+1}\left(\frac{-im\nu}{2\pi\tau_j}\right)^{1/2}\int_{\mathbf{R}^J}\exp\left(i\nu S(\gamma_\Delta)(x_{J+1},\dots,x_1,x_0)\right)\prod_{j=1}^{J} dx_j.$$

That is

$$\int_{\Omega} e^{i\nu S(\gamma)}\mathscr{D}\gamma = \lim_{|\Delta|\to 0} I(\Delta;\nu,b,a,x,y), \tag{1.1.17}$$

where

$$I(\Delta;\nu,b,a,x,y) = \prod_{j=1}^{J+1}\left(\frac{-im\nu}{2\pi\tau_j}\right)^{1/2}\int_{\mathbf{R}^J} e^{i\nu S(\gamma_\Delta)(x_{J+1},\dots,x_1,x_0)}\prod_{j=1}^{J} dx_j. \tag{1.1.18}$$

Feynman's original definition (1.1.17) is called the definition of the Feynman path integral by the time slicing approximation method or by the time slicing method and $I(\Delta;\nu,b,a,x,y)$ is called the piecewise classical approximation or time slicing approximation of the Feynman path integral associated with the division Δ.

Remark 1.1 From the mathematically rigorous point of view, the following questions arise:

Q1 Is the integral $I(\Delta;\nu,b,a,x,y)$ for fixed Δ finite?
Q2 Does the limit $\lim_{|\Delta|\to 0}$ in (1.1.17) exist?
Q3 Is it the correct propagator?

We will answer, at least for a small time interval, these questions under certain assumptions for $V(x)$ which will be given later.

Remark 1.2 Feynman [22] also discussed an integral of the form

$$\int_{\Omega} F(\gamma)e^{i\nu S(\gamma)}\mathscr{D}\gamma \tag{1.1.19}$$

involving the general functional $F(\gamma)$ of γ. This is called the Feynman path integral with integrand $F(\gamma)$.

Its time slicing approximation is

$$
\begin{aligned}
& I[F_\Delta](\Delta; \nu, b, a, x, y) \\
&= \prod_{j=1}^{J+1} \left(\frac{\nu m}{2\pi i \tau_j} \right)^{1/2} \int_{\mathbf{R}^J} e^{i\nu S(\gamma_\Delta)} F(\gamma_\Delta)(x_{J+1}, x_J, \ldots, x_1, x_0) \prod_{j=1}^{J} dx_j \\
&= \prod_{j=1}^{J+1} \left(\frac{\nu m}{2\pi i \tau_j} \right)^{1/2} \int_{\mathbf{R}^J} e^{i\nu S_\Delta(x_{J+1}, x_J, \ldots, x_1, x_0)} F_\Delta(x_{J+1}, x_J, \ldots, x_1, x_0) \prod_{j=1}^{J} dx_j,
\end{aligned}
\tag{1.1.20}
$$

where $F_\Delta = F(\gamma_\Delta)$, $S_\Delta = S(\gamma_\Delta)$, $x_{J+1} = x$ and $x_0 = y$, cf. Feynman [21], Feynman and Hibbs [22].

1.2 Examples

One can calculate explicitly the Feynman path integral if potential $V(x)$ is of special form. In such a case it gives the correct propagator.

1.2.1 The Free Motion

This is the case that $V(x) = 0$, that is

$$
L(\dot{x}, x) = \frac{m\dot{x}^2}{2}. \tag{1.2.1}
$$

The classical path starting from y at time a and reaching x at time b is the solution of the boundary value problem of Euler's equation:

$$
m\frac{d^2}{dt^2}\gamma(t) = 0, \qquad \gamma(a) = y, \quad \gamma(b) = x. \tag{1.2.2}
$$

Thus

$$
\gamma(t) = y + \frac{t-a}{b-a}(x-y). \tag{1.2.3}
$$

Let Δ be a division of time interval $[a, b]$

$$
\Delta : a = T_0 < T_1 < T_2 < \cdots < T_{J+1} = b. \tag{1.2.4}
$$

We set $\tau_j = T_j - T_{j-1}, \quad j = 1, 2, \ldots, J+1$, and

$$
|\Delta| = \max\{\tau_j;\, j = 1, 2, \ldots, J+1\}.
$$

Let $\{x_j\}_{j=1}^J$ be arbitrary J points in $\mathbf{R}$ and $x_0 = y$, $x_{J+1} = x$. The piecewise classical path $\gamma_\Delta(x_{J+1}, \dots, x_1, x_0)$ passes $x_j \in \mathbf{R}$ at time T_j and is a straight line in the interval $[T_{j-1}, T_j]$.

Therefore,

$$S(T_j, T_{j-1}, x_{j-1}, x_j) = \frac{1}{2}\int_{T_{j-1}}^{T_j} m\left(\frac{x_j - x_{j-1}}{T_j - T_{j-1}}\right)^2 dt = \frac{m(x_j - x_{j-1})^2}{2\tau_j}. \quad (1.2.5)$$

We have

$$S(\gamma_\Delta)(x_{J+1}, \dots, x_1, x_0) = \sum_{j=1}^{J+1} S(T_j, T_{j-1}, x_j, x_{j-1}) = m\sum_{j=1}^{J+1} \frac{(x_j - x_{j-1})^2}{2\tau_j}.$$

Thus

$$I(\Delta; \nu, b, a, x, y) = \prod_{j=1}^{J+1}\left(\frac{-i\nu m}{2\pi\tau_j}\right)^{1/2} \int_{\mathbf{R}^J} \mathrm{e}^{i\nu\sum_{j=1}^{J+1}\frac{m|x_j - x_{j-1}|^2}{2\tau_j}} \prod_{j=1}^{J} dx_j. \quad (1.2.6)$$

This integral on the right-hand side does not converge absolutely. However, it's meaning is given in the following way. We regard $x = (x_1, x_2, \dots, x_J)$ as a vector $x \in \mathbf{R}^J$, and rewrite

$$\sum_{j=1}^{J+1} \frac{|x_j - x_{j-1}|^2}{2\tau_j} = \frac{x_0^2}{2\tau_1} + \frac{x_{J+1}^2}{2\tau_{J+1}} + \frac{1}{2}x\cdot Hx - x\cdot\xi, \quad (1.2.7)$$

where $\xi = (\tau_1^{-1}x_0, 0, \dots, 0, \tau_{J+1}^{-1}x_{J+1}) \in \mathbf{R}^J$, the synbol $x \cdot \xi$ means the inner product of two vectors $x, \xi \in \mathbf{R}^J$ and H is the following $J \times J$ square matrix:

$$H = \begin{pmatrix} \frac{1}{\tau_1} + \frac{1}{\tau_2} & -\frac{1}{\tau_2} & 0 & 0 & \cdots \\ -\frac{1}{\tau_2} & \frac{1}{\tau_2} + \frac{1}{\tau_3} & -\frac{1}{\tau_3} & 0 & \cdots \\ 0 & -\frac{1}{\tau_3} & \frac{1}{\tau_3} + \frac{1}{\tau_4} & -\frac{1}{\tau_4} & \cdots \\ 0 & 0 & \cdots & \cdots & \cdots \end{pmatrix}. \quad (1.2.8)$$

Using positive parameter $\varepsilon > 0$, we define

$$I(\Delta; \nu, b, a, x, y) = \quad (1.2.9)$$

$$\prod_{j=1}^{J+1}\left(\frac{-i\nu m}{2\pi\tau_j}\right)^{1/2} \mathrm{e}^{i\nu m(\frac{x_0^2}{2\tau_1} + \frac{x_{J+1}^2}{2\tau_{J+1}})} \lim_{\varepsilon\to+0}\int_{\mathbf{R}^J} \mathrm{e}^{-\frac{1}{2}\varepsilon|x|^2 + i\nu(\frac{m}{2}x\cdot Hx - x\cdot\xi)} \prod_{j=1}^{J} dx_j.$$

Let A be a matrix. Then we write $\det A$ for the determinant of A. We use the formula for Gaussian integrals:

Lemma 1.1 *Let $x \cdot y$ denote the inner product of $x, y \in \mathbf{R}^n$ and A be a real symmetric regular $n \times n$ square matrix. Then we define the Fourier transform $F_\varepsilon(\xi)$ of the function*

$$\exp\Big(-\frac{1}{2}\varepsilon|y|^2 + \frac{i\nu}{2}y \cdot Ay\Big)$$

with a parameter $\varepsilon > 0$ by

$$F_\varepsilon(\xi) = \Big(\frac{\nu}{2\pi}\Big)^{n/2} \int_{\mathbf{R}^n} \mathrm{e}^{-\frac{1}{2}\varepsilon|y|^2 + i\nu(\frac{1}{2}y\cdot Ay - y\cdot\xi)} dy.$$

Then

$$F_\varepsilon(\xi) = \nu^{n/2}(\det(\varepsilon I - i\nu A))^{-1/2}\mathrm{e}^{-\frac{\nu^2}{2}\xi\cdot(\varepsilon I - i\nu A)^{-1}\xi}.$$

Definition 1.1 Let A be as in Lemma 1.1. We define

$$\begin{aligned}
&\Big(\frac{\nu}{2\pi}\Big)^{n/2} \int_{\mathbf{R}^n} \mathrm{e}^{i\nu(\frac{1}{2}y\cdot Ay - y\cdot\xi)} dy = \lim_{\varepsilon\to+0} F_\varepsilon(\xi) \qquad (1.2.10)\\
&= |\det A|^{-1/2} \exp\Big\{\frac{(n - 2\mathrm{Ind}(A))\pi i}{4}\Big\} \exp\Big(-\frac{i\nu\xi\cdot A^{-1}\xi}{2}\Big)\\
&= (\det A)^{-1/2} \exp\Big(-\frac{i\nu\xi\cdot A^{-1}\xi}{2}\Big).
\end{aligned}$$

Here $\mathrm{Ind}(A)$ is the number of negative eigenvalues of A.

We apply this to (1.2.9) after replacing A by mH. In this case $A = mH$ is positive definite and

$$\det H = \frac{\tau_1 + \tau_2 + \cdots + \tau_{J+1}}{\tau_1\tau_2\cdots\tau_{J+1}}. \qquad (1.2.11)$$

Let g_{ij} be the (i, j) element of the matrix $G = H^{-1}$. Then

$$g_{ij} = \begin{cases} \dfrac{(\tau_1 + \cdots + \tau_i)(\tau_{j+1} + \cdots + \tau_{J+1})}{\tau_1 + \tau_2 + \cdots + \tau_{J+1}}, & 1 \le i \le j \le J,\\[2mm] \dfrac{(\tau_1 + \cdots + \tau_j)(\tau_{i+1} + \cdots + \tau_{J+1})}{\tau_1 + \tau_2 + \cdots + \tau_{J+1}}, & 1 \le j \le i \le J. \end{cases} \qquad (1.2.12)$$

We have

$$\frac{1}{2}\xi\cdot G\xi = \frac{1}{2}\Big(\frac{x_0^2}{\tau_1} - \frac{x_0^2}{b-a} + \frac{2x_0x_{J+1}}{b-a} + \frac{x_{J+1}^2}{\tau_{J+1}} - \frac{x_{J+1}^2}{b-a}\Big),$$

where we have used the trivial equality:

$$\tau_1 + \tau_2 + \cdots + \tau_{J+1} = b - a.$$

Therefore, we have

$$I(\Delta;\nu,b,a,x,y)=\left(\frac{-i\nu m}{2\pi(b-a)}\right)^{1/2}\exp\frac{i\nu|x-y|^2}{2(b-a)}. \tag{1.2.13}$$

Since this is independent of division Δ,

$$\begin{aligned}K(b,a,x,y)&=\int_{\Omega_{x,y}}\exp\left(i\nu\int_a^b\frac{m\dot{\gamma}(t)^2}{2}dt\right)\mathscr{D}\gamma\\&=\left(\frac{-i\nu m}{2\pi(b-a)}\right)^{1/2}\exp\frac{i\nu m|x-y|^2}{2(b-a)}.\end{aligned}$$

It is not difficult to prove that

$$K(t,a,x,y)=\left(\frac{-i\nu m}{2\pi(t-a)}\right)^{1/2}\exp\frac{i\nu m|x-y|^2}{2(t-a)}$$

is the fundamental solution of the Schrödinger equation for the free motion

$$i\hbar\frac{\partial}{\partial t}u(t,x)=\frac{-\hbar^2}{2m}\frac{\partial^2}{\partial x^2}u(t,x). \tag{1.2.14}$$

1.2.2 The Harmonic Oscillator

This is the case of

$$V(x)=\frac{m\omega^2x^2}{2}$$

with a positive constant ω. The Lagrangian function is

$$L(\dot{x},x)=\frac{m}{2}\dot{x}^2-\frac{m\omega^2}{2}x^2. \tag{1.2.15}$$

The classical path starting from y at time a and reaching x at time b is the solution of Euler's equation:

$$\frac{d^2}{dt^2}\gamma(t)+\omega^2\gamma(t)=0,\quad \gamma(a)=y,\quad \gamma(b)=x. \tag{1.2.16}$$

In the case

$$\frac{\omega(b-a)}{\pi}\neq 0,\pm1,\pm2,\dots, \tag{1.2.17}$$

the boundary value problem (1.2.16) has the unique solution

$$\gamma(t) = \frac{x \sin \omega(t-a) + y \sin \omega(b-t)}{\sin \omega(b-a)}. \tag{1.2.18}$$

In the case

$$\frac{\omega(b-a)}{\pi} = 2n, \qquad n = 0, \pm 1, \pm 2, \pm 3, \ldots \tag{1.2.19}$$

equation (1.2.16) has no solution if $x \neq y$. If $x = y$, then for any constant c

$$\gamma(t) = y \cos \omega(t-a) + c \sin \omega(t-a) \tag{1.2.20}$$

is a solution of (1.2.16).

In the case

$$\frac{\omega(b-a)}{\pi} = 2n-1, \qquad n = 0, \pm 1, \pm 2, \pm 3, \ldots \tag{1.2.21}$$

(1.2.16) has no solution if $x \neq -y$. If $x = -y$, then the solution is of the form

$$\gamma(t) = y \cos \omega(t-a) + c \sin \omega(t-a), \tag{1.2.22}$$

where c is an arbitrary constant.

We discuss the case $0 < \omega(b-a) < \pi$. Calculation shows

$$S(b, a, x, y) = \frac{m\omega((x^2+y^2)\cos \omega(b-a) - 2xy)}{2 \sin \omega(b-a)}. \tag{1.2.23}$$

Let $|\Delta|$ be an arbitrary division as in (1.2.4) with $\tau_j \omega < \pi$ for every $j = 1, 2, \ldots, J+1$. Since

$$S(T_j, T_{j-1}, x_j, x_{j-1}) = \frac{m\omega((x_{j-1}^2 + x_j^2)\cos \omega\tau_j - 2x_{j-1}x_j)}{2 \sin \omega\tau_j}, \tag{1.2.24}$$

$$\begin{aligned} S(\gamma_\Delta)(x_{J+1}, \ldots, x_1, x_0) &= \sum_{j=1}^{J+1} S(T_j, T_{j-1}, x_j, x_{j-1}) \\ &= \frac{m}{2} x \cdot Hx - mx \cdot \xi + \frac{m\omega}{2}(x_0^2 \cot \omega\tau_1 + x_{J+1}^2 \cot \omega\tau_{J+1}), \end{aligned} \tag{1.2.25}$$

where $x = (x_1, x_2, \ldots, x_J) \in \mathbf{R}^J$, $\xi = (\omega x_0 / \sin\omega\tau_1, 0, \ldots, 0, \omega x_{J+1}/\sin\omega\tau_{J+1}) \in \mathbf{R}^J$ and

$$H = \omega \begin{pmatrix} \cot\omega\tau_1 + \cot\omega\tau_2 & -\frac{1}{\sin\omega\tau_2} & 0 & \cdots \\ -\frac{1}{\sin\omega\tau_2} & \cot\omega\tau_2 + \cot\omega\tau_3 & -\frac{1}{\sin\omega\tau_3} & \cdots \\ 0 & -\frac{1}{\sin\omega\tau_3} & \cot\omega\tau_3 + \cot\omega\tau_4 & \cdots \\ 0 & 0 & \cdots & \cdots \end{pmatrix}. \tag{1.2.26}$$

Calculation shows that

$$\det H = \frac{\omega^J \sin\omega(\tau_1 + \tau_2 + \cdots \tau_{J+1})}{\sin\omega\tau_1 \sin\omega\tau_2 \ldots \sin\omega\tau_{J+1}} \tag{1.2.27}$$

and that the (i, j) element g_{ij} of the matrix $G = H^{-1}$ is

$$g_{ij} = \frac{\sin\omega(\tau_1 + \cdots + \tau_i)\sin\omega(\tau_{j+1} + \cdots + \tau_{J+1})}{\omega\sin\omega(\tau_1 + \tau_2 + \cdots + \tau_{J+1})}, \quad \text{if } 1 \le i \le j \le J, \tag{1.2.28}$$

$$g_{ij} = \frac{\sin\omega(\tau_1 + \cdots + \tau_j)\sin\omega(\tau_{i+1} + \cdots + \tau_{J+1})}{\omega\sin\omega(\tau_1 + \tau_2 + \cdots + \tau_{J+1})}, \quad \text{if } 1 \le j \le i \le J. \tag{1.2.29}$$

Since $\xi = (\omega x_0/\sin\omega\tau_1, 0, \ldots, 0, \omega x_{J+1}/\sin\omega\tau_{J+1})$, we have

$$\begin{aligned} &-\frac{1}{2}(\xi \cdot H^{-1}\xi) + \frac{\omega}{2}(x_0^2 \cot\omega\tau_1 + x_{J+1}^2 \cot\omega\tau_{J+1}) \\ &= \frac{\omega((x_0^2 + x_{J+1}^2)\cos\omega(b-a) - 2x_0 x_{J+1})}{2\sin\omega(b-a)}. \end{aligned} \tag{1.2.30}$$

Using Lemma 1.1 and Definition 1.1, we have

$$\begin{aligned} I(\Delta; \nu, b, a, x, y) = &\left(\prod_{j=1}^{J+1} \frac{\sin\omega\tau_j}{\omega\tau_j}\right)^{1/2} \left(\frac{-i\nu m\omega}{2\pi\sin\omega(b-a)}\right)^{1/2} \\ &\times \exp\left\{\frac{i\nu m\omega((x^2 + y^2)\cos\omega(b-a) - 2xy)}{2\sin\omega(b-a)}\right\}. \end{aligned} \tag{1.2.31}$$

Since

$$\lim_{|\Delta| \to +0} \left(\prod_{j=1}^{J+1} \frac{\sin\omega\tau_j}{\omega\tau_j}\right)^{1/2} = 1, \tag{1.2.32}$$

we have

$$K(b,a,x,y) = \int_{\Omega} \exp\left\{i\nu\int_a^b \left(m\frac{\dot{\gamma}^2}{2} - m\omega^2\frac{\gamma^2}{2}\right)dt\right\}\mathscr{D}\gamma \tag{1.2.33}$$
$$= \lim_{|\Delta|\to+0} I(\Delta;\nu,b,a,x,y)$$
$$= \left(\frac{-i\nu m\omega}{2\pi\sin\omega(b-a)}\right)^{1/2}\exp\left\{\frac{i\nu m\omega((x^2+y^2)\cos\omega(b-a)-2xy)}{2\sin\omega(b-a)}\right\}.$$

It is not difficult to prove that this is the fundamental solution of the Schrödinger equation:

$$i\hbar\frac{\partial}{\partial t}K(t,s,x,y) = \left(\frac{-\hbar^2}{2m}\frac{\partial^2}{\partial x^2} + \frac{m\omega^2}{2}x^2\right)K(t,s,x,y). \tag{1.2.34}$$

Remark 1.3 If $0 < s < t_1 < t$, then every path starting y at time s and reaching x at time t should pass some point $z \in \mathbf{R}$ at time t_1. Therefore,

$$\int_{\mathbf{R}} K(t,t_1,x,z)K(t_1,s,z,y)dz = K(t,s,x,y). \tag{1.2.35}$$

This is called the evolution property. In fact, (1.2.35) is easily proved if $0 < s < t_1 < t < \omega^{-1}\pi$. For more detailed discussion see Theorem 6.5.

Remark 1.4 If $\pi < \omega(b-a) < 2\pi$, then $detH < 0$ and $Ind(H) = 1$ by (1.2.27). Thus $(detH)^{-1/2} = |detH|^{1/2}e^{-i\pi/2}$. Consequently,

$$K(b,a,x,y) =$$
$$\left(\frac{-i\nu m\omega}{2\pi|\sin\omega(b-a)|}\right)^{1/2}e^{-i\pi/2}\exp\left\{\frac{i\nu m\omega((x^2+y^2)\cos\omega(b-a)-2xy)}{2\sin\omega(b-a)}\right\}. \tag{1.2.36}$$

Remark 1.5 The evolution property holds even in the case $\omega|t-s| > \pi$. For instance, suppose that $0 < \pi < \omega t < 2\pi$. Then the evolution property gives

$$K(t,0,x,y) = \int_{\mathbf{R}} K\left(t,\frac{t}{2},x,z\right)K\left(\frac{t}{2},0,z,y\right)dz. \tag{1.2.37}$$

Similarly, we can discuss the case for all $t > 0$ if $\omega t \neq n\pi$ for $n = 1, 2, \ldots,$. See, for example, Maslov's work [49].

Remark 1.6 If $\pi^{-1}\omega(b-a)$ is an integer, (1.2.31) suggests that $I(\Delta;\nu,b,a,x,y)$ is not convergent. This corresponds to the fact the fundamental solution of (1.2.34) is not an ordinary function but delta function if $\pi^{-1}\omega(b-a)$ is an integer.

In these examples Gauss's integral Lemma 1.1 played an essential role. This is possible only if the potential $V(x)$ is a quadratic function of x. What can one say if $V(x)$ is not a quadratic function of x? From the next chapter we shall discuss this problem.

1.3 Notes

The Feynman path integral was first published in [21], cf. also [22] and [57]. However, it is very difficult to justify the Feynman path integral mathematically as we pointed out in Remark 1.1 on page xxx, for instance. The difficulty comes from the fact that there is no countably additive measure in the path space corresponding to the symbol

$$\int_{\Omega} \mathscr{D}\gamma. \tag{1.3.1}$$

Various attempts have been made to overcome the difficulty. We will introduce some of them briefly.

If we replace $\hbar$ with $-i\hbar$ in the equation (1.1.10), we obtain the heat equation. Gelfand and Yaglom [35] introduced a complex parameter $\hbar + i\delta$ with $\delta < 0$ in place of $\hbar$ in (1.1.17) and (1.1.18) and they proposed to treat the Feynman path integral (1.1.17) as an analytic continuation from the measure theoretic integration by the Wiener measure on the path space.

However, Cameron [13] pointed out that the Wiener kernel with non-real variance cannot define a countably additive measure even if the real part of the variance is positive. Avoiding this difficulty, he proposed the notion of the "sequential Wiener integral" in [13] and "Ilstow integrals" in [14] to define the Feynman path integral by analytic continuation of the integral of functionals including a complex parameter by the Wiener integral. The potentials he treated belong to some class of bounded functions. See also [15, 20, 61].

Itô [42] defined the notion "the generalized uniform complex measure" on a separable Hilbert space. Using this notion, Itô proved that the Feynman path integral exists and is the fundamental solution of the Schrödinger equation if potential $V(x)$ is one of the cases:

1. $V(x)$ is of the form

$$V(x) = \int_{\mathbf{R}} e^{i(x,y)}\, d\mu(y), \tag{1.3.2}$$

   where μ is a complex measure of absolutely bounded variation $\|\mu\|$ on $\mathbf{R}$.
2. $V(x) = Cx$ where C is a real constant.
3. $V(x) = Cx^2$ where C is a positive constant.

He discussed case 1 by using the following facts:

$$(2\pi i)^{-1/2}\int_{\mathbf{R}} e^{\frac{i}{2}|x|^2}V(x)dx = \int_{\mathbf{R}} e^{-\frac{i}{2}|y|^2}\,d\mu(y), \tag{1.3.3}$$

because for $x, y \in \mathbf{R}$,

$$(2\pi i)^{-1/2}\int_{\mathbf{R}} e^{\frac{i}{2}|x|^2}e^{i(x,y)}dx = e^{-\frac{i}{2}|y|^2}. \tag{1.3.4}$$

Notice that

$$\sup_x |V(x)| \leq \|\mu\| \tag{1.3.5}$$

and that the class of functions of the form (1.3.2) is closed under multiplication because the class of measures of bounded variation forms a Banach algebra with norm $\|x\|$ under convolution.

Hida [60] discussed path integrals by using the white noise calculus [37, 38].

Albeverio and Krohn [4, 8] (and its 2nd edition [9] with Mazzucchi) defined the notion of "Fresnel integrals" on a separable Hilbert space. Let $\mathscr{H}$ be a real separable Hilbert space with inner product (a, b) and norm $|a|$. Let $\mathscr{M}(\mathscr{H})$ be the Banach algebra of complex Borel measures of bounded total variation $\|x\|$ on $\mathscr{H}$ under convolution. They defined $\mathscr{F}(\mathscr{H})$ as the space of functions $f(x)$ on $\mathscr{H}$ of the form

$$f(x) = \int_{\mathscr{H}} e^{i(x,y)}\,d\mu(y), \tag{1.3.6}$$

for some $\mu \in \mathscr{M}(\mathscr{H})$. They called such $f(x)$ Fresnel integrable functions. Generalizing the relationship (1.3.3), they defined the oscillatory integral $\tilde{\int} f(x)\,dx$ of $f(x)$ on $\mathscr{H}$ by

$$\tilde{\int} f(x)\,dx = \int_{\mathscr{H}} e^{-\frac{i}{2}|y|^2}\,d\mu(y). \tag{1.3.7}$$

They called the left-hand side of this the "Fresnel integral of f". Since $\mathscr{M}(\mathscr{H})$ is the Banach algebra and (1.3.5) holds, $\mathscr{F}(\mathscr{H})$ is closed under multiplication.

They discussed in [8] Feynman path integral in the case that $V(x)$ equals the polynomial of x of degree 2 plus a Fresnel integrable function and proved that the Feynman integral gives the fundamental solution of the Schrödinger equation. They proved the semi-classical asymptotic formula also for the solution of the Schrödinger equation. The notion of Fresnel integrals is generalized in various ways, see [1, 2, 7]. In particular Albeverio and Mazzucchi in [5, 6, 50] generalized the notion of Fresnel integrals and treated potentials which are the sum of a real polynomial of x of even degree with positive leading coefficient and a function in $\mathscr{F}(\mathscr{H})$. This is remarkable because Yajima [68] proved that the fundamental solution of the1-dimensional Schrödinger equation is nowhere C^1 if there exist positive constants $C > 0$ and $\varepsilon > 0$ such that potential $V(x)$ satisfies

$$V(x) \geq C|x|^{2+\varepsilon}, \quad \text{as } |x| \to \infty. \tag{1.3.8}$$

Although it is operator theoretic but not pointwise, Nelson [53] treated very wild potentials. Suppose that the potential V is independent of t. Let $A = \frac{1}{2m}\left(\frac{\hbar}{i}\partial_x\right)^2$ and $B = V(x)$. Then A and B are self-adjoint operators and the Hamiltonian operator (1.1.9) is expressed as

$$H = A + B. \tag{1.3.9}$$

The evolution operator (1.1.11) is of the form

$$U(t,s) = \mathrm{e}^{-i\nu(t-s)(A+B)}, \tag{1.3.10}$$

By the Trotter–Kato product formula (Theorem VIII.29 of [55]),

$$U(t,s) = s - \lim_{n\to\infty}\left(\mathrm{e}^{-i\nu\frac{(t-s)}{n}A}\mathrm{e}^{-i\nu\frac{(t-s)}{n}B}\right)^n. \tag{1.3.11}$$

Here $s - \lim$ means the limit in the strong topology of operators in $L^2(\mathbf{R})$. Notice that for any $f \in L^2(\mathbf{R})$

$$\mathrm{e}^{-i\nu\frac{(t-s)}{n}A}\mathrm{e}^{-i\nu\frac{(t-s)}{n}B}f(x) = \left(\frac{-i\nu m}{2\pi(\frac{t-s}{n})}\right)^{1/2}\int_{\mathbf{R}} e^{i\nu(m\frac{(x-y)^2}{2(t-s)/n}+\frac{t-s}{n}V(y))}f(y)dy. \tag{1.3.12}$$

Thus we have, at least formally,

$$U(t,s) = s - \lim_{n\to\infty}\left(\mathrm{e}^{-i\nu\frac{(t-s)}{n}A}\mathrm{e}^{-i\nu\frac{(t-s)}{n}}B\right)^n f(x) \tag{1.3.13}$$

$$= s - \lim_{n\to\infty}\left(\frac{-i\nu m}{2\pi(t-s)/n}\right)^{n/2}\int_{\mathbf{R}^n} e^{i\nu S_n(x_{n+1},x_n,\ldots,x_2,x_1)}f(x_1)\prod_{j=1}^n dx_j, \tag{1.3.14}$$

where $x_{n+1} = x$ and

$$S_n(x_{n+1},x_n,\ldots,x_2,x_1) = \left(\frac{t-s}{n}\right)\sum_{j=1}^n\left(\frac{m}{2}\left(\frac{(x_{j+1}-x_j)}{(t-s)/n}\right)^2 + V(x_j)\right). \tag{1.3.15}$$

Formally,

$$\lim_{n\to\infty}\left(\frac{t-s}{n}\right)\sum_{j=1}^n\left(\frac{m}{2}\left(\frac{(x_{j+1}-x_j)}{(t-s)/n)}\right)^2 + V(x_j)\right) \tag{1.3.16}$$

$$= \int_s^t\left(\frac{m}{2}\left(\frac{d}{d\tau}x(\tau)\right)^2 + V(x(\tau))\right)d\tau.$$

In this way Nelson [53] regarded (1.3.13) as the Feynman path integral. Moreover, he considered m instead of $\hbar$ as a complex parameter. If $m = i\delta$, $\delta > 0$, then (1.3.13) converges to a well-defined integral by the Wiener measure as $n \to \infty$. Nelson discussed analytic continuation of (1.3.13) from purely imaginary values $m = i\delta$, $\delta > 0$ to positive m. Nelson [53] succeeded in treating the Feynman path integral for a very wide class of potentials V, even in the case of very irregular potentials.

From the next chapter we will discuss the Feynman path integral, under the restriction on $V(x)$ given in Assumption in the next Sect. 2.1, (1.1.17) and (1.1.18), where $S(\gamma_\Delta)(x_{J+1}, \ldots, x_1, x_0)$ is the action of the piecewise classical path γ_Δ corresponding to the division Δ. Feynman in [22] also defined time slicing approximation of path integral by using piecewise linear paths in place of piecewise classical paths. Kumano-go [46] treated path integrals with integrand $f(\gamma)$ as in (1.1.19) by means of piecewise linear approximation and discussed various operations, for instance, integration by parts, semi-classical approximation,..., etc., under the same assumption on the potential (See [30] for the corresponding piecewise classical path case).

If the non-zero magnetic field as well as $V(x)$ exist, another difficulty arises. Tsuchida [34, 62] treated this case. See also [44, 67].

Phase space path integral was also proposed by Feynman [57]. However, mathematical treatment is much more complicated than path integral on configuration space, because other difficulties such as uncertainty principle come in. See [41, 57]. Reference [3], Kumano-go [32, 47] discussed phase space path integral.

Although Feynman path integral is not an integral by countably additive measure, path integral for Dirac equation seems to have different feature. In fact, it is very interesting that Ichinose [40] proved that path integral for the Dirac equation of 1-space dimension can be constructed by a countably additive measure on path space.

As the content in this section is not complete, we strongly recommend readers to consult with, for instance, [9, 48, 50, 57] and references therein for more detailed information.

Chapter 2
Assumption on Potentials

Abstract We begin mathematically rigorous treatment of Feynman path integrals in the case of the general potential $V(x)$ by the time slicing approximation method. First we explain our assumption on potentials. Afterward we show some consequences of our assumption which will play important roles later.

2.1 Assumption on Potentials

Let $[a, b]$ be a time interval. We treat potentials which may depend on time, i.e. the potentials of the form $V(t, x)$. We assume that a potential $V(t, x)$ has the following properties:

Assumption 2.1 $V(t, x)$ is a real-valued continuous function of $(t, x) \in [a, b] \times \mathbf{R}$. It is an infinitely differentiable function of x if $t \in [a, b]$ is fixed and $\partial_x^\alpha V(t, x)$ is a continuous function of (t, x) for any α. Moreover, for any integer $m \geq 2$ there exists a positive constant v_m such that

$$|\partial_x^\alpha V(t, x)| \leq v_m \tag{2.1.1}$$

if α is an integer with $2 \leq \alpha \leq m$ and $(t, x) \in [a, b] \times \mathbf{R}$.

Remark 2.1 It follows from Assumption 2.1 that there exist positive constants v_0 and v_1 such that for all $(t, x) \in [a, b] \times \mathbf{R}$

$$|V(t, x)| \leq v_0(1 + |x|)^2, \tag{2.1.2}$$
$$|\partial_x V(t, x)| \leq v_1(1 + |x|). \tag{2.1.3}$$

Remark 2.2 Pauli [54] discussed the case $sup_x|V(x)|(1 + |x|^2)^{-1} < \infty$ in the way physicists do.

D. Fujiwara, *Rigorous Time Slicing Approach to Feynman Path Integrals*, Mathematical Physics Studies, DOI 10.1007/978-4-431-56553-6_2

2.2 Classical Paths

2.2.1 Action Functional and Function Spaces

The Lagrangian function is

$$L(t, \dot{x}, x) = \frac{1}{2}|\dot{x}|^2 - V(t, x) \tag{2.2.1}$$

and the action is

$$S(\gamma) = \int_a^b \frac{1}{2}\left|\frac{d}{dt}\gamma(t)\right|^2 - V(t, \gamma(t))\, dt.$$

In order to study classical paths under Assumption 2.1, we use the following function spaces. Let $H^1(a, b)$ denote the real L^2-Sobolev space of order 1, that is,

$$H^1(a, b) = \{f \in L^2(a, b) : f \text{ is absolutely continuous and } \frac{d}{dt} f(t) \in L^2(a, b)\}.$$

$H^1(a, b)$ is a Hilbert space equipped with the inner product

$$(f, g)_{H^1(a,b)} = \int_a^b f(t)g(t)\, dt + \int_a^b \frac{d}{dt} f(t)\frac{d}{dt} g(t)\, dt$$

of $f, g \in H^1(a, b)$. The norm of $f \in H^1(a, b)$ is denoted by $\|f\|_{H^1(a,b)}$:

$$\|f\|^2_{H^1(a,b)} = \int_a^b |f(t)|^2\, dt + \int_a^b \left|\frac{d}{dt} f(t)\right|^2 dt. \tag{2.2.2}$$

Let $C[a, b]$ be the space of all continuous functions on $[a, b]$. The space $C[a, b]$ is a Banach space equipped with the maximum norm $\|f\|_C = \max_{t\in[a,b]} |f(t)|$ for any $f \in C[a, b]$. The next lemma is well-known:

Lemma 2.1 *If $H^1(a, b) \ni f$, then $f(t)$ is a continuous function of t in $[a, b]$. The embedding map $H^1(a, b) \to C[a, b]$ is a continuous map. There exists a constant $C > 0$ such that for any $f \in H^1(a, b)$*

$$\|f\|_C \le C\|f\|_{H^1(a,b)}. \tag{2.2.3}$$

Proposition 2.1 *$S(\gamma)$ is well-defined for any $\gamma \in H^1(a, b)$. $S(\gamma)$ is a continuous functional of $f \in H^1(a, b)$.*

Proof Suppose $\gamma \in H^1(a, b)$. This implies $\int_a^b |\frac{d}{dt}\gamma(t)|^2 dt$ exists. By Lemma 2.1, $\gamma(t)$ is a continuous function of $t \in [a, b]$. Thus $V(t, \gamma(t))$ is a continuous function of $t \in [a, b]$ and integrable. Therefore, $S(\gamma)$ is well-defined.

Suppose $\{\gamma_n\}_n \subset H^1(a,b)$ is a sequence such that $\lim_{n\to\infty} \gamma_n = \gamma$ in $H^1(a,b)$. Then it is clear that

$$\lim_{n\to\infty} \int_a^b |\frac{d}{dt}\gamma_n(t)|^2\, dt = \int_a^b |\frac{d}{dt}\gamma(t)|^2\, dt.$$

By Lemma 2.1, $\gamma_n(t)$ converges uniformly to $\gamma(t)$ on $[a,b]$. This implies that $V(t,\gamma_n(t)) \to V(t,\gamma(t))$ uniformly. Consequently, $S(\gamma_n) \to S(\gamma)$ as $n \to \infty$. □

For any $x, y \in \mathbf{R}$ let $\mathscr{H}_{x,y}$ be the closed subset of $H^1(a,b)$ given by

$$\mathscr{H}_{x,y} = \{\gamma \in H^1(a,b) : \gamma(a) = y, \gamma(b) = x\}. \tag{2.2.4}$$

We wish to study $S(\gamma)$ restricted to $\mathscr{H}_{x,y}$. Let γ_0 be the straight line segment

$$\gamma_0(t) = \frac{t-a}{b-a}x + \frac{b-t}{b-a}y, \quad t \in [a,b]. \tag{2.2.5}$$

Then for any $\gamma \in \mathscr{H}_{x,y}$,

$$\gamma - \gamma_0 \in H_0^1(a,b)$$

where

$$H_0^1(a,b) = \{\xi \in H^1(a,b) : \xi(a) = \xi(b) = 0\}.$$

Abbreviating $H_0^1(a,b)$ to $\mathscr{H}$, we can write

$$\mathscr{H}_{x,y} = \mathscr{H} + \gamma_0. \tag{2.2.6}$$

We define a bilinear form on $\mathscr{H} \times \mathscr{H}$

$$(f,g)_{\mathscr{H}} = \int_a^b \frac{d}{dt}f(t)\frac{d}{dt}g(t)\, dt, \quad \text{for } f, g \in \mathscr{H} \tag{2.2.7}$$

and a seminorm $\|f\|_{\mathscr{H}} = (f,f)_{\mathscr{H}}^{1/2}$ associated with it. We adopt $(f,g)_{\mathscr{H}}$ as the inner product of f, g in $\mathscr{H}$.

Let $\omega = \pi(b-a)^{-1}$. For $n = 1, 2, \ldots,$ let $e_n(t) = 2^{1/2}(b-a)^{-1/2} \sin(n\omega(t-a))$ and $\varphi_n(t) = (n\omega)^{-1}e_n(t)$. Then it is clear from Fourier sine series expansion that $\{e_n : n = 1, 2, \ldots\}$ and $\{\varphi_n : m = 1, 2, \ldots\}$ are ortho-normal bases of $L^2(a,b)$ and $\mathscr{H}$, respectively.

The next proposition shows that $\| * \|_{\mathscr{H}}$ is a norm equivalent to $\| * \|_{H^1(a,b)}$ on $\mathscr{H}$.

Proposition 2.2 (Poincaré's inequality and Sobolev's inequality) *For any $f \in \mathscr{H}$, let $\|f\|_{L^2}$ be the norm of f in the space $L^2(a,b)$. Then the following inequality holds:*

$$\|f\|_{L^2} \le \frac{(b-a)}{\pi}\|f\|_{\mathscr{H}} \qquad \textit{Poincaré's inequality}, \tag{2.2.8}$$

$$\|f\|_{\mathscr{H}}^2 \leq \|f\|_{H^1(a,b)}^2 \leq \Big(1 + \frac{(b-a)^2}{\pi^2}\Big)\|f\|_{\mathscr{H}}^2.$$

$$\max_{t\in[a,b]} |f(t)| \leq \pi\sqrt{\frac{(b-a)}{3}}\|f\|_{\mathscr{H}} \qquad \textit{Sobolev's inequality.} \tag{2.2.9}$$

Proof For $f \in \mathscr{H}$ we have

$$f(t) = \sum_{n=1}^{\infty} a_n \varphi_n(t).$$

Then

$$\|f\|_{\mathscr{H}}^2 = \sum_{n=1}^{\infty} |a_n|^2 \qquad \text{and} \qquad \|f\|_{L^2}^2 = \sum_{n=1}^{\infty} |a_n|^2 (n\omega)^{-2}.$$

The Poincaré inequality follows from these. Now for $t \in [a, b]$,

$$|f(t)| \leq 2^{1/2}(b-a)^{-1/2} \sum_{n=1}^{\infty} |a_n|(n\omega)^{-1} \leq 2^{1/2}(b-a)^{-1/2}\|f\|_{\mathscr{H}} \Bigg(\sum_{n=1}^{\infty}(n\omega)^2\Bigg)^{1/2}.$$

Inequality (2.2.9) follows from this. □

The equality (2.2.6) implies that $\mathscr{H}_{x,y}$ is a differentiable manifold and the tangent space to $\mathscr{H}_{x,y}$ at a point $\gamma \in \mathscr{H}_{x,y}$ is identified with the Hilbert space $\mathscr{H}$. The functional $S(\gamma)$ is Fréchet differentiable at $\gamma \in \mathscr{H}_{x,y}$. Let $DS(\gamma)$ denote its Fréchet differential at γ. Its directional derivative in the direction of $\xi \in \mathscr{H}$ is

$$DS(\gamma)[\xi] = \int_a^b \frac{d}{dt}\gamma(t)\frac{d}{dt}\xi(t)\,dt - \int_a^b \partial_x V(t, \gamma(t))\xi(t)\,dt. \tag{2.2.10}$$

2.2.2 Classical Paths

For $x, y \in \mathbf{R}$, suppose that $\gamma \in \mathscr{H}_{x,y}$ is a classical path. Then by definition it is the critical (or stationary) point of $S(\gamma)$. In other words, it satisfies

$$DS(\gamma)[\xi] = 0 \tag{2.2.11}$$

for any $\xi \in \mathscr{H}$.

Let $g(t, s)$ be the Green function for the Dirichlet boundary value problem, i.e.

$$g(t,s) = \begin{cases} (b-t)(s-a)/(b-a) & \text{if } a \leq s \leq t \leq b, \\ (b-s)(t-a)/(b-a) & \text{if } a \leq t \leq s \leq b. \end{cases} \tag{2.2.12}$$

For any fixed $t \in [a, b]$, let g_t be the function of s defined by $g_t(s) = g(t, s)$. Then g_t is an element of $\mathscr{H}$. Equation (2.2.11) means that

$$0 = DS(\gamma)[g_t] = \int_a^b \frac{d}{ds}\gamma(s)\partial_s g(t, s)\, ds - \int_a^b \partial_x V(s, \gamma(s))g(t, s)\, ds. \quad (2.2.13)$$

Using γ_0 in (2.2.5), we decompose γ into two parts:

$$\gamma = \gamma_0 + \xi \quad (2.2.14)$$

where $\xi \in \mathscr{H}$.

Integration by parts gives that

$$\int_a^b \frac{d}{ds}(\gamma_0(s) + \xi(s))\partial_s g(t, s)\, ds = \xi(t).$$

Therefore, if γ in (2.2.14) is a classical path, then ξ satisfies the integral equation:

$$\xi(t) = \int_a^b g(t, s)\partial_x V(s, \gamma_0(s) + \xi(s))\, ds. \quad (2.2.15)$$

We look for the solution of the equation (2.2.15). The integral operator

$$Gf(t) = \int_a^b g(t, s) f(s)\, ds, \qquad \forall f \in L^1(a, b) \quad (2.2.16)$$

is called the Green operator.

The following well-known inequality is useful (cf. for example, [36] Theorem 275 or [19] p. 527).

Lemma 2.2 *Let (X, Σ, μ) be a positive measure space. For $1 \le p \le \infty$, let $\|f\|_{L^p(X)}$ denote*

$$\|f\|_{L^p(X)} = \left(\int_X |f(x)|^p\, d\mu(x)\right)^{1/p}$$

and $L^p(X)$ denote the Banach space of measurable functions $f(x)$ satisfying $\|f\|_{L^p(X)} < \infty$. Let $k(x, y)$ be a $\mu \times \mu$-measurable function on $X \times X$ and K be the integral operator:

$$Kf(x) = \int_X k(x, y) f(y)\, d\mu(y).$$

Suppose that there exists a constant $M > 0$ such that

$$\sup_x \int_X |k(x, y)|\, d\mu(y) \le M, \quad \sup_y \int_X |k(x, y)|\, d\mu(x) \le M.$$

Then for $1 \le p \le \infty$ *and for* $f \in L^p(X, \Sigma, \mu)$

$$\|Kf\|_{L^p(X)} \le M\|f\|_{L^p(X)}.$$

Proof Let $1 \le p \le \infty$ and leq q be such that $1/p + 1/q = 1$. Since $|k(x, y)| = |k(x, y)|^{1/p}|k(x, y)|^{1/q}$, by Hölder's inequality

$$\begin{aligned}\left|\int_X |k(x,y)||f(y)|\,d\mu(y)\right|^p &\le \left(\int_X |k(x,y)|\,d\mu(y)\right)^{p/q}\left(\int_X |k(x,y)||f(y)|^p\,d\mu(y)\right)\\ &\le M^{p/q}\left(\int_X |k(x,y)||f(y)|^p\,d\mu(y)\right).\end{aligned}$$

By interchanging order of integration, we obtain

$$\int_X\left(\int_X |k(x,y)||f(y)|^p\,d\mu(y)\right)d\mu(x) = \int_X\left(|f(y)|^p\int_X |k(x,y)|\,d\mu(x)\right)d\mu(y).$$

Consequently,

$$\|Kf\|^p_{L^p(X)} \le M^{p/q}\|f\|^p_{L^p(X)}\sup_y\int_X |k(x,y)|\,d\mu(x).$$

□

We apply the previous lemma to the Green operator G.

Lemma 2.3 *1. For* $s, t \in [a, b]$, $g(t, s) = g(s, t) \ge 0$.

2. *Let* $1 \le p \le \infty$ *and* $\|f\|_{L^p}$ *be the norm of* f *in* $L^p(a, b)$. *Then for any* $f \in L^p(a, b)$

$$\|Gf\|_{L^p} \le \frac{1}{8}(b-a)^2\|f\|_{L^p}. \tag{2.2.17}$$

3. *If* $f \in L^1(a, b)$, Gf *is a continuous function of* t *with* $Gf(a) = Gf(b) = 0$ *and*

$$\|Gf\|_C \le \frac{1}{4}(b-a)\|f\|_{L^1} \quad \text{for all } f \in L^1(a,b). \tag{2.2.18}$$

4. *If* $f \in C[a, b]$, *then* Gf *is of class* $C^2(a, b) \cap C[a, b]$ *and*

$$-\frac{d^2}{dt^2}Gf(t) = f(t) \quad \text{for } t \in (a,b), \tag{2.2.19}$$

$$Gf(a) = Gf(b) = 0. \tag{2.2.20}$$

Also

$$\left\|\frac{d}{dt}Gf\right\|_C \le \frac{1}{2}(b-a)\|f\|_C. \tag{2.2.21}$$

Proof The first statement is clear. Inequality (2.2.17) follows from the fact that

$$\int_a^b g(t,s)\,ds \le \frac{1}{8}(b-a)^2 \tag{2.2.22}$$

and from Lemma 2.2. Let $f \in L^1(a,b)$. Then $g(t,s)f(s)$ is a continuous function of t and $|g(t,s)f(s)| \le 4^{-1}|b-a||f(s)|$. It follows from the dominated convergence theorem that $Gf(t)$ is a continuous function of t. (2.2.18) follows from this inequality. If $f \in C[a,b]$, equalities (2.2.19) and (2.2.20) are well-known. Since

$$\partial_t g(t,s) = \begin{cases} -(s-a)/(b-a) & \text{if } a \le s \le t \le b \\ (b-s)/(b-a) & \text{if } a \le t \le s \le b, \end{cases} \tag{2.2.23}$$

$$\int_a^b |\partial_t g(t,s)|\,ds \le \frac{1}{2}(b-a).$$

Inequality (2.2.21) follows from this inequality. □

We now show the unique existence of the solution of equation (2.2.15) by the contraction mapping principle under the assumption that $|b-a|$ is small. Let $C_0 = \{f \in C[a,b] : f(a) = f(b) = 0\}$. Then C_0 is a closed subspace of $C[a,b]$ equipped with the induced norm $\|\mathrm{a}\|_C$. We define a continuous map:

$$\mathscr{F} : C_0 \ni \xi \to \mathscr{F}(\xi) = G\partial_x V(*, \gamma_0 + \xi) \in C_0;$$

here $\partial_x V(*, \gamma_0+\xi)$ is a function of t defined by

$$\partial_x V(*, \gamma_0+\xi)(t) = (\partial_x V)(t, \gamma_0(t)+\xi(t)).$$

The map $\mathscr{F}$ is Fréchet differentiable in C_0. Its directional derivative in the direction of $\eta \in C_0$ at ξ is

$$D\mathscr{F}(\xi)[\eta] = G\partial_x^2 V(*, \gamma_0+\xi)\eta. \tag{2.2.24}$$

Lemma 2.4 *For any $\xi \in C_0$ and $\eta \in C_0$, we have*

$$\|D\mathscr{F}(\xi)[\eta]\|_{C_0} \le \frac{v_2(b-a)^2}{8}\|\eta\|_{C_0}. \tag{2.2.25}$$

Here v_2 is the constant in Assumption 2.1.

Proof By assumption, $\partial_x^2 V(t, \gamma_0+\xi)\eta \in C[a,b]$ and Assumption 2.1 implies that

$$\|\partial_x^2 V(*, \gamma_0+\xi)\eta\|_C \le v_2\|\eta\|_C. \tag{2.2.26}$$

Thus (2.2.17) with $p = \infty$ gives the inequality (2.2.25). □

Lemma 2.5 *For any $\eta, \eta_1 \in C_0$ we have the estimate*

$$\|\mathscr{F}(\eta) - \mathscr{F}(\eta_1)\|_C \leq \frac{(b-a)^2 v_2}{8} \|\eta - \eta_1\|_C.$$

Proof

$$\begin{aligned}\|\mathscr{F}(\eta) - \mathscr{F}(\eta_1)\|_C &= \Big\| \int_0^1 D\mathscr{F}(\theta\eta + (1-\theta)\eta_1)[\eta - \eta_1]\, d\theta \Big\|_C \\ &\leq \frac{v_2(b-a)^2}{8} \|\eta - \eta_1\|_C.\end{aligned}$$

□

Definition 2.1 Choose $\delta > 0$ so small that

$$\frac{v_2 \delta^2}{8} < 1. \tag{2.2.27}$$

Lemma 2.6 *If $|b - a| \leq \delta$, then there exists one and only one element ξ^* in the space C_0 satisfying equation (2.2.15).*

Proof If $|b - a| \leq \delta$, then $\mathscr{F}$ is a contraction mapping on the Banach space C_0. It follows from this and the contraction mapping principle that its fixed point ξ^* exists and is unique in C_0. □

Theorem 2.2 *For any $x, y \in \mathbf{R}$ there exists one and only one critical point $\gamma^* \in \mathscr{H}_{x,y}$ of $S(\gamma)$ if $|b - a| \leq \delta$. The critical point γ^* is of class $C^2(a, b)$ and $\gamma^*(t)$ satisfies Euler's equation and boundary conditions:*

$$\frac{d^2}{dt^2}\gamma^*(t) + \partial_x V(t, \gamma^*(t)) = 0, \qquad \gamma^*(a) = y, \quad \gamma^*(b) = x. \tag{2.2.28}$$

Proof Uniqueness: Assume that γ is a classical path. Decompose it into two parts as in (2.2.14). Then ξ satisfies the equation (2.2.28) and ξ is in $\mathscr{H}$. The property $\xi \in \mathscr{H}$ implies that it is in C_0. Uniqueness follows from the previous lemma.

Existence: By the previous lemma the solution $\xi^* \in C_0$ of (2.2.28) exists. Since $\xi^* \in C_0$, Assumption 2.1 implies that $\partial_x V(s, \gamma_0(s) + \xi^*(s)) \in C$. It follows from equation (2.2.28) and Lemma 2.3 that $\xi^* \in C_0[a, b] \cap C^2(a, b) \subset \mathscr{H}$.

Therefore, $\gamma^* = \gamma_0 + \xi^*$ belongs to $\mathscr{H}_{x,y} \cap C^2(a, b)$. Thus for any $\eta \in \mathscr{H}$ it turns out from integration by parts and from (2.2.19) that

$$\int_a^b \frac{d}{dt}\gamma^*(t)\frac{d}{dt}\eta(t)\, dt = -\int_a^b \frac{d^2}{dt^2}\gamma^*(t)\eta(t)\, dt = \int_a^b \partial_x V(t, \gamma^*(t))\eta(t)\, dt.$$

Therefore,

$$DS(\gamma^*)[\eta] = 0.$$

Since $\eta \in \mathscr{H}$ is arbitrary, $\gamma^* \in \mathscr{H}_{xy}$ is the critical point of $S(\gamma)$ that is the classical path. □

The following facts are well-known.

Corollary 2.1 *If $|b - a| \leq \delta$, then $\gamma \in \mathscr{H}_{xy}$ is the critical point of $S(\gamma)$ if and only if $\gamma \in C^2(a, b) \cap C[a, b]$ and satisfies Euler's equation with boundary conditions (2.2.28).*

Corollary 2.2 *Suppose that $a < c < b$. Let $\gamma(t)$ be a path such that $\gamma(t)$ is continuous in $[a, b]$ and it is a classical path in each of $[a, c]$ and $[c, b]$. If $\frac{d}{dt}\gamma(c + 0) = \frac{d}{dt}\gamma(c - 0)$, then $\gamma(t)$ is of C^2-class and is a classical path in the whole interval $[a, b]$.*

Proof It follows from the assumption $\frac{d}{dt}\gamma(c + 0) = \frac{d}{dt}\gamma(c - 0)$ and Euler's equation (2.2.28) that $\frac{d^2}{dt^2}\gamma(c + 0) = \frac{d^2}{dt^2}\gamma(c - 0)$. Thus $\gamma(t)$ is of class $C^2(a, b)$ and satisfies Euler's equation (2.2.28). Therefore, it is a classical path in $[a, b]$. □

2.2.3 The Minimum of the Action

Proposition 2.3 *Assume that $|b - a| \leq \delta$. Then among all paths in $\mathscr{H}_{x,y}$, the classical path $\gamma^* \in \mathscr{H}_{x,y}$ attains the minimum of the action $S(\gamma)$.*

Proof For any $\gamma_1 \in \mathscr{H}_{x,y}$, $\gamma_1 - \gamma^* \in \mathscr{H}$. Let $\gamma_\theta = \theta\gamma_1 + (1 - \theta)\gamma^*$ for $0 \leq \forall\theta \leq 1$. Then $\gamma_\theta \in \mathscr{H}_{x,y}$. By Taylor's theorem there exists some θ with $0 < \theta < 1$ such that

$$S(\gamma_1) = S(\gamma^*) + DS(\gamma^*)[\gamma_1 - \gamma^*] + \frac{1}{2}\frac{d^2}{d\theta^2}S(\gamma_\theta). \tag{2.2.29}$$

Since γ^* is the classical path, $DS(\gamma^*)[\gamma_1 - \gamma^*] = 0$. Calculation shows

$$\frac{d^2}{d\theta^2}S(\gamma_\theta) = \|\gamma_1 - \gamma^*\|^2_{\mathscr{H}} - \int_a^b \partial_x^2 V(t, \gamma_\theta(t))(\gamma_1(t) - \gamma^*(t))^2\, dt.$$

It follows from Assumption 2.1 and the Poincaré inequality (2.2.8) in Proposition 2.2 that

$$\left| \int_a^b \partial_x^2 V(t, \gamma_\theta(t))(\gamma_1(t) - \gamma^*(t))^2 \, dt \right| \le \nu_2 \|\gamma_1(t) - \gamma^*(t)\|_{L^2}^2$$
$$\le \nu_2 \frac{(b-a)^2}{\pi^2} \|\gamma_1(t) - \gamma^*(t)\|_{\mathscr{H}}^2.$$

Thus it turns out from $\pi^2 > 8$, $(b-a) \le \delta$ and $\dfrac{\nu_2 \delta^2}{8} < 1$ that if $\gamma_1 \ne \gamma^*$, then

$$\frac{d^2}{d\theta^2} S(\gamma_\theta) \ge \left(1 - \nu_2 \frac{(b-a)^2}{\pi^2}\right) \|\gamma_1(t) - \gamma^*(t)\|_{\mathscr{H}}^2$$
$$\ge \left(1 - \nu_2 \frac{\delta^2}{8}\right) \|\gamma_1(t) - \gamma^*(t)\|_{\mathscr{H}}^2 > 0.$$

Thus if $\gamma_1 \ne \gamma^*$, then

$$S(\gamma_1) > S(\gamma^*).$$

The proposition is proved. □

In the same way we can prove

Remark 2.3 The action functional is a convex functional on $\mathscr{H}_{x,y}$ if x and y are fixed.

2.2.4 Comparison with the Straight Line

In the following we always assume Assumption 2.1. We fix δ as in Definition 2.1 and assume $|b-a| \le \delta$ in this subsection. Let $\gamma \in \mathscr{H}_{xy}$ be the classical path and γ_0 be the straight line given by (2.2.5). We use the decomposition $\xi = \gamma - \gamma_0 \in \mathscr{H}$ as before in (2.2.14). We write $\partial_x V(*, \gamma_0(*))$ for the function of $t \in [a, b]$ defined by

$$\partial_x V(*, \gamma_0(*)) : [a, b] \ni t \to \partial_x V(t, \gamma_0(t)) \in \mathbf{R}. \tag{2.2.30}$$

Proposition 2.4 *If $|b-a| \le \delta$, then there exist positive constants C_1, C_2 such that*

$$\|\xi\|_C \le C_1 |b-a|^2 \|\partial_x V(*, \gamma_0(*))\|_C, \tag{2.2.31}$$
$$\left\| \frac{d}{dt} \xi \right\|_C \le C_2 |b-a| \|\partial_x V(*, \gamma_0(*))\|_C. \tag{2.2.32}$$

Here we can choose

$$C_1 = \frac{1}{8}\left(1 - \frac{\delta^2 \nu_2}{8}\right)^{-1}, \quad C_2 = \frac{1}{2}\left(1 - \frac{\delta^2 \nu_2}{8}\right)^{-1}. \tag{2.2.33}$$

Thus one may say that C_1 and C_2 depend only on ν_2.

Proof By (2.2.28) we have

$$\|\xi\|_C \le \|G\partial_x V(*,\gamma_0(*)+\gamma_1(*))\|_C \le \frac{(b-a)^2}{8}\max_{t\in[a,b]}|\partial_x V(t,\gamma_0(t)+\xi(t))|.$$

On the other hand,

$$\begin{aligned}|\partial_x V(t,\gamma_0(t)+\xi(t))| &= \left|\partial_x V(t,\gamma_0(t)) + \int_0^1 \partial_x^2 V(t,\gamma_0(t)+\theta\xi(t))\xi(t)\,d\theta\right| \\ &\le |\partial_x V(t,\gamma_0(t))| + \nu_2|\xi(t)|.\end{aligned}$$

Thus

$$\|\xi\|_C \le \frac{|b-a|^2}{8}(\|\partial_x V(*,\gamma_0(*))\|_C + \nu_2\|\xi\|_C).$$

The condition $|b-a|\le\delta$ implies that

$$\left(1-\frac{\delta^2}{8}\nu_2\right)\|\xi\|_C \le \frac{(b-a)^2}{8}\|\partial_x V(*,\gamma_0(*))\|_C.$$

This and Definition 2.1 prove (2.2.31).

Differentiating (2.2.15) and using (2.2.21), we obtain that

$$\begin{aligned}\left\|\frac{d}{dt}\xi\right\|_C &\le \frac{|b-a|}{2}\|\partial_x V(*,\gamma_0(*)+\xi(*))\|_C \\ &\le \frac{|b-a|}{2}(\|\partial_x V(*,\gamma_0(*))\|_C + \nu_2\|\xi\|_C).\end{aligned}$$

Replacing $\|\xi\|_C$ here with its bound (2.2.31), we have (2.2.32). □

From here we shall deal with many inequalities. In doing so, we shall use the same letter C to express various positive constants if we don't have an interest in distinguishing each of them.

Corollary 2.3 *If $|b-a|\le\delta$, then there exists a positive constant C depending only on ν_1 and ν_2 such that*

$$\|\xi\|_C \le C|b-a|^2(1+|x|+|y|) \tag{2.2.34}$$

$$\left\|\frac{d}{dt}\xi\right\|_C \le C|b-a|(1+|x|+|y|). \tag{2.2.35}$$

Proof As $|\gamma_0(t)|\le|x|+|y|$, Assumption 2.1 implies

$$|\partial_x V(t,\gamma_0(t)+\xi(t))| \le \nu_1(1+|x|+|y|+\|\xi\|_C).$$

Corollary follows from Proposition 2.4 and these inequalities. □

Corollary 2.4 *If $|b-a| \le \delta$, then there exists a constant $C > 0$ depending only on v_0, v_1, v_2 such that for any $t \in [a, b]$*

$$|\gamma(t)| \le C(1+|x|+|y|),$$
$$\Big|\frac{d}{dt}\gamma(t)\Big| \le C(1+|x|+|y|).$$

So far we fixed end points x and y of paths. If x or y changes, then values $\gamma_0(t)$, $\xi(t)$ and $\frac{d}{dt}\xi(t)$ change. It is clear that

$$\partial_x \gamma_0(t) = \frac{b-t}{b-a}, \quad \partial_y \gamma_0(t) = \frac{t-a}{b-a}, \tag{2.2.36}$$
$$\partial_x^\alpha \partial_y^\beta \gamma_0(t) = 0, \quad \text{if } \alpha+\beta \ge 2. \tag{2.2.37}$$

Proposition 2.5 *Suppose that $|b-a| \le \delta$. Then there exist positive constants C_1 and C_2 depending only on v_2 such that if $\alpha+\beta = 1$, then*

$$\|\partial_x^\alpha \partial_y^\beta \xi\|_C \le C_1|b-a|^2, \tag{2.2.38}$$
$$\left\|\partial_x^\alpha \partial_y^\beta \frac{d}{dt}\xi\right\|_C \le C_2|b-a|. \tag{2.2.39}$$

The following choice suffices:

$$C_1 = \frac{v_2}{8}\left(1-\frac{\delta^2 v_2}{8}\right)^{-1}, \quad C_2 = \frac{v_2}{2}\left(1-\frac{\delta^2 v_2}{8}\right)^{-1}. \tag{2.2.40}$$

Proof We discuss the case of $\alpha = 0$ and $\beta = 1$. The case $\alpha = 1$ and $\beta = 0$ is similar. Let η denote $\partial_y \xi$. Then differentiating (2.2.15) by y, we have

$$\eta(t) = \int_a^b g(t,s)\partial_x^2 V(s,\gamma(s))\Big(\frac{b-s}{b-a}+\eta(s)\Big)ds. \tag{2.2.41}$$

From this follows

$$\|\eta\|_C \le \frac{|b-a|^2}{8}v_2(1+\|\eta\|_C).$$

Since $|b-a| \le \delta$,

$$\Big(1-\frac{\delta^2}{8}v_2\Big)\|\eta\|_C \le \frac{|b-a|^2}{8}v_2.$$

Estimate (2.2.38) follows from this together with the assumption $|b-a| \le \delta$.

Next we shall prove (2.2.39). We differentiate (2.2.41) by t and obtain

$$\frac{d}{dt}\eta(t) = \int_a^b \partial_t g(t,s)\partial_x^2 V(s,\gamma(s))\Big(\frac{b-s}{b-a} + \eta(s)\Big)ds. \tag{2.2.42}$$

Using Lemma 2.3 (2.2.21), we have

$$\Big\|\frac{d}{dt}\eta\Big\|_C \leq \frac{\nu_2|b-a|}{2}(1+\|\eta\|_C). \tag{2.2.43}$$

This together with (2.2.38) proves (2.2.39). □

Corollary 2.5 *Suppose that $|b-a| \leq \delta$. Then there exists a constant $C > 0$ depending only on ν_2 such that if $|\alpha| + |\beta| = 1$, for any $t \in [a,b]$*

$$|\partial_x^\alpha \partial_y^\beta \gamma(t)| \leq C,$$
$$|\partial_x^\alpha \partial_y^\beta \frac{d}{dt}\gamma(t)| \leq C.$$

We now discuss derivatives of higher order.

Proposition 2.6 *Assume that $|b-a| \leq \delta$. Let $m \geq 1$ be an integer. Then $\xi(t)$ and $\frac{d}{dt}\xi(t)$ are m times differentiable functions of $(x,y) \in \mathbf{R}^2$ if t is fixed. Moreover, there exists a positive constant C_m such that for any pair of non-negative integers α and β with the property $1 \leq |\alpha+\beta| \leq m$ we have*

$$\|\partial_y^\alpha \partial_x^\beta \xi\|_C \leq C_m(b-a)^2, \tag{2.2.44}$$
$$\|\partial_y^\alpha \partial_x^\beta \frac{d}{dt}\xi\|_C \leq C_m|b-a|. \tag{2.2.45}$$

We can choose C_m so that it depends only on $\nu_2, \nu_3, \dots, \nu_m$.

Proof For short write $\gamma^{\alpha,\beta}$ and $\xi^{\alpha,\beta}$ for $\partial_y^\alpha \partial_x^\beta \gamma$ and $\partial_y^\alpha \partial_x^\beta \xi$, respectively. We prove the proposition by induction on m. By Proposition 2.5 both (2.2.44) and (2.2.45) are true in the case $m = 1$. Assuming (2.2.44) is true in the case $|\alpha| + |\beta| \leq m-1$, we shall prove them for $|\alpha| + |\beta| = m$. Suppose $|\alpha| + |\beta| = m \geq 2$. Then $\gamma^{\alpha,\beta} = \xi^{\alpha,\beta}$. Differentiating (2.2.28), we have

$$\xi^{\alpha,\beta}(t) = \int_a^b g(t,s)\partial_x^2 V(s,\gamma(s))\xi^{\alpha,\beta}(s)\,ds + F(t). \tag{2.2.46}$$

Here $F(t)$ is the sum of terms of the form

$$\int_a^b g(t,s)\partial_x{}^k V(s,\gamma(s))\gamma^{\alpha_1,\beta_1}(s)\cdots\gamma^{\alpha_{k-1},\beta_{k-1}}(s)\,ds, \tag{2.2.47}$$

where

$$1 \le k \le m, \quad \alpha_1 + \alpha_2 + \cdots + \alpha_{k-1} = \alpha, \quad \beta_1 + \beta_2 + \cdots + \beta_{k-1} = \beta,$$

and $1 \le \alpha_j + \beta_j \le m - 1$ for $j = 1, 2, \ldots, k-1$. By the induction hypothesis, for $j = 1, 2, \ldots, k-1$,

$$|\gamma^{\alpha_j,\beta_j}(s)| \equiv |\gamma_0^{\alpha_j\beta_j}(s) + \xi^{\alpha_j\beta_j}(s)| \le 1 + C_{m-1}.$$

It follows from this and Lemma 2.3 that there exists a constant $C > 0$ depending only on C_{m-1} and $v_2, \ldots, v_m$ such that

$$\left| \int_a^b g(t,s)\partial_x{}^k V(s,\gamma(s))\gamma^{\alpha_1,\beta_1}(s)\cdots\gamma^{\alpha_{k-1},\beta_{k-1}}(s)\,ds \right| \le C(b-a)^2.$$

Hence, there exists a positive constant $C > 0$ depending only on m and $v_2, \ldots, v_m$ such that

$$|F(t)| \le C|b-a|^2.$$

This fact and (2.2.46) give

$$\|\xi^{\alpha,\beta}\|_C \le \frac{v_2|b-a|^2}{8}\|\xi^{\alpha,\beta}\|_C + C|b-a|^2.$$

It follows from this and $|b-a| \le \delta$ that there exists a constant $C > 0$ depending only on $v_2, \ldots, v_m$ such that

$$\|\partial_y^\alpha \partial_x^\beta \xi\|_C \le C(b-a)^2. \tag{2.2.48}$$

(2.2.44) for m has been proved. Therefore (2.2.44) is proved for all $m \ge 1$.

Similary, differentiating

$$\frac{d}{dt}\xi(t) = \int_a^b \partial_t g(t,s)\partial_x V(s,\gamma_0(s)+\xi(s))\,ds, \tag{2.2.49}$$

we can prove (2.2.45) by induction on m. □

Corollary 2.6 *Suppose that $|b-a| \le \delta$. For any m there exists a constant $C_m > 0$ such that as far as $m \ge |\alpha| + |\beta| \ge 1$,*

$$|\partial_x^\alpha \partial_y^\beta \gamma(t)| \le C_m$$
$$|\partial_x^\alpha \partial_y^\beta \dot{\gamma}(t)| \le C_m.$$

for any $t \in [a,b]$. We can choose C_m so that it depends only on $v_2, v_3, \ldots, v_m$.

The next lemma which follows from Corollary 2.6 and the chain rule will be used later.

Lemma 2.7 *Let $m, \lambda \geq 0$ be constant and $f(x, z_2, z_1, y)$ be a function of $(x, z_2, z_1, y) \in \mathbf{R}^4$. Suppose that for all $k \geq 0$ there exists a constant $A_k > 0$ such that if $|\alpha| + |\alpha_1| + |\alpha_2| + |\beta| \leq k$, then for any $(x, z_1, z_2, y) \in \mathbf{R}^4$*

$$|\partial_x^\alpha \partial_y^\beta \partial_{z_1}^{\alpha_1} \partial_{z_2}^{\alpha_2} f(x, z_1, z_2, y)| \leq A_k(1 + \lambda + |x| + |z_1| + |z_2| + |y|)^m.$$

If $|b - a| \leq \delta$, then for $a < T_1 < T_2 < b$ there exists a constant $C_k > 0$ such that

$$|\partial_x^\alpha \partial_y^\beta f(x, \gamma(T_2), \gamma(T_1), y)| \leq C_k A_k(1 + \lambda + |x| + |y|)^m. \tag{2.2.50}$$

Here we can choose C_k independent of T_1 and T_2.

2.3 Properties of the Classical Action

The difference of the classical path from the free motion is proved to be small under Assumption 2.1 if $|b - a| \leq \delta$. Using this fact, we show that the classical action has properties convenient for us to discuss time slicing approximation of Feynman path integrals.

Definition 2.2 We define

$$S(b, a, x, y) = \inf_{\gamma \in \mathscr{H}_{x,y}} S(\gamma). \tag{2.3.1}$$

As a consequence of Proposition 2.3 we have

Proposition 2.7 *Assume that $|b - a| \leq \delta$. Then*

$$S(b, a, x, y) = S(\gamma^*).$$

Here γ^* stands for the classical path starting y at time a and reaching x at time b.

2.3.1 Comparison with the Classical Action of the Free Motion

Suppose that $|b - a| \leq \delta$. Then Proposition 2.7 means

$$S(b, a; x, y) = \int_a^b \left(\frac{1}{2} \left| \frac{d}{dt} \gamma^*(t) \right|^2 - V(t, \gamma^*(t)) \right) dt, \tag{2.3.2}$$

where $\gamma^* \in \mathscr{H}_{x,y}$ is the classical path. Let γ_0 be as in (2.2.5). Using the expression $\gamma^* = \gamma_0 + \xi$, we shall divide $S(b, a, x, y)$ into two parts, the first of which is nothing but the classical action of the free motion.

Proposition 2.8 *Suppose $|b - a| \leq \delta$. Then*

$$S(b, a, x, y) = \frac{|x - y|^2}{2(b - a)} + (b - a)\phi(b, a, x, y), \tag{2.3.3}$$

where

$$\phi(b, a, x, y) = (b - a)^{-1} \int_a^b \left(\left|\frac{d}{dt}\xi(t)\right|^2 - V(t, \gamma_0(t) + \xi(t)) \right) dt. \tag{2.3.4}$$

There exists a positive constant $C > 0$ depending only on v_0, v_1, v_2 such that

$$|\phi(b, a, x, y)| \leq C(1 + |x|^2 + |y|^2). \tag{2.3.5}$$

Proof (2.3.3) follows from (2.3.2) and the fact

$$\int_a^b \frac{d}{dt}\gamma_0(t)\frac{d}{dt}\xi(t)dt = 0.$$

Owing to Corollary 2.3 we know

$$\int_a^b \left|\frac{d}{dt}\xi(t)\right|^2 dt \leq C|b - a|^3 v_1^2(1 + |x|^2 + |y|^2). \tag{2.3.6}$$

Assumption 2.1 together with Corollary 2.4 lead us to the estimate

$$\begin{aligned} \left|\int_a^b V(t, \gamma(t))dt\right| &\leq v_0 \int_a^b (1 + |\gamma(t)|^2)dt \\ &\leq C v_0 |b - a|(1 + |x|^2 + |y|^2). \end{aligned}$$

Therefore, there exists a positive constant C depending only on v_0, v_1, v_2 such that

$$|\phi(b, a, x, y)| \leq C(1 + |x|^2 + |y|^2).$$

□

2.3.2 Estimates of Derivatives of $\phi(b, a, x, y)$

If $|b-a| < \delta$, then $\xi(t)$ is a C^1-function of (t, x, y) and is a C^∞-function of (x, y) if t is fixed. This implies that $\phi(b, a, x, y)$ is of C^1-class in all variables and infinitely differentiable with respect to (x, y) if a, b are fixed.

Lemma 2.8 *Suppose that $|b-a| \leq \delta$. Then $S(b, a, x, y)$ is the generating function of the canonical mapping defined by Hamiltonian flow, i.e.*

$$\partial_y S(b, a, x, y) = -\frac{d}{dt}\gamma(a), \qquad \partial_x S(b, a, x, y) = \frac{d}{dt}\gamma(b). \tag{2.3.7}$$

Proof These are-well known formulae in classical mechanics. However, we give here a proof.

$$\partial_y S(b, a, x, y) = \int_a^b \left(\frac{d}{dt}\gamma(t)\partial_y \frac{d}{dt}\gamma(t) - \partial_x V(t, \gamma(t))\partial_y \gamma(t)\right) dt.$$

Since $\partial_y \frac{d}{dt}\gamma(t) = \frac{d}{dt}\partial_y \gamma(t)$, integration by parts gives

$$\partial_y S(b, a, x, y) = -\frac{d}{dt}\gamma(a) - \int_a^b \left(\frac{d^2{}^2}{dt}\gamma(t) + \partial_x V(t, \gamma(t))\right)\partial_y \gamma(t)\, dt. \tag{2.3.8}$$

Since $\gamma(t)$ satisfies Euler's equation as in Corollary 2.1, the second term of the right-hand side vanishes. Thus the first part of the lemma has been proved. Similar discussion proves the second part. □

Theorem 2.3 *Let δ be so small that $\delta^2 v_2/8 < 1$. If $|b-a| \leq \delta$, then*

$$S(b, a, x, y) = \frac{(x-y)^2}{2(b-a)} + (b-a)\phi(b, a, x, y). \tag{2.3.9}$$

There exist positive constants κ_0 and κ_1 such that.

$$|\phi(b, a, x, y)| \leq \kappa_0(1 + |x|^2 + |y|^2), \tag{2.3.10}$$

$$|\partial_y \phi(b, a, x, y)| \leq \kappa_1(1 + |x|^2 + |y|^2)^{1/2}, \tag{2.3.11}$$

$$|\partial_x \phi(b, a, x, y)| \leq \kappa_1(1 + |x|^2 + |y|^2)^{1/2}. \tag{2.3.12}$$

For any $m \geq 2$ there exists a positive constant κ_m such that

$$\max_{2 \leq |\alpha|+|\beta| \leq m} \sup_{(x,y)\in \mathbf{R}^2, a \leq s \leq t \leq b} |\partial_x^\alpha \partial_y^\beta \phi(t, s, x, y)| \leq \kappa_m < \infty. \tag{2.3.13}$$

If $m = 2$*, then we can choose that*

$$\kappa_2 \leq \frac{\nu_2}{2}\left(1 - \frac{\nu_2\delta^2}{8}\right)^{-1}. \tag{2.3.14}$$

In particular if δ *is so small that* $\frac{\nu_2\delta^2}{8} \leq 2^{-1}$*, then we can choose that* $\kappa_2 = \nu_2$*.*

Proof In view of Proposition 2.8, we have only to prove (2.3.11), (2.3.12) and (2.3.13). It follows from the previous lemma and definition of $\phi(b, a, x, y)$ that

$$\partial_y \phi(b, a, x, y) = -(b-a)^{-1}\frac{d}{dt}\xi(a), \quad \partial_x \phi(b, a, x, y) = (b-a)^{-1}\frac{d}{dt}\xi(b). \tag{2.3.15}$$

Thus (2.3.11) and (2.3.12) follow from (2.2.35). Now

$$\partial_y^2 \phi(b, a, x, y) = \frac{-1}{b-a}\partial_y \frac{d}{dt}\xi(a). \tag{2.3.16}$$

Proposition 2.5 gives

$$|\partial_y^2 \phi(b, a, x, y)| \leq \frac{\nu_2}{2}\left(1 - \frac{\nu_2\delta^2}{8}\right)^{-1}.$$

Similarly, if $|\alpha + \beta| = 2$, then

$$|\partial_x^\alpha \partial_y^\beta \phi(b, a, x, y)| \leq \frac{\nu_2}{2}\left(1 - \frac{\nu_2\delta^2}{8}\right)^{-1}.$$

For any pair of indices α, β we have from Proposition 2.6

$$|\partial_x^\alpha \partial_y^{\beta+1} \phi(b, a, x, y)| = \frac{1}{b-a}|\partial_x^\alpha \partial_y^\beta \frac{d}{dt}\xi(a)| \leq C.$$

Similarly, we can prove

$$|\partial_x^{\alpha+1} \partial_y^\beta \phi(b, a, x, y)| = \frac{1}{b-a}|\partial_x^{\alpha+1} \partial_y^\beta \frac{d}{dt}\xi(b)| \leq C.$$

□

Chapter 3
Path Integrals and Oscillatory Integrals

Abstract The time slicing approximation of a Feynman path integral does not converge absolutely. But it is expected to have a definite finite value, because the factor $\exp i\nu S(\gamma_\Delta)(x_{J+1}, x_J, \ldots, x_1, x_0)$ oscillates rapidly and as a consequence there occurs a large scale of cancellation. Such an integral is commonly treated by oscillatory integral techniques and is given a definite value under some conditions. We give an example of a sufficient condition for that in Sect. 3.2. Furthermore, in such a case the stationary phase method, which is given by Theorem 3.5 in Sect. 3.3, gives the value of the oscillatory integral asymptotically as $\nu \to \infty$.

3.1 What Does the Time Slicing Approximation of a Feynman Path Integral Look Like?

The potential $V(t, x)$ is assumed to satisfy Assumption 2.1 in Sect. 2.1. Let Δ be the division of time interval $[a, b]$ such that

$$\Delta : a = T_0 < T_1 < T_2 < \cdots < T_{J+1} = b.$$

We set $\tau_j = T_j - T_{j-1}$ for $j = 1, 2, \ldots, J+1$ and $|\Delta| = \max_{1 \le j \le J+1} \{\tau_j\}$. We assume

$$|\Delta| \le \delta, \tag{3.1.1}$$

where $\delta > 0$ is the constant with

$$\frac{v_2 \delta^2}{8} < 1. \tag{3.1.2}$$

Let $x, y \in \mathbf{R}$, and let $x_1, x_2, \ldots, x_J \in \mathbf{R}$ be arbitrary J points. We set $x_0 = y$ and $x_{J+1} = x$ as a convention. Then by virtue of Theorem 2.2 there exists one and only one classical path γ_j starting from x_{j-1} at time T_{j-1} and arriving at x_j at time T_j for $j = 1, 2, \ldots, J+1$. Connecting all these, we obtain the unique piecewise classical path $\gamma_\Delta(t; x, x_J, \ldots, x_1, y)$, where $t \in [a, b]$ is the time parameter, connecting points $(a, y) = (T_0, x_0), (T_1, x_1), (T_2, x_2), \ldots, (T_J, x_J), (T_{J+1}, x_{J+1}) = (b, x)$ in time-space $\mathbf{R}^2$. We will often abbreviate this to $\gamma_\Delta(x_{J+1}, x_J, \ldots, x_1, x_0)$ or $\gamma_\Delta(t)$.

D. Fujiwara, *Rigorous Time Slicing Approach to Feynman Path Integrals*,
Mathematical Physics Studies, DOI 10.1007/978-4-431-56553-6_3

We write $\Gamma(\Delta)$ for the space of all piecewise classical paths associated with Δ. $\Gamma(\Delta)$ is naturally identified with $\mathbf{R}^{J+2}$ by the correspondence:

$$\Gamma(\Delta) \ni \gamma_\Delta(x_{J+1}, x_J, \ldots, x_0) \to (x_{J+1}, x_J, \ldots, x_0) \in \mathbf{R}^{J+2}.$$

The action along $\gamma_\Delta(x_{J+1}, x_J, \ldots, x_0)$ is a function on $\mathbf{R}^{J+2}$:

$$\begin{aligned} S(\gamma_\Delta(x_{J+1}, x_J, \ldots, x_0)) &= \sum_{j=1}^{J+1} S(T_j, T_{j-1}, x_j, x_{j-1}) \\ &= \sum_{j=1}^{J+1} \int_{T_{j-1}}^{T_j} \left(\frac{1}{2} \left| \frac{d}{dt} \gamma_\Delta(t) \right|^2 - V(t, \gamma_\Delta(t)) \right) dt. \end{aligned} \tag{3.1.3}$$

We will often abbreviate $S(\gamma_\Delta(x_{J+1}, x_J, \ldots, x_0))$ as $S(x_{J+1}, x_J, \ldots, x_0)$ or $S(\gamma_\Delta)$.

The time slicing approximation of the Feynman path integral corresponding to the division Δ of the time interval is

$$I(\Delta; \nu, b, a, x, y) = \prod_{j=1}^{J+1} \left(\frac{-i\nu}{2\pi\tau_j} \right)^{\frac{1}{2}} \int_{\mathbf{R}^J} e^{i\nu S(x_{J+1}, \ldots, x_0)} \prod_{j=1}^{J} dx_j. \tag{3.1.4}$$

Here $\nu = \hbar^{-1}$, $x_0 = y$ and $x_{J+1} = x$.

In Remark 1.2 we introduced the integral $\int_\Omega F(\gamma) \exp S(\gamma) \mathscr{D}[\gamma]$ for a functional $F(\gamma)$ of path γ. Its time slicing approximation associated with the division Δ is defined by

$$\begin{aligned} & I[F_\Delta](\Delta; \nu, b, a, x_{J+1}, x_0) \\ &= \prod_{j=1}^{J+1} \left(\frac{-i\nu}{2\pi\tau_j} \right)^{\frac{1}{2}} \int_{\mathbf{R}^J} e^{i\nu S(x_{J+1}, \ldots, x_0)} F(\gamma_\Delta(x_{J+1}, \ldots, x_0)) \prod_{j=1}^{J} dx_j. \end{aligned} \tag{3.1.5}$$

Clearly, (3.1.4) is a special case $F \equiv 1$ of (3.1.5).

Since (3.1.2) and (3.1.1) are assumed, the discussions in the previous chapter are applied and

$$S(T_j, T_{j-1}, x_j, x_{j-1}) = \frac{(x_j - x_{j-1})^2}{2\tau_j} + \tau_j \phi(T_j, T_{j-1}, x_j, x_{j-1}). \tag{3.1.6}$$

The following estimate holds: for any integer $n \geq 2$

$$\max_{2 \leq |\alpha| + |\beta| \leq n} \sup_{(x,y) \in \mathbf{R}^2} |\partial_{x_j}^\alpha \partial_{x_{j-1}}^\beta \phi(t_j, t_{j-1}, x_j, x_{j-1})| \leq \kappa_n < \infty. \tag{3.1.7}$$

If $n = 2$, then

$$\kappa_2 \leq \left(1 - \frac{\nu_2 \delta^2}{8}\right)^{-1} \frac{\nu_2}{2}. \tag{3.1.8}$$

3.2 Oscillatory Integrals

3.2.1 Multi-index Notation

Before explaining oscillatory integral techniques we introduce some notational conventions which are convenient in the following discussions.

We write $|x|$ for $\sqrt{x_1^2 + x_2^2 + \cdots + x_n^2}$ if $x = (x_1, \ldots, x_n)$ is a vector in $\mathbf{R}^n$. $(a_{jk})_{jk=1}^n$ stands for the $n \times n$ square matrix of which the (j, k)-entry equals a_{jk}.

An n-tuple of non-negative integers $\alpha = (\alpha_1, \ldots, \alpha_n)$ is called a multi-index. If α is a multi-index, then $|\alpha| = \alpha_1 + \alpha_2 + \cdots + \alpha_n$ is called the length of α. Let $\alpha = (\alpha_1, \ldots, \alpha_n)$ and $\beta = (\beta_1, \ldots, \beta_n)$ be two multi-indices. Then the multi-index $\alpha + \beta$ is defined to be $\alpha + \beta = (\alpha_1 + \alpha_2, \ldots, \alpha_n + \beta_n)$. We define $\alpha \leq \beta$ if and only if $\alpha_j \leq \beta_j$ for all $j = 1, 2, \ldots, n$. We also write $\beta \geq \alpha$ if $\alpha \leq \beta$. If $\alpha \leq \beta$ and $\alpha \neq \beta$, then $\alpha < \beta$ which is equivalent to $\beta > \alpha$. The relation $\leq$ is an semi-order relation among multi-indices. If $\alpha \geq \beta$, then $\alpha - \beta = (\alpha_1 - \alpha_2, \ldots, \alpha_n - \beta_n)$ is also a multi-index. The symbol $\alpha!$ is the abbreviation of $\alpha_1!\alpha_2!\ldots\alpha_n!$. If $\alpha \geq \beta$, then the binomial symbol $\binom{\alpha}{\beta}$ means $\dfrac{\alpha!}{(\alpha-\beta)!\beta!}$. For n variables $x = (x_1, \ldots, x_n)$ the monomial $x_1^{\alpha_1} x_2^{\alpha_2} \ldots x_n^{\alpha_n}$ is denoted by x^α. Thus $(x + y)^\alpha = \sum_{\beta \leq \alpha} \binom{\alpha}{\beta} x^\beta y^{\alpha-\beta}$.

We write ∂_x for partial differential operator $\dfrac{\partial}{\partial x}$. Higher derivatives $\dfrac{\partial^{\alpha_1}}{\partial x_1^{\alpha_1}} \cdots \dfrac{\partial^{\alpha_n}}{\partial x_n^{\alpha_n}}$ are written as ∂_x^α. Thus Leibniz's formula for derivatives of the product of functions is written as

$$\partial_x^\alpha (f(x)g(x)) = \sum_{\beta \leq \alpha} \binom{\alpha}{\beta} \partial_x^\beta f(x) \partial_x^{\alpha-\beta} g(x).$$

Taylor's formula up to degree N of a function $f(x)$ of n variables $(x_1, \ldots, x_n)$ is

$$\begin{aligned} f(x) = & \sum_{|\alpha|<N} \frac{(x-a)^\alpha}{\alpha!} \partial_x^\alpha f(a) \\ & + N \sum_{|\alpha|=N} \frac{(x-a)^\alpha}{\alpha!} \int_0^1 (1-t)^{N-1} \partial_x^\alpha f(t(x-a)+a) dt. \end{aligned}$$

The multiple integral $\int \cdots \int f(x_1, \ldots, x_n)\, dx_1 dx_2 \ldots dx_n$ is written as $\int f(x_1, \ldots, x_n) dx$.

3.2.2 Definition of Oscillatory Integrals

An oscillatory integral is an integral of the form

$$I(a) = \int_{\mathbf{R}^n} a(x) e^{i\nu\phi(x)} dx. \tag{3.2.1}$$

Here ν is a large positive parameter and i is the imaginary unit. We always assume $\nu > 1$. Function $\phi(x)$ is a real-valued function of $x \in \mathbf{R}^n$. $\phi(x)$ is called the phase function, or simply the phase, and $a(x)$ is called the amplitude function, or the amplitude.

In addition to integrals of this type, we also use integrals of the following type:

$$I(a_\lambda)(y) = \int_{\mathbf{R}^n} a_\lambda(x, y) e^{i\nu\phi(x,y)} dx. \tag{3.2.2}$$

Here $y \in \mathbf{R}^m$ and $\lambda \in \mathbf{R}^{m'}$ are parameters. We regard (3.2.1) as the special case of (3.2.2) with $m = m' = 0$.

The formal integral (3.2.1) or (3.2.2) does not converge absolutely even in the simplest case $a_\lambda(x, y) \equiv 1$. It is treated in the following way.

Definition 3.1 Let $\{\omega_\varepsilon(x)\}_{\varepsilon>0}$ be a family of infinitely differentiable functions with the following properties:

1. $\lim_{\varepsilon \to 0} \omega_\varepsilon(x) = 1$ for any $x \in \mathbf{R}^n$.
2. For any multi-index α and for any non-negative integer N there exists a positive constant $C_{\alpha,N}$ independent of ε such that

$$|\partial_x^\alpha \omega_\varepsilon(x)| \leq C_{\alpha,N}(1 + |x|)^{-N}. \tag{3.2.3}$$

3. For any multi-index α with $|\alpha| \geq 1$,

$$\lim_{\varepsilon \to 0} \partial_x^\alpha \omega_\epsilon(x) = 0. \tag{3.2.4}$$

Using $\{\omega_\varepsilon(x)\}$ as above, we define

$$I_\varepsilon(a_\lambda)(y) = \int_{\mathbf{R}^n} \omega_\varepsilon(x) a_\lambda(x, y) e^{i\nu\phi(x,y)} dx.$$

Definition 3.2 We define

$$I(a_\lambda)(y) = \lim_{\varepsilon \to 0} \int_{\mathbf{R}^n} \omega_\varepsilon(x) a_\lambda(x, y) e^{i\nu\phi(x,y)} dx \tag{3.2.5}$$

if the following conditions hold:

1. For any family of functions $\{\omega_\epsilon(x)\}$ as above the integral $I_\varepsilon(y)$ converges absolutely if $\varepsilon > 0$.
2. The limit of the right-hand side of (3.2.5) exists and is independent of the choice of $\{\omega_\varepsilon\}$.

The next proposition is clear.

Proposition 3.1 *Suppose that $a_\lambda(x, y) \in C_0^\infty(\mathbf{R}^{m+n})$ if λ is fixed. Then*

$$I(a_\lambda)(y) = \int_{\mathbf{R}^n} a_\lambda(x, y) e^{i\nu\phi(x,y)} dx \tag{3.2.6}$$

is absolutely convergent, and $I(a_\lambda)(y) \in C_0^\infty(\mathbf{R}^m)$ if λ is fixed.

We shall discuss another sufficient condition for existence of $I(a_\lambda)(y)$.

Assumption 3.1 Let $x \in \mathbf{R}^n$ and $y \in \mathbf{R}^m$. We suppose that the following conditions are satisfied:

1. The real-valued function $\phi(x, y)$ is infinitely differentiable. If $|\alpha| + |\beta| \geq 2$, then there exists a positive constant $C_{\alpha\beta}$ such that

$$|\partial_x^\alpha \partial_y^\beta \phi(x, y)| \leq C_{\alpha\beta}. \tag{3.2.7}$$

2. Let $(\partial_{x_j} \partial_{x_k} \phi(x, y))_{jk=1}^n$ be the $n \times n$ matrix whose (j, k) entry is $\partial_{x_j} \partial_{x_k} \phi(x, y)$. Then there exists a positive constant c_0 such that for any $(x, y) \in \mathbf{R}^{n+m}$

$$|det(\partial_{x_j} \partial_{x_k} \phi(x, y))| \geq c_0 > 0. \tag{3.2.8}$$

3. The amplitude function $a_\lambda(x, y)$ which may depend on λ as well as (x, y) is infinitely differentiable with respect to (x, y) if λ is fixed. There exists a non-negative constant p such that for any α, β

$$\sup_{(x,y) \in \mathbf{R}^{n+m}} (1 + |\lambda| + |x| + |y|)^{-p} |\partial_x^\alpha \partial_y^\beta a_\lambda(x, y)| < \infty. \tag{3.2.9}$$

Remark 3.1 Suppose that $\phi(x, y)$ satisfies Assumption 3.1. Then for any α, β with $|\alpha| + |\beta| \leq 1$, there exists a positive constant $C_{\alpha\beta}$ such that

$$|\partial_x^\alpha \partial_y^\beta \phi(x, y)| \leq C_{\alpha\beta}(1 + |x| + |y|)^{2-|\alpha|-|\beta|}.$$

Remark 3.2 The parameter λ in (3.2.9) will be used in later discussion.

The second condition of Assumption 3.1 assures that the function $e^{i\nu\phi(x,y)}$ oscillates rapidly when x moves from one place to another. For the proof of this fact, we require Hadamard's global implicit function theorem.

Theorem 3.1 (Hadamard's global implicit function theorem) *Let $f : \mathbf{R}^n \to \mathbf{R}^n$ be a map of class C^1. If its differential $Df(x)$ is invertible at every $x \in \mathbf{R}^n$ and the norm of $Df(x)^{-1}$ as a linear map is bounded uniformly with respect to x. Then the map f is a global diffeomorphism of class C^1.*

Proof A sketch of proof is given at the end of this chapter. □

In the following, let $\zeta_j = \partial_{x_j}\phi(x, y)$, $j = 1, 2, \ldots, n$.

Lemma 3.1 *For each fixed $y \in \mathbf{R}^m$ define the map $\Phi_y : \mathbf{R}^n \ni x = (x_1, \ldots, x_n) \to \zeta = (\zeta_1, \ldots, \zeta_n) \in \mathbf{R}^n$. Then the following facts hold:*

1. *For fixed y the map Φ_y is a diffeomorphism of $\mathbf{R}^n$.*
2. *For fixed y, $x^* = \Phi_y^{-1}(0)$ is the only critical point of the function $\phi(x, y)$ with respect to x.*
3. *There exists a constant $C > 0$ independent of $y \in \mathbf{R}^m$ such that for any $x, x' \in \mathbf{R}^n$*

$$C^{-1}|x - x'| \le |\Phi_y(x) - \Phi_y(x')| \le C|x - x'|. \tag{3.2.10}$$

4. *For any $\alpha \neq 0$ there exists a positive constant C_α such that*

$$|\partial_x^\alpha \zeta| \le C_\alpha \quad \textit{and} \quad |\partial_\zeta^\alpha x| \le C_\alpha. \tag{3.2.11}$$

Proof The (j, k) entry of Jacobian matrix $J(x, y)$ of the map Φ_y at the point x is $\partial_{x_j}\partial_{x_k}\phi(x, y)$ and it is bounded because of condition 1 of Assumption 3.1. Condition 2 of Assumption 3.1 means that

$$0 < c_0 \le |\det J(x, y)|.$$

It follows from these and Theorem 3.1 that the map Φ_y is a global diffeomorphism of $\mathbf{R}^n$. It is clear that the critical point of the function $x \to \phi(x, y)$ coincides with $\Phi_y^{-1}(0)$. There exists one and only one critical point, because Φ_y is a diffeomorphism. In addition we know that for any $x, x' \in \mathbf{R}^n$,

$$\zeta_j(x, y) - \zeta_j(x', y) = \sum_{k=1}^{n}(x_k - x_k') \int_0^1 \partial_{x_j}\partial_{x_k}\phi(\theta x + (1 - \theta)x', y)\, d\theta. \tag{3.2.12}$$

Condition 1 of Assumption 3.1 implies that

$$|\zeta_j(x, y) - \zeta_j(x', y)| \le C|x - x'|,$$

with some constant C. The right-hand side of (3.2.10) follows from here.

As a consequence of (3.2.7) and (3.2.8) of Assumption 3.1, every entry of the inverse matrix $J(x, y)^{-1}$ of $J(x, y)$ is a uniformly bounded function of (ζ, y). Let $x(\zeta, y) = \Phi_y^{-1}(\zeta)$. Then there exists a positive constant C such that

$$|\partial_{\zeta_k} x_j(\zeta, y)| \le C,$$

because $\partial_{\zeta_k} x_j(\zeta, y)$ equals the (j, k) entry of the matrix $J(x, y)^{-1}$. This means that we can interchange the rôle of x and ζ in (3.2.12). Thus for all $j = 1, 2, \ldots, n$,

$$|x_j - x'_j| \le \sum_{k=1}^{n} |\zeta_k - \zeta'_k| \int_0^1 |\partial_{\zeta_k} x_j(\theta\zeta + (1-\theta)\zeta', y)|\, d\theta \le C|\zeta - \zeta'|.$$

The left-hand side of (3.2.10) follows from this. Similarly, we can prove (3.2.11). □

Lemma 3.2 *Let x^* be the critical point of $x \to \phi(x, y)$ and Φ_y be as above. Then*

$$C^{-1}|x - x^*| \le |\Phi_y(x)| \le C|x - x^*|. \tag{3.2.13}$$

$$C^{-1}|\Phi_y(0)| \le |x^*| \le C|\Phi_y(0)|. \tag{3.2.14}$$

$$|x^*| \le C(1 + |y|). \tag{3.2.15}$$

$$|\partial_y^\alpha x^*| \le C_\alpha \quad \textit{for any multi-index } \alpha \ne 0. \tag{3.2.16}$$

Proof Let $x' = x^*$ in inequality (3.2.10). Then we have (3.2.13), because $\Phi_y(x^*) = 0$. Set $x = 0$ in (3.2.13). Then we obtain (3.2.15). Let $\Phi_y(0) = (\zeta_1(0), \zeta_2(0), \ldots, \zeta_n(0))$ as above. Then $\zeta_j(0) = \partial_{x_j}\phi(0, y)$. Thus Taylor's formula leads us to

$$\zeta_j(0) - \partial_{x_j}\phi(0, 0) = \sum_{k=1}^{m} y_k \int_0^1 \partial_{y_k}\partial_{x_j}\phi(0, ty)\, dt.$$

This together with (3.2.7) give that there exists a constant C such that

$$|\zeta_j(0) - \partial_{x_j}\phi(0, 0)| \le C|y|.$$

This implies that with some positive constant C

$$|\Phi_y(0)| = |\zeta(0)| \le C(1 + |y|).$$

Inequality (3.2.15) follows from this and (3.2.14).

Since $\partial_{x_j}\phi(x^*, y) = 0$,

$$\sum_{k=1}^{n} \partial_{x_k}\partial_{x_j}\phi(x^*, y)\partial_{y_l}x_k^* = -\partial_{y_l}\partial_{x_j}\phi(x^*, y).$$

The inverse matrix Q of the Jacobi matrix $J(x^*, y)$ is uniformly bounded because of (3.2.7) and (3.2.8). Applying Q to both sides of the above equality, we have (3.2.16) in the case $|\alpha| = 1$. Similarly, we can prove (3.2.16) in the case $|\alpha| > 1$. □

Corollary 3.1 *Let $\zeta_j = \partial_{x_j}\phi(x, y)$ as above. Then for $\nu \geq 1$ there exists a constant $C > 0$ such that*

$$(1 + \nu|\zeta|^2)^{-1} \leq (1 + C\nu|x - x^*|^2)^{-1}. \tag{3.2.17}$$

Corollary 3.2 *Let $d(x, y) = |\det(\partial_{x_j}\partial_{x_k})\phi(x, y)|$ and $d(y) = d(x^*, y)^{-1/2}$. Then for any α, there exists a constant $C_\alpha > 0$ such that*

$$|\partial_y^\alpha d(y)| \leq C_\alpha. \tag{3.2.18}$$

Proof It follows from Assumption 3.1 that for any α, β there exists a constant $C_{\alpha,\beta} > 0$ with $\partial_x^\alpha \partial_y^\beta d(x, y)^{-1/2} \leq C_{\alpha,\beta}$. Hence estimate (3.2.18) follows from the chain rule and (3.2.16). □

To prove that $I(a_\lambda)$ has a definite meaning we make use of the technique of integration by parts. Let $D_{x_j} = \nu^{-1/2}\partial_{x_j}$ for $j = 1, 2, \ldots, n$.

Theorem 3.2 *Suppose that Assumption 3.1 holds. Then the oscillatory integral $I(a_\lambda)(y)$ has a definite value, which is given by the absolutely convergent integral:*

$$I(a_\lambda)(y) = \int_{\mathbf{R}^n} \left(M^{*p'+n} a_\lambda(x, y)\right) e^{i\nu\phi(x,y)}\, dx. \tag{3.2.19}$$

Here p' is an integer such that $p < p' \leq p + 1$ and M^* is the formal adjoint operator of

$$M = \frac{1 - i\sum_{j=1}^n \nu^{1/2}\zeta_j D_{x_j}}{1 + \nu|\zeta|^2}.$$

Moreover, there exists a constant $C > 0$ independent of y, λ such that for any $y \in \mathbf{R}^m$

$$\nu^{n/2}(1 + |\lambda| + |y|)^{-p}|I(a_\lambda)(y)| \leq C \sum_{|\alpha| \leq p'+n} \sup_{x \in \mathbf{R}^n} (1 + |\lambda| + |x| + |y|)^{-p} |\partial_x^\alpha a_\lambda(x, y)|. \tag{3.2.20}$$

Proof Let $\omega_\varepsilon(x)$ be as in Definition 3.1. It follows from condition 3 of Assumption 3.1 and (3.2.3) that $I_\varepsilon(a_\lambda)(y)$ is an absolutely convergent integral. Since

$$M e^{i\nu\phi(x,y)} = e^{i\nu\phi(x,y)},$$

it turns out from integration by parts that

$$I_\varepsilon(a_\lambda)(y) = \int_{\mathbf{R}^n} \omega_\varepsilon(x) a_\lambda(x, y) M\{e^{i\nu\phi(x,y)}\} dx = \int_{\mathbf{R}^n} M^*\{\omega_\varepsilon(x) a_\lambda(x, y)\} e^{i\nu\phi(x,y)} dx.$$

Here M^* stands for the formal adjoint operator of M:

$$M^* = \sum_{j=1}^{n} b_j(x, y) D_{x_j} + b_0(x, y),$$

where

$$b_j(x, y) = \frac{i\nu^{1/2}\zeta_j}{1+\nu|\zeta|^2}, \quad j = 1, 2, \ldots, n, \tag{3.2.21}$$

$$b_0(x, y) = \frac{1}{1+\nu|\zeta|^2} + \sum_{j=1}^{n} D_{x_j} b_j(x, y). \tag{3.2.22}$$

After repeating this process $p' + n$ times, we have

$$I_\varepsilon(a_\lambda)(y) = \int_{\mathbf{R}^n} M^{*p'+n}\{\omega_\varepsilon(x) a_\lambda(x, y)\} e^{i\nu\phi(x,y)}\, dx.$$

Let $[M^{*p'+n}, \omega_\epsilon] = M^{*p'+n}\omega_\epsilon - \omega_\epsilon M^{*p'+n}$ be the commutator. Then

$$\begin{aligned} I_\epsilon(a_\lambda)(y) = &\int_{\mathbf{R}^n} \omega_\epsilon(x) M^{*p'+n}(a_\lambda(x, y)) e^{i\nu\phi(x,y)}\, dx \\ &+ \int_{\mathbf{R}^n} \{[M^{*p'+n}, \omega_\epsilon] a_\lambda(x, y)\} e^{i\nu\phi(x,y)}\, dx. \end{aligned} \tag{3.2.23}$$

We look for the bound of integrands of these two integrals.

Lemma 3.3 *For $k = 1, 2, 3, \ldots$, let*

$$A_k = \sum_{|\alpha| \le k} \sup_{x \in \mathbf{R}^n} (1 + |\lambda| + |x| + |y|)^{-p} |\partial_x^\alpha a_\lambda(x, y)|. \tag{3.2.24}$$

Then there exists a positive constant C such that for any $(x, y) \in \mathbf{R}^{m+n}$

$$\left|M^{*p'+n} a_\lambda(x, y)\right| \le C(1 + \nu|\zeta|^2)^{-(p'-p+n)/2} A_{p'+n}(1 + |\lambda| + |y|)^p, \tag{3.2.25}$$

$$\left|[M^{*p'+n}, \omega_\epsilon] a_\lambda(x, y)\right| \le C(1 + \nu|\zeta|^2)^{-(p'-p+n)/2} A_{p'+n-1}(1 + |\lambda| + |y|)^p. \tag{3.2.26}$$

Moreover, for all $x \in \mathbf{R}^n$,

$$\lim_{\varepsilon \to 0} \left\{[M^{*p'+n}, \omega_\varepsilon] a_\lambda(x, y)\right\} = 0. \tag{3.2.27}$$

Assuming that this lemma is true for the moment, we continue the proof of Theorem. By property 1 of Definition 3.1 and (3.2.25),

$$|\omega_\epsilon(x)\big(M^{*p'+n}a_\lambda(x,y)\big)e^{i\nu\phi(x,y)}| \le C(1+\nu|\zeta|^2)^{-(p'-p+n)/2}A_{p'+n}(1+|\lambda|+|y|)^p.$$

Since $p'-p+n>n$, (3.2.17) and change of variables yield

$$\int_{\mathbf{R}^n}(1+\nu|\zeta|^2)^{-(p'-p+n)/2}\,dx \le \nu^{-n/2}C\int_{\mathbf{R}^n}(1+|x-x^*|^2)^{-(p'-p+n)/2}\,dx<\infty. \tag{3.2.28}$$

Since $\lim_{\epsilon\to 0}\omega_\epsilon(x)=1$, it follows from Lebesgue's convergence theorem that

$$\lim_{\epsilon\to+0}\int_{\mathbf{R}^n}\omega_\epsilon(x)\big(M^{*p'+n}a_\lambda(x,y)\big)e^{i\nu\phi(x,y)}\,dx=\int_{\mathbf{R}^n}\big(M^{*p'+n}a_\lambda(x,y)\big)e^{i\nu\phi(x,y)}\,dx.$$

The right-hand side of this equality is an absolutely convergent integral independent of choice of $\{\omega_\epsilon\}$ because of (3.2.25) and (3.2.28). It follows from Lebesgue's convergence theorem together with (3.2.26), (3.2.17) and (3.2.27) that

$$\lim_{\epsilon\to+0}\int_{\mathbf{R}^n}\{[M^{*p'+n},\omega_\epsilon]a_\lambda(x,y)\}e^{i\nu\phi(x,y)}\,dx=0.$$

Consequently, (3.2.19) has been proved. Making use of (3.2.25), (3.2.17) and (3.2.28), we obtain

$$\begin{aligned}|I(a_\lambda)(y)| &\le CA_{p'+n}(1+|\lambda|+|y|)^p\int_{\mathbf{R}^n}(1+\nu|\zeta|^2)^{-(p'-p+n)/2}dx\\ &= C\nu^{-n/2}A_{p'+n}(1+|\lambda|+|y|)^p.\end{aligned}$$

This proves (3.2.20). Proof of Theorem 3.2 is complete up to the proof of Lemma 3.3.

Proof of Lemma 3.3.

We claim that for any multi-index α there exists a positive constant C_α such that

$$|D_x^\alpha b_j(x)|\le C_\alpha(1+\nu|\zeta|^2)^{-1/2},\quad j=0,1,2,\dots,n. \tag{3.2.29}$$

First we prove (3.2.29) in the case $j\ge 1$ by induction on $|\alpha|$. Inequality (3.2.29) is clear if $\alpha=0$. Suppose that (3.2.29) is true in the case $0\le|\alpha|\le k-1$. Let α be a multi-index with $|\alpha|=k$. By definition (3.2.21),

$$(1+\nu|\zeta|^2)b_j=i\nu^{1/2}\zeta_j.$$

Apply D_x^α to this. Then Leibniz's rule gives

$$(1+\nu|\zeta|^2)D_x^\alpha b_j = -\sum_{\beta<\alpha}\binom{\alpha}{\beta}D_x^\beta b_j D_x^{\alpha-\beta}(1+\nu|\zeta|^2) + i\nu^{1/2}D_x^\alpha \zeta_j.$$

Since $|\beta| \le k-1$, we can use the induction hypothesis and

$$|D_x^\beta b_j| \le C_\beta(1+\nu|\zeta|^2)^{-1/2}.$$

By (3.2.7) of Assumption 3.1, function $\nu^{1/2}D_x^\alpha \zeta_j$ is uniformly bounded. Since $\alpha - \beta \neq 0$, there exists a positive constant C such that

$$|D_x{}^{\alpha-\beta}(1+\nu|\zeta|^2)| \le C(1+\nu|\zeta|^2)^{1/2}.$$

Consequently,

$$|(1+\nu|\zeta|^2)D_x^\alpha b_j| \le C(1+\nu|\zeta|^2)^{1/2}.$$

This proves (3.2.29) for $\alpha = k$. The induction process is complete and (3.2.29) in the case $1 \le j$ has been proved.

The case $j = 0$ is proved in the similar way. The claim is now proved.

It is clear that there exist functions $B_{\alpha,k}(x, y)$ and $B_k(x, y)$ such that

$$M^{*k} = \sum_{|\alpha|\le k} B_{\alpha,k}(x, y)D_x^\alpha + B_k(x, y). \tag{3.2.30}$$

We shall show that for any α and β

$$|D_x^\beta B_{\alpha,k}(x, y)| \le C(1+\nu|\zeta|^2)^{-k/2}, \quad |D_x^\beta B_k(x, y)| \le C(1+\nu|\zeta|^2)^{-k/2}, \tag{3.2.31}$$

with some constant $C > 0$.

We prove these by induction on k. In the case $k = 1$, (3.2.31) is true by the above claim (3.2.29). Suppose it is true in the case of k. Since

$$M^{*k+1} = \Big(\sum_{j=1}^n b_j D_{x_j} + b_0\Big)\Big(\sum_{|\alpha|\le k} B_{\alpha,k}(x)\,(D_x)^\alpha + B_k(x)\Big),$$

$B_{\alpha,k+1}$ and B_{k+1} are sums of $b_j B_{\beta,k}$, $b_j B_k$, $b_j D_{x_j} B_{\beta,k}$ and $b_j D_{x_j} B_k$ with β such that $|\beta| \le k$. It follows from (3.2.29) and the induction hypothesis that (3.2.31) for $k+1$ holds. The induction has been completed.

Now we can prove (3.2.25). By (3.2.30),

$$M^{*p'+n}a_\lambda(x, y) = \sum_{|\alpha|\le p'+n} B_{\alpha,p'+n}(x)D_x^\alpha a_\lambda(x, y) + B_{p'+n}(x)a_\lambda(x, y). \tag{3.2.32}$$

As a result of (3.2.9) of Assumption 3.1,

$$|B_{\alpha,p'+n}(x,y)D_x^\alpha a_\lambda(x,y)| \leq C(1+\nu|\zeta|^2)^{-(p'+n)/2}A_{|\alpha|}(1+|\lambda|+|x|+|y|)^p, \tag{3.2.33}$$

where A_k is the constant defined in (3.2.24). It is clear from (3.2.15) and (3.2.13) that with some constant $C > 0$,

$$(1+|\lambda|+|x|+|y|) \leq (1+|x-x^*|)(1+|\lambda|+|x^*|+|y|) \leq C(1+|\zeta|^2)^{1/2}(1+|\lambda|+|y|).$$

This and (3.2.33) give

$$|B_{\alpha,p'+n}(x,y)D_x^\alpha a_\lambda(x,y)| \leq C(1+\nu|\zeta|^2)^{-(p'-p+n)/2}A_{p'+n}(1+|\lambda|+|y|)^p, \tag{3.2.34}$$

because $\nu \geq 1$. Similarly, the following estimate holds:

$$|B_{p'+n}(x)a_\lambda(x,y)| \leq C(1+\nu|\zeta|^2)^{-(p'-p+n)/2}A_{p'+n}(1+|\lambda|+|y|)^p.$$

Therefore,

$$|M^{*p'+n}a_\lambda(x,y)| \leq C(1+\nu|\zeta|^2)^{-(p'-p+n)/2}A_{p'+n}(1+\lambda+|y|)^p. \tag{3.2.35}$$

We have proved (3.2.25).

Next we discuss (3.2.26). Notice that

$$[M^{*p'+n}, \omega_\epsilon] = \sum_{|\alpha|\leq p'+n} B_{\alpha,p'+n}(x)[D_x^\alpha, \omega_\epsilon]. \tag{3.2.36}$$

Here

$$[D_x^\alpha, \omega_\epsilon] = \sum_{\beta<\alpha} \binom{\alpha}{\beta} D_x^{\alpha-\beta}\omega_\epsilon(x)D_x^\beta. \tag{3.2.37}$$

Since $\alpha - \beta \neq 0$, by Definition 3.1

$$|D_x^{\alpha-\beta}\omega_\epsilon| \leq C. \tag{3.2.38}$$

Thus we can prove (3.2.26) in just the same way as we did in proving (3.2.25).

Since $\alpha - \beta \neq 0$, $\lim_{\epsilon\to 0} D_x^{\alpha-\beta}\omega_\epsilon = 0$. (3.2.27) follows from this. We have proved Lemma 3.3. □

Therefore, proof of Theorem 3.2 is now complete. □

Definition 3.3 Let $p \geq 0$. We define norms for $a_\lambda(x,y)$:

$$\|a_\lambda\|_{(p)} = \sup_{(x,y)\in\mathbf{R}^{n+m}} (1+|\lambda|+|x|+|y|)^{-p}|a_\lambda(x,y)|,$$

$$\|a_\lambda\|_{\{p,k\}} = \sum_{\alpha+\beta\leq k} \|\partial_x^\alpha\partial_y^\beta a_\lambda(x,y)\|_{(p)} \quad \text{for } k = 0, 1, 2, \ldots.$$

It is clear that the totality $\mathscr{B}_p(\mathbf{R}^{m+n})$ of functions $a_\lambda(x, y)$ that satisfy the condition 3 of Assumption 3.1 forms a Fréchet space equipped with norms $\|a\|_{\{p,k\}}$, $k = 0, 1, 2, \ldots$. It is clear that if $p_1 \geq p$, then $\mathscr{B}_p(\mathbf{R}^{m+n}) \subset \mathscr{B}_{p_1}(\mathbf{R}^{m+n})$ and the inclusion map is continuous.

Remark 3.3 If $p = 0$, then $\mathscr{B}_0(\mathbf{R}^{m+n})$ forms an algebra. It is clear that $\mathscr{B}_p(\mathbf{R}^{m+n})$ is a $\mathscr{B}_0(\mathbf{R}^{m+n})$ module.

Lemma 3.4 *Suppose that $a_\lambda \in \mathscr{B}_p(\mathbf{R}^{m+n})$. Let $\omega \in C_0^\infty(\mathbf{R}^{m+n})$ such that*

$$\omega(x, y) = \begin{cases} 1 & \text{if } |x| + |y| \leq 1 \\ 0 & \text{if } |x| + |y| \geq 2. \end{cases}$$

For any $k = 1, 2, 3, \ldots$, define $a_{\lambda,k}(x, y) = \omega(k^{-1}x, k^{-1}y)a_\lambda(x, y)$. Then $\{a_{\lambda,k}\}_k$ is a bounded sequence in $\mathscr{B}_p(\mathbf{R}^{m+n})$. If $p_1 > p$, then the sequence $\{a_{\lambda,k}\}_k$ converges to a_λ in $\mathscr{B}_{p_1}(\mathbf{R}^{m+n})$, i.e. for any fixed $l \geq 0$,

$$\lim_{k\to\infty} \|a_\lambda - a_{\lambda,k}\|_{\{p_1,l\}} = 0. \tag{3.2.39}$$

Proof Let $\omega_k(x, y) = \omega(k^{-1}x, k^{-1}y)$. Then for any (α, β) there exists a positive constant $C_{\alpha\beta}$ such that

$$\max_{(x,y)\in\mathbf{R}^{m+n}} |\partial_x^\alpha \partial_y^\beta \omega_k(x, y)| \leq C_{\alpha\beta} k^{-|\alpha|-|\beta|}.$$

Suppose $a_\lambda(x, y) \in \mathscr{B}_p(\mathbf{R}^{m+n})$. Then for any $l = 0, 1, 2, \ldots$, $\|a_{\lambda,k}\|_{\{p,l\}} \leq C_l$ with some constant C_l independent of k. This means that $\{a_{\lambda,k}\}_k$ is bounded in $\mathscr{B}_p(\mathbf{R}^{m+n})$.

Suppose $p_1 > p$. Since $\partial_x^\alpha \partial_y^\beta (a_\lambda(x, y) - a_{\lambda,k}(x, y)) = 0$ in $\{(x, y) \in \mathbf{R}^{m+n} : |x| + |y| \leq k\}$, for any α and β with $|\alpha| + |\beta| \leq l$

$$\begin{aligned}
&\sup_{(x,y)\in\mathbf{R}^{m+n}} (1 + |\lambda| + |x| + |y|)^{-p_1} |\partial_x^\alpha \partial_y^\beta (a_\lambda(x, y) - a_{\lambda,k}(x, y))| \\
&= \sup_{|x|+|y|\geq k} (1 + |\lambda| + |x| + |y|)^{p-p_1} (1 + \lambda + |x| + |y|)^{-p} |\partial_x^\alpha \partial_y^\beta (a_\lambda(x, y) - a_{\lambda,k}(x, y))| \\
&\leq (1 + k)^{p-p_1} \sup_{|x|+|y|\geq k} (1 + |\lambda| + |x| + |y|)^{-p} |\partial_x^\alpha \partial_y^\beta (a_\lambda(x, y) - a_{\lambda,k}(x, y))| \\
&\leq (1 + k)^{p-p_1} (\|a_\lambda\|_{\{p,l\}} + \|a_{\lambda,k}\|_{\{p,l\}}).
\end{aligned}$$

Therefore, $\lim_{k\to\infty} \|a_\lambda - a_{\lambda,k}\|_{\{p_1,l\}} = 0$ for any $l \geq 0$. □

Corollary 3.3 *Let $a_\lambda \in \mathscr{B}_p(\mathbf{R}^{m+n})$ and let $a_{\lambda,k}(x, y)$ be as in Lemma 3.4. If $p_1 > p$, then*

$$\lim_{k\to\infty} \sup_{y\in\mathbf{R}^m} (1 + |\lambda| + |y|)^{-p_1} |I(a_\lambda)(y) - I(a_{\lambda,k})(y)| = 0.$$

Proof This is a consequence of Lemma 3.4 and Theorem 3.2.

Definition 3.4 If $f_\lambda(y)$ is a function of y with parameter λ, for $p \geq 0$ we define

$$|||| f_\lambda ||||_{(p)} = \sup_{y \in \mathbf{R}^m} (1 + |\lambda| + |y|)^{-p} |f_\lambda(y)|. \tag{3.2.40}$$

We write $C_{(p)}(\mathbf{R}^m)$ for the Banach space of continuous functions $f_\lambda(y)$ such that $|||| f_\lambda ||||_{(p)} < \infty$.

Remark 3.4 It is clear that $C_{(p)}(\mathbf{R}^m)$ is a module over the algebra $C_{(0)}(\mathbf{R}^m)$.

The next corollary follows from (3.2.20) in Theorem 3.2.

Corollary 3.4 *Suppose that Assumption 3.1 holds. Then there exists a positive constant C independent of $a_\lambda(x, y)$ such that*

$$\nu^{n/2} |||| I(a_\lambda) ||||_{(p)} \leq C \sum_{|\alpha| \leq p+n+1} \| \partial_x^\alpha a_\lambda \|_{(p)}, \tag{3.2.41}$$

and $I(a_\lambda)(y)$ is a continuous function of y. The correspondence $a_\lambda(x, y) \to \nu^{n/2}$ $I(a_\lambda)(y)$ is a continuous linear map from $\mathscr{B}_p(\mathbf{R}^{m+n})$ to $C_{(p)}(\mathbf{R}^m)$.

Proof For any $a_\lambda \in \mathscr{B}_p(\mathbf{R}^{n+m})$ the value $\nu^{n/2} I(a_\lambda)(y)$ is well-defined and satisfies (3.2.41) because of (3.2.20) in Theorem 3.2.

We use a continuity argument to prove that $I(a_\lambda)(y)$ is a continuous function. Let $p_1 > p$. We can choose a sequence of functions $\{a_{\lambda,k}(x, y)\}_{k=1,2,\ldots} \subset C_0^\infty(\mathbf{R}^{m+n})$ such that $\{a_{\lambda,k}\}$ is a bounded sequence in $\mathscr{B}_p(\mathbf{R}^{m+n})$ and $\lim_{k\to\infty} a_{\lambda,k} = a_\lambda$ in $\mathscr{B}_{p_1}(\mathbf{R}^{n+m})$ by Lemma 3.4. Then $I(a_{\lambda,k})(y)$ is a continuous function of y and $\lim_{k\to\infty} \sum_{|\alpha| \leq p_1+n+1} \|\partial_x^\alpha (a_\lambda - a_{\lambda,k})\|_{(p_1)} = 0$. Applying (3.2.41) to $I(a_\lambda - a_{\lambda,k})$, we have

$$\begin{aligned} &|||| \nu^{n/2} I(a_\lambda) - \nu^{n/2} I(a_{\lambda,k}) ||||_{(p_1)} = \nu^{n/2} |||| I(a_\lambda - a_{\lambda,k}) ||||_{(p_1)} \\ &\leq C \sum_{|\alpha| \leq p_1+n+1} \|\partial_x^\alpha (a_\lambda - a_{\lambda,k})\|_{(p_1)} \quad \to 0 \quad \text{as } k \to \infty. \end{aligned}$$

The corollary has been proved. □

Lemma 3.5 (Integration by parts formula) *Suppose that Assumption 3.1 holds. Then the following equality of oscillatory integrals holds:*

$$\int_{\mathbf{R}^n} \partial_{x_j} \phi(x, y) a_\lambda(x, y) e^{i\nu\phi(x,y)} \, dx = -(i\nu)^{-1} \int_{\mathbf{R}^n} \partial_{x_j} a_\lambda(x, y) e^{i\nu\phi(x,y)} \, dx. \tag{3.2.42}$$

Proof Since $a_\lambda(x, y) \in \mathscr{B}_p(\mathbf{R}^{m+n})$, it is clear that $\partial_{x_j} a_\lambda(x, y) \in \mathscr{B}_p(\mathbf{R}^{m+n})$. As a consequence of Lemma 3.1 there exists a positive constant C such that

$$|\partial_{x_j} \phi(x, y)| \leq C|x - x^*| \leq C(|x| + |x^*|) \leq C(1 + |x| + |y|).$$

By (3.2.7), for any α and β with $\alpha+\beta \neq 0$ there exists a positive constant C such that

$$|\partial_x^\alpha \partial_y^\beta (\partial_{x_j}\phi(x, y))| \leq C_{\alpha\beta}.$$

It follows from these that $\partial_{x_j}\phi(x, y)a_\lambda(x, y) \in \mathscr{B}_{p+1}(\mathbf{R}^{m+n})$.

We use the continuity argument to prove (3.2.42). Let $p_1 > p$. Then by Lemma 3.4 we can choose a sequence $\{a_{\lambda,k}(x, y)\}_k \subset C_0^\infty(\mathbf{R}^{m+n})$ such that $a_{\lambda,k} \to a_\lambda$ in $\mathscr{B}_{p_1}(\mathbf{R}^{m+n})$. Then $\partial_{x_j}\phi\, a_{\lambda,k} \to \partial_{x_j}\phi\, a_\lambda$ in $\mathscr{B}_{p_1+1}(\mathbf{R}^{m+n})$ and $\partial_{x_j}a_{\lambda,k} \to \partial_{x_j}a_\lambda$ in $\mathscr{B}_{p_1}(\mathbf{R}^{m+n})$. Thus $I(\partial_{x_j}\phi\, a_{\lambda,k})(y) \to I(\partial_{x_j}\phi\, a_\lambda)(y)$ and $I(\partial_{x_j}a_{\lambda,k})(y) \to I(\partial_{x_j}a_\lambda)$ (y) for each y. Obviously equality (3.2.42) with a_λ replaced by $a_{\lambda,k}$ is true for each k. Tending $k \to \infty$, we prove (3.2.42). □

The next lemma is simple but very useful in the subsequent discussions.

Lemma 3.6 *There exist functions $q_{kl}(x, y) \in \mathscr{B}_0(\mathbf{R}^{m+n})$, for $k, l = 1, 2, \ldots, n$, such that*

$$x_k - x_k^* = \sum_{l=1}^{n} q_{kl}(x, y)\zeta_l(x, y). \tag{3.2.43}$$

Proof First we make a remark that the choice of $q_{kl}(x, y)$ is not unique. For example, we can use the following formula:

$$x_k - x_k^* = \sum_{l=1}^{n} \zeta_k \int_0^1 \partial_{\zeta_l} x_k(t\zeta_\lambda(x, y), y)\, dt.$$

However, we prefer the following construction. We define for $j, k = 1, 2, \ldots, n$,

$$h_{jk}(x, y) = \int_0^1 \partial_{x_k}\zeta_j(tx + (1-t)x^*, y)\, dt = \int_0^1 (\partial_{x_k}\partial_{x_j}\phi)(tx + (1-t)x^*, y)\, dt. \tag{3.2.44}$$

Then by Taylor's formula,

$$\zeta_j(x, y) = \sum_{k=1}^{n} h_{jk}(x, y)(x_k - x_k^*). \tag{3.2.45}$$

Let $H(x, y) = (h_{jk}(x, y))_{jk}^n$ be the $n \times n$ matrix whose (j, k) entry is $h_{jk}(x, y)$. Then it follows from Assumption 3.1 that $H(x^*, y)$ is invertible and that there exists a positive constant ϵ independent of y such that $H(x, y)^{-1}$ exists in $\{x;\, |x - x^*| \leq 3\epsilon\}$. We write $\tilde{q}_{jk}(x, y)$ for the (j, k) entry of $H(x, y)^{-1}$ there. Thus if $|x - x^*| \leq 3\epsilon$,

$$\tilde{q}_{jk}(x, y)h_{kl}(x, y) = \delta_{jl}, \tag{3.2.46}$$

where δ_{jl} is Kronecker's symbol, i.e. $\delta_{jl} = 0$ if $j \neq l$ and $\delta_{jj} = 1$. Let $\omega(x, y)$ be a real-valued $C_0^\infty(\mathbf{R}^n)$ function such that

$$\omega(x) = \begin{cases} 1 & \text{if } |x| \le \epsilon, \\ 0 & \text{if } |x| \ge 2\epsilon. \end{cases}$$

Since by Lemma 3.2 $\sum_{l=1}^{n} \zeta_l^2 > C|x - x^*|^2$ with some constant C independent of y, we can define, for any j, k and $(x, y) \in \mathbf{R}^{m+n}$,

$$q_{jk}(x, y) = \omega(x - x^*(y))\tilde{q}_{jk}(x, y) + \big(1 - \omega(x - x^*(y))\big)\frac{(x_j - x_j^*)\zeta_k(x, y)}{\sum_{l=1}^{n} \zeta_l(x, y)^2}. \tag{3.2.47}$$

Then for any j with $1 \le j \le n$,

$$\begin{aligned} &\sum_{k=1}^{n} q_{jk}(x, y)\zeta_k(x, y) \\ &= \omega(x - x^*(y)) \sum_{k=1}^{n} \tilde{q}_{jk}(x, y)h_{kl}(x, y)(x_l - x_l^*) + \big(1 - \omega(x - x^*(y))\big) \sum_{k=1}^{n} \frac{(x_j - x_j^*)\zeta_k(x, y)^2}{\sum_{l=1}^{n} \zeta_l(x, y)^2} \\ &= x_j - x_j^*. \end{aligned}$$

Moreover, it is clear that $q_{jk} \in \mathscr{B}_0(\mathbf{R}^{m+n})$. We have proved the lemma. □

Remark 3.5 Let $q_{jk}(x, y)$ and $h_{jk}(x, y)$ be as above. Then by construction

$$q_{jk}(x, y)h_{kl}(x, y) = \delta_{jl} \tag{3.2.48}$$

if $|x - x^*| < \epsilon$.

We shall discuss derivatives of $I(a_\lambda)(y)$.

Theorem 3.3 *Assume that $a_\lambda \in \mathscr{B}_p(\mathbf{R}^{m+n})$. Then the oscillatory integral $I(a_\lambda)(y)$ is continuously differentiable with respect to y. For any $j = 1, 2, \ldots, m$,*

$$\partial_{y_j} \int_{\mathbf{R}^n} a_\lambda(x, y)e^{i\nu\phi(x,y)}\, dx = \int_{\mathbf{R}^n} \partial_{y_j}(a_\lambda(x, y)e^{i\nu\phi(x,y)})\, dx. \tag{3.2.49}$$

Both sides are interpreted as oscillatory integrals.

Moreover, there exist $p_{jk}(x, y) \in \mathscr{B}_0(\mathbf{R}^{m+n})$, $k = 0, 1, 2, \ldots, n$, such that the function

$$a_j(x, y) = \partial_{y_j} a_\lambda(x, y) + p_{j0}(x, y)a_\lambda(x, y) + \sum_{k=1}^{n} p_{jk}(x, y)\partial_{x_k} a_\lambda(x, y) \tag{3.2.50}$$

satisfies

$$\partial_{y_j}\big(e^{-i\nu\phi(x^*,y)} I(a_\lambda)(y)\big) = e^{-i\nu\phi(x^*,y)} \int_{\mathbf{R}^n} a_j(x, y)e^{i\nu\phi(x,y)}\, dx. \tag{3.2.51}$$

For any α, β there exists a positive constant $C_{\alpha\beta}$ such that for any $(x, y) \in \mathbf{R}^{m+n}$

$$|\partial_x^\alpha \partial_y^\beta a_j(x, y)| \le C_{\alpha\beta} \sum_{|\alpha'| \le |\alpha|+1, \beta' \le \beta} |\partial_x^{\alpha'} \partial_y^{\beta'} a_\lambda(x, y)|. \tag{3.2.52}$$

We have

$$|||\nu^{n/2} \partial_{y_j} I(a_\lambda)|||_{(p)} \le C\Big(\sum_{|\alpha| \le p+n+2} \|\partial_x^\alpha a_\lambda\|_{(p)} + \sum_{|\alpha| \le p+n+1} \|\partial_x^\alpha \partial_{y_j} a_\lambda\|_{(p)} \Big). \tag{3.2.53}$$

Proof Suppose that $a_\lambda \in \mathscr{B}_p(\mathbf{R}^{m+n})$. We prove (3.2.49) by the continuity argument. Let $p_1 > p$. By Lemma 3.4, we can choose a sequence $\{a_{\lambda,k}\}_{k=1,2,\ldots} \subset C_0^\infty(\mathbf{R}^{m+n})$ such that $\{a_{\lambda,k}\}$ is bounded in $\mathscr{B}_p(\mathbf{R}^{m+n})$ and $a_{\lambda,k} \to a_\lambda$ in $\mathscr{B}_{p_1}(\mathbf{R}^{m+n})$ as $k \to \infty$. Since $a_{\lambda,k} \in C_0^\infty(\mathbf{R}^{m+n})$, $\partial_{y_j} I(a_{\lambda,k})(y)$ exists and

$$\partial_{y_j} I(a_{\lambda,k})(y) = \int_{\mathbf{R}^n} (\partial_{y_j} a_{\lambda,k}(x, y) + i\nu \partial_{y_j} \phi(x, y) a_{\lambda,k}(x, y)) e^{i\nu\phi(x,y)}\, dy. \tag{3.2.54}$$

It is clear that $\partial_{y_j} a_{\lambda,k}(x, y) \to \partial_{y_j} a_\lambda(x, y)$ in $\mathscr{B}_{p_1}(\mathbf{R}^{m+n})$ and that $\partial_{y_j} \phi(x, y) a_{\lambda,k}(x, y) \to \partial_{y_j} \phi(x, y) a_{\lambda,k}(x, y)$ in $\mathscr{B}_{p_1+1}(\mathbf{R}^{m+n})$ as $k \to \infty$. Therefore, as $k \to \infty$, the right-hand side of (3.2.54) converges uniformly on compact sets in y-space to

$$\int_{\mathbf{R}^n} (\partial_{y_j} a_\lambda(x, y) + i\nu \partial_{y_j} \phi(x, y) a_\lambda(x, y)) e^{i\nu\phi(x,y)}\, dy. \tag{3.2.55}$$

This implies that $\partial_{y_j} I(a_\lambda)(y)$ exists and is equal to (3.2.55). Therefore, (3.2.49) has been proved.

Using (3.2.49), we have

$$\begin{aligned} &\partial_{y_j} (e^{-i\nu\phi(x^*,y)} I(a_\lambda)(y)) \\ &= e^{-i\nu\phi(x^*,y)} \int_{\mathbf{R}^n} \Big(\partial_{y_j} a_\lambda(x, y) + i\nu(\partial_{y_j} \phi(x, y) - \partial_{y_j} \phi(x^*, y)) a_\lambda(x, y)\Big) e^{i\nu\phi(x,y)}\, dx. \end{aligned} \tag{3.2.56}$$

Taylor's formula gives

$$\partial_{y_j} \phi(x, y) - \partial_{y_j} \phi(x^*, y) = \sum_{l=1}^n h_{jl}(x, y)(x_l - x_l^*),$$

with

$$h_{jl}(x, y) = \int_0^1 \partial_{x_l} \partial_{y_j} \phi(tx + (1-t)x^*, y)\, dt.$$

$h_{jk}(x, y) \in \mathscr{B}_0(\mathbf{R}^{m+n})$ because of Assumption 3.1.

Applying Lemmas 3.6 and 3.5, we have

$$
\begin{aligned}
& i\nu \int_{\mathbf{R}^n} \left(\partial_{y_j}\phi(x, y) - \partial_{y_j}\phi(x^*, y)\right) a_\lambda(x, y) e^{i\nu\phi(x,y)}\, dx \\
&= i\nu \int_{\mathbf{R}^n} \Big(\sum_{k=1}^n \sum_{l=1}^n h_{jl}(x, y) q_{lk}(x, y) \partial_{x_k}\phi(x, y)\Big) e^{i\nu\phi(x,y)} a_\lambda(x, y)\, dx \\
&= -\int_{\mathbf{R}^n} \Big(\sum_{k=1}^n \sum_{l=1}^n \partial_{x_k}\big(h_{jl}(x, y) q_{lk}(x, y) a_\lambda(x, y)\big)\Big) e^{i\nu\phi(x,y)}\, dx.
\end{aligned}
$$

Equalities (3.2.51) and (3.2.50) follow from this and (3.2.56) if we define

$$
\begin{aligned}
a_j(x, y) &= \partial_{y_j} a_\lambda(x, y) - \sum_{k=1}^n \sum_{l=1}^n \partial_{x_k}(h_{jl}(x, y) q_{lk}(x, y) a_\lambda(x, y)), \\
p_{jk}(x, y) &= -\sum_{l=1}^n h_{jl}(x, y) q_{lk}(x, y), \quad \text{for } j = 1, 2, \ldots, n, \\
p_{j0}(x, y) &= -\sum_{k=1}^n \sum_{l=1}^n \partial_{x_k}(h_{jl}(x, y) q_{lk}(x, y)).
\end{aligned}
$$

Since $h_{jk} \in \mathscr{B}_0(\mathbf{R}^{m+n})$ and $q_{kl} \in \mathscr{B}_0(\mathbf{R}^{m+n})$, we have $p_{jk}(x, y) \in \mathscr{B}_0(\mathbf{R}^{m+n})$. Inequality (3.2.52) is a consequence of (3.2.50). (3.2.53) follows from (3.2.52) and Corollary 3.4. We have proved the theorem. □

Let $\mathscr{B}_p(\mathbf{R}^m)$ be the space of C^∞ functions $f_\lambda(y)$ of y with parameter λ such that $|||\partial_y^\beta f_\lambda|||_{(p)} < \infty$ for any multi-index β. This is a Fréchet space equipped with seminorms $\sum_{|\beta|\le k} |||\partial_y^\beta f_\lambda|||_{(p)}$, $k = 0, 1, 2, \ldots$.

Corollary 3.5 *Suppose that Assumption 3.1 holds. Then the function $I(a_\lambda)(y)$ is an infinitely differentiable function of y. For any multi-index β there exists a function $a_{\lambda\beta}(x, y) \in \mathscr{B}_p(\mathbf{R}^{m+n})$ such that*

$$
\partial_y^\beta (e^{-i\nu\phi(x^*,y)} I(a_\lambda)(y)) = e^{-i\nu\phi(x^*,y)} I(a_{\lambda\beta})(y). \tag{3.2.57}
$$

We can choose $a_{\lambda\beta}(x, y)$ in such a way that

$$
a_{\lambda\beta}(x, y) = \partial_y^\beta a_\lambda(x, y) + \sum_{\substack{|\alpha|+|\beta'|\le|\beta| \\ \beta'<\beta}} p_{\alpha\beta}(x, y) \partial_x^\alpha \partial_y^{\beta'} a_\lambda(x, y), \tag{3.2.58}
$$

where $p_{\alpha\beta}(x, y) \in \mathscr{B}_0(\mathbf{R}^{m+n})$. There exists a positive constant C such that

$$|||\partial_y^\beta(e^{-i\nu\phi(x^*,y)}I(a_\lambda)(y))|||_{(p)} \le C \sum_{\substack{\beta'\le\beta \\ |\alpha|\le|\beta|+n+p+2}} \|\partial_x^\alpha\partial_y^{\beta'}a_\lambda\|_{(p)}. \tag{3.2.59}$$

The correspondence $a_\lambda(x, y) \to e^{-i\nu\phi(x^*,y)}I(a_\lambda)(y)$ *is a continuous linear map from the Fréchet space* $\mathscr{B}_p(\mathbf{R}^{m+n})$ *to the Fréchet space* $\mathscr{B}_p(\mathbf{R}^m)$.

Proof Apply Lemma 3.2.49 repeatedly. Then we can prove that $I(a_\lambda)(y)$ is infinitely differentiable and that equations (3.2.57) and (3.2.58) hold. Applying Corollary 3.4 to $I(a_\beta)(y)$, we have (3.2.59), which means that the correspondence $a_\lambda(x, y) \to e^{-i\nu\phi(x^*,y)}I(a_\lambda)(y)$ is a continuous linear map from the Fréchet space $\mathscr{B}_p(\mathbf{R}^{m+n})$ to the Fréchet space $\mathscr{B}_p(\mathbf{R}^m)$. □

Corollary 3.6 *We can write that*

$$I(a_\lambda)(y) = e^{i\nu\phi(x^*,y)}b(y). \tag{3.2.60}$$

Then for any α *there exists a constant* $C_\alpha > 0$ *such that*

$$\begin{aligned}&\sup_y(1 + |\lambda| + |y|)^{-p}|\partial_y^\beta b(y)| \\ &\le C_\beta \sum_{\substack{\beta'\le\beta \\ |\alpha|\le|\beta|+n+p+2}} \sup_{x,y}(1 + |\lambda| + |x| + |y|)^{-p}|\partial_x^\alpha\partial_y^{\beta'}a_\lambda(x, y)|.\end{aligned} \tag{3.2.61}$$

Let $d(x, y) = |\det(\partial_{x_j}\partial_{x_k}\phi(x, y))_{jk}|$. *Then for any* β *there exists a constant* $C \ge 0$ *such that*

$$\begin{aligned}&\sup_y(1 + |\lambda| + |y|)^{-p}|\partial_y^\beta(d(x^*, y)^{-1/2}b(y))| \\ &\le C \sum_{\substack{\beta'\le\beta \\ |\alpha|\le|\beta|+n+p+2}} \sup_{x,y}(1 + |\lambda| + |x| + |y|)^{-p}|\partial_x^\alpha\partial_y^{\beta'}a_\lambda(x, y)|.\end{aligned} \tag{3.2.62}$$

Proof (3.2.61) is nothing but another expression of (3.2.59). Estimate (3.2.62) follows from the chain rule, (3.2.61) and (3.2.18). □

Next we shall discuss iterated integration of oscillatory integrals.

Let $A(x, y) = (\partial_{x_j}\partial_{x_k}\phi(x, y))_{j,k=1}^n$, $B(x, y) = (\partial_{y_k}\partial_{x_j}\phi(x, y))_{j=1,k=1}^{n,m}$ and $C(x, y) = (\partial_{y_k}\partial_{y_j}\phi(x, y))_{j,k=1}^n$. Then the Hessian matrix $Hess_{x,y}\phi(x, y)$ of $\phi(x, y)$ with respect to variables $(x, y) \in \mathbf{R}^{m+n}$ is the $(m+n) \times (m+n)$ square matrix with block expression:

$$Hess_{x,y}\phi(x, y) = \begin{pmatrix} A(x, y) & B(x, y) \\ {}^tB(x, y) & C(x, y) \end{pmatrix}.$$

The following simple fact plays a role.

Lemma 3.7 *Consider a* $(m+n) \times (m+n)$ *square matrix with block expression*

$$\begin{pmatrix} A & B \\ {}^tB & C \end{pmatrix},$$

where A *is an* $n \times n$ *square matrix,* B *is an* $n \times m$ *matrix,* C *is an* $m \times m$ *square matrix and* ${}^tB$ *is the transposed matrix of* B. *Suppose that* A *is invertible. Then*

$$\det \begin{pmatrix} A & B \\ {}^tB & C \end{pmatrix} = \det A \cdot \det(C - {}^tBA^{-1}B).$$

Proof Take the determinant of both sides of the following equality:

$$\begin{pmatrix} A & B \\ {}^tB & C \end{pmatrix} \begin{pmatrix} I & -A^{-1}B \\ 0 & I \end{pmatrix} = \begin{pmatrix} A & 0 \\ {}^tB & C - {}^tBA^{-1}B \end{pmatrix}.$$

□

Assumption 3.2 Let $\det Hess_{x,y}\phi(x, y)$ be the determinant of Hessian matrix of $\phi(x, y)$. There exists a positive constant c_1 such that for any $(x, y) \in \mathbf{R}^{m+n}$

$$|\det Hess_{x,y}\phi(x, y)| = \left|\det \begin{pmatrix} A(x, y) & B(x, y) \\ {}^tB(x, y) & C(x, y) \end{pmatrix}\right| > c_1 > 0.$$

Theorem 3.4 (Iterated integration) *Suppose that the above Assumption 3.2 holds in addition to Assumption 3.1. Let* $x^* = x^*(y)$ *denote the stationary point of* $x \to \phi(x, y)$ *if* y *is fixed. Then* $\phi^*(y) = \phi(x^*, y)$ *has following properties:*

1. *For any* $|\alpha| \geq 2$, $\sup_{y \in \mathbf{R}^m} |\partial_y^\alpha \phi^*(y)| < \infty$.
2. *There exists a positive constant* c_2 *such that*

$$|\det (\partial_{y_j}\partial_{y_k}\phi^*(y))_{jk=1}^m| \geq c_2 > 0. \tag{3.2.63}$$

Moreover, there holds equality between oscillatory integrals:

$$\int_{\mathbf{R}^{m+n}} a_\lambda(x, y)e^{i\nu\phi(x,y)}dxdy = \int_{\mathbf{R}^m} \left(e^{-i\nu\phi^*(y)}I(a)(y)\right) e^{i\nu\phi^*(y)}\, dy. \tag{3.2.64}$$

This may be written as

$$\int_{\mathbf{R}^{m+n}} a_\lambda(x, y)e^{i\nu\phi(x,y)}dxdy = \int_{\mathbf{R}^m} \left(\int_{\mathbf{R}^n} a_\lambda(x, y)e^{i\nu\phi(x,y)}\, dx\right) dy.$$

Proof Since $x^*(y)$ is the stationary point, for any $j, k = 1, 2, \dots, m$,

$$\partial_{y_k}\partial_{y_j}\phi^*(y) = \partial_{y_k}\partial_{y_j}\phi(x, y)\big|_{x=x^*} + \sum_{l=1}^{n} \partial_{x_l}\partial_{y_j}\phi(x, y)\big|_{x=x^*}\partial_{y_k}x_l^*.$$

It is clear from this and (3.2.16) in Lemma 3.2 and the chain rule that $\sup_{y\in\mathbf{R}^m} |\partial_y^\alpha\phi^*(y)| < \infty$ if $|\alpha| \geq 2$. Differentiating $\partial_{x_j}\phi(x^*, y) = 0$ with respect to y_k, we have

$$\sum_{l=1}^{n} \partial_{x_l}\partial_{x_j}\phi(x^*, y)\partial_{y_k}x_l^* = -\partial_{y_k}\partial_{x_j}\phi(x^*, y) \quad \text{for all } 1 \leq j \leq n, 1 \leq k \leq m.$$

Thus

$$(\partial_{y_k}\partial_{y_j}(\phi^*(y)))_{j,k=1}^{n} = C(x^*, y) - {}^tB(x^*, y)A(x^*, y)^{-1}B(x^*, y).$$

It follows from Lemma 3.7 and Assumption 3.2 that

$$c_0|\det A(x^*, y)|^{-1} \leq |\det(\partial_{y_j}\partial_{y_k}\phi^*(y))_{j,k}^{m}|.$$

By (3.2.7) of Assumption 3.1 there exists a constant $M > 0$ such that

$$|\partial_{x_j}\partial_{x_k}\phi(x, y)| \leq M. \tag{3.2.65}$$

This means that the length of each column vector of the matrix $A(x^*, y)$ does not exceed $n^{1/2}M$. Considering the n-dimensional volume of the fundamental domain of the lattice in $\mathbf{R}^n$ generated by these column vectors, we have Hadamard's inequality:

$$|\det A(x^*, y)| \leq n^{n/2}M^n. \tag{3.2.66}$$

Therefore,

$$c_0 n^{-n/2}M^{-n} \leq |\det(\partial_{y_j}\partial_{y_k}\phi^*(y))_{j,k}^{m}|. \tag{3.2.67}$$

This proves (3.2.63). Both properties of $\phi^*(y)$ have been proved.

Now we regard the right-hand side of (3.2.64) as an oscillatory integral with $\phi^*(y)$ as the phase function and $e^{-i\nu\phi^*(y)}I(a_\lambda)(y)$ as the amplitude function. We shall check that they satisfy Assumption 3.1 with n replaced by m, m replaced by 0 and x replaced by y. Properties 1 and 2 of $\phi^*(y)$ correspond to (3.2.7) and (3.2.8) of Assumption 3.1, respectively. Since $a_\lambda(x, y) \in \mathscr{B}_p(\mathbf{R}^{m+n})$, inequality (3.2.59) holds. This means (3.2.59) corresponds to (3.2.9) of Assumption 3.1 with a_λ replaced by $e^{-i\nu\phi^*(y)}I(a_\lambda)(y)$. Thus Assumption 3.1 is satisfied. It follows from Theorem 3.2 that the oscillatory integral

$$\int_{\mathbf{R}^m} \left(e^{-i\nu\phi^*(y)} I(a_\lambda)(y)\right) e^{i\nu\phi^*(x)}\, dy$$

has a definite meaning.

Now we prove equality (3.2.64) by the continuity argument. Let us define a smooth function $\omega(x)$ of $x \in \mathbf{R}^n$ and a smooth function $\chi(y)$ of $y \in \mathbf{R}^m$:

$$\omega(x) = \begin{cases} 1 & \text{if } |x| \le 1 \\ 0 & \text{if } |x| \ge 2 \end{cases}, \qquad \text{and} \qquad \chi(y) = \begin{cases} 1 & \text{if } |y| \le 1 \\ 0 & \text{if } |y| \ge 2. \end{cases} \tag{3.2.68}$$

For any positive integers k, l we set $\chi_k(y) = \chi(k^{-1}y)$ and $\omega_l(x) = \omega(l^{-1}x)$. Let us consider the double sequence

$$A_{kl} = \int_{\mathbf{R}^{m+n}} \chi_k(y)\omega_l(x) a_\lambda(x, y) e^{i\nu\phi(x,y)}\, dxdy \quad (j, k = 1, 2, \ldots,).$$

Since $\chi_k(y)\omega_l(x)a_\lambda(x, y)e^{i\nu\phi(x,y)}$ is integrable, Fubini's theorem is applied to the above integral. Thus

$$A_{kl} = \int_{\mathbf{R}^m} \chi_k(y) e^{i\nu\phi^*(y)} \left(e^{-i\nu\phi^*(y)} I(\omega_l a_\lambda)(y)\right) dy.$$

Owing to assumptions of Theorem 3.4 we can apply Theorem 3.2 with n replaced by $m + n$ and m replaced by 0. As a result the oscillatory integral on $\mathbf{R}^{m+n}$

$$A = \int_{\mathbf{R}^{m+n}} a_\lambda(x, y) e^{i\nu\phi(x,y)}\, dxdy \tag{3.2.69}$$

has a definite finite value, which we denote by A. That means

$$A = \lim_{\substack{k\to\infty \\ l\to\infty}} A_{kl} \tag{3.2.70}$$

exists. In other words, for any $\epsilon > 0$ there exists a natural number N such that if $k > N, l > N$ then

$$|A_{kl} - A| \le \epsilon. \tag{3.2.71}$$

On the other hand, by Theorem 3.2

$$\lim_{l\to\infty} \int_{\mathbf{R}^n} \omega_l(x) a_\lambda(x, y) e^{i\nu\phi(x,y)}\, dx = I(a_\lambda)(y) \tag{3.2.72}$$

exists. Moreover, if $p' > p$, then $\omega_l(x)a_\lambda(x, y) \to a_\lambda(x, y)$ in the topology of $\mathscr{B}_{p'}(\mathbf{R}^{m+n})$ and hence $e^{-i\nu\phi^*(y)} I(\omega_l a_\lambda)(y) \to e^{-i\nu\phi^*(y)} I(a_\lambda)(y)$ in $\mathscr{B}_{p'}(\mathbf{R}^m)$ as $l \to \infty$. Thus for each fixed k

$$\lim_{l\to\infty} A_{kl} = \lim_{l\to\infty} \int_{\mathbf{R}^m} \chi_k(y) \left(e^{-i\nu\phi^*(y)} I(\omega_l a_\lambda)(y)\right) e^{i\nu\phi^*(y)}\, dy$$
$$= \int_{\mathbf{R}^m} \chi_k(y) \left(e^{-i\nu\phi^*(y)} I(a_\lambda)(y)\right) e^{i\nu\phi^*(y)}\, dy.$$

Therefore, letting $l \to \infty$ in (3.2.71), we have

$$\left| \int_{\mathbf{R}^m} \chi_k(y) \left(e^{-i\nu\phi^*(y)} I(a_\lambda)(y)\right) e^{i\nu\phi^*(y)}\, dy - A \right| \le \epsilon$$

if $k > N$. Thus we have proved that

$$A = \lim_{k\to\infty} \int_{\mathbf{R}^m} \chi_k(y) \left(e^{-i\nu\phi^*(y)} I(a_\lambda)(y)\right) e^{i\nu\phi^*(y)}\, dy.$$

Equality (3.2.64) has been proved. □

3.3 Stationary Phase Method

In this section we always assume that Assumption 3.1 holds and we discuss the oscillatory integral

$$I(a_\lambda)(y) = \int_{\mathbf{R}^n} a_\lambda(x, y) e^{i\nu\phi(x,y)}\, dx. \tag{3.3.1}$$

A heuristic explanation of the reason why an oscillatory integral has a finite value is that a large-scale cancellation phenomenon takes place. If the length of the gradient vector $\operatorname{grad}_x \phi(x, y) = (\partial_{x_1}\phi(x, y), \partial_{x_2}\phi(x, y), \ldots, \partial_{x_n}\phi(x, y))$ is large at a point x, then the value of function $e^{i\nu\phi(x,y)}$ oscillates rapidly as x moves in the direction of that vector. This means that cancellation occurs here in the process of integration. Thus the contribution of the value of $a_\lambda(x, y)e^{i\nu\phi(x,y)}$ near x to the right-hand side of (3.3.1) is small, see, for example, Lemma 3.3. In other words, the principal contribution to $I(a_\lambda)(y)$ comes from the value of $a_\lambda(x, y)e^{i\nu\phi(x,y)}$ around the stationary point x^* of the phase function $\phi(x, y)$. Moreover, Taylor's formula

$$\phi(x, y) = \phi(x^*, y) + \frac{1}{2}((x - x^*), H(x - x^*)) + \cdots .$$

gives us a good approximation of $\phi(x, y)$ near x^*. Here H is the Hessian matrix of $\phi(x, y)$ at x^*, i.e. it is an $n \times n$ square matrix whose j, k entry equals $h_{jk}(x^*, y) = \partial_{x_j}\partial_{x_k}\phi(x^*, y)$ and the symbol (a, a) denotes the standard inner product of two vectors in $\mathbf{R}^n$. Therefore, the principal part of $I(a_\lambda)(y)$ is heuristically expected to be

$$a_\lambda(x^*, y)e^{i\nu\phi(x^*,y)} \int_{\mathbf{R}^n} e^{i\nu(z,H(x^*,y)z)/2}dz.$$

This phenomenon becomes more apparant as ν becomes larger. Thus we will be able to expect the asymptotic formula:

$$I(a_\lambda)(y) \approx e^{i\nu\phi(x^*,y)}(a_\lambda(x^*, y) + O(\nu^{-1})) \int_{\mathbf{R}^n} e^{i\nu(z,H(x^*,y)z)/2}dz,$$

as $\nu \to \infty$.

This is called the stationary phase method. We give a rigorous proof of this formula based on the integration by parts formula in the following subsection.

3.3.1 Application of Integration by Parts Formula

Let $h_{jk}(x, y)$ and $q_{jk}(x, y)$ be as in Lemma 3.6. Then we have

Lemma 3.8 *Let* $a_\lambda(x, y) \in \mathscr{B}_p(\mathbf{R}^{m+n})$. *Then* $\sum_{k=1}^n \partial_{x_k}(q_{jk}a_\lambda) \in \mathscr{B}_p(\mathbf{R}^{m+n})$ *and*

$$I((x_j - x_j^*)a_\lambda)(y) = -(i\nu)^{-1}I\Big(\sum_{k=1}^n \partial_{x_k}(q_{jk}a_\lambda)\Big)(y), \quad j = 1, 2, \dots, n. \quad (3.3.2)$$

Proof Using Lemma 3.6 and integration by parts formula, we have

$$I((x_j - x_j^*)a_\lambda)(y) = \sum_{k=1}^n I(q_{jk}(\partial_{x_k}\phi)a_\lambda)(y) = -(i\nu)^{-1}I\Big(\sum_{k=1}^n \partial_{x_k}(q_{jk}a_\lambda)\Big)(y).$$

Since $q_{jk} \in \mathscr{B}_0(\mathbf{R}^{m+n})$, it is clear that $\sum_{k=1}^n \partial_{x_k}(q_{jk}a_\lambda) \in \mathscr{B}_p(\mathbf{R}^{m+n})$. □

Lemma 3.9 *Let* $a_\lambda(x, y) \in \mathscr{B}_p(\mathbf{R}^{m+n})$ *and*

$$b_\lambda(x, y) = \sum_{k=1}^n \sum_{j=1}^n \partial_{x_k}(q_{jk}(x, y)\tilde{a}_{\lambda,j}(x, y)), \quad (3.3.3)$$

where

$$\tilde{a}_{\lambda,j}(x, y) = \int_0^1 \partial_{x_j}a_\lambda(tx + (1-t)x^*, y)\,dt.$$

Then $b_\lambda(x, y) \in \mathscr{B}_p(\mathbf{R}^{m+n})$ *and*

$$I(a_\lambda)(y) = a_\lambda(x^*, y)I(1)(y) - (i\nu)^{-1}I(b_\lambda)(y), \quad (3.3.4)$$

where $I(1)(y)$ is the oscillatory integral

$$I(1)(y) = \int_{\mathbf{R}^n} e^{i\nu\phi(x,y)}\, dx.$$

For any α, β there exists a constant $C > 0$ such that

$$\begin{aligned} &\sup_{x\in\mathbf{R}^n} (1+\lambda+|x|+|y|)^{-p} |\partial_x^\alpha \partial_y^\beta b_\lambda(x,y)| \\ &\le C \sum_{\substack{\beta'\le\beta \\ 1\le|\alpha'|\le|\alpha|+|\beta|+2}} \sup_{x\in\mathbf{R}^n} (1+\lambda+|x|+|y|)^{-p} |\partial_x^{\alpha'} \partial_y^{\beta'} a_\lambda(x,y)|. \end{aligned} \tag{3.3.5}$$

and for any β there exists a constant $C > 0$ such that

$$\begin{aligned} &\nu^{n/2}(1+\lambda+|y|)^{-p} |\partial_y^\beta (e^{-i\nu\phi(x^*,y)} I(b_\lambda)(y))| \\ &\le C \sum_{\substack{\beta'\le\beta \\ 1\le|\alpha|\le|\beta|+p+n+3}} \sup_{x\in\mathbf{R}^n} (1+\lambda+|x|+|y|)^{-p} |\partial_x^{\alpha} \partial_y^{\beta'} a_\lambda(x,y)|. \end{aligned} \tag{3.3.6}$$

Proof Taylor's theorem gives that

$$a_\lambda(x,y) = a_\lambda(x^*,y) + \sum_{l=1}^n (x_j - x_j^*)\tilde{a}_{\lambda j}(x,y).$$

Applying Lemma 3.8, we have

$$I(a_\lambda) = a_\lambda(x^*,y) I(1) - (i\nu)^{-1} I\Big(\sum_{j,k=1}^n \partial_{x_k}(q_{jk}\tilde{a}_{\lambda j})\Big).$$

(3.3.4) is proved. (3.3.5) and (3.3.6) follow from this and Theorem 3.2. □

Corollary 3.7 *Let $a_\lambda(x,y) \in \mathscr{B}_p(\mathbf{R}^{m+n})$. Suppose further that there exists $\rho > 0$ such that $a_\lambda(x,y) \equiv 0$ in $S_\rho = \{x; |x-x^*| \le \rho\}$. Then for any $N = 1, 2, \ldots,$ there exists $a_{\lambda,N}(x,y) \in \mathscr{B}_p(\mathbf{R}^{m+n})$ such that*

$$I(a_\lambda)(y) = \nu^{-N} I(a_{\lambda,N})(y). \tag{3.3.7}$$

Proof By assumption, $a_\lambda(x^*,y) = 0$. Thus Lemma 3.9 implies that

$$I(a_\lambda)(y) = -(i\nu)^{-1} I(b_\lambda)(y). \tag{3.3.8}$$

It is clear from (3.3.3) that $b_\lambda(x,y) \equiv 0$ for $x \in S_\rho$. Therefore, the same discussion leads us to the conclusion that if we define

$$c_\lambda(x, y) = \sum_{k=1}^{n}\sum_{j=1}^{n} \partial_{x_k}\left(q_{jk}(x, y)\int_0^1 \partial_{x_j} b_\lambda(tx + (1-t)x^*, y)\,dt\right),$$

then we have $c_\lambda(x, y) \in \mathscr{B}_p(\mathbf{R}^{m+n})$ and

$$I(a_\lambda)(y) = -(i\nu)^{-1} I(b_\lambda)(y) = (i\nu)^{-2} I(c_\lambda)(y).$$

Since $c_\lambda(x, y) \equiv 0$ for $x \in S_\rho$, we are able to repeat this process N times and we prove the corollary. □

The next lemma improves Lemma 3.9 one step further.

Lemma 3.10 *Suppose* $a_\lambda(x, y) \in \mathscr{B}_p(\mathbf{R}^{m+n})$. *Then*

$$I(a_\lambda)(y) = \left(a_\lambda(x^*, y) + (i\nu)^{-1} r_1(a_\lambda)(y)\right) I(1)(y) + (i\nu)^{-2} r_2(a_\lambda)(y).$$

Here

$$r_1(a_\lambda)(y) = \sum_{j,k,l,s=1}^{n} \frac{1}{2}\partial_{x_j} a_\lambda(x^*, y) q_{jl}(x^*, y)(\partial_{x_k}\partial_{x_l}\partial_{x_s}\phi)(x^*, y) q_{sk}(x^*, y)$$

$$- \sum_{j,k=1}^{n} \frac{1}{2}(\partial_{x_j}\partial_{x_k} a_\lambda)(x^*, y) q_{jk}(x^*, y),$$

$$r_2(a_\lambda)(y)$$

$$= \sum_{j,k,l,s=1}^{n} \{\partial_{x_j} a_\lambda(x^*, y) I(\partial_{x_s}(q_{ls}\tilde{q}_{jkkl}))(y) + \frac{1}{2}\partial_{x_j}\partial_{x_k} a_\lambda(x^*, y) I(\partial_{x_s}(q_{ls}\tilde{q}_{jkl}))(y)\}$$

$$+ \sum_{j,k,l,s=1}^{n} \{I(\partial_{x_s}(q_{ls}\partial_{x_k}(q_{jk}\tilde{a}_{jl})))(y) + \frac{1}{2} I(\partial_{x_s}(q_{ls} q_{jk}\tilde{a}_{jkl}))(y)\},$$

$$\tilde{a}_{jk}(x, y) = \int_0^1 (1-t)\partial_{x_j}\partial_{x_k} a_\lambda(tx + (1-t)x^*, y)\,dt,$$

$$\tilde{a}_{jkl}(x, y) = \int_0^1 (1-t)^2 \partial_{x_j}\partial_{x_k}\partial_{x_l} a_\lambda(tx + (1-t)x^*, y)\,dt,$$

$$\tilde{q}_{jkl}(x, y) = \int_0^1 \partial_{x_l} q_{jk}(tx + (1-t)x^*, y)\,dt,$$

$$\tilde{q}_{jkkl} = \int_0^1 \partial_{x_l}\partial_{x_k} q_{jk}(tx + (1-t)x^*, y)\,dt.$$

Proof We shall calculate $b_\lambda(x, y)$ in Lemma 3.9 further. For the sake of brevity, we write b for b_λ and a_j for $a_{\lambda,j}$ and so on, omitting λ. Expanding $\tilde{a}_j(x, y)$, we have

$$\tilde{a}_j(x, y) = \int_0^1 \partial_{x_j} a_\lambda(tx + (1-t)x^*, y)\,dt = \partial_{x_j} a_\lambda(x^*, y) + \sum_{l=1}^{n}(x_l - x_l^*)\tilde{a}_{jl}(x, y).$$

Similarly,

$$\tilde{a}_{jk}(x, y) = \frac{1}{2}\partial_{x_j}\partial_{x_k}a_\lambda(x^*, y) + \frac{1}{2}\sum_{l=1}^{n}(x_l - x_l^*)\tilde{a}_{jkl}(x, y).$$

Abbreviating $q_{jk}(x, y)$ as q_{jk} and so on,

$$\begin{aligned}
\partial_{x_k}(q_{jk}\tilde{a}_j(x, y)) &= \partial_{x_k}q_{jk}\partial_{x_j}a_\lambda(x^*, y) + q_{jk}\tilde{a}_{jk}(x, y) + \sum_{l=1}^{n}(x_l - x_l^*)\partial_{x_k}(q_{jk}\tilde{a}_{jl}) \\
&= \partial_{x_k}q_{jk}\partial_{x_j}a_\lambda(x^*, y) + \frac{1}{2}q_{jk}\partial_{x_j}\partial_{x_k}a_\lambda(x^*, y) \\
&\quad + \frac{1}{2}\sum_{l=1}^{n}(x_l - x_l^*)q_{jk}\tilde{a}_{jkl}(x, y) + \sum_{l=1}^{n}(x_l - x_l^*)\partial_{x_k}(q_{jk}\tilde{a}_{jl}).
\end{aligned}$$

Replacing this in (3.3.3), (3.3.4) and applying Lemma 3.8, we have

$$\begin{aligned}
I(a_\lambda) &= a_\lambda(x^*, y)I(1) \\
&\quad - (i\nu)^{-1}\sum_{j,k=1}^{n}\left(\partial_{x_j}a_\lambda(x^*, y)I(\partial_{x_k}q_{jk}) + \frac{1}{2}\partial_{x_j}\partial_{x_k}a_\lambda(x^*, y)I(q_{jk})\right) \\
&\quad + (i\nu)^{-2}\sum_{j,k,l,s=1}^{n}\left(I\left(\partial_{x_s}(q_{ls}\partial_{x_k}(q_{jk}\tilde{a}_{jl}))\right) + \frac{1}{2}I\left(\partial_{x_s}(q_{ls}q_{jk}\tilde{a}_{jkl})\right)\right).
\end{aligned}$$

We have further

$$\begin{aligned}
I(q_{jk}) &= q_{jk}(x^*, y)I(1) - (i\nu)^{-1}\sum_{l,s=1}^{n}I(\partial_{x_s}(q_{ls}\tilde{q}_{jkl})), \\
I(\partial_{x_k}q_{jk}) &= \partial_{x_k}q_{jk}(x^*, y)I(1) - (i\nu)^{-1}\sum_{s=1}^{n}I(\partial_{x_s}(q_{ls}\tilde{q}_{jkkl})).
\end{aligned}$$

Therefore,

$$\begin{aligned}
I(a_\lambda) &= a_\lambda(x^*, y)I(1) \\
&\quad - (i\nu)^{-1}\sum_{j,k=1}^{n}\left(\partial_{x_j}a_\lambda(x^*, y)\partial_{x_k}q_{jk}(x^*, y) + \frac{1}{2}\partial_{x_j}\partial_{x_k}a_\lambda(x^*, y)q_{jk}(x^*, y)\right)I(1) \\
&\quad + (i\nu)^{-2}\sum_{j,k,l,s=1}^{n}\left(\partial_{x_j}a_\lambda(x^*, y)I(\partial_{x_s}(q_{ls}\tilde{q}_{jkkl})) + \frac{1}{2}\partial_{x_j}\partial_{x_k}a_\lambda(x^*, y)I(\partial_{x_s}(q_{ls}\tilde{q}_{jkl}))\right) \\
&\quad + (i\nu)^{-2}\sum_{j,k,l,s=1}^{n}\left(I\left(\partial_{x_s}(q_{ls}\partial_{x_k}(q_{jk}\tilde{a}_{jl}))\right) + \frac{1}{2}I\left(\partial_{x_s}(q_{ls}q_{jk}\tilde{a}_{jkl})\right)\right).
\end{aligned} \tag{3.3.9}$$

Finally we calculate $\partial_{x_l} q_{jk}(x^*, y)$. By our construction (3.2.47), function $q_{jk}(x, y)$ is the (j, k) entry of the matrix $Q(x, y) = H(x, y)^{-1}$ in some neighbourhood of x^*. Thus $\partial_{x_l} q_{jk}(x, y)$ is the (j, k) entry of the matrix $-H(x, y)^{-1}\partial_{x_l} H(x, y)H(x, y)^{-1}$. Therefore, we have

$$\partial_{x_l} q_{jk}(x^*, y) = -\sum_{r,s=1}^{n} q_{jr}(x^*, y)\partial_{x_l} h_{rs}(x^*, y)q_{sk}(x^*, y).$$

It follows from (3.2.44) that

$$\partial_{x_l} h_{rs}(x^*, y) = (\partial_{x_l}\partial_{x_s}\partial_{x_r}\phi)(x^*, y)\int_0^1 t\, dt = \frac{1}{2}(\partial_{x_l}\partial_{x_s}\partial_{x_r}\phi)(x^*, y).$$

Replacing $\partial_{x_l} q_{jk}(x^*, y)$ in (3.3.9) by the right-hand side of this equation, we have proved Lemma 3.10. □

Corollary 3.8 *For any β there exists a positive constant C_β independent of $a_\lambda(x, y)$ such that*

$$\nu^{n/2}|||\partial_y^\beta e^{-i\nu\phi(x^*,y)} r_2(a_\lambda)(y)|||_{(p)} \le C_\beta \sum_{\substack{\beta' \le \beta \\ 1\le|\alpha|\le|\beta|+n+p+5}} \|\partial_x^\alpha \partial_y^{\beta'} a\|_{(p)}.$$

Proof Since $a \in \mathscr{B}_p(\mathbf{R}^{m+n})$ and $q_{jk} \in \mathscr{B}_0(\mathbf{R})^{m+n}$, it is clear that $\partial_{x_s}(q_{ls}\partial_{x_k}(q_{jk}\tilde{a}_{jl})) \in \mathscr{B}_p(\mathbf{R}^{m+n})$, $\partial_{x_s}(q_{ls}q_{jk}\tilde{a}_{jkl}) \in \mathscr{B}_p(\mathbf{R}^{m+n})$ and that $\partial_{x_s}(q_{ls}\tilde{q}_{jkkl}), \partial_{x_s}(q_{ls}\tilde{q}_{jkl}) \in \mathscr{B}_0(\mathbf{R})^{m+n}$. The corollary follows from these and Theorem 3.2. □

3.3.2 Stationary Phase Method

Now we discuss

$$I(1)(y) = \int_{\mathbf{R}^n} e^{i\nu\phi(x,y)}\, dx.$$

Let $H_0 = H(x^*, y)$ be the Hessian matrix of $\phi(x, y)$ at x^*, i.e., H_0 is an $n \times n$ matrix whose (j, k) entry is

$$h_{jk} = \partial_{x_k}\partial_{x_j}\phi(x^*, y).$$

Let (x, x') be the standard inner product of vectors x and x' in $\mathbf{R}^n$. Then the quadratic form $\frac{1}{2}(x, H_0 x)$ satisfies (3.2.7) and (3.2.8) in Assumption 3.1 on page xxx. For any $a_\lambda(x, y) \in \mathscr{B}_p(\mathbf{R}^{m+n})$ we define an oscillatory integral

$$I_0(a_\lambda)(y) = \int_{\mathbf{R}^n} a_\lambda(x, y)e^{i\nu\frac{1}{2}(x,H_0x)}\, dx.$$

By Lemma 1.1,

$$I_0(1)(y) = \int_{\mathbf{R}^n} e^{i\nu\frac{1}{2}(x, H_0 x)}\,dx = \left(\frac{2\pi}{\nu}\right)^{n/2} |\det H_0|^{-1/2} \exp\frac{i\pi}{4}(n - 2\mathrm{Ind}(H_0)).$$

We will compare $I(1)(y)$ with $I_0(1)(y)$. For that purpose we shall make use of Morse's lemma of the following form.

Lemma 3.11 (Morse's lemma) *Suppose that $\phi(x, y)$ satisfies conditions in Assumption 3.1. Let $x^* = x^*(y)$ be the stationary point of $\phi(x, y)$ with respect to x and $H_0 = H(x^*, y)$ be the Hessian matrix of $\phi(x^*, y)$ at x^*, i.e. $H_0 = (\partial_{x_k}\partial_{x_j}\phi(x^*, y))_{jk}^n$. Then there exists a constant $\delta > 0$ independent of y such that inside the sphere $U_y = \{x; |x - x^*(y)| \le \delta\}$ we can make a change of variables $U_y \ni x \to z = z(x, y) \in G_y \subset \mathbf{R}^n$ such that*

$$\phi(x, y) - \phi(x^*, y) = \frac{1}{2}(z(x, y), H_0 z(x, y)), \tag{3.3.10}$$

$$z(x^*, y) = 0, \quad \partial_x z(x^*, y) = I. \tag{3.3.11}$$

Here $\partial_x z(x^*, y)$ is the Jacobian matrix of the change of variables and I is the $n \times n$ unit matrix. Moreover, there exist positive constants C, C' such that

$$C|x - x^*| \le |z(x, y)| \le C'|x - x^*|, \tag{3.3.12}$$

and for arbitrary multi-indices α and β there exists a positive constant $C_{\alpha\beta}$ such that for any $y \in \mathbf{R}^m$ and $x \in U_y$

$$|\partial_x^\alpha \partial_y^\beta z(x, y)| \le C_{\alpha\beta}. \tag{3.3.13}$$

Let $x(z, y)$ denote x regarded as a function of $z \in G_y$. Then for any α, β

$$|\partial_z^\alpha \partial_y^\beta x(z, y)| \le C_{\alpha\beta} \quad \text{for any } z \in G_y. \tag{3.3.14}$$

There exists some $\rho > 0$ independent of y such that

$$\{z; |z| \le 3\rho\} \subset \cap_{y\in\mathbf{R}^m}\{z(x, y); x \in U_y\}.$$

A proof of Morse's lemma will be given at Sect. 3.3.3. For the time being we assume that it is true and prove the following proposition.

Proposition 3.2 *Let $J(x, y)$ be the Jacobian determinant $J(x, y) = \det \partial_x z(x, y)$ of the change of variables $x \to z$ and $q_{jk}(x, y)$ be as in Lemma 3.6. Define*

$$b_1(y) = -\frac{1}{2}\sum_{j,k=1}^n (\partial_{z_j}\partial_{z_k} J(x(z, y), y)^{-1}|_{z=0}) q_{jk}(x^*, y).$$

Then there exist functions $b_2(z, y)$, $b_3(z, y)$ of (z, y) and $b_4(x, y)$ of (x, y) such that

$$I(1)(y) = e^{i\nu\phi(x^*,y)} I_0(1)(y) \left(1 + (i\nu)^{-1} b_1(y)\right) \quad (3.3.15)$$
$$+ (i\nu)^{-2} e^{i\nu\phi(x^*,y)} I_0(b_2)(y) + (i\nu)^{-3} \left(e^{i\nu\phi(x^*,y)} I_0(b_3)(y) + I(b_4)(y)\right).$$

For any α, β there exists a positive constant $C_{\alpha\beta}$ such that

$$|\partial_z^\alpha \partial_y^\beta b_2(z, y)| \le C_{\alpha\beta} \max_{1\le|\alpha'|\le|\alpha|+3, \beta'\le\beta} \sup_{|z|\le 2\rho} |\partial_z^{\alpha'} \partial_y^{\beta'} J(x(z, y), y)^{-1}|, \quad (3.3.16)$$

$$|\partial_z^\alpha \partial_y^\beta b_3(z, y)| \le C_{\alpha\beta}, \quad (3.3.17)$$

$$|\partial_x^\alpha \partial_y^\beta b_4(x, y)| \le C_{\alpha\beta}. \quad (3.3.18)$$

Proof Let U_y, G_y and ρ be as in Morse's lemma 3.11. We set $V_{3\rho} = \{z \in \mathbf{R}^n;\ |z| \le 3\rho\}$ be the ball with radius $3\rho > 0$ in z-space. We have $V_{3\rho} \subset \cap_{y\in\mathbf{R}^m} \{z(x, y);\ x \in U_y\}$ by Lemma 3.11. For $x \in U_y$

$$\phi(x, y) = \phi(x^*, y) + \frac{1}{2}(z(x, y), H_0 z(x, y)).$$

Let $v(z)$ be a $C_0^\infty(\mathbf{R}^n)$ function such that $0 \le v(z) \le 1$ for any z and

$$v(z) = \begin{cases} 1 & \text{if } |z| \le \rho, \\ 0 & \text{if } |z| \ge 2\rho. \end{cases}$$

We define the following function $u(x, y)$ of $(x, y) \in \mathbf{R}^{m+n}$:

$$u(x, y) = \begin{cases} v(z(x, y)) & \text{if } x \in U_y, \\ 0 & \text{otherwise.} \end{cases}$$

Then $u(x, y)$ is a function in $C^\infty(\mathbf{R}^{m+n})$. Using this, we divide $I(1)(y)$ into two parts:

$$I(1)(y) = I_1(y) + I_2(y),$$

where

$$I_1(y) = \int_{\mathbf{R}^n} u(x, y) e^{i\nu\phi(x,y)}\, dx, \qquad I_2(y) = \int_{\mathbf{R}^n} (1 - u(x, y)) e^{i\nu\phi(x,y)}\, dx.$$

We discuss $I_1(y)$. Since Morse's lemma holds on the set where the integrand u_ρ of I_1 does not vanish, we change coordinates from x to z.

Let $G_y \ni z \to x(z, y)$ be the inverse map of $x \to z(x, y)$. Define a smooth function $f(z, y)$ of $(z, y) \in \mathbf{R}^{m+n}$ by

$$f(z, y) = \begin{cases} J(x(z, y), y)^{-1}v(z) & \text{for } |z| \le 3\rho, \\ 0 & \text{otherwise .} \end{cases}$$

Then

$$\begin{aligned} I_1(y) &= \int_{x\in U} u(x, y)e^{i\nu\phi(x,y)}\,dx \\ &= \int_{|z|\le 3\rho} u(x(z, y), y)J(x(z, y), y)^{-1}e^{i\nu\phi(x(z,y),y)}\,dz \\ &= e^{i\nu\phi(x^*,y)}\int_{\mathbf{R}^n} f(z, y)e^{i\nu\frac{1}{2}(z,H_0 z)}\,dz. \end{aligned}$$

The integral on the right-hand side is an oscillatory integral with phase $\frac{1}{2}(z, H_0 z)$ and amplitude $f(z, y)$. We shall apply Lemma 3.10 to this oscillatory integral. In this special case the phase function $\frac{1}{2}(z, H_0 z)$ is a quadratic form of z. Therefore, the map $\mathbf{R}^n \ni (z_1, z_2, \ldots, z_n) \to (\zeta_1, \zeta_2, \ldots, \zeta_n) \in \mathbf{R}^n$ defined by

$$\zeta_j = \sum_{k=1}^{n} h_{jk}z_k, \quad \text{for } j = 1, 2, \ldots, n,$$

is a linear mapping, i.e. h_{jk} which equals the (j, k) entry of H_0 is independent of z. Let q_{jk} be the (j, k) entry of the matrix H_0^{-1}, i.e. $q_{jk} = q_{jk}(x^*, y)$. Then

$$z_j = \sum_{k=1}^{n} q_{jk}\zeta_k.$$

As $J(x(0), y)^{-1} = 1$, it follows from Taylor's theorem that if $|z| \le 3\rho$,

$$J(x(z, y), y)^{-1} = 1 + \sum_{j=1}^{n} z_j \tilde{J}_j(z, y),$$

where

$$\tilde{J}_j(z, y) = \int_0^1 \partial_{z_j} J(x(tz), y)^{-1}\,dt, \quad \text{for } |z| \le 3\rho. \tag{3.3.19}$$

Define

$$f_j(z, y) = \begin{cases} \tilde{J}_j(z, y)v(z, y) & \text{if } |z| \le 3\rho, \\ 0 & \text{otherwise.} \end{cases}$$

Then $f_j(z, y)$ is a smooth function and

$$f(z, y) = v(z) + \sum_{j=1}^{n} z_j f_j(z, y), \quad (z, y) \in \mathbf{R}^{m+n}.$$

Applying Lemma 3.8, we have

$$e^{-i\nu\phi(x^*,y)} I_1(y) = I_0(v)(y) + \sum_{j=1}^{n} I_0(z_j f_j)(y)$$
$$= I_0(1)(y) + I_0(v-1)(y) - (i\nu)^{-1} \sum_{j=1}^{n} q_{jk} I_0(\partial_{z_k} f_j)(y).$$

As $1 - v(z) = 0$ in $\{z; |z| < \rho\}$, it follows from Corollary 3.7 that there exists a function $b_3(z, y) \in \mathscr{B}_0(\mathbf{R}^{m+n})$ such that

$$I_0(v-1)(y) = (i\nu)^{-3} I_0(b_3)(y).$$

For any α, β there exists a constant $C_{\alpha\beta}$ such that

$$\sup_{(z,y)\in\mathbf{R}^{m+n}} |\partial_z^\alpha \partial_y^\beta b_3(z, y)| \le C_{\alpha\beta}. \tag{3.3.20}$$

Since by (3.3.19)

$$\partial_{z_k} f_j(z, y)\big|_{z=0} = \partial_{z_k} \tilde{J}_j(z, y)\big|_{z=0} = \frac{1}{2} \partial_{z_k} \partial_{z_j} J(x(z, y), y)^{-1}\big|_{z=0},$$

it follows from Lemma 3.9 that there exist functions $b_{jk}(z, y)$ such that

$$I_0(\partial_{z_k} f_j)(y) = \frac{1}{2} \partial_{z_k} \partial_{z_j} J(x(z, y), y)^{-1}\big|_{z=0} I_0(1) - (i\nu)^{-1} I_0(b_{jk})(y).$$

For any α, β there exists a constant $C_{\alpha,\beta}$ such that

$$|\partial_z^\alpha \partial_y^\beta b_{jk}(z, x)| \le C_{\alpha\beta} \max_{1\le|\alpha'|\le|\alpha|+1, \beta'\le\beta} a \sup_{|z|\le 2\rho} |\partial_z^{\alpha'} \partial_y^{\beta'} \partial_{z_k} f_j(z, y)|$$
$$\le C_{\alpha\beta} \max_{1\le|\alpha'|\le|\alpha|+3, \beta'\le\beta} a \sup_{|z|\le 2\rho} |\partial_z^{\alpha'} \partial_y^{\beta'} J(x(z, y), y)^{-1}|.$$

Therefore, setting $b_2 = \sum_{jk} q_{jk} b_{jk}$, we can write

$$e^{-\nu\phi(x^*,y)} I_1(y) = I_0(1)(y) - (i\nu)^{-1} \frac{1}{2} \partial_{z_k} \partial_{z_j} J(x(z, y), y)\big|_{z=0} I_0(1)(y) + (i\nu)^{-2} I_0(b_2)(y) + (i\nu)^{-3} I_0(b_3)(y). \tag{3.3.21}$$

For any α, β there exists a constant $C_{\alpha,\beta} > 0$ such that

$$|\partial_z^\alpha \partial_y^\beta b_2(z, y)| \le C_{\alpha\beta} \max_{\substack{1\le|\alpha'|\le|\alpha|+3 \\ \beta'\le\beta}} a \sup_{|z|\le 2\rho} |\partial_z^{\alpha'} \partial_y^{\beta'} J^{-1}(x(z, y), y)|. \tag{3.3.22}$$

Now we discuss $I_2(y)$. Since $1 - u \in \mathscr{B}_0(\mathbf{R}^{m+n})$ and $1 - u(x, y) \equiv 0$ in some neighbourhood of 0, it follows from Lemma 3.10 that there exists an amplitude function $b_4(x, y) \in \mathscr{B}_0(\mathbf{R}^{m+n})$ such that

$$I_2(y) = (i\nu)^{-3} I(b_4)(y). \tag{3.3.23}$$

For any α, β there exists a positive constant C_β such that

$$|\partial_x^\alpha \partial_y^\beta b_4(x, y)| \le C_{\alpha\beta}. \tag{3.3.24}$$

Taking sum of (3.3.21) and (3.3.23), we have proved (3.3.15). Inequalities (3.3.16), (3.3.17) and (3.3.18) follow from (3.3.22), (3.3.20) and (3.3.24). Therefore, the proposition has been proved up to the proof of Morse's lemma. □

Now we can prove

Theorem 3.5 (Stationary phase method) *Suppose that Assumption 3.1 on page 43 holds. Let $q_{jk}(x, y)$ be as in Lemma 3.6. We write $Ind(H(x^*, y))$ for the number of negative eigenvalues of the Hessian matrix $H(x^*, y)$ and set*

$$A = (2\pi)^{n/2} |\det H(x^*, y)|^{-1/2} \exp \frac{i\pi}{4}(n - 2\mathrm{Ind}(H(x^*, y))). \tag{3.3.25}$$

Then

$$I(a_\lambda)(y) = e^{i\nu\phi(x^*,y)} \nu^{-n/2} \left(A a_\lambda(x^*, y) + (i\nu)^{-1} A R_1(a_\lambda)(y) + (i\nu)^{-2} R_2(a_\lambda)(y) \right). \tag{3.3.26}$$

Here

$$\begin{aligned} R_1(a_\lambda)(y) = &-\frac{1}{2} a_\lambda(x^*, y) \sum_{j,k=1}^n q_{jk}(x^*, y) \partial_{z_j} \partial_{z_k} J(x(z, y), y)^{-1}|_{z=0} \\ &+ \sum_{j,k,l,s=1}^n \frac{1}{2} \partial_{x_j} a_\lambda(x^*, y) q_{jl}(x^*, y) \partial_{x_k} \partial_{x_l} \partial_{x_s} \phi(x^*, y) q_{sk}(x^*, y) \\ &- \sum_{j,k=1}^n \frac{1}{2} \partial_{x_j} \partial_{x_k} a_\lambda(x^*, y) q_{jk}(x^*, y). \end{aligned} \tag{3.3.27}$$

For any β there exists a constant $C_\beta > 0$ such that

$$|||\partial_y^\beta R_2(a_\lambda)(y)|||_{(p)} \leq C_\beta \sum_{\substack{\beta' \leq \beta \\ |\alpha| \leq |\beta|+n+p+5}} \|\partial_x^\alpha \partial_y^{\beta'} a_\lambda\|_{(p)}. \tag{3.3.28}$$

Proof As a consequence of Lemma 3.10 and Proposition 3.2,

$$\begin{aligned} & e^{-i\nu\phi(x^*,y)} I(a_\lambda) \\ & = a_\lambda(x^*, y) I_0(1) \\ & + (i\nu)^{-1}\big(a_\lambda(x^*, y) b_1 I_0(1) + r_1(a_\lambda) I_0(1)\big) \\ & + (i\nu)^{-2}\big(a_\lambda(x^*, y) I_0(b_2) + r_1(a_\lambda) b_1 I_0(1) + e^{-i\nu\phi(x^*,y)} r_2(a_\lambda)\big) \\ & + (i\nu)^{-3}\big(a_\lambda(x^*, y) I_0(b_3) + a_\lambda(x^*, y) e^{-i\nu\phi(x^*,y)} I(b_4) + r_1(a_\lambda) I_0(b_2)\big) \\ & + (i\nu)^{-4}\big(r_1(a_\lambda) I_0(b_3) + r_1(a_\lambda) e^{-i\nu\phi(x^*,y)} I(b_4)\big). \end{aligned} \tag{3.3.29}$$

Define

$$\begin{aligned} R_1(a_\lambda) &= \nu^{n/2} I_0(1)\big(a_\lambda(x^*, y) b_1 + r_1(a_\lambda)\big), \\ R_2(a_\lambda) &= \nu^{n/2}\big[(a_\lambda(x^*, y) I_0(b_2) + r_1(a_\lambda) b_1 I_0(1) + e^{-i\nu\phi(x^*,y)} r_2(a_\lambda) \\ & \quad + (i\nu)^{-1}(a_\lambda(x^*, y) I_0(b_3) + a_\lambda(x^*, y) e^{-i\nu\phi(x^*,y)} I(b_4) + r_1(a_\lambda) I_0(b_2)) \\ & \quad + (i\nu)^{-2}(r_1(a_\lambda) I_0(b_3) + r_1(a_\lambda) e^{-i\nu\phi(x^*,y)} I(b_4))\big]. \end{aligned} \tag{3.3.30}$$

Then (3.3.26) and (3.3.27) hold.

We have only to prove inequality (3.3.28). It is clear that $b_1 \in \mathscr{B}_0(\mathbf{R}^m)$. Also $\nu^{n/2} I_0(b_2)(y)$, $\nu^{n/2} I_0(b_3)(y)$ and $\nu^{n/2} e^{-i\nu\phi(x^*,y)} I(b_4)(y) \in \mathscr{B}_0(\mathbf{R}^m)$, because $b_2, b_3, b_4 \in \mathscr{B}_0(\mathbf{R}^{m+n})$. Thus, for any β there exists a constant $C > 0$ such that for any $y \in \mathbf{R}^m$

$$\begin{gathered} |\partial_y^\beta b_1(y)| \leq C, \quad \nu^{n/2} |\partial_y^\beta I_0(b_2)(y)| \leq C, \\ \nu^{n/2} |\partial_y^\beta I_0(b_3)(y)| \leq C, \quad \nu^{n/2} |\partial_y^\beta e^{-i\nu\phi(x^*,y)} I(b_4)(y)| \leq C. \end{gathered}$$

On the other hand, $a_\lambda(x^*, y) \in \mathscr{B}_p(\mathbf{R}^m)$ and $r_1(a_\lambda) \in \mathscr{B}_p(\mathbf{R}^m)$. Thus for any β there exists a constant $C > 0$ such that

$$\begin{aligned} |||\partial_y^\beta a_\lambda(x^*, y)|||_{(p)} &\leq C \sum_{\beta' \leq \beta, |\alpha| \leq |\beta|} \|\partial_x^\alpha \partial_y^\beta a\|_{(p)}, \\ |||\partial_y^\beta r_1(a_\lambda)|||_{(p)} &\leq C \sum_{\beta' \leq \beta, 1 \leq |\alpha| \leq |\beta|+2} \|\partial_x^\alpha \partial_y^{\beta'} a\|_{(p)}. \end{aligned}$$

Hence

$$
\begin{aligned}
&\nu^{n/2}|||\partial_y^{\beta}\{a_\lambda(x^*,y)(I_0(b_3)(y)+e^{-i\nu\phi(x^*,y)}I(b_4)(y))\}|||_{(p)}\\
&\le C\sum_{\beta'\le\beta,|\alpha|\le|\beta|}\|\partial_x^{\alpha}\partial_y^{\beta'}a\|_{(p)},\\
&\nu^{n/2-1}|||\partial_y^{\beta}\{r_1(a_\lambda)(y)(I_0(b_3)(y)+e^{-i\nu\phi(x^*,y)}I(b_4))\}|||_{(p)}\\
&\quad\le C\nu^{-1}\sum_{\beta'\le\beta,1\le|\alpha|\le|\beta|+2}\|\partial_x^{\alpha}\partial_y^{\beta'}a\|_{(p)}.
\end{aligned}
$$

Similarly,

$$
\begin{aligned}
|||\partial_y^{\beta}\{r_1(a_\lambda)(y)b_1I_0(1)|||_{(p)} &\le C\sum_{\beta'\le\beta,1\le|\alpha|\le|\beta|+2}\|\partial_x^{\alpha}\partial_y^{\beta'}a\|_{(p)},\\
|||\partial_y^{\beta}a_\lambda(x^*,y)I_0(b_2)(y)|||_{(p)} &\le C\max_{\alpha\le|\beta|,\beta'\le\beta}\|\partial_x^{\alpha}\partial_y^{\beta'}a_\lambda(x^*,y)\|_{(p)},\\
|||r_1(a_\lambda)I_0(b_2)|||_{(p)} &\le C\max_{1\le|\beta|+p+4,\beta'\le\beta}\|\partial_x^{\alpha}\partial_y^{\beta'}a\|_{(p)}.
\end{aligned}
$$

Corollary 3.8 implies that

$$
\nu^{n/2}|||\partial_y^{\beta}\{e^{-i\nu\phi(x^*,y)}r_2(a_\lambda)(y)\}|||_p\le C\sum_{\substack{\beta'\le\beta\\1\le|\alpha|\le|\beta|+n+p+5}}\|\partial^{\alpha}\partial^{\beta'}a\|_{(p)}.
$$

Collecting all these inequalities, we obtain inequality (3.3.28). The theorem has been proved. □

Since $\dfrac{\partial}{\partial x}z(x,y)\big|_{x=x^*}$ is the unit matrix, calculation shows

Remark 3.6

$$
\begin{aligned}
&\partial_{z_j}\partial_{z_k}J(x(z,y),y)^{-1}|_{z=0}\\
&=-\partial_{x_j}\partial_{x_k}J(x,y)|_{x=x^*}+2(\partial_{x_j}J(x,y)|_{x=x^*})\,(\partial_{x_k}J(x,y)|_{x=x^*})\\
&\quad+\sum_{l=1}^{n}(\partial_{x_j}\partial_{x_k}z_l(x,y)|_{x=x^*})\,(\partial_{x_l}J(x,y)|_{x=x^*}).
\end{aligned}
$$

Pick $R(a_\lambda)(y)=AR_1(a_\lambda)(y)+(i\nu)^{-1}R_2(a_\lambda)(y)$ in (3.3.26). Then we can easily obtain the following corollary which is also usually called the "Stationary phase method".

Corollary 3.9 *Under the same assumption as in Theorem 3.5,*

$$
I(a_\lambda)(y)=e^{i\nu\phi(x^*,y)}\nu^{-n/2}\left(Aa_\lambda(x^*,y)+(i\nu)^{-1}R(a_\lambda)(y)\right). \tag{3.3.31}
$$

For any β there exists a positive constant C_β such that

$$|||\partial_y^\beta R(a_\lambda)(y)|||_{(p)} \leq C_\beta \sum_{\substack{\beta' \leq \beta \\ |\alpha| \leq |\beta|+n+p+5}} \|\partial_x^\alpha \partial_y^{\beta'} a_\lambda\|_{(p)}. \tag{3.3.32}$$

Remark 3.7 The information in (3.3.27), (3.3.28) and (3.3.32) on remainder terms as well as the principal term is important for our discussions later.

Remark 3.8 The stationary phase method is discussed in many literatures by various methods on various standpoints, cf. for example [17, 18, 39].

3.3.3 A Proof of Morse's Lemma

Finally we prove Morse's lemma 3.11. We write H_0 for $H(x^*, y)$. We shall look for a real symmetric matrix-valued function $S(x, y)$ such that the correspondence

$$x \to z = (I + H_0^{-1} S(x, y))(x - x^*)$$

is the required change of coordinates.

For that purpose we regard the totality $\mathscr{S}$ of real $n \times n$ symmetric matrix as a real vector space of dimension $n(n+1)/2$. We define a map F from $\mathscr{S}$ to itself:

$$F : \mathscr{S} \ni S \longrightarrow F(S) = (I + SH_0^{-1})H_0(I + H_0^{-1}S) \in \mathscr{S}.$$

Then $F(0) = H_0$ and the differential $dF(S)$ of F at S is the linear transformation in $\mathscr{S}$:

$$dF(S) : \mathscr{S} \ni X \longrightarrow dF(S)X = 2X + XH_0^{-1}S + SH_0^{-1}X \in \mathscr{S}.$$

Since $dF(0) = 2I$ is an isomorphism, it follows from the inverse function theorem that there exist a neighbourhood V of 0 in $\mathscr{S}$ and a neighbourhood W of $F(0) = H_0$ in $\mathscr{S}$ such that F is a diffeomorphism $V \to W$.

Let $H(x, y)$ be the symmetric matrix whose (j, k) entry equals

$$h_{jk}(x, y) = 2\int_0^1 (1-\theta)\partial_{x_j}\partial_{x_k}\phi(\theta x + (1-\theta)x^*, y)\, d\theta.$$

Since $H(x^*, y) = H_0$ for any y and there exists a positive constant C satisfying

$$\sup_{(x,y)\in \mathbf{R}^{m+n}} |\partial_{x_l} h_{jk}(x, y)| < C$$

owing to Assumption 3.1, there exists a positive constant ρ independent of y such that $H(x, y) \in W$ if $|x - x^*| < \rho$.

We set

$$S(x, y) = F^{-1}(H(x, y)) \in V \quad \text{for } |x - x^*| < \rho and y \in \mathbf{R}^m.$$

Using this $S(x, y)$, we define

$$z = (I + H_0^{-1} S(x, y))(x - x^*).$$

Then by Taylor's theorem,

$$\begin{aligned}\frac{1}{2}(z, H_0 z) &= \frac{1}{2}((x - x^*), (I + S(x, y)H_0^{-1})H_0(I + H_0^{-1}S(x, y))(x - x^*)) \\ &= \frac{1}{2}((x - x^*), F(S(x, y))(x - x^*)) = \frac{1}{2}((x - x^*), H(x, y)(x - x^*)) \\ &= \phi(x, y) - \phi(x^*, y).\end{aligned}$$

Since $S(x^*, y) = 0$, we have $z = 0$ if $x = x^*$. It is clear that $\partial z/\partial x\Big|_{x=x^*} = I$. Therefore, it follows from the inverse function theorem again that

$$x \longrightarrow z = (I + H_0^{-1} S(x, y))(x - x^*)$$

is a local coordinates in $|x - x^*| < \delta$ if $\delta > 0$ is sufficiently small.

We can easily prove (3.3.13) and (3.3.14) by virtue of Assumption 3.1.

We shall prove (3.3.12). We may assume the neighbourhood V of $0 \in \mathscr{S}$ is so small that for any $X \in V$

$$\|X\| < 2^{-1}\|H_0^{-1}\|^{-1}.$$

If $|x - x^*| < \rho$, then $S(x, y) \in V$ for any y by construction. This implies that $\|S(x, y)\| < 2^{-1}\|H_0^{-1}\|^{-1}$. Thus

$$\|H_0^{-1}S(x, y)\| < \|H_0^{-1}\|\|S(x, y)\| < \frac{1}{2}.$$

Therefore,

$$\begin{aligned}&\frac{1}{2}|x - x^*| \le (1 - \|H_0^{-1}S(x, y)\|)|x - x^*| \le |(I + H_0^{-1}S(x, y))(x - x^*)| \\ &= |z| \le |(I + H_0^{-1}S(x, y))(x - x^*)| \le (1 + \|H_0^{-1}S(x, y)\|)|x - x^*| \le \frac{3}{2}|x - x^*|.\end{aligned}$$

This proves (3.3.12). (3.3.13) and (3.3.14) are easy to prove. Morse's lemma is proved. □

Remark 3.9 Let $M(S)X = XH_0^{-1}S + SH_0^{-1}X$. Then $dF(S)X = (2I + M(S))X$ for any $X \in \mathscr{S}$. Let us write $\|S\|$ for the operator norm of a linear map S of $\mathbf{R}^n$ to itself. Assume that $\|S\| \le 2^{-1}\|H_0^{-1}\|^{-1}$. Then

$$\|M(S)X\| \le \|X\|.$$

This means that $L(S) = (2 + M(S))^{-1}$ exists and is the inverse map of $dF(S)$. Obviously $\|L(S)X\| \le \|X\|$. Therefore, we can construct $S(x, y)$ by Newton's iteration method if $|x - x^*|$ is sufficiently small, cf. [58].

3.4 A Proof of Hadamard's Global Implicit Function Theorem

Theorem 3.6 (Hadamard's global implicit function theorem) *Let $f : \mathbf{R}^n \to \mathbf{R}^n$ be a map of class C^1. If its differential $Df(x)$ is invertible at every $x \in \mathbf{R}^n$ and the norm of $Df(x)^{-1}$ as a linear map is bounded uniformly with respect to x, then the map f is a global diffeomorphism of class C^1.*

Proof We give a sketch of the proof. See Theorem 1.22 of Schwartz [58] for more detailed discussion. Let X and Y be the source space and the target space of f, i.e. $f : X \to Y$. By assumption

$$\sup_{x \in X} \|Df(x)^{-1}\| = K < \infty. \tag{3.4.1}$$

We first prove f is an onto mapping. Let $y_0 = f(0) \in Y$ and $y \in Y$ be an arbitrary point such that $y \neq y_0$. Consider the straight line segment in Y joining y_0 and y:

$$y(t) = (ty + (1 - t)y_0) \quad \text{for } 0 \le t \le 1. \tag{3.4.2}$$

Since $y_0 = f(0)$, by the usual implicit function theorem there exists a neighborhood U of 0 such that f maps U homeomorphically on to a neighborhood V of y_0. Thus there exist $\epsilon > 0$ and a continuous map $[0, \epsilon) \ni t \to x(t) \in X$ that satisfies $f(x(t)) = y(t)$. We prolong this map $x(t)$ as far as possible. Then we have a maximal subinterval J of $[0, 1]$ such that there exists a continuous map $J \ni t \to x(t) \in X$ with the property $y(t) = f(x(t))$.

We wish to prove that $J = [0, 1]$. Suppose that $a = \sup\{t : t \in J\}$. We claim $a \in J$. Since

$$\frac{d}{dt}x(t) = Df(x(t))^{-1}\frac{d}{dt}y(t), \tag{3.4.3}$$

$$|x(t) - x(s)| \le \int_s^t \|Df(x(r))^{-1}\| \left|\frac{d}{dr}y(r)\right| dr \le K|t - s||y - y_0|. \tag{3.4.4}$$

This means that $x(t)$ is Lipschitz continuous with Lipschitz constant K. Therefore, $\lim_{t \to a-0} x(t)$ exists in X. If we define $x(a) = \lim_{t \to a-0} x(t)$, then $x(t)$ is defined and continuous in $[0, a]$. We have also $f(x(a)) = \lim_{t\to a-0} f(x(t)) = \lim_{t\to a-0} y(t) = y(a)$. By the maximality of J, $[0, a] \subset J$.

Next we shall show that $a = 1$. Suppose that $a < 1$. Then by the usual local implicit function theorem, there exists a neighborhood U of $x(a)$ such that f maps U homeomorphically on to a neighborhood V of $y(a)$. We write the inverse image of $f|_U$ of $y \in V$ by $\left(f|_U\right)^{-1}(y)$. Therefore, there exists $\epsilon > 0$ such that

$$x(t) = \left(f|_U\right)^{-1}(y(t)) \tag{3.4.5}$$

is defined for $a < t \le a + \epsilon$. This contradicts maximality of J. Thus we have proved that $a = 1$. It follows from this that $f(x(1)) = y$. Therefore, the map f is onto.

It turns out that $f : X \to Y$ is a covering map. Since Y is simply connected, it follows from the monodromy argument that f is a diffeomorphism. The theorem is proved. □

Chapter 4
Statement of Main Results

Abstract Although the time slicing approximation of Feynman path integrals does not converge absolutely, it has a definite finite value if the potential satisfies Assumption 2.1 and if the time interval is short, because it is an oscillatory integral that satisfies Assumption 3.1. Furthermore, the time slicing approximation of Feynman path integrals converges to a limit as $|\Delta| \to 0$. The limit turns out to be the fundamental solution of the Schrödinger equation. The semi-classical asymptotic formula called Birkhoff's formula is proved from the standpoint of oscillatory integrals. In this chapter, these results as well as others are explained. Proofs will be given in subsequent chapters.

4.1 Time Slicing Approximation is an Oscillatory Integral

Let $[a, b]$ be the time interval and the potential $V(t, x)$ be assumed to satisfy Assumption 2.1, i.e. for any α the function $\partial_x^\alpha V(t, x)$ exists and is a continuous function on $[a, b] \times \mathbf{R}$ such that for $(t, x) \in [a, b] \times \mathbf{R}$ and $2 \leq |\alpha| \leq m$

$$|\partial_x^\alpha V(t, x)| \leq v_m. \tag{4.1.1}$$

We define a positive constant δ to be so small that

$$\frac{v_2 \delta^2}{8} < 1. \tag{4.1.2}$$

Let $[s, s'] \subset [a, b]$ with

$$|s' - s| \leq \delta$$

and let

$$\Delta : s = T_0 < T_1 < T_2 < \cdots < T_J < T_{J+1} = s' \tag{4.1.3}$$

be an arbitrary division of interval $[s, s']$. We use the symbols $\tau_j = T_j - T_{j-1}$, $|\Delta| = \max_{1 \leq j \leq J+1} \tau_j$.

D. Fujiwara, *Rigorous Time Slicing Approach to Feynman Path Integrals*, Mathematical Physics Studies, DOI 10.1007/978-4-431-56553-6_4

Let $x \in \mathbf{R}$ and $y \in \mathbf{R}$ be two points. For any $(x_1, \dots, x_J) \in \mathbf{R}^J$ we can define a piecewise classical path γ_Δ such that $\gamma_\Delta(T_j) = x_j$ for $j = 0, 1, 2, \dots, J+1$. Here and hereafter $x_0 = y$ and $x_{J+1} = x$ as usual. The action $S(\gamma_\Delta) = S_\Delta(x_{J+1}, x_J, \dots, x_1, x_0)$ is of the form

$$S_\Delta(x_{J+1}, \dots, x_0) = \sum_{j=1}^{J+1} S(T_j, T_{j-1}, x_j, x_{j-1}). \tag{4.1.4}$$

Here $S(T_j, T_{j-1}, x_j, x_{j-1})$ is the classical action:

$$\begin{aligned} S(T_j, T_{j-1}, x_j, x_{j-1}) &= \int_{T_{j-1}}^{T_j} L(t, \gamma_\Delta(t), \dot{\gamma_\Delta}(t))\, dt \\ &= \frac{(x_j - x_{j-1})^2}{2\tau_j} + \tau_j \phi(T_j, T_{j-1}, x_j, x_{j-1}). \end{aligned} \tag{4.1.5}$$

We often abbreviate $\phi(T_j, T_{j-1}, x_j, x_{j-1})$ to $\phi_j(x_j, x_{j-1})$. By Theorem 2.3, for each integer $m \geq 0$ there exists a constant $\kappa_m > 0$ independent of j such that for any α, β with $|\alpha| + |\beta| = m$ and for any $(x_j, x_{j-1}) \in \mathbf{R}^2$

$$|\partial_{x_j}^\alpha \partial_{x_{j-1}}^\beta \phi_j(x_j, x_{j-1})| \leq \kappa_m (1 + |x_j| + |x_{j-1}|)^{(2-m)_+}, \tag{4.1.6}$$

where $(2-m)_+ = max\{0, 2-m\}$.

The time slicing approximation of a Feynman path integral associated with Δ is

$$I(\Delta; \nu, s', s, x, y) = \prod_{j=1}^{J+1} \left(\frac{-i\nu}{2\pi\tau_j}\right)^{\frac{1}{2}} \int_{\mathbf{R}^J} e^{i\nu S_\Delta(x_{J+1}, x_J, \dots, x_1, x_0)} \prod_{j=1}^{J} dx_j. \tag{4.1.7}$$

We decompose the Hessian matrix $H_\Delta(x) = (h_{jk})_{jk=1}^J$ of $S_\Delta(x_{J+1}, x_J, \dots, x_1, x_0)$

$$h_{jk}(x_{J+1}, x_J, \dots, x_0) = \partial_{x_j} \partial_{x_k} S_\Delta(x_{J+1}, x_J, \dots, x_0), \quad j, k = 1, 2, \dots, J$$

into the sum of linear part H_Δ and non-linear part $W_\Delta(x)$, i.e.

$$H_\Delta(x) = H_\Delta + W_\Delta(x), \tag{4.1.8}$$

where

$$H_\Delta = \begin{pmatrix} \frac{1}{\tau_1} + \frac{1}{\tau_2} & -\frac{1}{\tau_2} & 0 & 0 & \cdots & 0 & 0 \\ -\frac{1}{\tau_2} & \frac{1}{\tau_2} + \frac{1}{\tau_3} & -\frac{1}{\tau_3} & 0 & \cdots & 0 & 0 \\ \cdots & \cdots & \cdots & \cdots & \cdots & \cdots & \cdots \\ \cdots & \cdots & \cdots & \cdots & \cdots & 0 & -\frac{1}{\tau_J} \\ 0 & 0 & 0 & \cdots & 0 & -\frac{1}{\tau_J} & \frac{1}{\tau_J} + \frac{1}{\tau_{J+1}} \end{pmatrix} \tag{4.1.9}$$

and the (j, k) entry w_{jk} of $W_\Delta(x)$ is given by

$$w_{jk} = \begin{cases} \partial_{x_j}^2(\tau_j\phi_j(x_j, x_{j-1}) + \tau_{j+1}\phi_{j+1}(x_{j+1}, x_j)) & \text{if } k = j, \\ \tau_j\partial_{x_{j-1}}\partial_{x_j}\phi_j(x_j, x_{j-1}) & \text{if } k = j-1, \\ \tau_{j+1}\partial_{x_{j+1}}\partial_{x_j}\phi_{j+1}(x_{j+1}, x_j) & \text{if } k = j+1, \\ 0 & \text{if } |j-k| \geq 2. \end{cases} \tag{4.1.10}$$

Definition 4.1 Let δ_0 be so small that $\kappa_2\delta_0^2 < 2^{-1}$ and $\delta_0 \leq \delta$.

Theorem 4.1 *Suppose $|s' - s| \leq \delta_0$. Then the phase function $S_\Delta(x_{J+1}, x_J, \ldots, x_1, x_0)$ satisfies conditions (3.2.7) and condition (3.2.8) in Assumption 3.1 if Δ is fixed. The time slicing approximation $I(\Delta; \nu, s', s, x, y)$ has a definite finite value as an oscillatory integral. It is of the following form:*

$$\begin{aligned} &I(\Delta; \nu, s', s, x, y) \\ &= \left(\frac{-i\nu}{2\pi(s'-s)}\right)^{1/2} e^{i\nu S(s',s,x,y)}\left(\det(I + H_\Delta^{-1}W_\Delta(x^*))\right)^{-1/2} a(\Delta; \nu, s', s, x, y). \end{aligned} \tag{4.1.11}$$

Furthermore, $a(\Delta; \nu, s', s, x, y)$ is of the form

$$a(\Delta; \nu, s', s, x, y) = 1 + \nu^{-1}r(\Delta; \nu, s', s, x, y).$$

For any α, β there exists a constant $C_{\Delta,\alpha,\beta} > 0$ such that

$$|\partial_x^\alpha\partial_y^\beta r(\Delta; \nu, s', s, x, y)| \leq C_{\Delta,\alpha,\beta}. \tag{4.1.12}$$

This theorem is proved through several lemmas.

Since

$$\det H_\Delta = \frac{\tau_1 + \cdots + \tau_{J+1}}{\tau_1\tau_2\cdots\tau_{J+1}} > 0,$$

it is invertible and we can write

$$H_\Delta(x) = H_\Delta(I + H_\Delta^{-1}W(x)). \tag{4.1.13}$$

We wish to prove that $H_\Delta^{-1}W(x)$ is small. Let g_{jk} be the (j, k) entry of the matrix $G_\Delta = H_\Delta^{-1}$. Then

$$g_{kl} = \begin{cases} \dfrac{(\tau_1 + \cdots + \tau_k)(\tau_{l+1} + \cdots + \tau_{J+1})}{\tau_1 + \tau_2 + \cdots + \tau_{J+1}}, & \text{if } 1 \leq k \leq l \leq J, \\ \dfrac{(\tau_1 + \cdots + \tau_l)(\tau_{k+1} + \cdots + \tau_L)}{\tau_1 + \tau_2 + \cdots + \tau_{J+1}}, & \text{if } 1 \leq l \leq k \leq J. \end{cases} \tag{4.1.14}$$

Definition 4.2 For any vector $\xi = (\xi_1, \xi_2, \ldots, \xi_J) \in \mathbf{R}^J$ we define the following two norms:

$$\|\xi\|_{l^\infty} = \max(|\xi_1|, \ldots, |\xi_J|), \quad \|\xi\|_{l^1} = \sum_{j=1}^{J} |\xi_j|. \tag{4.1.15}$$

Let $(G_\Delta \xi)_j$ be the j-th component of the vector $G_\Delta \xi$.

Lemma 4.1 *The following inequalities hold for any $\xi \in \mathbf{R}^J$.*

$$\begin{aligned}
\|W_\Delta \xi\|_{l^1} &\le 4\kappa_2(\tau_1 + \cdots + \tau_{J+1})\|\xi\|_{l^\infty}, \\
|(G_\Delta \xi)_j| &\le \min\{\tau_1 + \cdots + \tau_j, \tau_{j+1} + \cdots + \tau_{J+1}\}\|\xi\|_{l^1}, \\
\|G_\Delta \xi\|_{l^\infty} &\le \frac{\tau_1 + \cdots + \tau_{J+1}}{4}\|\xi\|_{l^1} \\
\|G_\Delta W_\Delta \xi\|_{l^\infty} &\le \kappa_2(\tau_1 + \cdots + \tau_{J+1})^2\|\xi\|_{l^\infty}.
\end{aligned}$$

Proof Let $T = \tau_1 + \cdots + \tau_{J+1}$. Then

$$\|W_\Delta \xi\|_{l^1} \le \|\xi\|_{l^\infty} \sum_{j,k=1}^{J} |w_{jk}| \le 4\kappa_2(\tau_1 + \cdots + \tau_{J+1})\|\xi\|_{l^\infty}.$$

The first inequality is proved. It is clear that if $k \le l$, then

$$\begin{aligned}
T^{-1}(\tau_1 + \cdots + \tau_k)(\tau_{l+1} + \cdots + \tau_{J+1}) &\le \min\{(\tau_1 + \cdots + \tau_k), (\tau_{l+1} + \cdots + \tau_{J+1})\} \\
&\le \min\{(\tau_1 + \cdots + \tau_k), (\tau_{k+1} + \cdots + \tau_{J+1})\}.
\end{aligned}$$

Hence,

$$|(G_\Delta \xi)_j| \le \|\xi\|_{l^1} \max_{k=1,\ldots,J} |g_{jk}| \le \|\xi\|_{l^1} \min\{(\tau_1 + \cdots + \tau_j), (\tau_{j+1} + \cdots + \tau_{J+1})\}.$$

The second inequality is proved. The third inequality follows from

$$(\tau_1 + \cdots + \tau_k)(\tau_{l+1} + \cdots + \tau_{J+1}) \le 4^{-1}T^2.$$

The last inequality follows from the first and the third inequalities. □

Lemma 4.2 *If $|s' - s| \le \delta_0$, then for any $(x_{J+1}, x_J, \ldots, x_1, x_0) \in \mathbf{R}^{J+2}$*

$$(1 - \kappa_2\delta_0^2)^J \le |\det(I + H_\Delta^{-1} W_\Delta(x))| \le (1 + \kappa_2\delta_0^2)^J, \tag{4.1.16}$$

$$\det(H_\Delta + W_\Delta(x)) \ge (1 - \kappa_2\delta_0^2)^J \frac{\tau_1 + \cdots + \tau_{J+1}}{\tau_1\tau_2\cdots\tau_{J+1}} > 0. \tag{4.1.17}$$

$H_\Delta(x)$ is a positive definite symmetric matrix.

Proof For any $\xi \in \mathbf{R}^J$

$$\|G_\Delta W_\Delta(x)\xi\|_{l^\infty} \le \kappa_2\delta_0^2\|\xi\|_{l^\infty}.$$

For θ with $0 \le \theta \le 1$ the spectre of $(I + \theta G W_\Delta(x))$ lies inside the circle $|z-1| \le \theta\kappa_2\delta_0^2$ of the complex plane. Thus (4.1.16) holds. Since

$$|\det(H_\Delta + \theta W_\Delta(x))| = |\det H_\Delta \det(I + \theta G_\Delta W_\Delta(x))| \ge (1-\theta\kappa_2\delta_0^2)^J \frac{\tau_1 + \cdots + \tau_{J+1}}{\tau_1\tau_2\cdots\tau_{J+1}},$$

the symmetric matrix $H_\Delta + \theta W_\Delta(x)$ is non-singular for any $\theta \in [0, 1]$, and $H_\Delta = H_\Delta + \theta W_\Delta(x)|_{\theta=0}$ is positive definite. Therefore, $H_\Delta(x) = H_\Delta + \theta W_\Delta(x)|_{\theta=1}$ is also positive definite. (4.1.17) follows from this and (4.1.16). The lemma is proved. □

Combining Lemmas 4.2 and 3.1, we have proved

Proposition 4.1 *Let $m = 2$, $n = J$. We regard $S_\Delta(x_{J+1}, x_J, \dots, x_1, x_0)$ as $\phi(x, y)$ with $y = (x_{J+1}, x_0)$, $x = (x_J, \dots, x_1)$. Then $S_\Delta(x_{J+1}, x_J, \dots, x_1, x_0)$ satisfies conditions (3.2.7) and (3.2.8) of Assumption 3.1. For any fixed $(x_{J+1}, x_0) \in \mathbf{R}^2$ there exists one and only stationary point $x^* = (x_J^*, \dots, x_1^*)$ of the function $\mathbf{R}^J \ni (x_1, \dots, x_J) \to S_\Delta(x_{J+1}, x_J, \dots, x_1, x_0) \in \mathbf{R}$. x^* is an infinitely differentiable function of (x_{J+1}, x_0) with values in $\mathbf{R}^J$.*

Proposition 4.2 *Suppose that $|s' - s| \le \delta$. Let $x = x_{J+1}$ and $y = x_0$. Let γ^* denote the only classical path with $\gamma^*(s) = y$ and $\gamma^*(s') = x$. Let $x^* = (x_J^*, x_{J-1}^*, \dots, x_1^*)$ be a stationary point of $(x_J, x_{J-1}, \dots, x_1) \to S_\Delta(x_{J+1}, x_J, x_{J-1}, \dots, x_1, x_0)$. Then $x_j^* = \gamma^*(T_j)$, $j = 1, 2, \dots, J$. and $S_\Delta(x, x_J^*, x_{J-1}^*, \dots, x_2^*, x_1^*, y) = S(s', s, x, y)$.*

Proof The stationary point $x^* = (x_J^*, x_{J-1}^*, \dots, x_2^*, x_1^*)$ satisfies:

$$\partial_{x_j} S(T_{j+1}, T_j, x_{j+1}^*, x_j^*) = \partial_{x_j}^* S(T_j, T_{j-1}, x_j^*, x_{j-1}^*).$$

Let γ_Δ^* be the piecewise classical path corresponding to x^*. Then the above equation and (2.8) means that

$$\frac{d}{dt}\gamma_\Delta^*(T_j + 0) = \frac{d}{dt}\gamma_\Delta^*(T_j - 0).$$

Then by Corollary 2.2 $\gamma_\Delta^* = \gamma^*$. Therefore, $x_j^* = \gamma^*(T_j)$, $j = 1, 2, \dots, J$. and

$$S_\Delta(x_{J+1}, x_J^*, x_{J-1}^*, \dots, x_2^*, x_1^*, x_0) = S(s', s, x_{J+1}, x_0).$$

□

Proof of Theorem 4.1 In the right-hand side of (3.2.2), let $n = J$, $x = (x_J, \dots, x_1)$, $m = 2$, $y = (x_{J+1}, x_0)$, $\phi(x, y) = S_\Delta(x_{J+1}, x_J, \dots, x_1, x_0)$, $p = 0$, $\lambda = 0$ and $a_\lambda = 1$. Then it turns out that $I(\Delta; \nu, s', s, x, y)$ is an oscillatory integral of type (3.2.2). We wish to show that Assumption 3.1 is satisfied. By (4.1.6), condition (3.2.7) of

Assumption 3.1 is satisfied. By Lemma 4.2, (3.2.8) is clearly satisfied if Δ is fixed. The amplitude function 1 obviously satisfies condition (3.2.9). Thus Assumption 3.1 is satisfied if Δ is fixed. We can apply stationary phase method Theorem 3.7 to this integral. By Proposition 2.7, the stationary value $S_\Delta(x_{J+1}, x_J^*, \ldots, x_1^*, x_0) = S(s', s, x_{J+1}, x_0)$. Notice in this case $a_\lambda \equiv 1$ and

$$A = (2\pi i)^{J/2}\left(\frac{\tau_{J+1}\cdots\tau_1}{(s'-s)}\right)^{-1/2}(\det(I + H_\Delta^{-1}W_\Delta(x^*)))^{-1/2}.$$

Using (3.3.26), we define

$$r(\Delta; \nu, s', s, x, y) = R_1(1)(y) + (i\nu)^{-1}A^{-1}R_2(1)(y).$$

Then Theorem 4.1 follows from Theorem 3.7. □

In Remark 1.2 of Sect. 1.1.4 we mentioned Feynman path integrals of the form

$$\int_\Omega F(\gamma)\exp\left(i\nu S(\gamma)\right)\mathscr{D}\gamma$$

involving general functional $F(\gamma)$ of γ. Now we discuss definition of the time slicing approximation of Feynman path integrals of this type.

Let $F(\gamma)$ be a functional of paths γ. We define the time slicing approximation of a Feynman path integral with integrand $F_\lambda(\gamma)$ by the oscillatory integral

$$\begin{aligned} &I[F_\Delta](\Delta; \nu, s', s, x, y) \qquad (4.1.18)\\ &= \prod_{j=1}^{J+1}\left(\frac{\nu}{2\pi i\tau_j}\right)^{d/2}\int_{\mathbf{R}^J} e^{i\nu S(\gamma_\Delta)}F(\gamma_\Delta)\prod_{j=1}^{J} dx_j\\ &= \prod_{j=1}^{J+1}\left(\frac{\nu}{2\pi i\tau_j}\right)^{d/2}\int_{\mathbf{R}^J} e^{i\nu S_\Delta(x_{J+1},x_J,\ldots,x_1,x_0)}F_\Delta(x_{J+1}, x_J, \ldots, x_1, x_0)\prod_{j=1}^{J} dx_j, \end{aligned}$$

where $x_{J+1} = x$ and $x_0 = y$, cf. Feynman [21], Feynman and Hibbs [22]. By this notation

$$I[1](\Delta; \nu, s', s, x, y) = I(\Delta; \nu, s', s, x, y).$$

We discuss $I[F_\Delta](\Delta; \nu, s', s, x, y)$ for functionals $F(\gamma)$ with the following property:

Assumption 4.2 Let m be a non-negative constant. For any non-negative integer K there exist positive constants A_K and X_K such that for any division Δ and for any indices α_j, $j = 0, 1, 2, \ldots, J+1$ satisfying $|\alpha_j| \le K$ there holds the inequality

$$\left|\left(\prod_{k=0}^{J+1} \partial_{x_j}^{\alpha_j}\right) F(\gamma_\Delta(x_{J+1}, x_J, \ldots, x_1, x_0))\right| \tag{4.1.19}$$
$$\leq A_K X_K^{J+2}(1 + |x_{J+1}| + |x_J| + \cdots + |x_1| + |x_0|)^m.$$

Theorem 4.3 *Suppose that $|s' - s| \leq \delta_0$. If $F(\gamma)$ satisfies the above assumption, then for fixed division Δ $I[F_\Delta](\Delta; \nu, s', s, x, y)$ has definite finite value. It is of the form*

$$I[F_\Delta](\Delta; \nu, s', s, x, y)$$
$$= \left(\frac{-i\nu}{2\pi(s' - s)}\right)^{1/2} \left(\det(I + H_\Delta^{-1} W_\Delta(x^*))\right)^{-1/2} e^{i\nu S(s', s, x_{J+1}, x_0)}$$
$$\times \left(F(\gamma^*) + (i\nu)^{-1} r(\Delta; \nu, s', s, x_{J+1}, x_0)\right).$$

For any α, β there exists a positive constant $C_{\Delta\alpha\beta}$ such that

$$|\partial_{x_{J+1}}^\alpha \partial_{x_0}^\beta r(\Delta; \nu, s', s, x_{J+1}, x_0)| \leq C_{\Delta\alpha\beta}(1 + |x_{J+1} + |x_0|)^m.$$

Proof If $F(\gamma)$ satisfies Assumption 4.2 above, then $F(\gamma_\Delta)$ satisfies Condition 3.2.9 with $p = m$ and $\lambda = 0$. It follows from Lemma 4.2 and Assumption 4.2 that $I[F_\Delta](\Delta; \nu, s', s, x, y)$ satisfies Assumption 3.1 if $|s' - s| \leq \delta_0$. The present theorem follows from Theorem 3.7. □

Remark 4.1 How does the constant $C_{\Delta\alpha\beta}$ behave when $|\Delta| \to 0$? The question will be answered in the next section.

4.2 Statement of Main Results

4.2.1 Convergence of Time Slicing Approximation of Feynman Path Integrals as $|\Delta| \to 0$

We discuss what happens if $|\Delta| \to 0$. Proofs are given in the subsequent chapters, because they are long. Suppose that $[s, s'] \subset [a, b]$ with $|s' - s| \leq \delta_0$. We set

$$D(\Delta; s', s, x, y) = \det(I + H_\Delta^{-1} W_\Delta(x^*)). \tag{4.2.1}$$

Then by Theorem 4.1 the time slicing approximation of Feynman path integrals is of the form

$$I(\Delta; \nu, s', s, x, y)$$
$$= \left(\frac{-i\nu}{2\pi(s' - s)}\right)^{1/2} e^{i\nu S(s', s, x, y)} D(\Delta; s', s, x, y)^{-1/2} \left(1 + (i\nu)^{-1} r(\Delta; \nu, s', s, x, y)\right).$$

We begin with

Lemma 4.3 *Suppose that* $0 < |s' - s| \le \delta_0$. *Define* $d(\Delta; s', s, x, y)$ *by*

$$D(\Delta; s', s, x, y) = 1 + (s' - s)^2 d(\Delta; s', s, x, y).$$

Then for any $k \ge 0$

$$\sup_{\substack{|s'-s|\le\delta,\\ |\Delta|\le\delta}} \max_{|\alpha|+|\beta|\le k} \sup_{(x,y)\in\mathbf{R}^2} |\partial_x^\alpha \partial_y^\beta d(\Delta; s', s, x, y)| = C_k < \infty.$$

When $|\Delta| \to 0$, we have the following

Theorem 4.4 *Suppose that* $|s - s'| \le \delta_0$. *Then the following limit exists:*

$$D(s', s, x, y) = \lim_{|\Delta|\to 0} D(\Delta; s', s, x, y). \tag{4.2.2}$$

More precisely, for any α, β *there exists a positive constant* $C_{\alpha\beta}$ *such that for* $(x, y) \in \mathbf{R}^2$

$$|\partial_x^\alpha \partial_y^\beta (D(s', s, x, y) - D(\Delta; s', s, x, y)| \le C_{\alpha\beta}|\Delta|(s' - s). \tag{4.2.3}$$

We can write

$$D(s', s, x, y) = 1 + (s' - s)^2 d(s', s, x, y). \tag{4.2.4}$$

For any α β *there exists a positive constant* $C_{\alpha\beta}$ *such that for* $(x, y) \in \mathbf{R}^2$

$$|\partial_x^\alpha \partial_y^\beta d(s', s, x, y)| \le C_{\alpha\beta}. \tag{4.2.5}$$

Theorem 4.4 will be proved in Sect. 5.2.

The next theorem improves Theorem 4.1.

Theorem 4.5 *Suppose that the potential* $V(t, x)$ *satisfies Assumption 2.1. Let* δ_0 *be so small that* $\kappa_2\delta_0^2 < 1$. *Suppose that* $|s' - s| \le \delta_0$. *Then there exists a constant* $\delta_1 > 0$ *such that if* $|\Delta| \le \delta_1$, *then*

$$\begin{aligned} &I(\Delta; S, \nu, s', s, x, y) \\ &= \left(\frac{-i\nu}{2\pi(s' - s)}\right)^{1/2} \exp\{i\nu S(s', s, x, y)\} k(\Delta; \nu, s', s, x, y) \end{aligned}$$

with

$$k(\Delta; \nu, s', s, x, y) = D(\Delta; s', s, x, y)^{-1/2}(1 + \nu^{-1} r(\Delta; \nu, s', s, x, y)). \tag{4.2.6}$$

Moreover, for any α, β *there exists a positive constant* $C_{\alpha,\beta}$ *independent of* ν *and of* Δ *such that for any* $(x, y) \in \mathbf{R}^2$

$$|\partial_x^\alpha \partial_y^\beta r(\Delta; \nu, s', s, x, y)| \leq C_{\alpha\beta}(s' - s)^2. \tag{4.2.7}$$

This theorem will be proved in Sect. 5.4.

When $|\Delta| \to 0$, we have

Theorem 4.6 (Convergence of Feynman path integral) *Suppose* $0 < |s' - s| \leq \delta_0$. *Then the following limit exists:*

$$K(\nu, s', s, x, y) = \lim_{|\Delta| \to 0} I(\Delta; \nu, s', s, x, y). \tag{4.2.8}$$

More precisely, $K(\nu, s', s, x, y)$ *is of the form*

$$K(\nu, s', s, x, y) = \left(\frac{-i\nu}{2\pi(s' - s)}\right)^{1/2} \exp\{i\nu S(s', s, x, y)\} k(\nu, s', s, x, y). \tag{4.2.9}$$

For any α, β *there exist constants* $C_{\alpha\beta} > 0$ *and* $\delta_1 > 0$ *such that if* $|\Delta| < \delta_1$, *then for* $(x, y) \in \mathbf{R}^2$

$$|\partial_x^\alpha \partial_y^\beta \big(k(s', s, \nu, x, y) - k(\Delta; \nu, s', s, x, y)\big)| \leq C_{\alpha\beta}|\Delta||s' - s|. \tag{4.2.10}$$

Proof Given in Sect. 5.4.

Corollary 4.1 *There exists a function* $r(\nu, s', s, x, y)$ *such that*

$$k(\nu, s', s, x, y) = D(s', s, x, y)^{-1/2}(1 + \nu^{-1} r(\nu, s', s, x, y)). \tag{4.2.11}$$

For any α, β *there exist constants* $C_{\alpha\beta} > 0$ *and* $\delta_1 > 0$ *independent of* ν *and of* Δ *such that if* $|\Delta| \leq \delta_1$, *then for* $(x, y) \in \mathbf{R}^2$

$$|\partial_x^\alpha \partial_y^\beta \big(r(\nu, s', s, x, y) - r(\Delta; \nu, s', s, x, y)\big)| \leq C_{\alpha\beta}|\Delta|(s' - s)^2.$$

For any α,β *there exists a positive constant* $C_{\alpha\beta}$ *independent of* ν, s', s *such that for* $(x, y) \in \mathbf{R}^2$

$$|\partial_x^\alpha \partial_y^\beta r(\nu, s', s, x, y)| \leq C_{\alpha\beta}|s' - s|^2. \tag{4.2.12}$$

Remark 4.2 If we adopt the symbols in Chap. 1, then Theorem 4.6 states that

$$\lim_{|\Delta| \to 0} \prod_{j=1}^{J+1} \left(\frac{-i\nu}{2\pi\tau_j}\right)^{1/2} \int_{\mathbf{R}^J} \exp(i\nu S(\gamma_\Delta)) \prod_{j=1}^{J} dx_j = K(\nu, s', s, x, y), \tag{4.2.13}$$

that is,

$$\int_\Omega \exp(i\hbar^{-1} S(\gamma)) \mathscr{D}\gamma = K(\nu, s', s, x, y). \tag{4.2.14}$$

In fact, (4.2.10) implies that the convergence occurs in a very strong topology. Moreover, it follows from (4.2.11) together with (4.2.12) that the semi-classical asymptotic formula for Feynman path integrals is true. This fact was proved earlier in [25].

We now consider convergence of path integrals with integrand $F(\gamma)$. We discuss (4.1.18):

$$
\begin{aligned}
& I[F_\Delta](\Delta; \nu, s', s, x, y) \qquad (4.2.15)\\
&= \prod_{j=1}^{J+1} \left(\frac{\nu}{2\pi i \tau_j}\right)^{d/2} \int_{\mathbf{R}^J} e^{i\nu S(\gamma_\Delta)} F(\gamma_\Delta) \prod_{j=1}^{J} dx_j \\
&= \prod_{j=1}^{J+1} \left(\frac{\nu}{2\pi i \tau_j}\right)^{d/2} \int_{\mathbf{R}^J} e^{i\nu S_\Delta(x_{J+1}, x_J, \dots, x_1, x_0)} F_\Delta(x_{J+1}, x_J, \dots, x_1, x_0) \prod_{j=1}^{J} dx_j,
\end{aligned}
$$

and we discuss convergence of Feynman path integral with integrand $F(\gamma)$

$$
\int_\Omega \exp(i\hbar^{-1} S(\gamma)) F(\gamma) \mathscr{D}\gamma = \lim_{|\Delta|\to 0} I[F_\Delta](\Delta; \nu, s', s, x, y). \qquad (4.2.16)
$$

Let γ_Δ denote a piecewise classical path associated with Δ such that $\gamma_\Delta(T_j) = x_j$. We assume the following assumption:

Assumption 4.7 Let m be a non-negative constant. For any integer $K \geq 0$ there exist constants $A_K > 0$ and $X_K \geq 1$ such that for any division Δ and for any indices α_j satisfying $|\alpha_j| \leq K$, $j = 0, 1, 2, \dots, J+1$

$$
\begin{aligned}
& \left| \left(\prod_{k=0}^{J+1} \partial_{x_j}^{\alpha_j} \right) F(\gamma_\Delta(x_{J+1}, x_J, \dots, x_1, x_0)) \right| \qquad (4.2.17)\\
& \qquad \leq A_K X_K^{J+2} (1 + |x_{J+1}| + |x_J| + \dots + |x_1| + |x_0|)^m.
\end{aligned}
$$

Our discussion requires additional assumptions on $F(\gamma)$ that are rather complicated. In order to describe them, we now introduce a symbol.

Let Δ be a division of interval $[s, s']$ as (4.1.3). We call a division Δ' of the interval $[s, s']$ coarser than the division Δ if Δ is a refinement of Δ'. Let $0 = j_0 < j_1 < \dots < j_p < j_{p+1} = J+1$ be any subsequence of $\{0, 1, \dots, J, J+1\}$. Then

$$
\Delta' : s = T_0 = T_{j_0} < T_{j_1} < \dots < T_{j_p} < T_{j_{p+1}} = T_{J+1} = s' \qquad (4.2.18)
$$

is a division of the interval $[s, s']$ coarser than Δ. For any fixed x_{j_n} and $x_{j_{n+1}}$ we define $(x^*_{j_{n+1}-1}, \dots, x^*_{j_n+1})$ as the stationary point of the function

$$
(x_{j_{n+1}-1}, \dots, x_{j_n+1}) \to \sum_{j=j_n+1}^{j_{n+1}} S(T_j, T_{j-1}, x_j, x_{j-1}).
$$

Since x_j^*, $j = j_n + 1, \dots, j_{n+1} - 1$, is a function of $(x_{j_{n+1}}, x_{j_n})$ we write $x_j^* = x_j^*(x_{j_{n+1}}, x_{j_n})$ for it.

For any function $a(x_{J+1}, x_J, \dots, x_0)$ of $(x_{J+1}, x_J, \dots, x_0) \in \mathbf{R}^{J+2}$ we define a new function $\iota_{\Delta'}^{\Delta} a(x_{J+1}, x_{j_p}, \dots, x_{j_1}, x_0)$ of $(x_{J+1}, x_{j_p}, \dots, x_{j_1}, x_0) \in \mathbf{R}^{p+2}$ by

$$\begin{aligned}&\iota_{\Delta'}^{\Delta} a(x_{J+1}, x_{j_p}, \dots, x_{j_1}, x_0) \\ &= a(x_{J+1}, x_J, \dots, x_k, \dots, x_0)\Big|_{\substack{x_k = x_k^*(x_{j_n}, x_{j_{n-1}}) \\ j_{n-1} < k < j_n, n=1,2,\dots,p+1}} .\end{aligned} \tag{4.2.19}$$

If $a(x_{J+1}, x_J, \dots, x_0) = F(\gamma_\Delta)(x_{J+1}, x_J, \dots, x_0)$ with some functional $F(\gamma)$, then by definition

$$\iota_{\Delta'}^{\Delta} F(\gamma_\Delta) = F(\gamma_{\Delta'}). \tag{4.2.20}$$

In particular,

$$\iota_{\Delta'}^{\Delta} S(x_{J+1}, x_{j_p}, \dots, x_{j_1}, x_0) = \sum_{n=1}^{p+1} S(T_{j_n}, T_{j_{n-1}}, x_{j_n}, x_{j_{n-1}}). \tag{4.2.21}$$

For any integer $k = 1, 2, \dots, J + 1$ we define a division $\Delta(k)$ coarser than Δ as follows:

$$\Delta(k) : s = T_0 < T_k < T_{k+1} < \cdots < T_{J+1} = s'. \tag{4.2.22}$$

In particular $\Delta(1) = \Delta$ and $\Delta(J + 1) = [T_0, T_{J+1}] = [s, s']$. Thus

$$\iota_{\Delta(J+1)}^{\Delta} a(x_{J+1}, x_0) = a(x_{J+1}, x_J, \dots, x_1, x_0)\Big|_{\{x_j = \gamma^*(T_j), 1 \le j \le J\}} .$$

Here γ^* is the classical path on $[s, s']$ with $\gamma^*(T_{J+1}) = x_{J+1}$ and $\gamma^*(T_0) = x_0$. We write $a(\gamma^*)$ for $\iota_{\Delta(J+1)}^{\Delta} a$ for simplicity.

Let Δ be a division of the interval $[s, s']$ as in (4.1.3) and let $(x_{J+1}, x_J, \dots, x_0) \in \mathbf{R}^{J+2}$.

Kumano-go [46] introduced a sufficient condition of $F(\gamma)$ for the Feynman path integral (4.2.16) to converge.

Assumption 4.8 There exist a non-negative constant m and a bounded non-negative Borel measure ρ on $[s, s']$. For any integer $K \ge 0$ there exist constants $A_K > 0$ and $X_K \ge 1$ with the following property: Let Δ be an arbitrary division of $[s, s']$ as (4.1.3) and Δ' be as (4.2.18) coarser than Δ and for any indices α_{j_n} satisfying $|\alpha_{j_n}| \le K$, $n = 0, 1, 2, \dots, p + 1$

$$\begin{aligned}&\left| \left(\prod_{n=0}^{p+1} \partial_{x_{j_n}}^{\alpha_{j_n}} \right) \left(\iota_{\Delta'}^{\Delta} \partial_{x_k}^{\beta} F_\Delta \right)(x_{J+1}, x_{j_p}, \dots, x_{j_1}, x_0) \right| \\ &\le A_K X_K^{p+2} \rho([T_{k-1}, T_{k+1}])(1 + |x_{J+1}| + |x_{j_p}| + \cdots + |x_{j_1}| + |x_0|)^m,\end{aligned} \tag{4.2.23}$$

for all $k = 1, 2, \ldots, J$ and $\beta = 1, 2$.

Example 4.1 Cf. [46]. Let $\rho(t)$ be a function of bounded variation on $[s, s']$ and $f(t, x)$ be a continuous function of $(t, x) \in [s, s'] \times \mathbf{R}$. Also let $f(t, x)$ be an infinitely differentiable function of x if t is fixed. Suppose that for any α there exists a positive constant C_α such that

$$|\partial_x^\alpha f(t, x)| \le C_\alpha (1 + |x|)^m, \tag{4.2.24}$$

with some $m \ge 0$ independent of α and (t, x). Define

$$F(\gamma) = \int_s^{s'} f(t, \gamma(t))\, d\rho(t). \tag{4.2.25}$$

Then $F(\gamma)$ satisfies Assumptions 4.7 and 4.8.

The next theorem gives the answer to the question in Remark 4.1.

Theorem 4.9 *Suppose that the potential $V(t, x)$ fulfills Assumption 2.1 and the functional $F(\gamma)$ satisfies Assumptions 4.7 and 4.8. Let δ_0 be so small that $\kappa_2 \delta_0^2 < 1$. If $|s - s'| \le \min\{\delta_0, 1\}$, then there exists a positive constant δ_1 such that if $|\Delta| \le \delta_1$, then*

$$I[F_\Delta](\Delta; \nu, s', s, x, y) \tag{4.2.26}$$

$$= \left(\frac{-i\nu}{2\pi(s' - s)} \right)^{1/2} \exp\{i\nu S(s', s, x, y)\} \tag{4.2.27}$$

$$\times D(\Delta; s', s, x, y)^{-1/2} \big(F(\gamma^*) + \nu^{-1} R_\Delta[F_\Delta](\nu, s', s, x, y) \big). \tag{4.2.28}$$

Here γ^* is the classical path starting from y at time s and reaching x at time s'. Moreover, for any non-negative integer K there exist a positive constant C_K and a positive integer $M(K)$ independent of ν and Δ such that for $(x, y) \in \mathbf{R}^2$

$$\begin{aligned} &|\partial_x^\alpha \partial_y^\beta R_\Delta[F_\Delta](\nu, s', s, x, y)| \\ &\le C_K A_{M(K)} |s - s'| (|s - s'| + \rho([s, s'])) (1 + |x| + |y|)^m, \end{aligned} \tag{4.2.29}$$

as far as $|\alpha|, |\beta| \le K$.

Proof Will be given in Sect. 5.4.

We set

$$k[F_\Delta](\Delta; \nu, s', s, x, y) = D(\Delta; s', s, x, y)^{-1/2} \big(F(\gamma^*) + \nu^{-1} R_\Delta[F_\Delta](\nu, s', s, x, y) \big). \tag{4.2.30}$$

When $|\Delta| \to 0$, we have

Theorem 4.10 *Assume that functional $F(\gamma)$ satisfies Assumptions 4.7 and 4.8 above and $|s'-s| \le \delta_0$. Then the following limit exists:*

$$K[F](\nu, s', s, x, y) = \lim_{|\Delta|\to 0} I[F_\Delta](\Delta; \nu, s', s, x, y). \tag{4.2.31}$$

$K[F](\nu, s', s, x, y)$ is of the form

$$K[F](\nu, s', s, x, y) = \left(\frac{\nu}{2\pi i(s'-s)}\right)^{1/2} e^{i\nu S(\gamma^*)} k(F; \nu, s', s, x, y). \tag{4.2.32}$$

More precisely, for any integer $K \ge 0$ there exist constants $C_{\alpha\beta} > 0$, $\delta_1 > 0$ and an integer $M(K) \ge 0$ such that if $|\Delta| \le \delta_0$, then for $(x, y) \in \mathbf{R}^2$

$$|\partial_x^\alpha \partial_y^\beta \left(k[F_\Delta](\Delta; \nu, s', s, x, y) - k(F; \nu, s', s, x, y)\right)| \tag{4.2.33}$$
$$\le C_{\alpha\beta} A_{M(K)} |\Delta| (\rho([s, s'] + (s'-s))(1 + |x| + |y|)^m, \tag{4.2.34}$$

as far as $|\alpha| \le K$ and $|\beta| \le K$.

Proof Will be given in Sect. 5.4.

Corollary 4.2 *$k(F; \nu, s', s, x, y)$ is of the form*

$$k(F; \nu, s', s, x, y) = D(s', s, x, y)^{-1/2} \left(F(\gamma^*) + \nu^{-1} R[F](\nu, s', s, x, y)\right), \tag{4.2.35}$$

and as far as $|\alpha| \le K$ and $|\beta| \le K$, for $(x, y) \in \mathbf{R}^2$

$$|\partial_x^\alpha \partial_y^\beta R_\Delta[F](\nu, s', s, x, y)| \tag{4.2.36}$$
$$\le C_K A_{M(K)} |s-s'|(|s-s'| + \rho([s, s']))(1 + |x| + |y|)^m,$$
$$|\partial_x^\alpha \partial_y^\beta \left(R_\Delta[F_\Delta](\nu, s', s, x, y) - R[F](\nu, s', s, x, y)\right)| \tag{4.2.37}$$
$$\le C_K A_{M(K)} |\Delta| |s-s'|(|s-s'| + \rho([s, s']))(1 + |x| + |y|)^m.$$

Remark 4.3 We may write

$$\int_\Omega e^{i\nu S(\gamma)} F(\gamma) \mathscr{D}[\gamma] = K[F](\nu, s', s, x, y). \tag{4.2.38}$$

Remark 4.4 Corollary 4.2 together with (4.2.36) imply the semi-classical asymptotic formula.

Remark 4.5 While the case $F(\gamma) \equiv 1$ had been proved in [28, 34], from Theorem 4.9 to Corollary 4.2 were proved first by Kumano-go [46].

Remark 4.6 In the case $F(\gamma) \equiv 1$, the corresponding result was discussed in [62] if a non-zero magnetic field exists.

Detailed proofs of results stated in this subsection will be given in Sects. 5.2 and 5.4 in Chap. 5.

4.2.2 The Feynman Path Itegral and the Schrödinger Equation

The case $F(\gamma) \equiv 1$ is particularly important, because in this case $\int_\Omega e^{i\nu S(\gamma)} \mathscr{D}[\gamma] = K(\nu, t, s, x, y)$ is expected to be the fundamental solution of the Schrödinger equation. [21]. This conjecture is in fact true as stated below. Moreover, we can prove the semi-classical asymptotic formula of the fundamental solution of the Schrödinger equation. This proof is completely different from the usual discussion originated by Birkhoff [11].

We shall discuss the integral transformations:

Definition 4.3 Suppose that $[s, t] \subset [a, b]$ and $|t - s| \leq \delta_0$. Then we define for any $\varphi \in C_0^\infty(\mathbf{R})$

$$I(\Delta; \nu, t, s)\varphi(x) = \int_{\mathbf{R}} I(\Delta; \nu, t, s, x, y)\varphi(y)dy, \tag{4.2.39}$$

$$K(\nu, t, s)\varphi(x) = \int_{\mathbf{R}} K(\nu, t, s, x, y)\varphi(y)dy. \tag{4.2.40}$$

We shall treat these transformations on the Hilbert space $L^2(\mathbf{R})$. We express the norm of a linear operator A on $L^2(\mathbf{R})$ by writing $\|A\|$.

Proposition 4.3 *Suppose that* $|t - s| \leq \delta_0$. $I(\Delta; \nu, t, s)$ *and* $K(\nu, t, s)$ *can be extended as bounded linear operators on* $L^2(\mathbf{R})$. *There exists a positive constant* C *independent of* ν, t *and* s *such that*

$$\|I(\Delta; \nu, t, s)\| \leq C, \quad \|K(\nu, t, s)\| \leq C. \tag{4.2.41}$$

Proof of Proposition 4.3 is given in Proposition 6.1.

Theorem 4.11 *Suppose that* $|t - s| \leq \delta_0$. *Then the operator* $I(\Delta; \nu, t, s)$ *converges to* $K(\nu, t, s)$ *in uniform operator topology as* $|\Delta| \to 0$. *More precisely, for sufficiently small positive* δ_1 *there exists a positive constant independent of* Δ, ν, t *and* s *such that if* $|\Delta| < \delta_1$

$$\|I(\Delta; \nu, t, s) - K(\nu, t, s)\| \leq C|t - s||\Delta|. \tag{4.2.42}$$

Proof of this theorem will be given in Theorem 6.4 of Sect. 6.1.

Next we shall discuss the relation between the Feynman path integral and the fundamental solution of the Schrödinger equation. Proofs will be given in Chap. 6.

Theorem 4.12 *Suppose* $|t-s| \le \delta_0$. *Define*

$$e(t,s,x,y) = \left(\frac{1}{2\pi(t-s)}\right)^{1/2} D(t,s,x,y)^{-1/2}. \qquad (4.2.43)$$

Then it satisfies the transport equation:

$$\partial_t e(t,s,x,y) + \partial_x S(t,s,x,y)\partial_x e(t,s,x,y) + \frac{1}{2}\partial_x^2 S(t,s,x,y) e(t,s,x,y) = 0.$$

Proof Given in Theorem 6.6.

Remark 4.7 1. Our proof given in Theorem 6.6 of Theorem 4.12 is different from the standard method by Birkhoff [11, 49].
2. This theorem implies that $D(t,s,x,y)(t-s)$ is **Morette–Van Vleck** determinant, cf. [51, 63].

Let $H(t)$ be the Hamiltonian operator:

$$H(t) = \frac{1}{2}\left(-i\nu^{-1}\partial_x\right)^2 + V(t,x). \qquad (4.2.44)$$

The next theorem is the main objective of Feynman that the Feynman path integral gives the propagator of the Schrödinger equation.

Theorem 4.13 *Suppose that* $|t-s| \le \delta_0$. *For any* $f \in C_0^\infty(\mathbf{R})$ *the* $L^2(\mathbf{R})$*-valued function* $t \to K(\nu,t,s)f$ *is strongly differentiable. It satisfies*

$$i\nu^{-1}\frac{d}{dt}K(\nu,t,s)f = H(t)K(\nu,t,s)f, \qquad (4.2.45)$$

$$s-\lim_{|t-s|\to 0} K(\nu,t,s)f = f. \qquad (4.2.46)$$

Proof Given in Theorem 6.8 of Sect. 6.4.

Remark 4.8 In other words, this implies that $\int_\Omega e^{i\nu S(\gamma)}\mathscr{D}[\gamma]$ is in fact the propagator as Feynman [21] conjectured.

Corollary 4.3 $K(\nu,t,s)f(x)$ *is the classical solution of the Scrödinger equation*

$$i\nu^{-1}\frac{\partial}{\partial t}K(\nu,t,s)f = \left[\frac{1}{2}\left(-i\nu^{-1}\frac{\partial}{\partial x}\right)^2 + V(t,x)\right]K(\nu,t,s)f(x), \qquad (4.2.47)$$

if $f \in C_0^\infty$.

Proof Given in Corollary 6.2 of Sect. 6.4.

Remark 4.9 Taking this Corollary and Theorem 4.12 into account, we can say that the equality (4.2.11) implies that the semi-classical asymptotic formula of Birkhoff holds. There are many works discussing semi-classical asymptotic formulas from the Feynman path integrals, cf. [21, 22, 51, 57, 61]. See also the beautiful result of [66].

The next theorem explains the meaning of $D(t, s, x, y)$.

Theorem 4.14 *Suppose that* $|t - s| \leq \delta_0$. *Then* $D(t, s, x, y)$ *is an infinite dimensional determinant that is closely related to the Jacobi differential operator, More precisely,*

$$D(t, s, x, y) = \det(JG). \tag{4.2.48}$$

Here G is the Green operator that we used in Sect. 2.2 and J is the Jacobi differential operator

$$Jx(t) = \left[-\frac{d^2}{dt^2} - \partial_x^2 V(t, \gamma^*) \right] x(t), \tag{4.2.49}$$

at the classical orbit $\gamma^*(t)$ *such that* $\gamma^*(s) = y$ *and* $\gamma^*(t) = x$.

Proof Given in Theorems 6.9 and 6.12.

Remark 4.10 Formally, we can write

$$D(t, s, x, y) = \det\left[\left(-\frac{d^2}{dt^2} - \partial_x^2 V(t, \gamma^*) \right) \left(-\frac{d^2}{dt^2} \right)^{-1} \right].$$

We can write also

$$D(t, s, x, y) = \det \delta^2 S(\gamma^*),$$

where $\delta^2 S(\gamma^*)$ is the second variation, i.e. the Fr'echet derivative in $\mathscr{H}$ of order 2 of $S(\gamma)$ at $\gamma = \gamma^*$. See Theorem 6.9 for more detail.

Concerning the semi-classical asymptotic formula we can prove the following sharper facts:

Theorem 4.15 *Suppose* $|t - s| \leq \delta_0$. *Define*

$$v_0(t, x, y) = \left(\frac{\nu}{2\pi i(t-s)} \right)^{1/2} D(t, s, x, y)^{-1/2}$$

$$v_1(t, s, x, y) = -\left(\frac{\nu}{2\pi i(t-s)} \right)^{1/2} (t-s) D(t, s, x, y)^{-1/2} p(t, s, x, y),$$

where

$$(t-s)p(t, s, \nu, x, y) = -\frac{1}{2} \int_s^t D(t, s, \gamma_{x,y}^*(t), y)^{1/2} \partial_z^2 \left(D(t, s, z, y)^{-1/2} \right) \Big|_{z=\gamma^*(t)} dt. \tag{4.2.50}$$

Then

$$\int_{\Omega_{x,y}} e^{i\nu S(\gamma)} \mathscr{D}\gamma = e^{i\nu S(t,s,x,y)}(\nu_0(t,s,x,y) + (i\nu)^{-1}\nu_1(t,s,x,y) + \nu^{-2}\nu_2(\nu,t,s,x,y)).$$

For any $K \geq 0$ *there exists a positive constant* $C_K > 0$ *independent of* $(x, y) \in \mathbf{R}^2$ *and* ν *such that*

$$|\partial_x^\alpha \partial_y^\beta \nu_2(\nu, t, s, x, y)| < C_K. \tag{4.2.51}$$

$\nu_1(t, s.x.y)$ *satisfies the second transport equation:*

$$\frac{\delta}{\delta t}\nu_1(t,s,x,y) + \frac{1}{2}\partial_x^2 S(t,s,x,y)\nu_1(t,s,x,y) + \frac{1}{2}\partial_x^2 \nu_0(t,s,x,y) = 0.$$

The function $\nu_1(t, s, x, y)$ *coincides with the second term in Birkhoff's asymptotic expansion formula* [11, 49].

The proof will be given in Theorem 6.14 and Proposition 6.15.

Remark 4.11 The second term of semi-classical asymptotics is discussed in [29, 31], if $F(\gamma)$ is not necessarily a constant.

Detailed proofs of results stated in this subsection are given in Chaps. 5 and 6.

Chapter 5
Convergence of Feynman Path Integrals

Abstract In this chapter we prove the main results stated in the previous chapter. Our main tool is the stationary phase method of oscillatory integrals over a space of large dimension. In this chapter we show how to use this tool, while the proof of it is given in Part II.

5.1 Some Properties of Determinant of Hessian

Let Δ be the division of the interval $[s, s'] \subset [a, b]$ as in (4.1.3).

We discuss properties of $D(\Delta; s', s, x, y)$ in this section. The following proposition is well-known but is very useful, cf. Lemma 7.4 of Morse [52].

Proposition 5.1 *Let $\phi(x, y) \in C^\infty(\mathbf{R}^m \times \mathbf{R}^n)$. Suppose that there is a C^∞-map $y^\# : \mathbf{R}^m \ni x \longrightarrow y^\#(x) \in \mathbf{R}^n$ such that for any $x \in \mathbf{R}^m$,*

$$\partial_y \phi(x, y^\#(x)) = 0, \tag{5.1.1}$$

$$\det \operatorname{Hess}_y \phi(x, y)\big|_{y=y^\#(x)} \neq 0, \tag{5.1.2}$$

and that the function

$$\phi^\# : \mathbf{R}^m \ni x \longrightarrow \phi(x, y^\#(x)) \in \mathbf{R}$$

is stationary at the point $x = x^*$. Then $(x^*, y^*) = (x^*, y^\#(x^*)) \in \mathbf{R}^m \times \mathbf{R}^n$ is a stationary point of function $\phi(x, y)$ and the following equality holds:

$$\det \operatorname{Hess}_{(x^*,y^*)} \phi = \det \operatorname{Hess}_{x^*} \phi^\# \times \det \operatorname{Hess}_y \phi(x, y)\big|_{(x,y)=(x^*,y^*)}. \tag{5.1.3}$$

Proof We abbreviate $\partial_x \phi(x, y)\big|_{y=y^\#(x)}$ to $(\partial_x \phi)(x, y^\#(x))$ and $\partial_y \phi(x, y)\big|_{y=y^\#(x)}$ to $(\partial_y \phi)(x, y^\#(x))$. As a consequence of (5.1.1),

$$\partial_x \phi^\#(x) = (\partial_x \phi)(x, y^\#(x)) + (\partial_y)\phi(x, y^\#(x))\partial_x y^\#(x) = (\partial_x \phi)(x, y^\#(x)). \tag{5.1.4}$$

Replacing $x = x^*$ in this equality, we have

D. Fujiwara, *Rigorous Time Slicing Approach to Feynman Path Integrals*,
Mathematical Physics Studies, DOI 10.1007/978-4-431-56553-6_5

$$(\partial_x \phi)(x^*, y^*) = 0.$$

Therefore, (x^*, y^*) is a stationary point of $\phi(x, y)$.

The Hessian at (x^*, y^*) is decomposed into the following block expression:

$$\mathrm{Hess}_{(x^*,y^*)}\phi = \begin{pmatrix} \partial_x^2 \phi(x, y) & \partial_y \partial_x \phi(x, y) \\ \partial_x \partial_y \phi(x, y) & \partial_y^2 \phi(x, y) \end{pmatrix} \Big|_{x=x^*, y=y^*}.$$

Apply Lemma 3.7 to this expression. Then

$$\begin{aligned} \det \mathrm{Hess}_{(x^*,y^*)}\phi &= \det[\mathrm{Hess}_{x^*}\phi^{\#} - \partial_x \partial_y \phi(x^*, y^*)(\mathrm{Hess}_y \phi(x^*, y^*))^{-1} \partial_x \partial_y \phi(x^*, y^*)] \\ &\quad \times \det \mathrm{Hess}_y \phi(x, y)\Big|_{y=y^{\#}, x=x^*}. \end{aligned} \tag{5.1.5}$$

Differentiate (5.1.1) and (5.1.4) with respect to x. Then

$$\partial_x \partial_y \phi(x, y^{\#}(x)) + \partial_y^2 \phi(x, y^{\#}(x)) \partial_x y^{\#}(x) = 0. \tag{5.1.6}$$

$$\partial_x^2 \phi^{\#}(x) = \partial_x^2 \phi(x, y^{\#}(x)) + \partial_y \partial_x \phi(x, y^{\#}(x)) \partial_x y^{\#}(x). \tag{5.1.7}$$

Eliminating $\partial_x y^{\#}(x)$ from these and (5.1.6) by virtue of (5.1.2), we obtain

$$\partial_x^2 \phi^{\#}(x) = \partial_x^2 \phi(x, y^{\#}(x)) - \partial_y \partial_x \phi(x, y^{\#}(x)) \partial_y^2 \phi(x, y^{\#}(x))^{-1} \partial_x \partial_y \phi(x, y^{\#}(x)).$$

Setting $x = x^*$ in this equation and replacing that in (5.1.5), we have equality (5.1.3). We have proved the proposition. □

We apply this proposition to the phase function

$$S(x_{J+1}, x_J, \ldots, x_0) = \sum_{j=1}^{J+1} S(T_j, T_{j-1}, x_j, x_{j-1}).$$

Here $S(T_j, T_{j-1}, x_j, x_{j-1})$ is the classical action (2.3.2). We often abbreviate it as $S_j(x_j, x_{j-1})$. Let j, l be two integers such that $0 \le j < l \le J + 1$. Then for any fixed $(x_l, x_{j-1}) \in \mathbf{R}^2$ let $(x_{l-1}^*(x_l, x_j), x_{l-2}^*(x_l, x_j), \ldots, x_j^*(x_l, x_{j-1}))$ be the critical point of

$$(x_{l-1}, \ldots, x_j) \to S_l(x_l, x_{l-1}) + S_{l-1}(x_{l-1}, x_{l-2}) + \cdots + S_j(x_j, x_{j-1}).$$

By Proposition 4.2 the critical value $S_{l,j}(x_l, x_{j-1})$ of this function equals the classical action, i.e.,

$$S_{l,j}(x_l, x_{j-1}) = S(T_l, T_{j-1}, x_l, x_{j-1}).$$

For any integers $0 < j < l$ we define

$$D_{x^*_{l-1},\dots,x^*_j}\Big(\sum_{k=j}^{l} S_k; x_l, x_{j-1}\Big) = \Big(\frac{\tau_l + \cdots + \tau_j}{\tau_l \cdots \tau_j}\Big) \det Hess\Big(\sum_{k=j}^{l} S_k(x_k, x_{k-1})\Big), \tag{5.1.8}$$

Here *Hess* means the Hessian at the critical point $(x^*_{l-1}(x_l, x_{j-1}), \dots, x^*_j(x_l, x_{j-1}))$. In the case $l = J + 1$ and $j = 1$, then it is clear that

$$D_{x^*_J,\dots,x^*_1}\Big(\sum_{k=1}^{J+1} S_k; x, y\Big) = D(\Delta; s', s, x, y) = \det(I + H_\Delta^{-1} W^*_\Delta(x^*)). \tag{5.1.9}$$

In the case $j = 1$ and $l = k$ for any $k = 1, 2, \dots, J$

$$D_{x^*_{k-1},\dots,x^*_1}\Big(\sum_{j=1}^{k} S_j; x_k, x_0\Big) = D(\Delta_k; x_k, x_0), \tag{5.1.10}$$

where Δ_k is the division of $[s, T_k]$:

$$\Delta_k : s = T_0 < T_1 < \cdots < T_k. \tag{5.1.11}$$

Lemma 5.1 *If* $1 < l \leq J$, *then*

$$\begin{aligned} &D_{x^*_l, x^*_{l-1},\dots,x^*_1}\Big(\sum_{k=1}^{l+1} S_k; x_{l+1}, x_0\Big) \\ &= D_{x^*_l}\Big(S_{l+1} + S_{l,1}; x_{l+1}, x_0\Big) D_{x^*_{l-1}\dots x^*_1}\Big(\sum_{k=1}^{l} S_k; x_l, x_0\Big)\Big|_{x_l = x^*}. \end{aligned} \tag{5.1.12}$$

If $1 \leq j \leq k < l$, *then*

$$\begin{aligned} &D_{x^*_l,\dots,x^*_j}(x_{l+1}, x_{j-1}) \\ &= D_{x^*_k}(S_{l+1,k+1} + S_{k,j}; x_{l+1}, x_{j-1}) \\ &\times D_{x^*_l,\dots,x^*_{k+1}}(S_{l+1.k+1}; x_{l+1}, x_k)\Big|_{x_k = x^*_k} D_{x^*_{k-1},\dots,x^*_j}(S_{k,j}; x_k, x_{j-1})\Big|_{x_k = x^*_k}. \end{aligned} \tag{5.1.13}$$

Proof Let $(x^*_l, x^*_{l-1}, \dots, x^*_1)$ be the critical point of the function defined by

$$\mathbf{R}^l \ni (x_l, x_{l-1}, \dots, x_1) \longrightarrow \sum_{k=1}^{l+1} S_k(x_k, x_{k-1}) \in \mathbf{R}.$$

We divide the independent variables $(x_l, x_{l-1}, \dots, x_1)$ into two groups $\{x_l\}$ and $\{x_{l-1}, \dots, x_1\}$ and apply Proposition 5.1. Fixing x_l, x_0, we write $(x^\#_{l-1}, \dots, x^\#_1)$ for the stationary point of the function

$$\mathbf{R}^{l-1} \ni (x_{l-1}, \ldots, x_2, x_1) \longrightarrow \sum_{k=1}^{l} S_k(x_k, x_{k-1}) \in \mathbf{R}.$$

Then $(x_{l-1}^{\#}, \ldots, x_1^{\#})$ is a function of x_l and x_0. As mentioned before,

$$\sum_{k=1}^{l} S_k(x_k, x_{k-1})\Big|_{x_j = x_j^{\#}, 1 \le j \le l-1} = S_{l,1}(x_l, x_0).$$

It follows from this and Proposition 5.1 that

$$\det\left[\mathrm{Hess}_{x_l^*, x_{l-1}^*, \ldots, x_1^*}\Big(\sum_{k=1}^{l+1} S_k(x_k, x_{k-1})\Big)\right] \tag{5.1.14}$$

$$= \det\left[\mathrm{Hess}_{x_l^*}(S_{l+1} + S_{l,1})\right] \det\left[\mathrm{Hess}_{x_{l-1}^{\#}(x_l^*, x_0), \ldots, x_1^{\#}(x_l^*, x_0)}\Big(\sum_{k=1}^{l} S_k(x_k, x_{k-1})\Big)\right].$$

Since $x_k^{\#}(x_l^*, x_0) = x_k^*$ for $k = 1, 2, \ldots, l-1$, by (5.1.8) we have

$$\det\left[\mathrm{Hess}_{x_l^*}(S_{l+1} + S_{l,1})\right] = \frac{\tau_{l+1} + \tau_l + \cdots + \tau_1}{\tau_{l+1}(\tau_l + \cdots + \tau_1)} D_{x_l^*}(S_{l+1} + S_{l,1}; x_{l+1}, x_0)$$

and

$$\det\left[\mathrm{Hess}_{x_{l-1}^*, \ldots, x_1^*}\Big(\sum_{k=1}^{l} S_k(x_k, x_{k-1})\Big)\right] = \frac{\tau_l + \cdots + \tau_1}{\tau_l \tau_{l-1} \cdots \tau_1} D_{x_{l-1}^* \ldots x_1^*}\Big(\sum_{k=1}^{l} S_k; x_l, x_0\Big)\Big|_{x_l = x^*}.$$

(5.1.12) has been proved.

Similarly, we can prove (5.1.13) by regrouping independent variables $\{x_l, x_{l-1}, \ldots, x_j\}$ into two groups $\{x_k\}$ and $\{x_l, x_{l-1}, \ldots, x_{k+1}, x_{k-1}, \ldots, x_j\}$ and applying Proposition 5.1. We have

$$\det\left[\mathrm{Hess}_{x_l^*, x_{l-1}^*, \ldots, x_{k+1}^*}\Big(\sum_{n=j}^{l+1} S_n; x_{l+1}, x_{j-1}\Big)\right] \tag{5.1.15}$$

$$= \det\left[\mathrm{Hess}_{x_k^*}(S_{l+1,k+1} + S_{k,j}; x_{l+1}, x_{j-1})\right]$$

$$\times \det\left[\mathrm{Hess}_{x_l^{\#}, x_{l-1}^{\#}, \ldots, x_{k+1}^{\#}}\Big(\sum_{n=k+1}^{l+1} S_n; x_{l+1}, x_k\Big)\right]\Big|_{x_k = x_k^*} \det\left[\mathrm{Hess}_{x_{k-1}^{\#}, \ldots, x_j^{\#}}\Big(\sum_{n=j}^{k} S_n; x_k, x_{j-1}\Big)\right]\Big|_{x_k = x_k^*}$$

(5.1.13) follows from this. □

Lemma 5.2 *If* $|s' - s| \le \delta_0$, *then*

$$D(\Delta; s', s, x_{J+1}, x_0) = \prod_{k=2}^{J+1} D_{x_{k-1}^*}(S_k + S_{k-1,1}; x_k, x_0)\big|_{x_k = x_k^*}. \tag{5.1.16}$$

Here $(x_J^*, \ldots, x_k^*, \ldots, x_1^*)$ *is the stationary point of* $S(x_{J+1}, x_J, \ldots, x_0)$.

Proof By (5.1.9), equality (5.1.16) is nothing but the following equality in the case $l = J + 1$:

$$\begin{aligned} &D_{(x_{l-1}^*, \ldots, x_1^*)}(S_l + \cdots + S_1; x_l, x_0) \\ &= \prod_{k=2}^{l} D_{x_{k-1}^*}(S_k + S_{k-1,1}; x_k, x_0)\big|_{x_k = x_k^*(x_l, x_0), 1 \le k \le l-1}. \end{aligned} \tag{5.1.17}$$

We prove (5.1.17) by induction on l. Equality (5.1.17) is clearly true if $l = 2$. Assuming that equality (5.1.17) is true in the case of l, we shall prove it in the case of $l + 1$ if $l \le J$. Assume that $l \le J$. We fix x_{l+1}, x_0. We apply the induction hypothesis (5.1.17) to the second factor on the right-hand side of (5.1.12). Then we have

$$\begin{aligned} &D_{x_l^*, \ldots, x_1^*}\Big(\sum_{k=1}^{l+1} S_k; x_{l+1}, x_0\Big) \\ &= D_{x_l^*}(S_{l+1} + S_{l,1}; x_{l+1}, x_0) \prod_{k=2}^{l} D_{x_{k-1}^*}(S_k + S_{k-1,1}; x_k, x_0)\big|_{x_k = x_k^*}. \end{aligned}$$

This proves (5.1.8) in the case of $l+1$. Thus the induction is now complete. Equality (5.1.17) in the case of $J + 1$ has been proved. We have completed the proof of the lemma. □

Proposition 5.2 *Suppose that* $|s' - s| \le \delta_0$. *Then for any* $0 \le j < k < l \le J + 1$

$$D_{x_k^*}(S_{l,k+1} + S_{k,j}; x_l, x_{j-1}) = 1 + (T_l - T_k)(T_k - T_{j-1})g(x_l, x_{j-1}). \tag{5.1.18}$$

For any $m \ge 0$ *there exists a constant* $\tilde{\kappa}_{m+2} > 0$ *independent of* j, k, l *and of* κ_j *for* $j > m + 2$ *such that*

$$\sup_{(x_l, x_{j-1}) \in \mathbf{R}^2} |\partial_{x_l}^{\alpha} \partial_{x_{j-1}}^{\beta} g(x_l, x_{j-1})| \le \tilde{\kappa}_{m+2} \tag{5.1.19}$$

if $0 \le |\alpha| + |\beta| \le m$. *We can so choose that* $\tilde{\kappa}_2 = \kappa_2$.

Proof Let x_k^* be the critical point of the function $x_k \to S(T_l, T_k, x_l, x_k) + S(T_k, T_{j-1}, x_k, x_{j-1})$. Define

$$g(x_l, x_{j-1}) = \frac{(T_l - T_k)}{(T_l - T_{j-1})} \partial_{x_k}^2 \phi(T_l, T_k, x_l, x_k)\Big|_{x_k = x_k^*} + \frac{(T_k - T_{j-1})}{(T_l - T_{j-1})} \partial_{x_k}^2 \phi(T_k, T_{j-1}, x_k, x_{j-1})\Big|_{x_k = x_k^*}.$$

Then

$$\det \mathrm{Hess}_{x_k^*}(S_{l,k+1} + S_{k,j}; x_l, x_{j-1}) = \frac{T_l - T_{j-1}}{(T_l - T_k)(T_k - T_{j-1})} \Big(1 + (T_l - T_k)(T_k - T_{j-1}) g(x_l, x_{j-1}) \Big).$$

This implies (5.1.18). Also

$$|g(x_l, x_{j-1})| = \frac{(T_l - T_k)}{(T_l - T_{j-1})} \kappa_2 + \frac{(T_k - T_{j-1})}{(T_l - T_{j-1})} \kappa_2 \le \kappa_2.$$

(5.1.19) follows from the chain rule and Theorem 2.3. □

Lemma 5.3 *Suppose* $0 \le \gamma < 1$. *Let* $C \ge -\gamma^{-1} \log(1 - \gamma)$. *Then for* $0 \le t \le \gamma$

$$e^{-Ct} \le (1 - t) \le 1 + t \le e^t. \tag{5.1.20}$$

In particular, if $\gamma \le 2^{-1}$, *then we can choose* $C = 2$.

Proof Let $C \ge -\gamma^{-1} \log(1 - \gamma)$. Then $C \ge \sum_{n=0}^{\infty} \frac{\gamma^n}{n+1} > 1$, because $0 < \gamma < 1$. Define $g(t) = e^{-Ct} - (1 - t)$ for $0 \le t$. Then $g'(0) = 1 - C < 0$, $g''(t) = C^2 > 0$, $g(0) = 0$ and $g(\gamma) = e^{-C\gamma} - (1 - \gamma) \le 0$. This implies that $g(t) \le 0$ for $0 \le t \le \gamma$. If $\gamma = 2^{-1}$, then we can choose $C = 2$, because $-\gamma^{-1} \log(1 - \gamma) = 2 \log 2 < 2$. □

The following is a simple but useful conclusion of Lemma 5.2.

Lemma 5.4 *Suppose that* $0 < |s' - s| \le \delta_0$. *We define* $d(\Delta; s', s, x, y)$ *by*

$$D(\Delta; s', s, x, y) = 1 + (s' - s)^2 d(\Delta; s', s, x, y). \tag{5.1.21}$$

Define $\rho_0 > 0$ *such that* $\rho_0 \delta_0 \kappa_2 \le 2^{-1}$. *If* $|\Delta| \le \rho_0$, *then there exists a constant* $C_0 > 0$ *such that*

$$|d(\Delta; s', s, x, y)| \le C_0. \tag{5.1.22}$$

We can choose $C_0 = 2\kappa_2 e^{2\kappa_2 \delta^2}$. *Moreover, for any* $K \ge 0$ *there exist constants* $\rho_K > 0$ *and* $C_K > 0$ *such that if* $|\Delta| \le \rho_K$

$$\sup_{\substack{|s'-s| \le \delta_0, \\ |\Delta| \le \rho_K}} \max_{|\alpha| \le K, |\beta| \le K} \sup_{(x,y) \in \mathbf{R}^2} |\partial_x^\alpha \partial_y^\beta d(\Delta; s', s, x, y)| = C_K < \infty. \tag{5.1.23}$$

Proof By virtue of Lemma 5.2,

$$D_{x_J^*,\dots,x_1^*}(S_J+\cdots+S_1;x_{J+1},x_0)=\prod_{k=2}^{J+1} D_{x_{k-1}^*}(S_k+S_{k-1,1};x_k,x_0)\Big|_{x_k=x^*}, \quad (5.1.24)$$

here $(x_J^*,\dots,x_1^*)$ is the stationary point of $(x_J,\dots,x_1)\to\sum_{k=1}^{J+1} S_k(x_k,x_{k-1})$.

By Lemma 5.2, for each $k=2,3,\dots,J+1$, there exists a function $g_{k,0}(x_k,x_0)$ such that

$$D_{x_{k-1}^*}(S_k+S_{k-1,1};x_k,x_0)\Big|_{x_k=x^*}=1+\tau_k(\tau_{k-1}+\cdots+\tau_1)g_{k,0}(x_k^*,x_0) \quad (5.1.25)$$

with $|g_{k,0}(x_k,x_0)|\le\kappa_2$. Let ρ_0 be so small as $\rho_0\delta_0\kappa_2\le 2^{-1}$. Suppose that $|\Delta|\le\rho_0$. Then

$$|\tau_k(\tau_{k-1}+\cdots+\tau_1)g_{k,0}(x_k^*,x_0)|\le\tau_k|s'-s|\kappa_2\le\rho_0\delta_0\kappa_2\le 2^{-1}.$$

Applying Lemma 5.3 to (5.1.25), we obtain

$$\begin{aligned}e^{-2\tau_k|s'-s|\kappa_2}\le 1-\tau_k|s'-s|\kappa_2\le D_{x_{k-1}^*}(S_k+S_{k-1,1};x_k,x_0)&\le 1+\tau_k|s'-s|\kappa_2\\ &\le e^{\tau_k|s'-s|\kappa_2}.\end{aligned}$$

It follows from this and (5.1.24) that

$$e^{-2|s'-s|\kappa_2\sum_{k=2}^{J+1}\tau_k}\le D_{x_J^*,\dots,x_1^*}(S_J+\cdots+S_1;x_{J+1},x_0)\le e^{|s'-s|\kappa_2\sum_{k=2}^{J+1}\tau_k}.$$

Hence

$$e^{-2|s'-s|^2\kappa_2}-1\le D_{x_J^*,\dots,x_1^*}(S_J+\cdots+S_1;x_{J+1},x_0)-1\le e^{|s'-s|^2\kappa_2}-1.$$

(5.1.22) with $C=2\kappa_2 e^{2\kappa_2\delta^2}$ follows from this, because $1-e^{-x}\le e^x-1\le xe^x$ for $0<x$.

Now we prove (5.1.23) with $K\ge 0$. By (5.1.19) we can apply Lemma 2.7 to $g_k(x_k^*,x_0)$. Then for any α and β with $|\alpha|\le K$, $|\beta|\le K$ there exists a constant $\kappa_K^*>0$ such that

$$|\partial_{x_{J+1}}^{\alpha}\partial_{x_0}^{\beta}g(x_k^*,x_0)|\le\kappa_K^*.$$

Define $\rho_K>0$ so small that $\kappa_{2K+2}^*\rho_K\delta_0<2^{-1}$. If $\tau_k\le\rho_K$, then by Lemma 5.2,

$$|\partial_x^{\alpha}\partial_y^{\beta}\tau_k(\tau_{k-1}+\cdots+\tau_1)g_{k,0}(x_k^*,x_0)|\le\kappa_{2K+2}^*\tau_k|s'-s|\le\kappa_{2K+2}^*\rho_K\delta_0\le 2^{-1}.$$

We apply (5.1.29) in the following lemma with $a\equiv 1$, $\varepsilon_j(x)=\kappa_{2K+2}^*\tau_j(\tau_{j-1}+\cdots+\tau_1)$ and $\gamma=2^{-1}$ to (5.1.24). Then it turns out that (5.1.23) is true. Lemma 5.4 has been proved up to the proof of the next lemma. □

Lemma 5.5 *Let M, K and d be natural numbers with the property $d \leq M$. Let $(x_1, \dots, x_M)$ be a point in $\mathbf{R}^M$. Suppose each of J functions $f_j(x_{j,1}, \dots, x_{j,d})$, $1 \leq j \leq J$, depends on at most d variables among M variables $(x_1, \dots, x_M)$. Suppose that there are a positive constant γ with $0 < \gamma < 1$ and functions $\varepsilon_j(x)$ with $0 \leq \varepsilon_j(x) < \gamma$ such that for any $|\alpha_{j_u}| \leq K$, $u = 1, 2, \dots, d$, the following inequalities hold:*

$$\Big|\prod_{u=1}^{d} \partial_{x_{j_u}}^{\alpha_{j_u}} f_j(x_{j_1}, \dots, x_{j_d})\Big| \leq \varepsilon_j(x), \quad j = 1, \dots, J. \tag{5.1.26}$$

Then there exists a positive constant C depending only on K and d but independent of J and M such that for any differential operator $\mathscr{L} = \prod_{j=1}^{M} \partial_{x_j}^{\alpha_j}$ with $\alpha_{j_u} \leq K$, $u = 1, 2, \dots, d$, and any function $a(x) \in C^\infty(\mathbf{R}^M)$ the following inequalities hold:

$$\begin{aligned}&\Big|\mathscr{L}\Big[a(x)\prod_{j=1}^{J}(1 + f_j(x_{j_1}, \dots, x_{j_d}))\Big] - \mathscr{L}\big[a(x)\big]\prod_{j=1}^{J}(1 + f_j(x_{j_1}, \dots, x_{j_d}))\Big| \\ &\leq \|a\|_K(x)\Big(\prod_{j=1}^{J}(1 + C\varepsilon_j(x)) - 1\Big).\end{aligned} \tag{5.1.27}$$

Here for $x \in \mathbf{R}^M$ the norm $\|a\|_K(x)$ is defined by

$$\|a\|_K(x) = \max_{\alpha_j \leq K, j=1,2,\dots,M} \Big|\Big(\prod_{j=1}^{M} \partial_{x_j}^{\alpha_j}\Big)a(x)\Big|. \tag{5.1.28}$$

Moreover, there exists a constant C_1 which may depend also on γ but not on J and M such that

$$\begin{aligned}&\Big|\mathscr{L}\Big[\prod_{j=1}^{J}(1 + f_j(x_{j_1}, \dots, x_{j_d})) - 1 - \sum_{j=1}^{J} f_j(x_{j_1}, \dots, x_{j_d})\Big]\Big| \\ &\leq C_1\Big(\sum_{j=1}^{J} \varepsilon_j(x)\Big)^2 \exp C_1\Big(\sum_{j=1}^{J} \varepsilon_j(x)\Big).\end{aligned} \tag{5.1.29}$$

In particular, if $\gamma = 2^{-1}$, then we can choose $C = C_1 = 2^{Kd+1}$.

Proof Suppose that $|\alpha_j| \leq K$ for $j = 1, \dots, M$. We shall show by induction on J that (5.1.27) holds with $C = 2^{kd+1}$. First we wish to show (5.1.27) in the case $J = 1$ holds with $C = 2^{Kd} - 1$. We may assume without loss of generality that the function f_1 depends only on variables $(x_1, x_2, \dots, x_d)$ but does not depend on other variables $(x_{d+1}, \dots, x_M)$. We set $\mathscr{L}_0 = \prod_{j=1}^{d} \partial_{x_j}^{\alpha_j}$ and $\mathscr{L}_1 = \prod_{j=d+1}^{M} \partial_{x_j}^{\alpha_j}$. Then $\mathscr{L} = \mathscr{L}_0\mathscr{L}_1$ and $\mathscr{L}_1 f_1(x_1, \dots, x_d) = 0$. Thus

$$\mathscr{L}\big[(1+f_1(x_1,\dots,x_d))a(x)\big] - (1+f_1(x_1,\dots,x_d))\mathscr{L}\big[a(x)\big]$$
$$= \mathscr{L}_0\big[f_1(x_1,\dots,x_d)\mathscr{L}_1(a(x))\big] - f_1(x_1,\dots,x_d)\mathscr{L}_0\mathscr{L}_1\big[(a(x)\big].$$

It follows from Leibnitz's rule that

$$\mathscr{L}_0\big[f_1(x_1,\dots,x_d)\mathscr{L}_1(a(x))\big] - f_1(x_1,\dots,x_d)\mathscr{L}_0\mathscr{L}_1\big[a(x)\big]$$
$$= \sum{}' \binom{\alpha_1}{\beta_1}\cdots\binom{\alpha_d}{\beta_d}\prod_{j=1}^{d}\partial_{x_j}^{\beta_j}\mathscr{L}_1(a(x))\prod_{j=1}^{d}\partial_{x_j}^{\alpha_j-\beta_j}f_1(x_1,\dots,x_d),$$

where the symbol $\sum'$ signifies summation over all multi-indices β satisfying $\beta < \alpha$. It is clear from the definition of $\|a\|_K(x)$ and the assumption of the lemma that

$$\Big|\prod_{j=1}^{d}\partial_{x_j}^{\beta_j}\mathscr{L}_1(a(x))\Big| \le \|a\|_K(x) \quad\text{and}\quad \Big|\prod_{j=1}^{d}\partial_{x_j}^{\alpha_j-\beta_j}f_1(x_1,\dots,x_d)\Big| \le \varepsilon_1(x).$$

Therefore,

$$|\mathscr{L}\big[a(x)(1+f_1(x_1,\dots,x_d))\big] - \mathscr{L}\big[a(x)\big](1+f_1(x_1,\dots,x_d))|$$
$$\le (2^{Kd}-1)\varepsilon_1(x)\|a\|_K(x), \tag{5.1.30}$$

because

$$\sum{}' \binom{\alpha_1}{\beta_1}\cdots\binom{\alpha_d}{\beta_d} = 2^{\alpha_1+\alpha_2+\cdots+\alpha_d} - 1 \le 2^{Kd}-1.$$

We have proved (5.1.27) for $J=1$ with $C = 2^{Kd}-1$.

Before proceeding further, we notice that

$$|\mathscr{L}\big[a(x)(1+f_1(x_1,\dots,x_d))\big]| \le (1+2^{Kd}\varepsilon_1(x))\|a\|_K(x). \tag{5.1.31}$$

This follows from (5.1.30) and

$$|\mathscr{L}\big[a(x)\big](1+f_1(x_1,\dots,x_d))| \le \|a\|_K(x)(1+\varepsilon_1(x)).$$

In particular

$$\|a(1+f_1(x_1,\dots,x_d))\|_K(x) \le (1+2^{Kd}\varepsilon_1(x))\|a\|_K(x). \tag{5.1.32}$$

Next assuming that (5.1.27) in the case of J is true with $C = 2^{Kd+1}$, we shall prove it in the case of $J+1$ with $C = 2^{Kd+1}$. In the following discussion the important point is the upper bound d of the number of variables on which each f_j depends and it is not important that each f_j depends on which variables. So we write simply $f_j(x)$ instead of $f_j(x_{j_1},\dots,x_{j_d})$ to avoid unnecessary complexity of notation. We write

$$
\mathscr{L}\Big[a(x)\prod_{j=1}^{J+1}(1+f_j(x))\Big]-\mathscr{L}[a(x)]\prod_{j=1}^{J+1}(1+f_j(x))=I_1(x)+I_2(x),
$$

$$
\begin{aligned}
I_1(x)&=\mathscr{L}\Big[a(x)(1+f_{J+1}(x))\prod_{j=1}^{J}(1+f_j(x))\Big]\\
&\quad-\mathscr{L}\Big[a(x)(1+f_{J+1}(x))\Big]\prod_{j=1}^{J}(1+f_j(x)),
\end{aligned}
$$

$$
I_2(x)=\Big(\mathscr{L}\Big[a(x)(1+f_{J+1}(x))\Big]-\mathscr{L}[a(x)](1+f_{J+1}(x))\Big)\prod_{j=1}^{J}(1+f_j(x)).
$$

Since (5.1.31) holds, we can use induction hypothesis (5.1.27) for J with $a(x)$ replaced by $a(x)(1+f_{J+1}(x))$. Hence by (5.1.31),

$$
\begin{aligned}
|I_1(x)|&\le \|a(1+f_{J+1})\|_K(x)\Big(\prod_{j=1}^{J}(1+C\varepsilon_j(x))-1\Big)\\
&\le \|a\|_K(x)(1+2^{Kd}\varepsilon_{J+1}(x))\Big(\prod_{j=1}^{J}(1+C\varepsilon_j(x))-1\Big).
\end{aligned}
$$

We use (5.1.30) with f_1 replaced with f_{J+1}. Then

$$
|I_2(x)|\le (2^{Kd}-1)\varepsilon_{J+1}(x)\|a\|_K(x)\prod_{j=1}^{J}(1+\varepsilon_j(x)).
$$

Since $C=2^{Kd+1}>1$,

$$
\begin{aligned}
&|I_1(x)+I_2(x)|\\
&\quad\le \|a\|_K(x)\Big((1+2^{Kd+1}\varepsilon_{J+1}(x))\prod_{j=1}^{J}(1+C\varepsilon_j(x))-(1+2^{Kd}\varepsilon_{J+1}(x))\Big)\\
&\quad\le \|a\|_K(x)\Big((1+2^{Kd+1}\varepsilon_{J+1}(x))\prod_{j=1}^{J}(1+C\varepsilon_j(x))-1\Big).
\end{aligned}
$$

Therefore, (5.1.27) for $J+1$ is true. The induction has been completed. We have proved (5.1.27).

Next we shall prove (5.1.29) by induction on J. In the case of $J = 1$ (5.1.29) is clear. Let $C_0 = 2^{Kd+1}$. Assuming (5.1.29) in the case of J, we shall prove it in the case of $J+1$. We write

$$\mathscr{L}\Big[\prod_{j=1}^{J+1}(1+f_j(x)) - 1 - \sum_{j=1}^{J+1} f_j(x)\Big] = I_1(x) + I_2(x) + I_3(x),$$

where

$$I_1(x) = \mathscr{L}\Big[\prod_{j=1}^{J}(1+f_j(x)) - 1 - \sum_{j=1}^{J} f_j(x)\Big],$$

$$I_2(x) = \mathscr{L}\Big[f_{J+1}\prod_{j=1}^{J}(1+f_j(x))\Big] - \mathscr{L}\Big[f_{J+1}(x)\Big]\prod_{j=1}^{J}(1+f_j(x)),$$

$$I_3(x) = \mathscr{L}\Big[f_{J+1}(x)\Big]\prod_{j=1}^{J}(1+f_j(x)) - \mathscr{L}\Big[f_{J+1}(x)\Big].$$

By the induction hypothesis

$$|I_1(x)| \le C_1\Big[\sum_{j=1}^{J}\varepsilon_j(x)\Big]^2 \exp\Big\{C_1\sum_{j=1}^{J}\varepsilon_j(x)\Big\}.$$

Applying (5.1.27) with a replaced with f_{J+1}, we have

$$\begin{aligned}|I_2(x)| &\le \varepsilon_{J+1}(x)\Big(\prod_{j=1}^{J}(1+C_0\varepsilon_j(x)) - 1\Big)\\ &\le \varepsilon_{J+1}(x)\Big(\exp\Big\{C_0\sum_{j=1}^{J}\varepsilon_j(x)\Big\} - 1\Big).\end{aligned}$$

Since $e^t - 1 \le te^t$ for any $t > 0$,

$$|I_2(x)| \le C_0\varepsilon_{J+1}(x)\Big(\sum_{j=1}^{J}\varepsilon_j(x)\Big)\exp\Big\{C_0\sum_{j=1}^{J}\varepsilon_j(x)\Big\}.$$

Now we treat $I_3(x)$. As $|\mathscr{L}(f_{J+1}(x))| \le \varepsilon_{J+1}(x)$,

$$|I_3| \le \varepsilon_{J+1}(x)|\Big(\prod_{j=1}^{J}(1+f_j(x)) - 1\Big)|.$$

By Lemma 5.3, there exists a constant $C_2 > 1$ such that $e^{-C_2\varepsilon_j(x)} \leq (1 - \varepsilon_j(x)) < 1 - |f_j(x)| \leq 1 + f_j(x) \leq e^{\varepsilon_j(x)}$. It follows from this that

$$\Big|\prod_{j=1}^{J}(1 + f_j(x)) - 1\Big| \leq \exp\Big\{C_2 \sum_{j=1}^{J} \varepsilon_j(x)\Big\} - 1.$$

Hence

$$\begin{aligned}|I_3(x)| &\leq \varepsilon_{J+1}(x)\Big(\exp\{C_2 \sum_{j=1}^{J} \varepsilon_j(x)\} - 1\Big) \\ &\leq C_2\varepsilon_{J+1}(x)\Big(\sum_{j=1}^{J} \varepsilon_j(x)\Big) \exp\Big\{C_2 \sum_{j=1}^{J} \varepsilon_j(x)\Big\}.\end{aligned}$$

Setting $C_1 \geq \max(C_0, C_2)$, we have

$$\begin{aligned}&\Big|\mathscr{L}\Big\{\prod_{j=1}^{J+1}(1 + f_j(x)) - 1 - \sum_{j=1}^{J+1} f_j(x)\Big\}\Big| \\ &\leq C_1\Big(\sum_{j=1}^{J} \varepsilon_j(x)\Big)^2 \exp\{C_1 \sum_{j=1}^{J} \varepsilon_j(x)\} \\ &\quad + 2C_1\varepsilon_{J+1}(x)\Big(\sum_{j=1}^{J} \varepsilon_j(x)\Big) \exp\{C_1 \sum_{j=1}^{J} \varepsilon_j(x)\} \\ &\leq C_1\Big(\sum_{j=1}^{J+1} \varepsilon_j(x)\Big)^2 \exp\{C_1 \sum_{j=1}^{J+1} \varepsilon_j(x)\}.\end{aligned}$$

We have proved (5.1.29) in the case of $J + 1$. If $\gamma = 2^{-1}$, then $C_2 = 2$ is enough by Lemma 5.3. It turns out that we can choose $C_1 = 2^{Kd+1}$.

We have proved Lemma 5.5. Proof of Lemma 5.4 has been completed. □

5.2 Existence of $\lim_{|\Delta|\to 0} D(\Delta; s', s, x, y)$

Let δ_0 be $\delta_0^2\kappa_2 < 2^{-1}$ as in Definition 4.1, and

$$\Delta : s = T_0 < T_1 < \cdots < T_J < T_{J+1} = s' \tag{5.2.1}$$

be a division of the interval $[s, s']$. In order to prove that $\lim_{|\Delta|\to 0} D(\Delta; s', s, x, y)$ exists we discuss Cauchy's condition of convergence.

Lemma 5.6 *Suppose* $[s, s'] \subset [a, b]$ *with* $|s' - s| \le \delta_0$ *and* $K \ge 0$. *Let* Δ' *be an arbitrary refinement of* Δ. *Define* $d(\Delta'/\Delta; x, y)$ *by the equality*

$$\frac{D(\Delta'; s', s, x, y)}{D(\Delta; s', s, x, y)} = 1 + |\Delta|(s' - s)d(\Delta'/\Delta; x, y). \tag{5.2.2}$$

Then for any $K \ge 0$ *there exist constants* $\rho_K > 0$ *and* $C_K > 0$ *such that if* $|\Delta| \le \rho_K$ *and* $|\alpha| \le K$ *and* $|\beta| \le K$, *then*

$$\sup_{(x,y)\in\mathbf{R}^2} |\partial_x^\alpha \partial_y^\beta d(\Delta'/\Delta; x, y)| \le C_K. \tag{5.2.3}$$

Proof First we shall discuss a special simple type of refinement of Δ. Let n be any one of $n = 1, \dots, J + 1$. We divide the n-th subinterval $I_n = [T_{n-1}, T_n]$ into smaller intervals and denote such division of I_n by $\Delta_\flat$, i.e.

$$\Delta_\flat : T_{n-1} = T_{n,0} < T_{n,1} < \cdots < T_{n,p_n+1} = T_n. \tag{5.2.4}$$

Adding these new dividing points of $\Delta_\flat$ to Δ and keeping other $[T_{j-1}, T_j]$ unchanged, we get a refinement $\tilde{\Delta}$ of Δ. In other words, $\tilde{\Delta}$ is the same as Δ except for the division $\Delta_\flat$ of $[T_{n-1}, T_n]$. We set $\sigma_k = T_{n,k} - T_{n,k-1}$, $k = 1, 2, \dots, p_{n+1}$. We claim that

$$\frac{D(\tilde{\Delta}; s', s, x, y)}{D(\Delta; s', s, x, y)} = 1 + \tau_n^2 d(\tilde{\Delta}/\Delta; x, y) \tag{5.2.5}$$

and that for any $K \ge 0$ there exist positive constants $\tilde{\rho}_K > 0$ and $C_K^* >$ such that if $|\Delta| \le \tilde{\rho}_K$, then for any α, β with $|\alpha| \le K$, $|\beta| \le K$

$$\left|\partial_x^\alpha \partial_y^\beta d(\tilde{\Delta}/\Delta; x, y)\right| \le C_K^*. \tag{5.2.6}$$

Let $\gamma_{\tilde{\Delta}}$ be an arbitrary piecewise classical path associated with the division $\tilde{\Delta}$. We set

$$x_j = \gamma_{\tilde{\Delta}}(T_j), \quad j = 0, 1, \dots, J + 1$$

and

$$y_k = \gamma_{\tilde{\Delta}}(T_{n,k}), \quad k = 0, 1, \dots, p_n + 1.$$

Here we understand $x_n = y_{p_n+1}$ and $x_{n-1} = y_0$. We write y for $(y_{p_n}, y_{p_n-1}, \dots, y_1)$. Let γ_Δ be the piecewise classical path associated with Δ that passes the points x_j at time T_j for all $j = 0, 1, 2, \dots, J + 1$. Then clearly $\gamma_{\tilde{\Delta}}(t) = \gamma_\Delta(t)$ if $t \notin [T_{n-1}, T_n]$.

The phase function $S_\Delta(x_{J+1}, \dots, x_0)$ associated with Δ is

$$S_\Delta(x_{J+1}, \dots, x_0) = \sum_{j=1}^{J+1} S_j(x_j, x_{j-1}),$$

where

$$S_j(x_j, x_{j-1}) = \int_{T_{j-1}}^{T_j} L(t, \gamma_\Delta(t), \dot{\gamma}_\Delta(t))\, dt.$$

The phase function associated with the division $\tilde{\Delta}$ is

$$\begin{aligned} &S_{\tilde{\Delta}}(x_{J+1}, \dots, x_{n+1}, y, x_n, \dots, x_1, x_0) \\ &= \int_s^{s'} L(t, \gamma_{\tilde{\Delta}}(t), \dot{\gamma}_{\tilde{\Delta}}(t))\, dt \\ &= \sum_{j=1}^{n-1} S_j(x_j, x_{j-1}) + \sum_{j=n+1}^{J+1} S_j(x_j, x_{j-1}) + S_{\Delta_\flat}(x_n, y, x_{n-1}), \end{aligned}$$

where

$$S_{\Delta_\flat}(x_n, y, x_{n-1}) = \int_{T_{n-1}}^{T_n} L(t, \gamma_{\tilde{\Delta}}(t), \dot{\gamma}_{\tilde{\Delta}}(t))\, dt.$$

Note that $S_{\Delta_\flat}(x_n, y, x_{n-1})$ can be regarded as the phase function associated with division $\Delta_\flat$ of the small interval $[T_{n-1}, T_n]$.

Let $y^\#(x_n, x_{n-1}) = (y^\#_{p_n}(x_n, x_{n-1}), y^\#_{p_n-1}(x_n, x_{n-1}), \dots, y^\#_1(x_n, x_{n-1}))$ be the stationary point of the function

$$(y_{p_n}, y_{p_n-1}, \dots, y_1) \longrightarrow S_{\Delta_\flat}(x_n, y, x_{n-1}) = S_{\Delta_\flat}(x_n, y_{p_n}, y_{p_n-1}, \dots, y_1, , x_{n-1})$$

for any fixed x_n and x_{n-1}. Then it is clear that $y^\#(x_n, x_{n-1}) = \gamma_\Delta(T_{n,k})$ for $k = 1, 2, \dots, p_n$ and

$$S_{\Delta_\flat}(x_n, y^\#(x_n, x_{n-1}), x_{n-1}) = S_n(x_n, x_{n-1}).$$

Let $x^* = (x^*_J, \dots, x^*_1)$ and $y^* = (y^*_{n,p_n}, \dots, y^*_{n,1})$ be the stationary point of

$$(x, y) = (x_J, \dots, x_n, y, x_{n-1}, \dots, x_1) \longrightarrow S_{\tilde{\Delta}}(x_{J+1}, x_J, \dots, x_n, y, x_{n-1}, \dots, x_1, x_0)$$

with fixed (x_{J+1}, x_0). Then it is clear that $y^\#_k(x^*_n, x^*_{n-1}) = y^*_k$ for $k = 1, 2, \dots, p_n$.

Now we calculate $D(\tilde{\Delta}; s', s, x, y)$. On one hand by definition

$$\begin{aligned} &\det Hess_{x^*,y^*} S_{\tilde{\Delta}}(x_{J+1}, x_0) \\ &= (s' - s)^{-1} \left(\prod_{j=n+1}^{J+1} \tau_j \right) \left(\prod_{j=1}^{n-1} \tau_j \right) \left(\prod_{k=1}^{p_n+1} \sigma_k \right) D(S_{\tilde{\Delta}}; x_{J+1}, x_0). \end{aligned}$$

On the other hand, it follows from Proposition 5.1 that

$$\begin{aligned}&\det Hess_{x^*,y^*} S_{\tilde{\Delta}}(x_{J+1}, x_0)\\&= \det Hess_{x^*}(S_\Delta; x_{J+1}, x_0)) \det Hess_{y^*}(S_{\Delta_\flat}; x_n, x_{n-1})\big|_{x_n=x_n^*, x_{n-1}=x_{n-1}^*}.\end{aligned}$$

Since $S_{\Delta_\flat}(x_n, y_{n,p_n}, \ldots, y_{n,1}, x_{n-1})$ is the phase function corresponding to division $\Delta_\flat$ of subinterval $[T_{n-1}, T_n]$, we can write

$$\det Hess_{y^*}(S_{\Delta_\flat}; x_n, x_{n-1}) = \tau_n{}^{-1}\left(\prod_{k=1}^{p_n+1} \sigma_k\right) D(S_{\Delta_\flat}; x_n, x_{n-1}).$$

Thus

$$\begin{aligned}&(s'-s)^{-1}\tau_n{}^{-1}\left(\prod_{j=1}^{J+1}\tau_j \prod_{k=1}^{p_n+1}\sigma_k\right) D(S_{\tilde{\Delta}}; x_{J+1}, x_0)\\&= \det Hess_{x^*,y^*} S_{\tilde{\Delta}}(x_{J+1}, x_0)\\&= \left[(s'-s)^{-1}\left(\prod_{j=1}^{J+1}\tau_j\right) D(S_\Delta; x_{J+1}, x_0)\right]\\&\quad\times\left[\tau_n{}^{-1}\left(\prod_{k=1}^{p_n+1}\sigma_k\right) D(S_{\Delta_\flat}; x_n, x_{n-1})\big|_{x_n=x_n^*, x_{n-1}=x_{n-1}^*}\right].\end{aligned}$$

It turns out from these that

$$D(S_{\tilde{\Delta}}; x_{J+1}, x_0) = D(S_\Delta; x_{J+1}, x_0) D(S_{\Delta_\flat}; x_n, x_{n-1})\Big|_{x_n=x_n^*, x_{n-1}=x_{n-1}^*}. \tag{5.2.7}$$

By Lemma 5.4, for any $K \geq 0$ there exist constants $\tilde{\rho}_k > 0$ and C_K^* such that if $|\Delta| \leq \tilde{\rho}_K$, then $|\Delta^\flat| \leq \tilde{\rho}_K$. Thus by Lemma 5.4,

$$D(S_{\Delta_\flat}; x_n, x_{n-1}) = 1 + \tau_n^2 d(\Delta_\flat, x_n, x_{n-1}) \tag{5.2.8}$$

and for any α, β with $|\alpha| \leq K$, $|\beta| \leq K$

$$|\partial_{x_n}^\alpha \partial_{x_{n-1}}^\beta d(\Delta_\flat, x_n, x_{n-1})| \leq C_K^*. \tag{5.2.9}$$

Therefore, we define

$$d(\tilde{\Delta}/\Delta; x, y) = d(\Delta_\flat, x_n, x_{n-1})\big|_{x_n=x_n^*, x_{n-1}=x_{n-1}^*}.$$

Then (5.2.5) follows from (5.2.7) and (5.2.6) follows from (5.2.9) and Lemma 2.7. Therefore we have proved the claim.

Next we prove Lemma 5.6 in full generality. For $K \geq 0$, let $\tilde{\rho}_K$, C_K^* be as above and Δ be such that $|\Delta| \leq \tilde{\rho}_K$. Let Δ' be an arbitrary refinement of Δ. We shall construct a sequence of divisions $\Delta_1, \Delta_2, \ldots \Delta_J$ in the following way.

Let Δ_1 be the division of which the dividing points are the union of all dividing points of Δ and dividing points of Δ' inside subinterval $[T_0, T_1]$. So the division Δ_1 is different from Δ only in the subinterval $[T_0, T_1]$. Thus we can apply the claim above and obtain

$$\frac{D(\Delta_1; s', s, x_{J+1}, x_0)}{D(\Delta; s', s, x_{J+1}, x_0)} = 1 + \tau_1^2 d(\Delta_1/\Delta; x_{J+1}, x_0)$$

and for any α, β with $|\alpha| \leq K$, $|\beta| \leq K$ such that

$$\left| \partial_{x_{J+1}}^{\alpha} \partial_{x_0}^{\beta} d(\Delta_1/\Delta; x_{J+1}, x_0) \big|_{x_1 = x_1^*,} \right| \leq C_K^*.$$

Let Δ_2 be the division whose dividing points are the union of dividing points of Δ_1 and dividing points inside subinterval $[T_1, T_2]$. Division Δ_2 is different from Δ_1 only in the interval $[T_1, T_2]$. We apply the claim again and obtain

$$\frac{D(\Delta_2; s', s, x_{J+1}, x_0)}{D(\Delta_1; s', s, x_{J+1}, x_0)} = 1 + \tau_2^2 d(\Delta_2/\Delta_1; x_{J+1}, x_0)$$

and for any α, β $|\alpha| \leq K$, $|\beta| \leq K$

$$\left| \partial_{x_{J+1}}^{\alpha} \partial_{x_0}^{\beta} d(\Delta_2/\Delta_1; x_{J+1}, x_0) \right| \leq C_K^*.$$

Starting from $\Delta_0 = \Delta$ and repeating this process, we find a sequence of divisions $\Delta_1, \Delta_2, \ldots, \Delta_J, \Delta_{J+1}$ such that Δ_n consists of all dividing points of Δ_{n-1} and the part of Δ' inside $[T_{n-1}, T_n]$. By construction it is clear that $\Delta_{J+1} = \Delta'$. Since Δ_n is a refinement of Δ_{n-1} and they are different only in $[T_{n-1}, T_n]$, it follows from the claim above that for any $n = 1, 2, \ldots, J+1$

$$\frac{D(\Delta_n; s', s, x_{J+1}, x_0)}{D(\Delta_{n-1}; s', s, x_{J+1}, x_0)} = 1 + \tau_n^2 d(\Delta_n/\Delta_{n-1}; x_{J+1}, x_0) \tag{5.2.10}$$

and for any α, β with $|\alpha| \leq K$, $|\beta| \leq K$ there exists a positive constant $C_{\alpha\beta}$ such that

$$\left| \partial_{x_{J+1}}^{\alpha} \partial_{x_0}^{\beta} d(\Delta_n/\Delta_{n-1}; x_{J+1}, x_0) \right| \leq C_K^*. \tag{5.2.11}$$

As $\Delta_{J+1} = \Delta'$,

$$\frac{D(\Delta'; s', s, x_{J+1}, x_0)}{D(\Delta; s', s, x_{J+1}, x_0)} = \prod_{n=1}^{J+1} (1 + \tau_n^2 d(\Delta_n/\Delta_{n-1}; x_{J+1}, x_0)). \tag{5.2.12}$$

Let ρ_K be so small that $\rho \le \tilde{\rho}_K$ and $\rho_K^2 C_K^* \le 2^{-1}$. Then it follows from (5.2.10) and Lemma 5.5 that if $|\Delta| \le \rho_K$, then

$$\prod_{n=1}^{J+1}(1 + \tau_n^2 d(\Delta_n/\Delta_{n-1}; x_{J+1}, x_0)) = 1 + \Big(\sum_{n=1}^{J+1} \tau_n^2\Big) f(x_{J+1}, x_0).$$

For any α, β with $|\alpha| \le K$, $|\beta| \le K$, there exists a constant C_K such that

$$|\partial_{x_{J+1}}^\alpha \partial_{x_0}^\beta f(x_{J+1}.x_0)| \le C_K.$$

Define

$$d(\Delta'/\Delta; x_{J+1}, x_0) = \frac{\sum_{n=1}^{J+1} \tau_n^2}{|\Delta|(s'-s)} f(x_{J+1}, x_0).$$

Then (5.2.2) and (5.2.3) hold, because $\sum_{n=1}^{J+1} \tau_n^2 \le |\Delta|(s'-s)$. Lemma 5.6 has been proved. □

Theorem 5.1 *Suppose that* $|s'-s| \le \delta_0$. *When* $|\Delta| \to 0$, $\{D(\Delta; s', s, , x, y)\}_\Delta$ *satisfies Cauchy's condition of convergence. More precisely, for any* $K \ge 0$ *there exist constants* $\rho_K > 0$ *and* $C_K > 0$ *such that if divisions* Δ *and* Δ_1 *satisfy* $|\Delta| \le \rho_K$ *and* $|\Delta_1| \le \rho_K$, *then for any* α, β *with* $|\alpha| \le K$, $|\beta| \le K$,

$$|\partial_x^\alpha \partial_y^\beta D(\Delta_1; s', s, x, y) - \partial_x^\alpha \partial_y^\beta D(\Delta; s', s, x, y)| \le C_K(|\Delta| + |\Delta_1|). \quad (5.2.13)$$

Proof Let Δ' be a common refinement of Δ and Δ_1. By Lemma 5.6, for any $K \ge 0$ there exists $\rho_K > 0$ such that if Δ and Δ_1 satisfy $|\Delta| \le \rho_K$ and $|\Delta_1| \le \rho_K$, then

$$\frac{D(\Delta'; s', s, x, y)}{D(\Delta; s', s, x, y)} = 1 + |\Delta_1|(s'-s) d(\Delta'/\Delta; x, y),$$
$$\frac{D(\Delta'; s', s, x, y)}{D(\Delta_1; s', s, x, y)} = 1 + |\Delta_1|(s'-s) d(\Delta'/\Delta_1; x, y).$$

Thus

$$\begin{aligned}
&D(\Delta; s', s, x, y) - D(\Delta_1; s', s, x, y)\\
&= D(\Delta; s', s, x, y) - D(\Delta'; s', s, x, y) + D(\Delta'; s', s, x, y) - D(\Delta_1; s', s, x, y)\\
&= D(\Delta; s', s, x, y)\left(1 - \frac{D(\Delta'; s', s, x, y)}{D(\Delta; s', s, x, y)}\right)\\
&\quad - D(\Delta_1; s', s, x, y)\left(1 - \frac{D(\Delta'; s', s, x, y)}{D(\Delta_1; s', s, x, y)}\right)\\
&= -D(\Delta; s', s, x, y)(s'-s)|\Delta| d(\Delta'/\Delta, s', s, x, y)\\
&\quad + D(\Delta_1; s', s, x, y)(s'-s)|\Delta_1| d(\Delta'/\Delta_1, s', s, x, y).
\end{aligned}$$

It follows from this, Lemmas 5.4 and 5.6 that if $|\Delta| \le \rho_K$, $|\Delta_1| \le \rho_K$, then for α, β with $|\alpha| \le K$, $|\beta| \le K$

$$|\partial_x^\alpha \partial_y^\beta (D(\Delta; s', s, x, y) - D(\Delta_1; s', s, x, y))| \le C_{\alpha,\beta}(|\Delta| + |\Delta_1|)(s' - s).$$

□

Since $\{D(s', s, , x, y)\}_\Delta$ forms a Cauchy net, the following theorem clearly follows from the above theorem.

Theorem 5.2 *Suppose that* $|s - s'| \le \delta_0$. *Then the following limit exists in the space* $\mathscr{B}(\mathbf{R}^2)$:

$$D(s', s, x, y) = \lim_{|\Delta| \to 0} D(\Delta; s', s, x, y). \tag{5.2.14}$$

More precisely, for any $K \ge 0$ *there exist constants* $\rho_K > 0$ *and* $C_K > 0$ *such that if* $|\Delta| \le \rho_K$, *then for* α, β *with* $|\alpha| \le K$, $|\beta| \le K$

$$|\partial_x^\alpha \partial_y^\beta (D(s', s, x, y) - D(\Delta; s', s, x, y))| \le C_K |\Delta||s' - s|. \tag{5.2.15}$$

We can write

$$D(s', s, x, y) = 1 + (s' - s)^2 d(s', s, x, y). \tag{5.2.16}$$

For any α β *with* $|\alpha| \le K$, $|\beta| \le K$ *there exists* $C_K > 0$ *such that*

$$|\partial_x^\alpha \partial_y^\beta d(s', s, x, y)| \le C_K. \tag{5.2.17}$$

Remark 5.1 Interesting properties of $D(s', s, x, y)$ will be discussed later.

5.3 Stationary Phase Method over a Space of Large Dimension

Our main theorem, Theorem 4.6, will be proved in the next section. Our proof is based on the stationary phase method of oscillatory integrals over a space of large dimension. As a preparation we collect in this section some facts that we shall use in the next section.

Suppose that $V(t, x)$ fulfills Assumption 2.1 and $[s, s'] \subset [a, b]$ with $|s' - s| \le \delta_0$. Let γ^* be the unique classical path starting y at time s and arriving at x at time s'. Let Δ be a division of $[s, s']$ as (5.2.1). We write $S(x_{J+1}, x_J, \dots, x_1, x_0)$ for $S(\gamma_\Delta)(x_{J+1}, x_J, \dots, x_1, x_0)$.

Let $a_\lambda(x_{J+1}, x_J, \dots, x_1, x_0)$ be a C^∞ function of $(x_{J+1}, x_J, \dots, x_0) \in \mathbf{R}^{J+2}$ with a parameter $\lambda \in \mathbf{R}^n$. We discuss the oscillatory integral

$$I(\Delta; S, a_\lambda, \nu)(x_{J+1}, x_0) \tag{5.3.1}$$
$$= \prod_{j=1}^{J+1} \left(\frac{-i\nu}{2\pi\tau_j}\right)^{\frac{1}{2}} \int_{\mathbf{R}^J} e^{i\nu S(x_{J+1}, x_J, \ldots, x_1, x_0)} a_\lambda(x_{J+1}, x_J, \ldots, x_1, x_0) \prod_{j=1}^{J} dx_j.$$

Remark 5.2 We do not exclude the case $n = 0$, i.e. a_λ is independent of λ. In that case we have only to set $\lambda = 0$ in the following.

Remark 5.3 We do not necessarily assume that there is a functional $F(\gamma)$ such that $a = F(\gamma_\Delta)$ in this section. If $F(\gamma)$ is a functional of path γ and $a(x_{J+1}, x_J, \ldots, x_1, x_0) = F_\Delta(x_{J+1}, x_J, \ldots, x_1, x_0)$, then

$$I(\Delta; S, a, \nu)(x_{J+1}, x_0) = I[F_\Delta](\Delta, \nu, s', s, x_{J+1}, x_0).$$

We assume

Assumption 5.3 Let m be a non-negative constant. Assume $|s' - s| \leq \delta_0$. For any non-negative integer K there exist positive constants A_K and $X_K \geq 1$ such that for any indices α_j satisfying $|\alpha_j| \leq K, j = 0, 1, 2, \ldots, J+1$ there holds the inequality

$$\left| \left(\prod_{k=0}^{J+1} \partial_{x_j}^{\alpha_j} \right) a_\lambda(x_{J+1}, x_J, \ldots, x_1, x_0) \right| \leq A_K X_K^{J+2} \Big(1 + |\lambda| + \sum_{j=0}^{J+1} |x_j|\Big)^m. \tag{5.3.2}$$

Moreover, for any division

$$\Delta' : T_0 = T_{j_0} < T_{j_1} < \cdots < T_{j_n} < T_{j_{n+1}} = T_{J+1}$$

which is coarser than Δ corresponding to subsequence $0 = j_0 < j_1 < \cdots < j_n < j_{n+1} = J + 1$,

$$\left| \left(\prod_{k=0}^{n+1} \partial_{x_{j_k}}^{\alpha_{j_k}} \right) \iota_{\Delta'}^{\Delta} a_\lambda(x_{j_{n+1}}, x_{j_n}, \ldots, x_{j_1}, x_{j_0}) \right| \leq A_K X_K^{n+2} \Big(1 + |\lambda| + \sum_{k=0}^{n+1} |x_{j_k}|\Big)^m, \tag{5.3.3}$$

as far as $|\alpha_{j_k}| \leq K, 0 \leq k \leq n+1$.

We have the following

Theorem 5.4 (Stationary phase method over a space of large dimension) *Suppose that $a_\lambda(x_{J+1}, x_J, \ldots, x_1, x_0)$ satisfies Assumption 5.3. Let δ_0 be the positive constant as in Definition 4.1. Then there holds the following equality:*

$$I(\Delta; S, a_\lambda, \nu)(x_{J+1}, x_0) \tag{5.3.4}$$
$$= \left(\frac{-\nu i}{2\pi(s'-s)}\right)^{1/2} \exp\{i\nu S(s', s, x_{J+1}, x_0)\}$$
$$\times D(\Delta; s', s, x_{J+1}, x_0)^{-1/2}(\iota^{\Delta}_{\Delta(J+1)} a_\lambda(x_{J+1}, x_0) + \nu^{-1} r_\lambda(\Delta; \nu, s', s, x_{J+1}, x_0)).$$

The remainder $r_\lambda(\Delta; \nu, x_{J+1}, x_0)$ satisfies the following estimates. For any non-negative integer k there exists a positive constant C_k and a natural number $M(k)$ such that for any (x_{J+1}, x_0)

$$\left|\partial^{\alpha_0}_{x_0} \partial^{\alpha_{J+1}}_{x_{J+1}} r_\lambda(\Delta; \nu, x_{J+1}, x_0)\right| \tag{5.3.5}$$
$$\le A_{M(k)} C_k X^2_{M(k)}(s'-s)(1 + |\lambda| + |x_{J+1}| + |x_0|)^m \exp\{C_k X^2_{M(k)}(s'-s)\},$$

if $|\alpha_0|, |\alpha_{J+1}| \le k$. Here $M(k)$ and C_k do not depend on Δ, (x_{J+1}, x_0) and ν. We may choose δ_0 so that $4\kappa_2\delta_0^2 < 1/2$.

Theorem 5.4 is proved in Part II. See Theorem 8.4 and Corollary 8.8 in Chap. 8.

Corollary 5.1 *Under the same assumption as in the theorem above we can write*

$$I(\Delta; S, a_\lambda, \nu)(x_{J+1}, x_0) \tag{5.3.6}$$
$$= \left(\frac{-\nu i}{2\pi(s'-s)}\right)^{1/2} \exp\{i\nu S(s', s, x_{J+1}, x_0)\}$$
$$\times D(\Delta; s', s, x_{J+1}, x_0)^{-1/2} \tilde{a}(\Delta; s', s, \nu, x_{J+1}, x_0).$$

For any non-negative integer k there exists a positive constant C_k such that for any (x_{J+1}, x_0) we have

$$\left|\partial^{\alpha_0}_{x_0} \partial^{\alpha_{J+1}}_{x_{J+1}} \tilde{a}_\lambda(\Delta; \nu, s', s, x_{J+1}, x_0)\right| \tag{5.3.7}$$
$$\le C_k A_{M(k)} (1 + |\lambda| + |x_{J+1}| + |x_0|)^m e^{\nu^{-1} C_k X^2_{M(k)}(s'-s)},$$

if $|\alpha_0|, |\alpha_{J+1}| \le k$. Here $M(k)$ is the same as above.

Remark 5.4 In the following we replace $C_k e^{\nu^{-1} C_k X^2_{M(k)} \delta_0}$ by C_k. Thus we use

$$\left|\partial^{\alpha_0}_{x_0} \partial^{\alpha_{J+1}}_{x_{J+1}} \tilde{a}_\lambda(\Delta; \nu, s', s, x_{J+1}, x_0)\right| \le C_k A_{M(k)} (1 + |\lambda| + |x_{J+1}| + |x_0|)^m e^{\nu^{-1} C_k X^2_{M(k)}(s'-s)}, \tag{5.3.8}$$

instead of (5.3.7).

The next theorem explains further the structure of the remainder term of the previous theorem.

Theorem 5.5 *Suppose that $|s'-s| \le \min\{1, \delta_0\}$ and $a_\lambda(x_{J+1}, x_J, \dots, x_1, x_0)$ satisfies Assumption 5.3. Then*

$$\begin{aligned} &I(\Delta; S, a_\lambda, \nu)(x_{J+1}, x_0) \\ &= \left(\frac{-i\nu}{2\pi(s'-s)}\right)^{1/2} e^{i\nu S(s',s,x_{J+1},x_0)} D(\Delta; s', s, x_{J+1}, x_0)^{-1/2} \\ &\times \left(\iota^{\Delta}_{\Delta(J+1)} a_\lambda(x_{J+1}, x_0) + \nu^{-1} i(s'-s) p(\Delta; s', s, x_{J+1}, x_0)\right) \\ &+ \left(\frac{-i\nu}{2\pi(s'-s)}\right)^{1/2} e^{i\nu S(s',s,x_{J+1},x_0)} \nu^{-1}(s'-s)^2 r(\Delta; \nu, s', s, x_{J+1}, x_0) \end{aligned} \tag{5.3.9}$$

with

$$\begin{aligned} &p(\Delta; s', s, x_{J+1}, x_0) \\ &= \sum_{k=1}^{J} \frac{(T_k - s)\tau_{k+1}}{2(s'-s)(T_{k+1}-s)} \iota^{\Delta(k)}_{\Delta(J+1)} \left(D_{x^*_{k-1},\dots,x^*_1}\Big(\sum_{j=1}^{k} S_j; x_k, x_0\Big)^{1/2} \right. \\ &\left. \times \partial^2_{x_k} \Big(D_{x^*_{k-1},\dots,x^*_1}\Big(\sum_{j=1}^{k} S_j; x_k, x_0\Big)^{-1/2} \iota^{\Delta}_{\Delta(k)} a_\lambda\Big) \right) (x_{J+1}, x_0). \end{aligned} \tag{5.3.10}$$

Also $r(\Delta; \nu, x_{J+1}, x_0)$ satisfies the following estimate. For any $K \ge 0$ there exists $M(K)$ and C_K such that

$$(1+|\lambda|+|x_{J+1}|+|x_0|)^{-m} |\partial^{\alpha_{J+1}}_{x_{J+1}} \partial^{\alpha_0}_{x_0} r(\Delta; \nu, s', s, x_{J+1}, x_0)| \le C_K A_{M(K)} X^2_{M(K)}, \tag{5.3.11}$$

if multi-indices α_0, α_{J+1} satisfy $|\alpha_0| \le K$ and $|\alpha_{J+1}| \le K$.

Proof of Theorem 5.5 is given in Part II. See Theorem 8.7 in Chap. 8.

5.4 Proof of Convergence of Feynman Path Integrals

Since $F(\gamma) \equiv 1$ satisfies Kumano-go's condition, Assumption 4.8, proof of Theorem 4.5 is reduced to that of Theorem 4.9 and proof of Theorem 4.6 is reduced to proof of Theorem 4.10. Therefore, we shall prove Theorems 4.9 and 4.10. Our discussions are based on the stationary phase method of oscillatory integrals over a space of large dimension discussed in the previous section.

5.4.1 Proof of Theorems 4.5 and 4.9

Suppose that $V(t,x)$ fulfills Assumption 2.1. In this section we always assume that $\kappa_2\delta_0^2 < 1$, $8^{-1}v_2\delta^2 < 1$ and $|s'-s| \le \min\{1, \delta_0\}$. Let x, y be arbitrary points in $\mathbf{R}^2$ and γ^* be the unique classical path satisfying $\gamma^*(s) = y$ and $\gamma^*(s') = x$.

For technical reasons, we discuss the case that the functional $F(\gamma)$ may depend also on parameter λ. Let $F_\lambda(\gamma)$ be a functional of paths. Let

$$\Delta : s = T_0 < T_1 < \cdots < T_J < T_{J+1} = s' \tag{5.4.1}$$

be an arbitrary division of the interval $[s, s']$. We write $F_{\lambda,\Delta} = F_\lambda(\gamma_\Delta)$. $S(\gamma_\Delta)(x_{J+1}, x_J, \ldots, x_1, x_0)$ has the unique stationary point $x^* = (x_J^*, \ldots, x_1^*)$, $x_j^* = \gamma^*(T_j)$ and its stationary value equals $S(\gamma^*)$. We discuss time slicing approximation of the Feynman path integral

$$\begin{aligned} &I[F_{\lambda,\Delta}](\Delta; \nu, s', s, x, y) \qquad (5.4.2)\\ &= \prod_{j=1}^{J+1} \left(\frac{\nu}{2\pi i\tau_j}\right)^{1/2} \int_{\mathbf{R}^J} e^{i\nu S(\gamma_\Delta)} F_\lambda(\gamma_\Delta) \prod_{j=1}^{J} dx_j \\ &= \prod_{j=1}^{J+1} \left(\frac{\nu}{2\pi i\tau_j}\right)^{1/2} \int_{\mathbf{R}^J} e^{i\nu S_\Delta(x_{J+1}, x_J, \ldots, x_1, x_0)} F_{\lambda,\Delta}(x_{J+1}, x_J, \ldots, x_1, x_0) \prod_{j=1}^{J} dx_j, \end{aligned}$$

where $x_{J+1} = x$ and $x_0 = y$.

Remark 5.5 (4.2.15) is the special case $\lambda = 0$ of (5.4.2).

As the functional F_λ may depend on parameters, we modify assumptions for F.

Assumption 5.6 Let m be a non-negative constant. For any integer $K \ge 0$ there exist constants $A_K > 0$ and $X_K \ge 1$ such that for any division Δ and for any indices α_j satisfying $|\alpha_j| \le K, j = 0, 1, 2, \ldots, J+1$

$$\begin{aligned} &\left| \left(\prod_{k=0}^{J+1} \partial_{x_j}^{\alpha_j} \right) F_\lambda(\gamma_\Delta(x_{J+1}, x_J, \ldots, x_1, x_0)) \right| \qquad (5.4.3)\\ &\le A_K X_K^{J+2} (1 + |\lambda| + |x_{J+1}| + |x_J| + \cdots + |x_1| + |x_0|)^m. \end{aligned}$$

Kumano-go's condition [46] is also modified.

Assumption 5.7 There exist a non-negative constant m and a bounded non-negative Borel measure ρ on $[s, s']$. For any integer $K \ge 0$ there exist constants $A_K > 0$ and $X_K \ge 1$ with the following property: Let Δ be an arbitrary division of $[s, s']$ as (4.1.3) and Δ' be as (4.2.18) coarser than Δ and for any indices α_{j_n} satisfying $|\alpha_{j_n}| \le K$, $n = 0, 1, 2, \ldots, p+1$

$$\left|\left(\prod_{n=0}^{p+1} \partial_{x_{j_n}}^{\alpha_{j_n}}\right)(\iota_{\Delta'}^{\Delta}\partial_{x_k}^{\beta}F_{\lambda,\Delta})(x_{J+1}, x_{j_p}, \ldots, x_{j_1}, x_0)\right|$$
$$\leq A_K X_K^{p+2}\rho([T_{k-1}, T_{k+1}])(1 + |\lambda| + |x_{J+1}| + |x_{j_p}| + \cdots + |x_{j_1}| + |x_0|)^m, \quad (5.4.4)$$

for all $k = 1, 2, \ldots, J$ and $\beta = 1, 2$.

We first show that Kumano-go's assumption, Assumption 5.7, leads us to a good estimate of remainder term (5.3.10) of Theorem 5.5.

In the following we often define a_λ for $F_\lambda(\gamma_\Delta)$. The next proposition follows from Assumption 5.7.

Proposition 5.3 *Suppose that $F_\lambda(\gamma)$ satisfies Assumption 5.7 as well as Assumption 5.6. Then for any indices α_{J+1}, α_0, satisfying $|\alpha_{J+1}|, |\alpha_0| \leq K$ there holds the inequality*

$$\left|\partial_{x_{J+1}}^{\alpha_{J+1}}\partial_{x_0}^{\alpha_0}\iota_{\Delta(J+1)}^{\Delta(k)}\left(\partial_{x_k}^{\beta}\iota_{\Delta(k)}^{\Delta}a_\lambda\right)(x_{J+1}, x_0)\right|$$
$$\leq A_K X_K^2\rho([T_0, T_{k+1}])(1 + |\lambda| + |x_{J+1}| + |x_0|)^m, \quad (5.4.5)$$

for all $k = 1, 2, \ldots, J$ and $\beta = 1, 2$.

The next theorem implies Theorem 4.9 as a special case of $\lambda = 0$.

Theorem 5.8 *Suppose that Assumption 5.7 is satisfied as well as Assumption 5.6. Let δ_0 be so small that $\kappa_2\delta_0^2 < 1$. If $|s - s'| \leq \min\{\delta_0, 1\}$, then there exists a positive constant δ such that for any Δ with $|\Delta| \leq \delta$*

$$\begin{aligned} &I[F_{\lambda,\Delta}](\Delta; \nu, s', s, x, y) \quad (5.4.6)\\ &= I(\Delta; S, a_\lambda, \nu)(x, y)\\ &= \left(\frac{-i\nu}{2\pi(s'-s)}\right)^{1/2} \exp\{i\nu S(s', s, x, y)\}\\ &\times D(\Delta; s', s, x, y)^{-1/2}(F_\lambda(\gamma^*)(x, y) + \nu^{-1}R[F_\lambda](\Delta; \nu, s', s, x, y)). \end{aligned}$$

Moreover, for any non-negative integer K there exist a positive constant C_K and a positive integer $M(K)$ independent of ν and of Δ such that if $|\alpha|, |\beta| \leq K$, then

$$\begin{aligned} &|\partial_x^\alpha\partial_y^\beta R[F_\lambda](\Delta; \nu, s', s, x_{J+1}, x_0)| \quad (5.4.7)\\ &\leq C_K A_{M(K)}|s - s'|(|s - s'| + \rho([s, s']))(1 + |\lambda| + |x_{J+1}| + |x_0|)^m. \end{aligned}$$

Proof We look for an estimate of $p(\Delta; s', s, x_{J+1}, x_0)$ in Theorem 5.5 by using the previous proposition. We have

$$
\begin{aligned}
&\iota^{\Delta(k)}_{\Delta(J+1)}\Big[D_{x^*_{k-1},\dots,x^*_1}\Big(\sum_{j=1}^{k} S_j; x_k, x_0\Big)^{1/2} \\
&\quad\times \partial^2_{x_k}\Big(D_{x^*_{k-1},\dots,x^*_1}\Big(\sum_{j=1}^{k} S_j; x_k, x_0\Big)^{-1/2}\iota^{\Delta}_{\Delta(k)}a_\lambda\Big)\Big](x_{J+1}, x_0) \\
&= \iota^{\Delta(k)}_{\Delta(J+1)}\left[D_{x^*_{k-1},\dots,x^*_1}\Big(\sum_{j=1}^{k} S_j; x_k, x_0\Big)^{1/2}\right](x_{J+1}, x_0) \\
&\quad\times \iota^{\Delta(k)}_{\Delta(J+1)}\Big[\partial^2_{x_k}\big(D_{x^*_{k-1},\dots,x^*_1}\Big(\sum_{j=1}^{k} S_j; x_k, x_0\Big)^{-1/2}\iota^{\Delta}_{\Delta(k)}a_\lambda\big)\Big](x_{J+1}, x_0).
\end{aligned}
$$

It follows from Proposition 5.3 that for $\beta = 1$ and $\beta = 2$

$$
(1+|\lambda|+|x_{J+1}|+|x_0|)^{-m}\left|\partial^{\alpha_0}_{x_0}\partial^{\alpha_{J+1}}_{x_{J+1}}\iota^{\Delta(k)}_{\Delta(J+1)}\partial^{\beta}_{x_k}\iota^{\Delta}_{\Delta(k)}a_\lambda(x_{J+1}, x_0)\right| \le \rho([s, T_{k+1}])A_K X_K^2,
$$

as far as $|\alpha_0| \le K$, $|\alpha_{J+1}| \le K$. Also

$$
(1+|\lambda|+|x_{J+1}|+|x_0|)^{-m}\left|\partial^{\alpha_0}_{x_0}\partial^{\alpha_{J+1}}_{x_{J+1}}\iota^{\Delta(k)}_{\Delta(J+1)}\iota^{\Delta}_{\Delta(k)}a_\lambda(x_{J+1}, x_0)\right| \le A_K X_K^2, \quad (5.4.8)
$$

because $\iota^{\Delta(k)}_{\Delta(J+1)}\iota^{\Delta}_{\Delta(k)} = \iota^{\Delta}_{\Delta(J+1)}$.

Since $s < T_1 < T_2 < \cdots < T_k$ is a division of interval $[s, T_k]$, we can apply Lemmas 5.4 and 4.2 of Sect. 4.1 to this case. Thus there exists a function $d(T_k, s, x_k, x_0) \in \mathscr{B}(\mathbf{R}^2)$ such that

$$
\Big(D_{x^*_{k-1},\dots,x^*_1}\Big(\sum_{j=1}^{k} S_j; x_k, x_0\Big)\Big)^{-1/2} = 1 + (T_k - s)^2 d(T_k, s, x_k, x_0),
$$

and for any $K \ge 0$ there exists C_K independent of k such that

$$
\max_{|\alpha|\le K, |\alpha_0|\le K}\ \sup_{(x,y)\in\mathbf{R}^2} |\partial^{\alpha_k}_{x_k}\partial^{\alpha_0}_{x_0} d(T_k, s, x_k, x_0)| < C_K.
$$

It follows from this that for any $K \ge 0$ there exists $C_K > 0$ such that as far as $|\alpha_0| \le K$, $|\alpha_{J+1}| \le K$ and $2 \ge |\beta| > 0$,

$$
\left|\partial^{\alpha_0}_{x_0}\partial^{\alpha_{J+1}}_{x_{J+1}}\iota^{\Delta(k)}_{\Delta(J+1)}\partial^{\beta}_{x_k}\Big(D_{x^*_{k-1},\dots,x^*_1}\Big(\sum_{j=1}^{k} S_j; x_k, x_0\Big)\Big)^{-1/2}\right| \le C_K(T_k - s)^2.
$$

Similarly, for any $K \ge 0$ there exists a positive constant C_K such that as far as $|\alpha_0| \le K$ and $|\alpha_{J+1}| \le K$,

$$\left| \partial_{x_0}^{\alpha_0} \partial_{x_{J+1}}^{\alpha_{J+1}} \iota_{\Delta(J+1)}^{\Delta(k)} \Big(D_{x_{k-1}^*,\dots,x_1^*} \big(\sum_{j=1}^{k} S_j; x_k, x_0 \big) \Big)^{-1/2} \right| \le C_K.$$

It turns out from these and Leibniz's formula that there exists a function $f_k(s', s, x_{J+1}, x_0)$ such that

$$\begin{aligned} &\iota_{\Delta(J+1)}^{\Delta(k)} D_{x_{k-1}^*,\dots,x_1^*} \Big(\sum_{j=1}^{k} S_j; x_k, x_0 \Big)^{1/2} \\ &\qquad \times \partial_{x_k}^2 \Big(D_{x_{k-1}^*,\dots,x_1^*} \Big(\sum_{j=1}^{k} S_j; x_k, x_0 \Big)^{-1/2} \iota_{\Delta(k)}^{\Delta} a_\lambda(x_{J+1}, \dots, x_k, x_0) \Big) \\ &= (\rho([s, T_{k+1}]) + (T_k - s)^2) f_k(s', s, x_{J+1}, x_0). \end{aligned}$$

The function $f_k(s', s, x_{J+1}, x_0)$ satisfies the following estimates: For any integer $K \ge 0$ there exists a constant $C_K > 0$ such that

$$\left| \partial_{x_{J+1}}^{\alpha} \partial_{x_0}^{\beta} f_k(s', s, x_{J+1}, x_0) \right| \le C_K X_K^2 A_K (1 + |\lambda| + |x_{J+1}| + |x_0|)^m, \tag{5.4.9}$$

as far as $|\alpha| \le K$ and $\beta| \le K$.

We apply the above estimate to the result of the stationary phase method, Theorem 5.5. Notice that

$$\begin{aligned} &\sum_{k=1}^{J} \frac{(T_k - s)\tau_{k+1}}{2(s' - s)(T_{k+1} - s)} (\rho([s, T_{k+1}]) + (T_k - s)^2) \\ &\le \sum_{k=1}^{J} \frac{\tau_{k+1}}{2(s' - s)} (\rho([s, s']) + (s' - s)^2) \le 2^{-1} (\rho([s, s']) + (s' - s)^2). \end{aligned}$$

It follows from this and (5.4.9) that

$$\begin{aligned} &p(\Delta; s', s, x_{J+1}, x_0) \\ &= \sum_{k=1}^{J} \frac{(T_k - s)\tau_{k+1}}{2(s' - s)(T_{k+1} - s)} (\rho([s, T_{k+1}]) + (T_k - s)^2) f_k(s', s, x_{J+1}, x_0). \\ &= (\rho([s, s']) + |s' - s|) q(\Delta, s', s, x_{J+1}, x_0). \end{aligned} \tag{5.4.10}$$

Here for any non-negative integer K there exists $C_K > 0$ such that as far as $|\alpha| \le K$ and $|\beta| \le K$

$$|\partial_{x_{J+1}}^{\alpha} \partial_{x_0}^{\beta} q(\Delta, s', s, x_{J+1}, x_0)| \le C_K X_K^2 A_K (1 + |\lambda| + |x_{J+1}| + |x_0|)^m. \tag{5.4.11}$$

Using Theorem 5.5, we set

$$R[F_\lambda](\Delta;\nu,s',s,x_{J+1},x_0)$$
$$= i(s'-s)p(\Delta;s',s,x_{J+1},x_0)+(s'-s)^2D(\Delta;s',s,x_{J+1},x_0)^{1/2}r(\Delta;\nu,s',s,x_{J+1},x_0).$$

Then by Theorem 5.5 we have

$$I(\Delta;S,a_\lambda,\nu)(x_{J+1},x_0)$$
$$=\left(\frac{\nu}{2\pi i(s'-s)}\right)^{1/2}D(\Delta;s',s,x_{J+1},x_0)^{-1/2}e^{i\nu S(s',s,x_{J+1},x_0)}$$
$$\times\left(\iota^{\Delta}_{\Delta(J+1)}a_\lambda(x_{J+1},x_0)+\nu^{-1}R[F_\lambda](\Delta;\nu,s',s,x_{J+1},x_0)\right).$$

Moreover, combining (5.3.11), (5.4.10) and (5.4.11), for any $K\geq 0$ we can find a constant C_K and an integer $M(K)$ such that

$$|\partial_x^\alpha\partial_y^\beta R[F_\lambda](\Delta;\nu,s',s,x_{J+1},x_0)|$$
$$\leq C_K A_{M(K)}|s-s'|(|s-s'|+\rho([s,s']))(1+|\lambda|+|x_{J+1}|+|x_0|)^m,$$

as far as $|\alpha|,|\beta|\leq K$. Therefore, Theorem 5.8 has been proved. □

Proof of Theorem 4.9

We can now prove Theorem 4.9. For any division Δ as above we set

$$a_\lambda(x_{J+1},x_J,\ldots,x_0)=F(\gamma_\Delta)(x_{J+1},x_J,\ldots,x_0).$$

Then by Remark 5.3,

$$I[F_\Delta](\Delta;\nu,s',s,x,y)=I(\Delta;S,a_\lambda,\nu)(x,y).$$

Suppose $F(\gamma)$ satisfies Assumption 4.7. Then a_λ is independent of λ and satisfies Assumption 5.3, because $\iota^{\Delta}_{\Delta'}F(\gamma_\Delta)=F(\gamma_{\Delta'})$ if Δ' is a division coarser than Δ.

Suppose further $F(\gamma)$ satisfies Assumption 4.8. Then a_λ satisfies Proposition 5.3. Thus under the assumption in Theorem 4.9 we can apply Theorem 5.8 with $\lambda=0$ to $I(\Delta;S,a_\lambda,\nu)(x,y)=I[F_\Delta](\Delta;\nu,s',s,x,y)$. Then we obtain the conclusion of Theorem 4.9. □

Proof of Theorem 4.5

Let $F(\gamma)\equiv 1$. Then $F(\gamma)$ satisfies Assumptions 4.7 and 4.8 with $\rho=0$. Therefore, Theorem 4.5 follows from Theorem 4.9 with $\rho=0$. Theorem 4.5 has been proved. □

5.4.2 Proof of Theorems 4.6 and 4.10

Since $F(\gamma) \equiv 1$ satisfies the assumptions in Theorem 4.10, Theorem 4.6 easily follows from Theorem 4.10. We shall prove Theorem 4.10.

Consider an arbitrary division Δ of time interval $[s, s']$ as follows:

$$\Delta : s = T_0 < T_1 < \cdots < T_J < T_{J+1} = s'. \tag{5.4.12}$$

It was proved by Theorem 5.8 that we can write

$$\begin{aligned} &I[F_{\lambda,\Delta}](\Delta; \nu, s', s, x, y) \\ &= \left(\frac{-i\nu}{2\pi(s'-s)}\right)^{1/2} \exp\{i\nu S(s', s, x, y)\} k[(F_{\lambda,\Delta}](\Delta; \nu, s', s, x, y). \end{aligned} \tag{5.4.13}$$

Here

$$k[F_{\lambda,\Delta}](\Delta; \nu, s', s, x, y) = D(\Delta; s', s, x, y)^{-1/2}(F_\lambda(\gamma^*) + \nu^{-1} R[F_\lambda](\Delta; \nu, s', s, x, y)). \tag{5.4.14}$$

Theorem 4.10 is a special case $\lambda = 0$ of the next theorem.

Theorem 5.9 *Let $\delta_0 > 0$ be so small that $\kappa_2\delta_0 < 1$. Assume that functional $F_\lambda(\gamma)$ satisfies Assumptions 5.6 and 5.7 above and $|s' - s| \le \delta_0$. Then the following limit exists:*

$$K[F_\lambda](\nu, s', s, x, y) = \lim_{|\Delta|\to 0} I[F_{\lambda,\Delta}](\Delta; \nu, s', s, x, y). \tag{5.4.15}$$

$K[F_\lambda](\nu, s', s, x, y)$ is of the form

$$K[F_\lambda](\nu, s', s, x, y) = \left(\frac{\nu}{2\pi i(s'-s)}\right)^{1/2} e^{i\nu S(\gamma^*)} k[F_\lambda](\nu, s', s, x, y). \tag{5.4.16}$$

More precisely, for any integer $K \ge 0$ there exist constants $C_{\alpha\beta} > 0$, $\delta_1 > 0$ and an integer $M(K) \ge 0$ such that if $|\Delta| \le \delta_1$, then for $(x, y) \in \mathbf{R}^2$

$$\begin{aligned} &|\partial_x^\alpha \partial_y^\beta \left(k[F_{\lambda,\Delta}](\Delta; \nu, s', s, x, y) - k[F_\lambda](\nu, s', s, x, y)\right)| \\ &\le C_K A_{M(K)} |\Delta| (\rho([s, s'] + (s' - s))(1 + |\lambda| + |x| + |y|)^m, \end{aligned} \tag{5.4.17}$$

as far as $|\alpha| \le K$ and $|\beta| \le K$.

Remark 5.6 We may write

$$\int_\Omega e^{i\nu S(\gamma)} F_\lambda(\gamma) \mathscr{D}[\gamma] = K[F_\lambda](\nu, s', s, x, y). \tag{5.4.18}$$

We shall prove Theorem 5.9 by showing that $k[F_{\lambda,\Delta}](\Delta; \nu, s', s, x, y)$ satisfies Cauchy's condition for convergence as $|\Delta| \to 0$. More precisely, we shall prove

Lemma 5.7 *Let Δ^* be any refinement of Δ. Then for any integer $k \geq 0$ there exists a positive constant C_k and a non-negative integer $M(k)$ independent of Δ, ν such that*

$$
\begin{aligned}
&|\partial_x^\alpha \partial_y^\beta (k[F_{\lambda,\Delta^*}](\Delta^*; \nu, s', s, x, y) - k[F_{\lambda,\Delta}](\Delta; \nu, s', s, x, y))| \\
&\leq C_k A_{M(k)} |\Delta| (\rho([s, s']) + |\Delta|)(1 + |\lambda| + |x| + |y|)^m,
\end{aligned}
\tag{5.4.19}
$$

as far as $|\alpha| \leq k$, $|\beta| \leq k$.

In the following we shall abbreviate $I[F_{\lambda,\Delta}](\Delta; \nu, s', s, x, y)$ to $I[F_{\lambda,\Delta}](\Delta; x, y)$ for the sake of simplicity.

The proof of Lemma 5.7 is divided into a few steps. First we shall discuss a special simple type of refinement of Δ. Let n be any one of $n = 1, 2, \ldots, J + 1$. We divide the n-th subinterval $[T_{n-1}, T_n]$ of Δ into smaller sub-subintervals and denote the division of $[T_{n-1}, T_n]$ by δ, i.e.,

$$
\delta : T_{n-1} = T_{n,0} < T_{n,1} < \cdots < T_{n,p_n+1} = T_n. \tag{5.4.20}
$$

We define a refinement Δ' of Δ by using the union of $\{T_j\}_{j=0}^{J+1}$ and $\{T_{n,k}\}_{k=1}^{p_n}$ as the set of dividing points of Δ'. In other words, Δ' is the same as Δ except for the division δ of $[T_{n-1}, T_n]$. We set $\sigma_k = T_{n,k} - T_{n,k-1}$ for $k = 1, 2, \ldots, p_n + 1$.

Lemma 5.8 *Suppose that $F_\lambda(\gamma)$ fulfills Assumptions 5.6 and 5.7. Let Δ and Δ' be as above. Then for any non-negative integer K there exist a positive constant C_K and a non-negative integer $M(K)$ independent of Δ and ν such that if $|\alpha|, |\beta| \leq K$ we have*

$$
\begin{aligned}
&|\partial_x^\alpha \partial_y^\beta \left(k[F_{\lambda,\Delta'}](\Delta'; \nu, s', s, x, y) - k[F_{\lambda,\Delta}](\Delta; \nu, s', s, x, y)\right)| \\
&\leq C_K A_{M(K)} X_{M(K)}^4 \tau_n \Big(\rho([T_{n-1}, T_n]) + \tau_n\Big)(1 + |\lambda| + |x| + |y|)^m.
\end{aligned}
\tag{5.4.21}
$$

Proof We consider an arbitrary piecewise classical path $\gamma_{\Delta'}$ associated with the division Δ' and we write

$$
\begin{aligned}
y_k &= \gamma_{\Delta'}(T_{n,k}), \quad \text{for} \quad 0 \leq k \leq p_n + 1, \\
x_j &= \gamma_{\Delta'}(T_j), \quad \text{for} \quad 0 \leq j \leq J + 1,
\end{aligned}
$$

where we assume that $y_0 = x_{n-1}$, $y_{p_n+1} = x_n$. We abbreviate the block of variables $(y_{p_n}, \ldots, y_1)$ to $y_{[p_n,1]}$. Similarly for any pair of integers $k \geq j \geq 0$ we write $x_{[k,j]}$ for $(x_k, \ldots, x_j)$. As a special case we set $x_{[k,k]} = x_k$. Let

$$S_{n,k}(y_k, y_{k-1}) = S(T_{n,k}, T_{n,k-1}, y_k, y_{k-1}) \tag{5.4.22}$$
$$= \int_{T_{n,k-1}}^{T_{n,k}} L\Big(t, \frac{d}{dt}\gamma_{\Delta'}(t), \gamma_{\Delta'}(t)\Big)\, dt \quad (k = 1, \ldots, p_n + 1).$$

Then the action $S(\gamma_{\Delta'})$ becomes

$$S(\gamma_{\Delta'}) = S_{\Delta'}(x_{[J+1,n]}, y_{[p_n,1]}, x_{[n-1,0]}) \tag{5.4.23}$$
$$= \sum_{j=1, j\neq n}^{J+1} S_j(x_j, x_{j-1}) + \sum_{k=1}^{p_n+1} S_{n,k}(y_k, y_{k-1}).$$

We set the latter term of (5.4.23) as

$$S_\delta(x_n, y_{[p_n,1]}, x_{n-1}) = \sum_{k=1}^{p_n+1} S_{n,k}(y_k, y_{k-1}), \tag{5.4.24}$$

and we set

$$F_{\lambda,\Delta'}(x_{[J+1,n]}, y_{[p_n,1]}, x_{[n-1,0]}) = F_\lambda(\gamma_{\Delta'}(x_{[J+1,n]}, y_{[p_n,1]}, x_{[n-1,0]})). \tag{5.4.25}$$

We distinguish two groups of variables $(x_J, \ldots, x_1)$ and $(y_{p_n}, \ldots, y_1)$. Integrating with respect to $(y_{p_n}, \ldots, y_1)$ prior to integration with respect to $(x_J, \ldots, x_1)$, we define $F_{\lambda,\Delta/\Delta'}(x_{J+1}, x_J, \ldots, x_1, x_0)$ by the equality

$$\left(\frac{\nu}{2\pi i\tau_n}\right)^{1/2} e^{i\nu S_n(x_n,x_{n-1})} F_{\lambda,\Delta/\Delta'}(x_{J+1}, x_J, \ldots, x_1, x_0) \tag{5.4.26}$$
$$= \prod_{k=1}^{p_n+1}\left(\frac{\nu}{2i\pi\sigma_k}\right)^{1/2} \int_{\mathbf{R}^{p_n}} e^{i\nu S_\delta(x_n, y_{p_n}, \ldots, y_1, x_{n-1})}$$
$$\times F_{\lambda,\Delta'}(x_{[J+1,n]}, y_{p_n}, \ldots, y_1, x_{[n-1,0]}) \prod_{k=1}^{p_n} dy_k.$$

Using (5.4.2) and (5.4.26), we have

$$I[F_{\lambda,\Delta'}](\Delta'; x, y) = \prod_{j=1}^{J+1}\left(\frac{\nu}{2\pi i\tau_j}\right)^{1/2}$$
$$\times \int_{\mathbf{R}^J} e^{i\nu S_\Delta(x_{J+1}, x_J, \ldots, x_1, x_0)} F_{\lambda,\Delta/\Delta'}(x_{J+1}, x_J, \ldots, x_1, x_0) \prod_{j=1}^{J} dx_j. \tag{5.4.27}$$

In other words,

$$I[F_{\lambda,\Delta'}](\Delta'; x, y) = I[F_{\lambda,\Delta/\Delta'}](\Delta; x, y). \tag{5.4.28}$$

Now we shall compare $F_{\lambda,\Delta}(x_{J+1}, \dots, x_0)$ with $F_{\lambda,\Delta/\Delta'}(x_{J+1}, \dots, , x_0)$. We regard $F_{\lambda,\Delta'}(x_{[J+1,n+1]}, y_{p_n+1}, \dots, y_0, x_{[n-2,0]})$ as a function of $(y_{p_n+1}, \dots, y_1, y_0)$ with parameters $x_{J+1}, \dots, x_{n+1}$ and $x_{n-2}, \dots, x_0$ as well as λ. Since $F_\lambda(\gamma)$ satisfies Assumptions 5.6 and 5.7, the function

$$(y_{p_n+1}, \dots, y_1) \longrightarrow F_{\lambda,\Delta'}(x_{[J+1,n+1]}, y_{p_n+1}, \dots, y_0, x_{[n-2,0]})$$

with parameters satisfies Assumptions 5.6 and 5.7. We can apply Theorem 5.8 to the integral (5.4.26) with respect to variables $\{y_{p_n}, \dots, y_1\}$ and we obtain

$$\begin{aligned}
&\prod_{k=1}^{p_n+1}\left(\frac{\nu}{2i\pi\sigma_k}\right)^{1/2}\int_{\mathbf{R}^{p_n}} e^{i\nu S_\delta(x_n, y_{p_n},\dots,y_1,x_{n-1})} F_{\lambda,\Delta'}(x_{[J+1,n]}, y_{p_n}, \dots, y_1, x_{[n-1,0]}) \prod_{k=1}^{p_n} dy_k. \\
&= \left(\frac{\nu}{2\pi\tau_n}\right)^{1/2} e^{i\nu S_n(x_n,x_{n-1})} D(\delta; x_n, x_{n-1})^{-1/2} \qquad (5.4.29)\\
&\quad \times \Big(F_{\lambda,\Delta}(x_{J+1}, x_J, \dots, x_n, x_{n-1}, \dots, x_0) + \nu^{-1} R_\delta[F_{\lambda,\Delta'}](\nu, s', s, x_{J+1}, x_J, \dots, x_1, x_0)\Big).
\end{aligned}$$

Here the principal term is

$$\begin{aligned}
F_{\lambda,\Delta'}(x_{[J+1,n]}, y^*_{[p_n,1]}, x_{[n-1,0]}) &= \iota_\Delta^{\Delta'} F_{\lambda,\Delta'}(x_{J+1}, x_J, \dots, x_n, x_{n-1}, \dots, x_0) \\
&= F_{\lambda,\Delta}(x_{J+1}, x_J, \dots, x_n, x_{n-1}, \dots, x_0).
\end{aligned}$$

and $R_\delta[F_{\lambda,\Delta'}](\nu, s', s, x_{J+1}, x_J, \dots, x_1, x_0)$ denotes the remainder. It follows from (5.4.7) that for any K there exist a positive constant C_K and a non-negative integer $M(K)$ independent of Δ such that

$$\begin{aligned}
&|\partial_{x_n}^{\alpha_n}\partial_{x_{n-1}}^{\alpha_{n-1}} R_\delta[F_{\lambda,\Delta'}](\nu, s', s, x_{J+1}, x_J, \dots, x_1, x_0)| \qquad (5.4.30)\\
&\le C_K \tau_n(\tau_n + \rho([T_{n-1}, T_n]))A_{M(K)} \\
&\quad \times (1 + |\lambda| + |x_{[J+1,n+1]}| + |x_n| + |x_{n-1}| + |x_{[n-2,0]}|)^m,
\end{aligned}$$

as far as $|\alpha_n| \le K$, $|\alpha_{n-1}| \le K$. Here $|x_{[J+1,n+1]}| = \sum_{j=n+1}^{J+1} |x_j|$ and $|x_{[n-2,0]}| = \sum_{j=0}^{n-2} |x_j|$. Substitute (5.4.29) in (5.4.28), we have

$$\begin{aligned}
&I[F_{\lambda,\Delta'}](\Delta'; x, y) = I[F_{\lambda,\Delta/\Delta'}](\Delta; x, y) \qquad (5.4.31)\\
&= I[D(\delta; x_n, x_{n-1})^{-1/2} F_{\lambda,\Delta}](\Delta; x, y) + \nu^{-1} I[D(\delta; x_n, x_{n-1})^{-1/2} R_\delta[F_{\lambda,\Delta'}]](\Delta; x, y).
\end{aligned}$$

This implies that

$$
\begin{aligned}
&I[F_{\lambda,\Delta'}](\Delta'; x, y) - I[F_{\lambda,\Delta}](\Delta; x, y) \qquad (5.4.32)\\
&= I[F_{\lambda,\Delta/\Delta'} - F_{\lambda,\Delta}](\Delta; x, y)\\
&= I[(D(\delta; x_n, x_{n-1})^{-1/2} - 1)F_{\lambda,\Delta}](\Delta; x, y) + \nu^{-1} I[D(\delta; x_n, x_{n-1})^{-1/2} R_\delta[F_{\lambda,\Delta'}]](\Delta; x, y).
\end{aligned}
$$

We will show that both terms of the right hand side of (5.4.32) are small.

We will show that $I[(D(\delta; x_n, x_{n-1})^{-1/2} - 1)F_{\lambda,\Delta}](\Delta; x, y)$ is small. It follows from Lemmas 4.2 and 5.4 that we can write

$$D(\delta; x_n, x_{n-1})^{-1/2} - 1 = \tau_n^2 f(x_n, x_{n-1}),$$

and for any $K \geq 0$ there exists a positive constant C_K

$$|\partial_{x_n}^{\alpha_n} \partial_{x_{n-1}}^{\alpha_{n-1}} f(x_n, x_{n-1})| \leq C_K$$

if $|\alpha_{n-1}| \leq K$ and $|\alpha_n| \leq K$. It follows from this and Assumption 4.7 that for any $K \geq 0$ there exists a constant C_K and

$$
\begin{aligned}
&\left| \left(\prod_{k=0}^{J+1} \partial_{x_j}^{\alpha_j} \right) (D(\delta; x_n, x_{n-1})^{-1/2} - 1) F_\lambda(\gamma_\Delta(x_{J+1}, x_J, \ldots, x_1, x_0)) \right| \qquad (5.4.33)\\
&\leq C_K \tau_n^2 A_K X_K^{J+2} (1 + |\lambda| + |x_{J+1}| + |x_J| + \cdots + |x_1| + |x_0|)^m.
\end{aligned}
$$

holds as far as $|\alpha_j| \leq K, j = 0, 1, \ldots, J+1$.

Let $j_0 = 0 < j_1 < \cdots < j_s < j_{s+1} = J+1$ be any subsequence of $\{0, 1, \ldots, J, J+1\}$. Then $\Delta^\flat : s = T_0 < T_{j_1} < \cdots < T_{j_s} < T_{j_{s+1}} = s'$ is a division of the interval $[s, s']$ coarser than Δ. Since Lemma 5.4, Corollary 2.6 and Assumption 4.7 hold, for any non-negative integer K there exists a positive constant C_K such that if $|\alpha_{j_k}| \leq K$

$$
\begin{aligned}
&\left| \left(\prod_{k=0}^{s+1} \partial_{x_{j_k}}^{\alpha_{j_k}} \right) \iota_{\Delta^\flat}^{\Delta} ((D(\delta; x_n, x_{n-1})^{-1/2} - 1) F_{\lambda,\Delta})(x_{j_{s+1}}, x_{j_s}, \ldots, x_{j_1}, x_0) \right|\\
&\leq C_K \tau_n^2 A_K X_K^{s+2} (1 + |\lambda| + |x_{j_{s+1}}| + \cdots + |x_0|)^m.
\end{aligned}
$$

This means that $(D(\delta; x_n, x_{n-1})^{-1/2} - 1) F_\lambda(\gamma_\Delta(x_{J+1}, x_J, \ldots, x_1, x_0))$ satisfies Assumption 5.3. Therefore, the stationary phase method on the space of large dimension, Theorem 5.4, leads us to the following conclusion: we can write

$$I_\Delta[(D(\delta; x_n, x_{n-1})^{-1/2} - 1)F_{\lambda,\Delta}](\nu, x, y) = \left(\frac{-i\nu}{2\pi(s'-s)} \right)^{1/2} e^{\nu S(s',s,x,y)} g_1(\nu, s', s, x, y), \qquad (5.4.34)$$

and for any non-negative integer K there exist a positive constant C_K and a non-negative integer $M(K)$ such that

$$\left|\partial_x^\alpha \partial_y^\beta g_1(\nu, s', s, x, y)\right| \leq C_K A_{M(K)} \tau_n^2 (1 + |\lambda| + |x| + |y|)^m \tag{5.4.35}$$

if α, β satisfy $|\alpha| \leq K$ and $|\beta| \leq K$.

Next we will show that $I[D(\delta; x_n, x_{n-1})^{-1/2} R_\delta[F_{\lambda,\Delta'}]](\Delta; x, y)$ is small. For that purpose we shall prove that we can apply the stationary phase method over a space of large dimension when we integrate with respect to $x_J, \ldots, x_{n+1}$ and $x_{n-2}, \ldots, x_1$. Let $j_0 = 0 < j_1 < \cdots < j_a < j_{a+1} = n-1$ be an arbitrary subsequence of $0, 1, \ldots, n-1$ and $k_0 = n < k_1 < \ldots, k_b < k_{b+1} = J+1$ be an arbitrary subsequence of $n, n+1, \ldots, J+1$. Then

$$\begin{aligned}\Delta^\sharp : T_0 = T_{j_0} < T_{j_1} < \cdots < T_{j_a} < T_{j_{a+1}} = T_{n-1} = T_{n,0} < T_{n,1} < T_{n,2} < \cdots < T_{n,p_n} \\ < T_{n,p_n+1} = T_n = T_{k_0} < T_{k_1} < \cdots < T_{k_b} < T_{k_{b+1}} = T_{J+1}\end{aligned}$$

is a division of $[s, s']$ coarser than Δ', and

$$\begin{aligned}\Delta^\flat : T_0 = T_{j_0} < T_{j_1} < \cdots < T_{j_a} < T_{j_{a+1}} = T_{n-1} \\ < T_n = T_{k_0} < T_{k_1} < \cdots < T_{k_b} < T_{k_{b+1}} = T_{J+1}\end{aligned}$$

is a division of $[s, s']$ coarser than Δ.

Moreover, for any α_j, for $j = j_1, j_2, \ldots, j_a$ and $j = k_1, \ldots, k_b$ let $P(\partial)$ be the differential operator

$$P(\partial) = \prod_{i=1}^{a} \partial_{x_{j_i}}^{\alpha_{j_i}} \prod_{i=1}^{b} \partial_{x_{k_i}}^{\alpha_{k_i}}.$$

Now we operate $P(\partial)\iota_{\Delta^\flat}^{\Delta}$ on both sides of (5.4.29). Then

$$\begin{aligned}P(\partial)\iota_{\Delta^\flat}^{\Delta} \prod_{k=1}^{p_n+1} \left(\frac{\nu}{2i\pi\sigma_k}\right)^{1/2} \int_{\mathbf{R}^{p_n}} e^{i\nu S_\delta(x_n, y_{p_n}, \ldots, y_1, x_{n-1})} \\ \times F_{\lambda,\Delta'}(x_{[J+1,n]}, y_{p_n}, \ldots, y_1, x_{[n-1,0]}) \prod_{k=1}^{p_n} dy_k. \\ = \left(\frac{\nu}{2\pi\tau_n}\right)^{1/2} e^{i\nu S_n(x_n, x_{n-1})} D(\delta; x_n, x_{n-1})^{-1/2} \\ \times \Big(P(\partial)\iota_{\Delta^\flat}^{\Delta} F_{\lambda,\Delta}(x_{J+1}, x_J, \ldots, x_n, x_{n-1}, \ldots, x_0) \\ + \nu^{-1} P(\partial)\iota_{\Delta^\flat}^{\Delta} R_\delta[F_{\lambda,\Delta'}](\nu, x_{J+1}, x_J, \ldots, x_1, x_0)\Big).\end{aligned} \tag{5.4.36}$$

Since x_j with $j \neq n-1, n$ are independent of $(y_{p_n+1}, y_{p_n}, \ldots, y_1, y_0)$ and x_n, x_{n-1}, we have

$$\iota^{\Delta}_{\Delta^\flat}\int_{\mathbf{R}^{p_n}} e^{i\nu S_\delta(x_n,y_{p_n},\dots,y_1,x_{n-1})} F_{\lambda,\Delta'}(x_{[J+1,n]}, y_{p_n},\dots,y_1,x_{[n-1,0]})\prod_{k=1}^{p_n} dy_k.$$

$$= \int_{\mathbf{R}^{p_n}} e^{i\nu S_\delta(x_n,y_{p_n},\dots,y_1,x_{n-1})}$$

$$\times\, \iota^{\Delta'}_{\Delta^\sharp} F_{\lambda,\Delta'}(x_{k_{b+1}}, x_{l_b},\dots,x_{k_1},x_n,y_{p_n},\dots,y_1,x_{n-1},x_{j_a},\dots,x_{j_0})\prod_{k=1}^{p_n} dy_k.$$

$$= \int_{\mathbf{R}^{p_n}} e^{i\nu S_\delta(x_n,y_{p_n},\dots,y_1,x_{n-1})}$$

$$\times\, F_{\lambda,\Delta^\sharp}(x_{k_{b+1}}, x_{l_b},\dots,x_{k_1},x_n,y_{p_n},\dots,y_1,x_{n-1},x_{j_a},\dots,x_{j_0})\prod_{k=1}^{p_n} dy_k. \quad (5.4.37)$$

Here we have used the equality

$$\iota^{\Delta'}_{\Delta^\sharp} F_{\lambda,\Delta'} = F_{\lambda,\Delta^\sharp}. \quad (5.4.38)$$

Similarly, we can operate $P(\partial)$ under the integral sign.

$$P(\partial)\iota^{\Delta}_{\Delta^\flat}\int_{\mathbf{R}^{p_n}} e^{i\nu S_\delta(x_n,y_{p_n},\dots,y_1,x_{n-1})} F_{\lambda,\Delta'}(x_{[J+1,n]}, y_{p_n},\dots,y_1,x_{[n-1,0]})\prod_{k=1}^{p_n} dy_k.$$

$$= \int_{\mathbf{R}^{p_n}} e^{i\nu S_\delta(x_n,y_{p_n},\dots,y_1,x_{n-1})}$$

$$\times\, P(\partial)F_{\lambda,\Delta^\sharp}(x_{k_{b+1}}, x_{l_b},\dots,x_{k_1},x_n,y_{p_n},\dots,y_1,x_{n-1},x_{j_a},\dots,x_{j_0})\prod_{k=1}^{p_n} dy_k. \quad (5.4.39)$$

It turns out from (5.4.36) and (5.4.39) that

$$\prod_{k=1}^{p_n+1}\left(\frac{\nu}{2i\pi\sigma_k}\right)^{1/2}\int_{\mathbf{R}^{p_n}} e^{i\nu S_\delta(x_n,y_{p_n},\dots,y_1,x_{n-1})} \quad (5.4.40)$$

$$\times\, P(\partial)(F_{\lambda,\Delta^\sharp})(x_{k_{b+1}},\dots,x_{k_1},x_n,y_{p_n},\dots,y_1,x_{n-1},x_{j_a},\dots,x_{j_0})\prod_{k=1}^{p_n} dy_k.$$

$$= \left(\frac{\nu}{2\pi i\tau_n}\right)^{1/2} e^{i\nu S_n(x_n,x_{n-1})} D(\delta; x_n, x_{n-1})^{-1/2}$$

$$\times \Big(P(\partial)F_{\lambda,\Delta^\flat}(x_{k_{b+1}},\dots,x_{k_1},x_n,x_{n-1},x_{j_a},\dots,x_{j_0})$$

$$+\,\nu^{-1}P(\partial)\big(\iota^{\Delta}_{\Delta^\flat}R_\delta[F_{\lambda,\Delta'}]\big)(\nu, x_{k_{b+1}},\dots,x_{k_1},x_n,x_{n-1},x_{j_a},\dots,x_{j_0})\Big).$$

On the other hand, the left-hand side of this is an oscillatory integral with amplitude $P(\partial)(F_{\lambda,\Delta^\sharp})$. Therefore, we wish to apply Theorem 5.8 to the left-hand side of equality (5.4.40). We shall show that $P(\partial)(F_{\lambda,\Delta^\sharp})$ satisfies the assumptions of Theorem 5.8.

Owing to Assumption 4.7, function $P(\partial)F_{\lambda,\Delta^\sharp}$ satisfies the following estimate:

$$\left|\prod_{i=0}^{p_{n+1}} \partial_{y_{j_i}}^{\beta_{j_i}} P(\partial)F_{\lambda,\Delta^\sharp}(x_{k_{b+1}}, x_{k_b}, \dots, x_{k_1}, y_{p_{n+1}}, y_{p_n}, \dots, y_1, y_0, x_{j_a}, \dots, x_{j_0})\right|$$
$$\le A_K X_K^{a+b+2+p_n}\left(1+|\lambda|+\sum_{i=0}^{a}|x_{j_i}|+\sum_{i=1}^{b+1}|x_{k_i}|+|y_{p_0}|+\cdots+|y_{p_{n+1}}|\right)^m \quad (5.4.41)$$

if $|\alpha_j| \le K$ for $j = j_i, (i = 1, \dots, a), j = k_i, i = 1, \dots, b$ and $|\beta_{p_i}| \le K$ for $i = 0, \dots, n+1$. Thus if we regard λ, $x_{j_i}, i = 0, \dots, a$ and $x_{k_i}, i = 0, \dots, b$ as parameters, then $P(\partial)F_{\lambda,\Delta^\sharp}(x_{k_{b+1}}, x_{k_b}, \dots, x_{k_1}, y_{p_{n+1}}, y_{p_n}, \dots, y_1, y_0, x_{j_a}, \dots, x_{j_0})$ satisfies Assumption 5.6.

Let $0 = l_0 < l_1 < \cdots < l_q < l_{q+1} = p_{n+1}$ be any subsequence of $0 < 1 < \cdots < p_n < p_{n+1}$. Then $\delta^\# : T_{n,p_0} = T_{n,0} = T_{n,l_0} < T_{n,l_1} < \cdots < T_{n,l_q} < T_{n,l_{q+1}} = T_{n,p_n+1}$ is a division of $[T_n, T_{n+1}]$ coarser than δ. The union of dividing points of $\delta^\#$ and of $\Delta^\#$ makes a new division $\Delta^{\#\#}$ of $[s, s']$ coarser than $\Delta^\#$. By Assumption 5.7, for any $K \ge 0$

$$\left|\left(\prod_{i=0}^{q_{n+1}} \partial_{y_{l_i}}^{\beta_{l_i}}\right)\iota_{\Delta^{\#\#}}^{\Delta^\#}\partial_{y_k}^{\beta_k} P(\partial)F_{\lambda,\Delta^\sharp}(x_{k_{b+1}}, x_{k_b}, \dots, x_{k_1}, y_{l_{q+1}}, y_{l_q}, \dots, y_{l_1}, y_0, x_{j_a}, \dots, x_{j_0})\right|$$
$$\le A_K X_K^{a+b+2+p_n}\rho([T_{n,k-1}, T_{n,k+1}])\left(1+|\lambda|+\sum_{i=0}^{a}|x_{j_i}|+\sum_{i=1}^{b+1}|x_{k_i}|+\sum_{i=0}^{q+1}|y_{l_i}|\right)^m. \quad (5.4.42)$$

Hence if we regard λ, $x_{j_i}, i = 0, \dots, a$ and $x_{k_i}, i = 0, \dots, b$ as parameters, then $P(\partial)F_{\lambda,\Delta^\sharp}(x_{k_{b+1}}, x_{k_b}, \dots, x_{k_1}, y_{p_{n+1}}, y_{p_n}, \dots, y_1, y_0, x_{j_a}, \dots, x_{j_0})$ satisfies Assumption 5.7 as well as Assumption 5.6.

Therefore, we can apply Theorem 5.8 to the oscillatory integral with amplitude function $P(\partial)F_{\lambda,\Delta^\sharp}$ on the left-hand side of (5.4.43) if we regard λ, $x_{j_i}, i = 0, \dots, a$ and $x_{k_i}, i = 0, \dots, b$ as parameters. We obtain

$$\prod_{k=1}^{p_n+1}\left(\frac{\nu}{2i\pi\sigma_k}\right)^{1/2}\int_{\mathbf{R}^{p_n}} e^{i\nu S_\delta(x_n, y_{p_n}, \dots, y_1, x_{n-1})} \quad (5.4.43)$$
$$\times P(\partial)(F_{\lambda,\Delta^\sharp})(x_{k_{b+1}}, \dots, x_{k_1}, x_n, y_{p_n}, \dots, y_1, x_{n-1}, x_{j_a}, \dots, x_{j_0})\prod_{k=1}^{p_n} dy_k$$
$$= \left(\frac{\nu}{2\pi i\tau_n}\right)^{1/2} e^{i\nu S_n(x_n, x_{n-1})} D(\delta; x_n, x_{n-1})^{-1/2}$$

$$\times\Big(\big(\iota_{\Delta^\flat}^{\Delta^\sharp}P(\partial)F_{\lambda,\Delta^\sharp}\big)(x_{k_{b+1}},x_b,\ldots,x_{k_1},x_n,x_{n-1},x_{j_a},\ldots,x_{j_0})$$
$$+\nu^{-1}R_\delta[P(\partial)F_{\lambda,\Delta^\sharp}](\nu,x_{k_{b+1}},x_b,\ldots,x_{k_1},x_n,x_{n-1},x_{j_a},\ldots,x_{j_0})\Big).$$

Notice that

$$\big(\iota_{\Delta^\flat}^{\Delta^\sharp}P(\partial)F_{\lambda,\Delta^\sharp}\big)=\big(P(\partial)F_{\lambda,\Delta^\flat}\big),$$

because $(y^*_{p_n},\ldots,y^*_1)$ depends only on (x_n,x_{n-1}). Therefore, comparing (5.4.43) with (5.4.40), we have equality

$$P(\partial)\big(\iota_{\Delta^\flat}^{\Delta}R_\delta[F_{\lambda,\Delta'}]\big)(\nu,x_{k_{b+1}},\ldots,x_{k_1},x_n,x_{n-1},x_{j_a},\ldots,x_{j_0})$$
$$=R_\delta[P(\partial)F_{\lambda,\Delta^\sharp}](\nu,x_{k_{b+1}},\ldots,x_{k_1},x_n,x_{n-1},x_{j_a},\ldots,x_{j_0}). \quad (5.4.44)$$

Since $R_\delta[P(\partial)F_{\lambda,\Delta^\sharp}](\nu,x_{k_b},\ldots,x_{k_1},x_n,x_{n-1},x_{j_a},\ldots,x_{j_0})$ is the remainder term of (5.4.43), it turns out from Theorem 5.8 that for any $K\geq 0$ there exist a constant C_K and a non-negative integer $M(K)$ such that

$$\big|\partial_{x_n}^{\alpha_n}\partial_{x_{n-1}}^{\alpha_{n-1}}R_\delta[P(\partial)F_{\lambda,\Delta^\sharp}](\nu,x_{k_b},\ldots,x_{k_1},x_n,x_{n-1},x_{j_a},\ldots,x_{j_0})\big| \quad (5.4.45)$$
$$\leq C_KA_{M(K)}X_{M(K)}^{a+b+2}\tau_n(\rho([T_{n-1},T_n])+\tau_n)$$
$$\times\Big(1+|\lambda|+|x_n|+|x_{n-1}|+\sum_{i=0}^{a}|x_{j_i}|+\sum_{i=1}^{b+1}|x_{k_i}|\Big)^m.$$

By (5.4.44), this means

$$\big|\partial_{x_{n+1}}^{\alpha_{n+1}}\partial_{x_n}^{\alpha_n}P(\partial)\big(\iota_{\Delta^\flat}^{\Delta}R_\delta[F_{\lambda,\Delta'}]\big)(\nu,x_{k_b},\ldots,x_{k_1},x_{n+1},x_n,x_{j_a},\ldots,x_{j_0})\big| \quad (5.4.46)$$
$$\leq C_KA_{M(K)}X_{M(K)}^{a+b+2}\tau_n(\rho([T_{n-1},T_n])+\tau_n)$$
$$\times\Big(1+|\lambda|+|x_n|+|x_{n-1}|+\sum_{i=0}^{a}|x_{j_i}|+\sum_{i=1}^{b+1}|x_{k_i}|\Big)^m.$$

It turns out from (5.4.46) that the function

$$(x_{J+1},x_J,\ldots,x_{n+1},x_n)\to R_\delta[F_{\lambda,\Delta'}](\nu,x_{J+1},x_J,\ldots,x_n,x_{n-1},\ldots,x_1,x_0)$$

satisfies Assumption 5.3 with parameter $\lambda,x_{n-1},x_{j_b},\ldots,x_{j_0}$. Similarly,

$$(x_n,x_{n-1},\ldots,x_1,x_0)\to R_\delta[F_{\lambda,\Delta'}](\nu,x_{J+1},x_J,\ldots,x_1,x_0)$$

also satisfies Assumption 5.3 with suitable parameters.

Consider oscillatory integral

$$\prod_{j=1,j\neq n}^{J+1}\left(\frac{\nu}{2\pi i\tau_j}\right)^{1/2}\int_{\mathbf{R}^{J-2}} e^{i\nu(\sum_{j=1,j\neq n}^{J+1} S_j(x_j,x_{j-1}))} \tag{5.4.47}$$
$$\times R_\delta[F_{\lambda,\Delta'}](\nu, x_{J+1}, x_J, \ldots, x_{n+1}, x_n, x_{n-1}, x_{n-2}, \ldots, x_1, x_0)\prod_{\substack{j=1,\\ j\neq n-1,n}}^{J} dx_j.$$

We can apply Corollary 5.1 to this integral twice in succession, first to integration with respect to $(x_{n-2}, \ldots, x_1)$ and second to integration with respect to $(x_J, \ldots, x_{n+1})$. As a consequence

$$\prod_{j=1,j\neq n}^{J+1}\left(\frac{\nu}{2\pi i\tau_j}\right)^{1/2}\int_{\mathbf{R}^{J-2}} e^{i\nu(\sum_{j=1,j\neq n}^{J+1} S_j(x_j,x_{j-1}))}$$
$$\times R_\delta[F_{\lambda,\Delta'}](\nu, x_{J+1}, x_J, \ldots, x_{n+1}, x_n, x_{n-1}, x_{n-2}, \ldots, x_1, x_0)\prod_{j=1,j\neq n-1,n}^{J} dx_j$$
$$= \left(\frac{\nu}{2\pi i(T_{J+1}-T_n)}\right)^{1/2}\left(\frac{\nu}{2\pi i(T_{n-1}-T_0)}\right)^{1/2} e^{i\nu(S(T_{J+1},T_n,x_{J+1},x_n)+S(T_{n-1},T_0,x_{n-1},x_0))}$$
$$\times D_{x_J^*\ldots x_{n+1}^*}\left(\sum_{j=n+1}^{J+1} S_j; x_{J+1}, x_n\right)^{-1/2} D_{x_{n-2}^*\ldots x_1^*}\left(\sum_{j=1}^{n-1} S_j; x_{n-1}, x_0\right)^{-1/2}$$
$$\times f(\nu, x_{J+1}, x_n, x_{n-1}, x_0). \tag{5.4.48}$$

Here $f(\nu, x_{J+1}, x_n, x_{n-1}, x_0)$ satisfies the following inequality: For any K there exist a positive constant C_K and a non-negative integer $M(K)$ such that

$$\left|\partial_{x_{J+1}}^{\alpha_{J+1}}\partial_{x_n}^{\alpha_n}\partial_{x_{n-1}}^{\alpha_{n-1}}\partial_{x_0}^{\alpha_0} f(\nu, x_{J+1}, x_n, x_{n-1}, x_0)\right| \tag{5.4.49}$$
$$\leq C_K A_{M(K)} X_{M(K)}^4 \tau_n(\rho([T_{n-1}, T_n]) + \tau_n)(1 + |\lambda| + |x_{J+1}| + |x_n| + |x_{n-1}| + |x_0|)^m$$

if $|\alpha_{J+1}| \leq K$, $|\alpha_n| \leq K$, $|\alpha_{n-1}| \leq K$, $|\alpha_0| \leq K$.

Recall

$$I[D(\delta; x_n, x_{n-1})^{-1/2} R_\delta[F_{\lambda,\Delta'}]](\Delta; x, y)$$
$$= \left(\frac{\nu}{2\pi i\tau_n}\right)^{1/2}\int_{\mathbf{R}^2} e^{i\nu S_n(x_n,x_{n-1})} D(\delta; x_n, x_{n-1})^{-1/2}$$
$$\times\left(\prod_{j=1,j\neq n}^{J+1}\left(\frac{\nu}{2\pi i\tau_j}\right)\int_{\mathbf{R}^{J-2}} e^{i\nu(\sum_{j=1,j\neq n}^{J+1} S_j(x_j,x_{j-1}))}\right.$$
$$\left.\times R_\delta[F_{\lambda,\Delta'}](\nu, x_{J+1}, x_J, \ldots, x_{n+1}, x_n, x_{n-1}, x_{n-2}, \ldots, x_1, x_0)\prod_{j=1,j\neq n-1,n}^{J} dx_j\right) dx_{n-1}dx_n.$$

Using (5.4.48), we have

$$
\begin{aligned}
& I[D(\delta; x_n, x_{n-1})^{-1/2} R_\delta[F_{\lambda,\Delta'}]](\Delta; x, y) \\
&= \left(\frac{\nu}{2\pi i \tau_n}\right)^{1/2} \int_{\mathbf{R}^2} e^{i\nu S_n(x_n, x_{n-1})} D(\delta; x_n, x_{n-1})^{-1/2} \\
&\times \left(\frac{\nu}{2\pi i (T_{J+1} - T_n)}\right)^{1/2} \left(\frac{\nu}{2\pi i (T_{n-1} - T_0)}\right)^{1/2} e^{i\nu(S(T_{J+1}, T_n, x_{J+1}, x_n) + S(T_{n-1}, T_0, x_{n-1}, x_0))} \\
&\times D_{x_J^* \ldots x_{n+1}^*}\Big(\sum_{j=n+1}^{J+1} S_; x_{J+1}, x_n\Big)^{-1/2} D_{x_{n-2}^* \ldots x_1^*}\Big(\sum_{j=1}^{n-1} S_j; x_{n-1}, x_0\Big)^{-1/2} \\
&\quad \times f(\nu, x_{J+1}, x_n, x_{n-1}, x_0)\, dx_n dx_{n-1}.
\end{aligned}
$$

Apply Corollary 3.5 to the right-hand side and use Lemma 5.4. Then we can write

$$
\begin{aligned}
& I[D(\delta; x_n, x_{n-1})^{-1/2} R_\delta[F_{\lambda,\Delta'}]](\Delta; x, y) \\
&= \left(\frac{\nu}{2\pi i (s' - s)}\right)^{1/2} e^{i\nu S(s', s, x, y)} g_2(\nu, s', s, x, y).
\end{aligned}
\tag{5.4.50}
$$

Here for any integer $K \geq 0$ there exist a constant C_K and an integer $M(K)$ such that

$$
\begin{aligned}
& |\partial_{x_{J+1}}^{\alpha_{J+1}} \partial_{x_0}^{\alpha_0} g_2(\nu, s', s, x, y)| \\
& \leq C_K A_{M(K)} X_{M(K)}^4 \tau_n (\rho([T_{n-1}, T_n]) + \tau_n)(1 + |\lambda| + |x_{J+1}| + |x_0|)^m.
\end{aligned}
\tag{5.4.51}
$$

Using (5.4.34) and (5.4.50), we can write

$$
k[F_{\lambda,\Delta'}](\Delta'; \nu, s', s, x, y) - k[F_{\lambda,\Delta}](\Delta; \nu, s', s, x, y)) = g_1(\nu, s', s, x, y) + g_2(\nu, s', s, x, y).
$$

It follows from (5.4.35) and (5.4.51) that for any integer $K \geq 0$ there exist a positive constant C_K and integer $M(K) \geq 0$ such that

$$
\begin{aligned}
& \left|\partial_x^\alpha \partial_y^\beta (k[F_{\lambda,\Delta'}](\Delta'; \nu, s', s, x, y) - k[F_{\lambda,\Delta}](\Delta; \nu, s', s, x, y))\right| \\
& \leq C_K A_{M(K)} \tau_n (\rho([T_{n-1}, T_n]) + \tau_n)(1 + |\lambda| + |x| + |y|)^m
\end{aligned}
$$

as far as $|\alpha| \leq K$ and $|\beta| \leq K$. This proves inequality (5.4.21).

Lemma 5.8 has been proved. □

We shall prove Lemma 5.7 in full generality.

Proof Let Δ^* be an arbitrary refinement of Δ. We shall construct a sequence of divisions $\Delta_1, \Delta_2, \ldots \Delta_J$ in the following way.

Starting from $\Delta_0 = \Delta$, let Δ_1 be the division of which the dividing points are the union of all dividing points of Δ and dividing points of Δ^* inside the subinterval

$[T_0, T_1]$. So the division Δ_1 is different from Δ only in the subinterval $[T_0, T_1]$. Thus we can apply Lemma 5.8 and obtain

$$\left|\partial_x^\alpha \partial_y^\beta (k[F_{\lambda,\Delta_1}](\Delta_1; \nu, s', s, x, y) - k[F_{\lambda,\Delta_0}](\Delta_0; \nu, s', s, x, y))\right|$$
$$\leq C_K A_{M(K)} X_{M(K)} \tau_1 (\rho([T_0, T_1]) + \tau_1)(1 + |\lambda| + |x_{J+1}| + |x_0|)^m.$$

Let Δ_2 be the division whose dividing points are union of dividing points of Δ_1 and dividing points of Δ^* inside subinterval $[T_1, T_2]$. Division Δ_2 is different from Δ_1 only in the interval $[T_1, T_2]$. We apply Lemma 5.8 again and obtain

$$\left|\partial_x^\alpha \partial_y^\beta \left(k[F_{\lambda,\Delta_2}](\Delta_2; \nu, s', s, x, y) - k[F_{\lambda,\Delta_1}](\Delta_1; \nu, s', s, x, y)\right)\right|$$
$$\leq C_K A_{M(K)} \tau_2 (\rho([T_1, T_2]) + \tau_2)(1 + |\lambda| + |x_{J+1}| + |x_0|)^m.$$

Starting from $\Delta_0 = \Delta$ and repeating this process, we find a sequence of divisions $\Delta_0 = \Delta$, Δ_1, Δ_2, ..., Δ_J, Δ_{J+1} such that Δ_n consists of all dividing points of Δ_{n-1} and the part of Δ^* inside $[T_{n-1}, T_n]$. By construction it is clear that $\Delta_{J+1} = \Delta^*$. Since Δ_n is a refinement of Δ_{n-1} and they are different only in $[T_{n-1}, T_n]$, it follows from the Lemma 5.8 that for any $n = 1, 2, \ldots, J + 1$

$$\left|\partial_x^\alpha \partial_y^\beta \left(k[F_{\lambda,\Delta_n}](\Delta_n; \nu, s', s, x, y) - k[F_{\lambda,\Delta_{n-1}}](\Delta_{n-1}; \nu, s', s, x, y)\right)\right|$$
$$\leq C_K A_{M(K)} \tau_n (\rho([T_{n-1}, T_n]) + \tau_n)(1 + |\lambda| + |x_{J+1}| + |x_0|)^m.$$

Therefore,

$$\begin{aligned}
&\left|\partial_x^\alpha \partial_y^\beta \left(k[F_{\lambda,\Delta^*}](\Delta^*; \nu, s', s, x, y) - k[F_{\lambda,\Delta}](\Delta; \nu, s', s, x, y)\right)\right| \\
&= \left|\sum_{n=1}^{J+1} \partial_x^\alpha \partial_y^\beta \left(k[F_{\lambda,\Delta_n}](\Delta_n; \nu, s', s, x, y) - k[F_{\lambda,\Delta_{n-1}}](\Delta_{n-1}; \nu, s', s, x, y)\right)\right| \\
&\leq \sum_{n=1}^{J+1} C_K A_{M(K)} \tau_n (\rho([T_{n-1}, T_n]) + \tau_n)(1 + |\lambda| + |x_{J+1}| + |x_0|)^m \\
&\leq C_K A_{M(K)} |\Delta| \left(\sum_{n=1}^{J+1} (\rho([T_{n-1}, T_n]) + \tau_n) \right) (1 + |\lambda| + |x_{J+1}| + |x_0|)^m \\
&\leq C_K A_{M(K)} |\Delta| (\rho([s, s']) + |s' - s|)(1 + |\lambda| + |x_{J+1}| + |x_0|)^m.
\end{aligned}$$

This proves (5.4.19). Lemma 5.7 has been proved. □

Since we have proved that $\left\{k[(F_{\lambda,\Delta}](\Delta; \nu, s', s, x, y)\right\}_\Delta$ satisfies Cauchy's condition for convergence as $|\Delta| \to 0$, we have completed the proof of Theorem 5.9. □

Corollary 5.2 *$k[F_\lambda](\nu, s', s, x, y)$ is of the form*

$$k[F](\nu, s', s, x, y) = D(s', s, x, y)^{-1/2} \left(F_\lambda(\gamma^*) + \nu^{-1} R[F_\lambda](\nu, s', s, x, y)\right), \tag{5.4.52}$$

and as far as $|\alpha| \le K$ and $|\beta| \le K$, for $(x, y) \in \mathbf{R}^2$

$$\begin{aligned} &|\partial_x^\alpha \partial_y^\beta R[F_\lambda](\nu, s', s, x, y)| &(5.4.53)\\ &\le C_K A_{M(K)} |s - s'|(|s - s'| + \rho([s, s']))(1 + |\lambda| + |x| + |y|)^m, \\ &|\partial_x^\alpha \partial_y^\beta \left(R[F_\lambda](\Delta; \nu, s', s, x, y) - R[F_\lambda](\nu, s', s, x, y)\right)| &(5.4.54)\\ &\le C_K A_{M(K)} |s - s'|(|s - s'| + \rho([s, s']))(1 + |\lambda| + |x| + |y|)^m. \end{aligned}$$

Proof Since $k[F_{\lambda,\Delta}](\Delta; \nu, s', s, x, y) - k[F_\lambda](\nu, s', s, x, y) = R[F_\lambda](\Delta; \nu, s', s, x, y) - R[F_\lambda](\nu, s', s, x, y)$, (5.4.54) follows from (5.4.17). Letting $|\Delta| \to 0$, (5.4.53) follows from (5.4.7). Corollary is proved. □

As a special case of $\lambda = 0$, Theorem 4.10, a fortiori, Theorem 4.6 have been proved.

In other words, the Feynman path integral converges and the semi-classical asymptotic formula is true under our assumptions.

Chapter 6
Feynman Path Integral and Schrödinger Equation

Abstract In this chapter we prove Feynman's conjecture that the integral transform defined by the Feynman path integral is in fact the fundamental solution of the Schrödinger equation. The main tool is the L^2-boundedness theorem proof of which is left to Chap. 8 in Part II. By the way we shall prove that the main term of the semi-classical asymptotic of the fundamental solution of the Schrödinger equation satisfies the transport equations. At the end we obtain the second term of the semi-classical asymptotic and prove that it satisfies the second transport equation. Our discussion of this is different from the usual method originated by Birkhoff (Bull Am Math Soc 39:681–700 (1933) [11]). Our method enables us to obtain the bound of the remainder term.

6.1 Convergence of Time Slicing Approximation of Feynman Path Integral as a Sequence of Operators

6.1.1 L^2 Boundedness Theorem

The proof of Theorem 4.11 is given in this section. In this section the symbol $\|A\|$ expresses the norm of an element of $L^2(\mathbf{R})$ or the norm of a linear operator on $L^2(\mathbf{R})$.

As a preparation we recall the L^2-boundedness theorem of oscillatory integral transformations, [10, 23]. Let us treat the oscillatory integral transformation

$$T(\nu)f(x) = \int_{\mathbf{R}} a(x, y)e^{i\nu\phi(x,y)} f(y)\,dy \tag{6.1.1}$$

under the following assumptions:

Assumption 6.1 $\phi(x, y)$ is a real-valued C^∞ function of $(x, y) \in \mathbf{R}^2$. There exists a positive constant γ such that for any $(x, y) \in \mathbf{R}^2$ the following estimate holds:

$$0 < \gamma \le |\partial_x \partial_y \phi(x, y)|. \tag{6.1.2}$$

D. Fujiwara, *Rigorous Time Slicing Approach to Feynman Path Integrals*, Mathematical Physics Studies, DOI 10.1007/978-4-431-56553-6_6

For any integer $m \geq 0$ there exists a positive constant κ_m such that for any (x, y)

$$|\partial_x^{\alpha+1} \partial_y^{\beta+1} \phi(x, y)| \leq \kappa_{m+2}$$

if $|\alpha| + |\beta| = m$.

Assumption 6.2 For any integer $m \geq 0$ there exists a positive constant A_m such that for any (x, y)

$$|\partial_x^\alpha \partial_y^\beta a(x, y)| \leq A_m$$

if $|\alpha| + |\beta| \leq m$.

Theorem 6.3 *Suppose $\phi(x, y)$ and $a(x, y)$ satisfy the above assumptions. Then there exists an integer $k > 0$ such that for any $f \in \mathscr{C}_0^\infty(\mathbf{R})$*

$$\|T(\nu) f\| \leq C \nu^{-1/2} A_k \|f\|, \tag{6.1.3}$$

with some constant $C > 0$ independent of $\nu \geq 1$ and of f. We can choose C so that it depends only on $\gamma, \kappa_0, \kappa_1, \ldots, \kappa_k$. By the continuity argument $T(\nu)$ is extended as a bounded linear operator from $L^2(\mathbf{R})$ to $L^2(\mathbf{R})$. If we write the extended operator as the same symbol $T(\nu)$, then inequality (6.1.3) is valid for any $f \in L^2(\mathbf{R})$, cf. [10, 23].

The proof of this theorem is found in Chap. 9 in Part II. We can see that $k = 10$ is enough.

Let $S(s', s, x, y)$ be the classical action function as in Theorem 2.3 in Sect. 2.3, and let κ_m be as in Theorem 2.3. Then we have

Lemma 6.1 *Let $a(\nu, s', s, x, y)$ be a function. Suppose that for any non-negative integer m there exists a positive constant A_m independent of s', s, x, y and of ν with $\nu \geq 1$ satisfying*

$$|\partial_x^\alpha \partial_y^\beta a(\nu, s', s, x, y)| \leq A_m$$

if $|\alpha| + |\beta| \leq m$. Define for $f \in C_0^\infty(\mathbf{R})$

$$U(\nu, s', s) f(x) = \left(\frac{-i\nu}{2\pi(s' - s)} \right)^{1/2} \int_{\mathbf{R}} a(\nu, s', s, x, y) \exp\{i\nu S(s', s, x, y)\} f(y) dy.$$

Suppose that $\delta_0 > 0$ is so small that $\delta_0^2 \kappa_2 < 1$ and $0 < |s' - s| \leq \delta_0$. Then there exists an integer $k > 0$ and a positive constant C independent of $\nu \geq 1$ and s, s' such that for any $f \in \mathscr{C}_0^\infty(\mathbf{R})$

$$\|U(\nu, s', s) f\| \leq C A_k \|f\|. \tag{6.1.4}$$

$U(\nu, s', s)$ is extended as a bounded linear operator from $L^2(\mathbf{R})$ to $L^2(\mathbf{R})$. If we write the extended operator as $U(\nu, t, s)$ again, then inequality (6.1.4) is valid for any $f \in L^2(\mathbf{R})$.

Proof We set $\mu = \nu/(s'-s)$ and $\psi(s',s,x,y) = (s'-s)S(s',s,x,y)$. Then $\nu S(s',s,x,y) = \mu\psi(s',s,x,y)$ and

$$U(\nu,s',s)f(x) = \left(\frac{-\mu i}{2\pi}\right)^{1/2} \int_{\mathbf{R}} a(\nu,s',s,x,y)e^{i\mu\psi(s',s,x,y)}f(y)dy. \tag{6.1.5}$$

$$\psi(s',s,x,y) = \frac{(x-y)^2}{2} + (s'-s)^2\phi(s',s,x,y). \tag{6.1.6}$$

Owing to Proposition 2.8 there exists a positive constant C such that

$$|\phi(s',s,x,y)| \le C(1+|x|^2+|y|^2). \tag{6.1.7}$$

$\phi(s',s,x,y)$ is a function of (x,y) of class $C^\infty(\mathbf{R}^2)$ if s and s' are fixed. For any integer $m \ge 2$ there exists a positive constant κ_m such that

$$|\partial_x^\alpha \partial_y^\beta \phi(s',s,x,y)| \le \kappa_m \tag{6.1.8}$$

if $2 \le |\alpha|+|\beta| \le m$. In particular, in the case $m=2$ we can so choose that $\kappa_2 \le \left(1-\frac{v_2\delta^2}{8}\right)^{-1}\frac{v_2}{2}$. Thus

$$|\partial_x\partial_y\psi(s',s,x,y)| > 1-(t-s)^2\kappa_2 > 1-\delta_0^2\kappa_2 > 0. \tag{6.1.9}$$

For any $m = |\alpha|+|\beta|$

$$|\partial_x^{1+\alpha}\partial_y^{1+\beta}\psi(s',s,x,y)| < (s'-s)^2\kappa_{m+2}.$$

Therefore, ψ satisfies Assumption 6.1 for the phase function. $a(\nu,s',s,x,y)$ satisfies Assumption 6.2 with A_m independent of s',s and of ν. Apply Theorem 6.3. Then there exist a positive integer k and a constant $C>0$ independent of μ,s',s such that

$$\|U(\nu,s',s)f\| \le CA_k\|f\|. \tag{6.1.10}$$

Lemma 6.1 is now proved. □

6.1.2 Convergence as a Sequence of Operators

Let $\delta_0 > 0$ be as in the previous subsection. Assume that $|s'-s| \le \delta_0$. Let

$$\Delta : s = T_0 < T_1 < \cdots < T_J < T_{J+1} = s' \tag{6.1.11}$$

be an arbitrary division of the interval $[s,s']$ as (5.4.1).

In Definition 4.3 the following integral transformations are defined for any $\varphi \in C_0^\infty(\mathbf{R})$:

$$I(\Delta; \nu, s', s)\varphi(x) = \int_{\mathbf{R}} I(\Delta; \nu, s', s, x, y)\varphi(y)dy,$$
$$K(\nu, s', s)\varphi(x) = \int_{\mathbf{R}} K(\nu, s', s, x, y)\varphi(y)dy.$$

Let us first prove Proposition 4.3 by showing:

Proposition 6.1 *Suppose that $|s - s'| \le \delta_0$. Then there exists a positive constant C independent of ν, s', s and Δ such that*

$$\|I(\Delta; \nu, s', s)\| \le C,$$
$$\|K(\nu, s', s)\| \le C.$$

Proof By Theorem 4.5, if $|s - s'| \le \delta_0$, then

$$I(\Delta; \nu, s', s, x, y) = \left(\frac{-i\nu}{2\pi(s' - s)}\right)^{1/2} \exp\{i\nu S(s', s, x, y)\}k(\Delta; \nu, s', s, x, y),$$

where

$$k(\Delta; \nu, s', s, x, y) = D(\Delta; s', s, x, y)^{-1/2}(1 + \nu^{-1}r(\Delta; \nu, s', s, x, y)). \quad (6.1.12)$$

Moreover, it follows from (4.2.7) in Theorem 4.5 that for any α, β there exists a positive constant $C_{\alpha,\beta}$ such that

$$|\partial_x^\alpha \partial_y^\beta k(\Delta; \nu, s', s, x, y)| \le C_{\alpha,\beta}. \quad (6.1.13)$$

This means that $k(\Delta; \nu, s', s, x, y)$ satisfies Assumption 6.2 for the amplitude function.

Thus we can apply Lemma 6.1 to the operator $I(\Delta; \nu, s', s)$ and obtain

$$\|I(\Delta; \nu, s', s)\| \le C$$

with some positive constant C independent of Δ, ν and of (s', s). This proves the first half of Proposition 6.1.

It follows from Theorem 4.6 that for any α, β

$$|\partial_x^\alpha \partial_y^\beta k(\nu, s', s, x, y)| \le C_{\alpha,\beta} \quad (6.1.14)$$

with some positive constant $C_{\alpha\beta}$ independent of ν, s, s'. Using Lemma 6.1 once more, we obtain

$$\|K(\nu, t, s)\| \leq C,$$

where C is independent of ν and s', s.

Proposition 6.1 has been proved. □

Proposition 4.3 has also been proved.

Next we shall discuss the case $|\Delta| \to 0$ and prove Theorem 4.11 by showing

Theorem 6.4 *There exists a positive constant C independent of* Δ, ν, (s, s') *such that*

$$\|I(\Delta; \nu, s', s) - K(\nu, s', s)\| \leq C|\Delta|(s' - s). \tag{6.1.15}$$

Proof By definition

$$\begin{aligned}&(I(\Delta; \nu, s', s) - K(\nu, s', s))f(x) \\ &= \left(\frac{-i\mu}{2\pi}\right)^{1/2} \int_{\mathbf{R}} (k(\Delta; \nu, s', s, x, y) - k(\nu, s', s, x, y))e^{\mu\psi(s',s,x,y)} f(y)dy.\end{aligned} \tag{6.1.16}$$

Theorem 4.6 says that for any non-negative integer m there exists a constant C_m independent of Δ, ν and of $(s', s)'$ such that

$$|\partial_x^\alpha \partial_y^\beta (k(\Delta; \nu, s', s, x, y) - k(\nu, s', s, x, y))| \leq C_{\alpha,\beta}|s' - s||\Delta|. \tag{6.1.17}$$

Applying Lemma 6.1 again, we obtain (6.1.15). Theorem 6.4 has been proved. □

Theorem 4.11 has been proved.

Remark 6.1 Theorem 4.11 was first proved in [24] by a different method. Another approach is given in [27].

6.2 Evolution Property of $K(\nu, t, s)$

We assume that $\delta_0^2\kappa_2 < 1$. Let

$$K(\nu, s', s, x, y) = \left(\frac{-\nu i}{2\pi(s' - s)}\right)^{1/2} \exp(i\nu S(s', s, x, y))k(\nu, t, s, x, y)$$

be as in Theorem 4.6.

Proposition 6.2 *Let $f(x)$ be a function uniformly bounded together with its derivatives of all order. Then the integral*

$$g(\nu, t, s)(x) = \int_{\mathbf{R}} K(\nu, t, s, x, y) f(y)\, dy \tag{6.2.1}$$

is well-defined as an oscillatory integral with parameters if $|t-s| \leq \delta_0$. And

$$\lim_{|t-s|\to 0} K(\nu, t, s) f(x) = f(x) \tag{6.2.2}$$

uniformly on every compact set of $x \in \mathbf{R}$.

Proof Suppose $\delta_0^2 \kappa_2 < 1$. We set $\mu = \nu(t-s)^{-1}$ and $\psi(t, s, x, y) = (t-s)S(t, s, x, y)$. Then

$$K(\nu, t, s) f(x) = \left(\frac{-i\mu}{2\pi}\right)^{1/2} \int_{\mathbf{R}} k(\nu, t, s, x, y) f(y) e^{i\mu\psi(t,s,x,y)} dy. \tag{6.2.3}$$

If $|t-s| \leq \delta_0$, then

$$|\partial_y^2 \psi(t, s, x, y)| = |1 + (t-s)^2 \partial_y^2 \phi(t, s, x, y)| \geq 1 - \kappa_2 \delta_0^2 > 0. \tag{6.2.4}$$

Suppose further $m \geq 2$. Then

$$\max_{2\leq|\beta|\leq m} \sup_{(x,y)} |\partial_y^\beta \psi(t, s, x, y)| \leq 1 + |t-s|^2 |\partial_y^\beta \phi(t, s, x, y)| \leq \kappa_m' < \infty \tag{6.2.5}$$

and $k(\nu, t, s, x, y)$ is given in Corollary 4.1. This means that the integral on the right hand side of (6.2.3) satisfies Assumption 3.1. Therefore, $g(\nu, t, s)(x)$ is well-defined and the stationary phase method, Corollary 3.9 of Theorem 3.7, gives

$$\begin{aligned} K(\nu, t, s) f(x) = {} & \left(1 + (t-s)^2 \partial_y^2 \phi(t, s, x, y^*)\right)^{-1/2} e^{i\mu\psi(t,s,x,y^*)} \\ & \times \left(k(\nu, t, s, x, y^*) f(y^*) + \mu^{-1} \tilde{r}(\mu, t, s, x)\right). \end{aligned} \tag{6.2.6}$$

Here y^* is the stationary point of the phase, i.e. it is the unique solution to

$$y^* - x + (t-s)^2 \partial_y \phi(t, s, x, y^*) = 0, \tag{6.2.7}$$

and $\tilde{r}(\mu, t, s, x)$ is the remainder. There exist a constant $C > 0$ and an integer $K(0) \geq 0$ such that

$$|\tilde{r}(\mu, t, s, x)| \leq C \max_{\alpha \leq K(0)} \sup_{y \in \mathbf{R}} |\partial_y^\alpha k(\nu, t, s, x, y) f(y)| \tag{6.2.8}$$

The constants C and $K(0)$ are independent of t, s, x, ν, because κ_m' and $(1-\kappa_2\delta_0^2)^{-1}$ are independent of t, s, x, ν.

By Corollary 4.1,

$$k(\nu, s', s, x, y) = D(s', s, x, y)^{-1/2}(1 + \nu^{-1}r(\nu, s', s, x, y)). \tag{6.2.9}$$

Moreover, for any α, β there exists a positive constant $C_{\alpha\beta}$ independent of ν, s', s such that for $(x, y) \in \mathbf{R}^2$

$$|\partial_x^\alpha \partial_y^\beta r(\nu, s', s, x, y)| \le C_{\alpha\beta}|s' - s|^2. \tag{6.2.10}$$

It follows from Theorem 5.2 and (6.2.10) that there exists a constant $C \ge 0$ such that

$$|\tilde{r}(\mu, t, s, x)| \le C \max_{\alpha \le K(0)} \sup_{y \in \mathbf{R}} |\partial_y^\alpha k(\nu, t, s, x, y) f(y)| \le C. \tag{6.2.11}$$

Thus

$$\mu^{-1}\tilde{r}(\mu, t, s, x) \to 0$$

uniformly if $t - s$ tends to 0.

Apply (2.3.13) of Theorem 2.3 to (6.2.7). Then

$$|y^* - x| \le |t - s|^2(|\partial_y \phi(t, s, x, x)| + \kappa_2|y^* - x|).$$

Since $|t - s| \le \delta_0$,

$$(1 - \kappa_2\delta_0^2)|y^* - x| \le |t - s|^2|\partial_y \phi(t, s, x, x)|.$$

Thus

$$|y^* - x| \le (1 - \kappa_2\delta_0^2)^{-1}|t - s|^2|\partial_y \phi(t, s, x, x)|. \tag{6.2.12}$$

If $|x| \le N$ for some $N > 0$, then it follows from (6.2.12) that y^* remains bounded and $|x - y^*|$ converges uniformly to 0. It follows from Theorem 2.3, (6.2.11), Theorem 5.2 and Corollary 4.1 that

$$\lim_{|t-s|\to 0} \left(1 + (t - s)^2\partial_y^2\phi(t, s, x, y^*)\right)^{-1/2} = 1,$$
$$\lim_{|t-s|\to 0} k(\nu, t, s, x, y^*) = 1,$$
$$\lim_{|t-s|\to 0} f(y^*) = f(x)$$

uniformly as far as x is bounded. On the other hand, we have from (6.2.12)

$$\left|\frac{(x - y^*)^2}{t - s}\right| \le (1 - \kappa_2\delta_0^2)^{-2}(t - s)^3|\partial_y \phi(t, s, x, x)|^2 \to 0, \quad (|t - s| \to 0).$$

Therefore, if ν is fixed and $|t-s| \to 0$

$$\mu\psi(t,s,x,y^*) = \nu S(t,s,x,y^*) = \nu\left(\frac{(x-y^*)^2}{t-s} + (t-s)\phi(t,s,x,y^*)\right) \to 0$$

uniformly if x is bounded. This implies that

$$\lim_{|t-s|\to 0} e^{i\mu\psi(t,s,x,y^*)} = 1$$

converges uniformly on every compact set.

It follows from all these that

$$\lim_{|t-s|\to 0} K(\nu,t,s)f(x) = f(x)$$

uniformly on every compact set with respect to x. Proposition 6.2 has been proved. □

We shall prove that operator $K(\nu,t,s)$ is strongly continuous in t. Recall that we have written $\|f\|_C = \sup_{x\in\mathbf{R}} |f(x)|$ for a continuous function $f(x)$.

Lemma 6.2 *Suppose $\delta_0^2\kappa_2 < 1$ and $0 < |t-s| \le \delta_0$. Let $p \ge 0$ be an integer and $a(\nu,t,s,x,y)$ be a function. For any integer $k \ge 0$ there exists a constant $A_k > 0$ with*

$$|\partial_y^\alpha a(\nu,t,s,x,y)| \le A_k(1+|x|+|y|)^p, \tag{6.2.13}$$

as far as $|\alpha| \le k$. Let $F(\nu,t,s)$ be an integral transformation defined for $f \in \mathbf{C}_0^\infty(\mathbf{R})$ by

$$F(\nu,t,s)f(x) = \left(\frac{\nu}{2\pi i(t-s)}\right)^{1/2} \int_{\mathbf{R}} a(\nu,t,s,x,y)e^{i\nu S(t,s,x,y)} f(y)\,dy. \tag{6.2.14}$$

Then there exists a constant $C > 0$ independent of (t,s) with $0 < t-s \le \delta_0$ such that for any $x \in \mathbf{R}$

$$|F(\nu,t,s)f(x)| \le C(1+|x|)^p A_{p+2} \max_{|\beta|\le p+2} \|\partial_y^\beta f\|_C. \tag{6.2.15}$$

Moreover, let $B = 2(1-\kappa_2\delta_0^2)^{-1}(1+2\kappa_1\delta_0^2)$ and choose a constant $R > 1$ so that $f(y) = 0$ for $|y| \ge R$. If $|x| > BR$, then for any integer n there exists a constant $C_n > 0$ such that

$$|F(\nu,t,s)f(x)| \le C_n A_n \left(\frac{t-s}{\nu}\right)^{n-1/2} (1-\kappa_2\delta_0^2)^{-n} R \max_{|\beta|\le n} \|\partial_y^\beta f\|_C\ |x|^{p-n}. \tag{6.2.16}$$

We can choose C_n independent of (t, s) as far as $0 < t - s \le \delta_0$.

Proof Let $\mu = \nu(t-s)^{-1}$ and $\psi(t, s, x, y) = (t-s)S(t, s, x, y)$ as before. Then

$$F(\nu, t, s)f(x) = \left(\frac{\mu}{2\pi i}\right)^{1/2} \int_{\mathbf{R}} a(\nu, t, s, x, y)e^{i\mu\psi(t,s,x,y)} f(y)\,dy, \tag{6.2.17}$$

where

$$\psi(t, s, x, y) = \frac{1}{2}|x-y|^2 + (t-s)^2\phi(t, s, x, y).$$

(6.2.15) follows from Theorem 3.3.

In the case $|x| \ge BR$, we prove (6.2.16). We define for any x, y

$$\zeta = \partial_y\psi(t, s, x, y) = (y-x) + (t-s)^2\partial_y\phi(t, s, x, y).$$

By the mean value theorem and Theorem 2.3, for any x, y

$$|\partial_y\phi(t, s, x, y)| \le |\partial_y\phi(t, s, 0, y)| + \kappa_2|x| \le \kappa_1(1+|y|) + \kappa_2|x| \le 2\kappa_1 R + \kappa_2|x|.$$

Let $B = 2(1-\kappa_2\delta_0^2)^{-1}(1+2\kappa_1\delta_0^2)$. If $|x| > BR$ and $f(y) \ne 0$, then it follows from this that

$$\begin{aligned}|\zeta| &\ge |x| - |y| - 2\delta^2\kappa_1 R - \kappa_2(t-s)^2|x| \ge (1-\kappa_2|t-s|^2)|x| - (1+2\kappa_1\delta_0^2)R \\ &\ge (1-\kappa_2|t-s|^2)|x| - 2^{-1}(1-\kappa_2\delta_0^2)BR \ge 2^{-1}(1-\kappa_2|t-s|^2)|x|.\end{aligned}$$

Let $L = (i\zeta)^{-1}\partial_y$. Then $Le^{i\mu\psi} = \mu e^{i\mu\psi}$. Thus

$$\begin{aligned}F(\nu, t, s)f(x) &= \left(\frac{\mu}{2\pi i}\right)^{1/2} \int_{|y|\le R} e^{i\mu\psi(t,s,x,y)} a(\nu, t, s, x, y)f(y)dy \\ &= \left(\frac{\mu}{2\pi i}\right)^{1/2} \int_{|y|\le R} \mu^{-1}L(e^{i\mu\psi(t,s,x,y)})a(\nu, t, s, x, y)f(y)dy \\ &= \left(\frac{\mu}{2\pi i}\right)^{1/2} \mu^{-1}\int_{|y|\le R} e^{i\mu\psi(t,s,x,y)}L^*(a(\nu, t, s, x, y)f(y))dy.\end{aligned}$$

Here $L^* = -i\zeta^{-1}\partial_y + i\zeta^{-2}\partial_y\zeta$ is the formal adjoint of L. Repeat this process n times. Then

$$\begin{aligned}&F(\nu, t, s)f(x) \\ &= \left(\frac{\mu}{2\pi i}\right)^{1/2} \mu^n \int_{|y|\le R} e^{i\mu\psi(t,s,x,y)}(L^*)^n(a(\nu, t, s, x, y)f(y))dy.\end{aligned} \tag{6.2.18}$$

Since $\partial_y\zeta = 1 + (t-s)^2\partial_y^2\phi$, for any $\alpha > 0$ there exists a constant $C_\alpha > 0$ such that

$$|\partial_y^\alpha \zeta| \le C_\alpha.$$

If $f(y) \neq 0$ and $|x| > BR$, then there exists some constant $C_n > 0$ such that

$$\begin{aligned}|(L^*)^n(a(\nu,t,s,x,y)f(y))| &\le C_n \max_{|\alpha|\le n} |\partial_y^\alpha(a(y)f(y))||\zeta|^{-n} \\ &\le C_n 2^n A_n 2^n \max_{|\alpha\le n} \|\partial_y^\alpha f\|_C (1-\kappa_2|t-s|^2)^{-n}\,|x|^{-n}(3|x|)^p.\end{aligned}$$

Here we used Leibniz's rule and the inequalities $B > 2$ and $R > 1$. It follows from this and (6.2.18) that if $|x| \ge BR$, then

$$|F(\nu,t,s)f(x)| \le C_n A_n \max_{|\alpha|\le n} \|\partial_y^\alpha f\|_C \left(\frac{\nu}{|t-s|}\right)^{-n+1/2} (1-\kappa_2|t-s|^2)^{-n}|x|^{p-n}2R.$$

Changing constant C_n, we have proved (6.2.16). □

Corollary 6.1 *Suppose $|t-s|$, $f(y)$, B and R are as in Lemma 6.2. If $|x| \le BR$, then*

$$|K(\nu,t,s)f(x)| \le A_{p+2} \max_{|\beta|\le p+2} \|\partial_y^\beta f\|_C.$$

If $|x| \ge BR$, then for any integer n there exists a constant $C_n > 0$ such that

$$\begin{aligned}&|K(\nu,t,s)f(x)| \\ &\le C_n\left(\frac{t-s}{\nu}\right)^{n-1/2}(1-\kappa_2|t-s|^2)^{-n}BR|x|^{-n}\max_{|\beta|\le n}\|\partial_y^\beta f\|_C.\end{aligned} \tag{6.2.19}$$

Proof Replace $a(\nu,t,s,x,y)$ with $k(\nu,t,s,x,y)$ in Lemma 6.2 with $p = 0$. □

Proposition 6.3 *Suppose $\delta_0^2\kappa_2 < 1$ and $0 < t-s \le \delta_0$. Let $p \ge 0$ and $a(\nu,t,s,x,y)$ be a continuous function of (t,s,x,y) with the following property: For any α there exists a constant $C_\alpha > 0$ such that*

$$|\partial_y^\alpha a(\nu,t,s,x,y)| \le C_\alpha(1+|x|+|y|)^p. \tag{6.2.20}$$

Define an integral transformation $F(\nu,t,s)$ by

$$F(\nu,t,s)f(x) = \left(\frac{\nu}{2\pi i(t-s)}\right)^{1/2} \int_{\mathbf{R}} a(\nu,t,s,x,y)e^{i\nu S(t,s,x,y)} f(y)\,dy, \tag{6.2.21}$$

for $f \in C_0^\infty(\mathbf{R})$ and $0 < t-s \le \delta_0$. Then

$$(1+|x|^2)F(\nu,t,s)f(x) \in L^2(\mathbf{R}).$$

Moreover, the correspondence $(t, s) \to (1 + |x|^2)F(\nu, t, s)f$ *is an* $L^2(\mathbf{R})$*-valued strongly continuous function.*

Proof Notice that if $f \in C_0^\infty(\mathbf{R})$, then $F(\nu, t, s)f(x)$ is a continuous function of (t, s, x). Let $(t, s) \in \mathbf{R}^2$ satisfy $0 < t - s \le \delta_0$ and $(t_1, s_1), (t_2, s_2), \ldots$ be any sequence such that $\lim_{n\to\infty}(t_n, s_n) = (t, s)$. Then $\lim_{n\to\infty} F(\nu, t_n, s_n)f(x) = F(\nu, t, s)f(x)$ for each $x \in \mathbf{R}$.

We may assume $\rho < t_n - s_n \le \delta_0$ for some $\rho > 0$. As $f \in C_0^\infty(\mathbf{R})$, let $R > 1$ be so that $f(y) = 0$ for $|y| \ge R$. Let $B = 2(1 - \kappa_2\delta_0^2)^{-1}(1 + 2\kappa_1\delta_0^2)$. If $\rho < t - s \le \delta_0$, then by Lemma 6.2 with $n = p + 3$, there exists a constant $C > 0$ independent of t, s and x such that

$$|(1 + |x|^2)F(\nu, t, s)f(x)| \le Cq(x), \tag{6.2.22}$$

where

$$q(x) = \begin{cases} A_{p+3}(1 + BR)^{p+3} \max_{|\alpha|\le p+3} \|\partial^\alpha f\|_C & \text{if } |x| \le BR \\ \left(\frac{\delta_0}{\nu}\right)^{p+3/2} (1 - \kappa_2\delta_0^2)^{-3-p} \frac{1+|x|^2}{|x|^3} R \max_{|\alpha|\le 3+p} \|\partial_y^\alpha f\|_C & \text{if } |x| \ge BR. \end{cases} \tag{6.2.23}$$

Since $q \in L^2(\mathbf{R})$, $(1 + |x|^2)F(\nu, t, s)f(x) \in L^2(\mathbf{R})$. Moreover, for any $x \in \mathbf{R}$ and $n = 1, 2, \ldots$

$$(1 + |x|^2)^2|F(\nu, t_n, s_n)f(x) - F(\nu, t, s)f(x)|^2 \le 4C^2q(x)^2.$$

Since $q \in L^2((R)$, it turns out from Lebesgue's bounded convergence theorem that

$$\lim_{n\to\infty} \|(1 + |x|^2)F(\nu, t_n, s_n)f - (1 + |x|^2)F(\nu, t, s)f\|_{L^2}^2 = 0. \tag{6.2.24}$$

This proves $L^2(\mathbf{R})$-strong continuity in the set $0 < t - s \le \delta_0$ of $(1+|x|^2)F(\nu, t, s)f$ for $f \in C_0^\infty(\mathbf{R})$. □

Proposition 6.4 *For any* $f \in C_0^\infty(\mathbf{R})$,

$$s - \lim_{|t-s|\to 0} K(\nu, t, s)f = f. \tag{6.2.25}$$

Proof Suppose $f(x) = 0$ outside $\{x; |x| \le R\}$. Let $B = 2(1 - \kappa_2\delta_0^2)^{-1}(1 + 2\kappa_1\delta_0^2)$ and

$$\chi(x) = \begin{cases} 1 & \text{if } |x| \le BR, \\ 0 & \text{otherwise.} \end{cases}$$

Then

$$K(\nu, t, s)f(x) = \chi(x)K(\nu, t, s)f(x) + (1 - \chi(x))K(\nu, t, s)f(x). \tag{6.2.26}$$

If $|x| \le BR$, then $\lim_{|t-s|\to 0} \chi(x)K(\nu, t, s)f(x) = f(x)$ uniformly owing to Proposition 6.2. Therefore,

$$s - \lim_{|t-s|\to 0} \chi(x)K(\nu, t, s)f(x) = f(x). \tag{6.2.27}$$

If $|x| > BR$, then we can apply (6.2.19) in Corollary 6.1 with $n = 1$. There exists a constant $C > 0$ such that

$$|K(\nu, t, s)f(x)| \le C(1 - \kappa_2\delta_0^2)^{-1} \left(\frac{|t-s|}{\nu}\right)^{1/2} \max_{|\alpha|\le 1} \|\partial_y^\alpha f\|_C |x|^{-1} BR.$$

Therefore,

$$\begin{aligned}
&\int_{\mathbf{R}} |(1-\chi(x))K(\nu, t, s)f(x)|^2 dx \\
&\le C\left(\frac{|t-s|}{\nu}\right)(1-\kappa_2\delta_0^2)^{-1} \max_{|\alpha|\le 1} \|\partial_y^\alpha f\|_C |x|^{-1} BR \int_{|x|>BR} |x|^{-2} dx \\
&\le C\left(\frac{|t-s|}{\nu}\right)(1-\kappa_2\delta_0^2)^{-1} \max_{|\alpha|\le 1} \|\partial_y^\alpha f\|_C.
\end{aligned}$$

Therefore,

$$s - \lim_{|t-s|\to 0} (1-\chi(x))K(\nu, t, s)f = 0.$$

Consequently, it follows from this, (6.2.26) and (6.2.27) that

$$s - \lim_{|t-s|\to 0} K(\nu, t, s)f = f \tag{6.2.28}$$

if $f \in C_0^\infty(\mathbf{R})$. □

The family of bounded operators $\{K(\nu, t, s)\}$ enjoys the following property:

Theorem 6.5 (Evolution property) *If $s < t' < t$, then*

$$K(\nu, t, t')K(\nu, t', s) = K(\nu, t, s). \tag{6.2.29}$$

For any $f \in L^2(\mathbf{R})$

$$s - \lim_{t\to s+0} K(\nu, t, s)f = f. \tag{6.2.30}$$

Define $K(\nu, s, s)f = f$ for any $f \in L^2(\mathbf{R})$. Then for any $f \in L^2(\mathbf{R})$, $K(\nu, t, s)f$ is a strongly continuous $L^2(\mathbf{R})$-valued function of (t, s) in $0 \le t - s \le \delta_0$.

Proof Since $s < t' < t$, $[s, t] = [s, t'] \cup [t', t]$. Let

$$\Delta_1 : s = t_0 < t_1 < \cdots < t_L = t'$$

and

$$\Delta_2 : t' = t_L < t_{L+1} < t_{L+2} < \cdots < t_{L+M} = t$$

be divisions of $[s, t']$ and $[t', t]$ respectively. Then $\Delta = \Delta_1 \cup \Delta_2$ makes a division of $[s, t]$. This implies that

$$I(\Delta; \nu, t, s) = I(\Delta_2; \nu, t, t')I(\Delta_1; \nu, t', s). \tag{6.2.31}$$

Let $|\Delta_1| \to 0$ and $|\Delta_2| \to 0$. Then $|\Delta| \to 0$. Thus

$$\lim_{|\Delta_1|\to 0} I(\Delta_2; \nu, t, t') = K(\nu, t, t'),$$
$$\lim_{|\Delta_2|\to 0} I(\Delta_1; \nu, t', s) = K(\nu, t', s),$$
$$\lim_{\substack{|\Delta_1|\to 0 \\ |\Delta_2|\to 0}} I(\Delta; \nu, t, s) = K(\nu, t, s)$$

in the uniform operator topology of bounded linear operators in $L^2(\mathbf{R})$. Equality (6.2.29) follows from these and (6.2.31).

If $g \in C_0^\infty(\mathbf{R})$, then by Propositions 6.3 and 6.4, $K(\nu, t, s)g$ is strongly continuous function of (t, s) in the set $0 \le t - s \le \delta_0$. We use the Ascoli–Arzelà principle to prove strong continuity of $K(\nu, t, s)f$ for arbitrary $f \in L^2(\mathbf{R})$. Let (t, s) be a point with $0 \le t - s \le \delta_0$. Let (t_n, s_n), $n = 1, 2, 3, \ldots$, be any sequence such that $0 \le t_n - s_n \le \delta_0$ and $(t_n, s_n) \to (t, s)$ as $n \to \infty$. For any $\epsilon > 0$ there exists a function $g \in C_0^\infty(\mathbf{R})$ which satisfies

$$\|f - g\| \le \epsilon.$$

Then by Proposition 4.3

$$\begin{aligned}
&\|K(\nu, t_n, s_n)f - K(\nu, t, s)f\| \\
&\le \|K(\nu, t_n, s_n)(f - g)\| + \|K(\nu, t_n, s_n)g - K(\nu, t, s)g\| + \|K(\nu, t, s)(g - f)\| \\
&\le C\epsilon + \|K(\nu, t, s)g - g\| + C\epsilon.
\end{aligned}$$

Let $n \to \infty$. Then it follows from (6.2.28) that

$$\limsup_{n\to 0} \|K(\nu, t_n, s_n)f - K(\nu, t, s)f\| \le 2C\epsilon.$$

This proves

$$\lim_{n\to\infty} \|K(\nu, t_n, s_n)f - K(\nu, t, s)f\| = 0,$$

because $\epsilon > 0$ is arbitrarily small. Strong continuity of $K(\nu, t, s)f$ has been proved. □

6.3 Transport Equation and Determinant of the Hessian

Although it is well-known that the classical action $S(t, s, x, y)$ is the solution of the Hamilton–Jacobi equation, we give a proof.

Proposition 6.5 *$S(t, s, x, y)$ satisfies the following Hamilton-Jacobi equations:*

$$\partial_t S(t, s, x, y) = -\frac{1}{2}(\partial_x S(t, s, x, y))^2 - V(t, x) \tag{6.3.1}$$

$$\partial_s S(t, s, x, y) = +\frac{1}{2}(\partial_y S(t, s, x, y))^2 + V(s, y) \tag{6.3.2}$$

Proof By definition

$$S(t, s, x, y) = \int_s^t L(\tau, \gamma(\tau), \dot{\gamma}(t))d\tau. \tag{6.3.3}$$

Differentiate both sides along the classical path $\gamma(\tau)$. Then

$$\partial_t S(t, s, x, y) + \dot{\gamma}(t)\partial_x S(t, s, x, y) = L(t, \gamma(t), \dot{\gamma(t)}). \tag{6.3.4}$$

Since $\dot{\gamma}(t) = \partial_x S(t, s, x, y)$, we obtain (6.3.1). Similarly, (6.3.2) is proved. □

The next theorem proves Theorem 4.12.

Theorem 6.6 *We define a function $p(t, s, x, y)$ by*

$$p(t, s, x, y) = \left(\frac{1}{2\pi(t-s)}\right)^{\frac{1}{2}} D(t, s, x, y)^{-1/2}. \tag{6.3.5}$$

Then $p(t, s, x, y)$ satisfies transport equations

$$\partial_t p(t, s, x, y) + \partial_x S(t, s, x, y)\partial_x p(t, s, x, y) + \frac{1}{2}\partial_x^2 S(t, s, x, y)p(t, s, x, y) = 0, \tag{6.3.6}$$

$$\partial_s p(t, s, x, y) - \partial_y S(t, s, x, y)\partial_x p(t, s, x, y) - \frac{1}{2}\partial_y^2 S(t, s, x, y)p(t, s, x, y) = 0. \tag{6.3.7}$$

Proof Let $\gamma(t)$ be the classical path such that $\gamma(t) = x$ and $\gamma(s) = y$. We use the symbol

$$D_t = \partial_t + \dot{\gamma}(t)\partial_x = \partial_t + \partial_x S(t, s, x, y)\partial_x \tag{6.3.8}$$

for the operation of differentiation along the curve $(t, \gamma(t))$ in the time-space $\mathbf{R}^2$. Using this we can rewrite (6.3.6) as

$$D_t p(t,s,x,y) = -\frac{1}{2}\partial_x^2 S(t,s,x,y)p(t,s,x,y). \tag{6.3.9}$$

This is equivalent to the equation

$$D_t q(t,s,x,y) = \partial_x^2 S(t,s,x,y)q(t,s,x,y), \tag{6.3.10}$$

for $q(t,s,x,y) = p(t,s,x,y)^{-2} = (t-s)D(t,s,x,y)$. In other words, we have only to prove

$$\frac{d}{dt}q(t,s,\gamma(t),y) = \partial_x^2 S(t,s,x,y)\Big|_{x=\gamma(t)} q(t,s,\gamma(t),y). \tag{6.3.11}$$

We prove (6.3.11). Suppose $\gamma(\tau)$ is defined in $[s,u]$ for $u > t$. We write $\gamma(u)$ as w, i.e., three points y, x, z are on the same classical path γ. We shall calculate

$$q(u,s,\gamma(u),y) - q(t,s,\gamma(t),y), \quad t < u.$$

Let

$$\Delta : s = T_0 < T_1 < \cdots < T_{J+1} = t$$

be a division of $[0,t]$. Set $u - t = \epsilon$. Let

$$\Delta_\epsilon : t = T_{J+1} < T_{J+2} < \cdots < T_{J+M+1} = u$$

be a division of $[t,u]$. Then $\Delta_u = \Delta \cup \Delta_\epsilon$ is a division of $[s,u]$. Let γ_{Δ_u} be a piecewise classical path associated with Δ_u such that $\gamma_{\Delta_u}(s) = x_0$ and $\gamma_{\Delta_u}(u) = w$. We set $\gamma_{\Delta_u}(T_j) = x_j$. We understand $x_0 = y$, $x_{J+M+1} = w$. $D(\Delta_u; u, s, w, y)$ is obtained from the determinant of Hessian of the action $\sum_{j=1}^{J+M+1} S_j(x_j, x_{j-1})$ with respect to $(x_{J+M}, \ldots, x_1)$ at the stationary point $(x^*_{J+M}, \ldots, x^*_1)$. It is clear that $x^*_j = \gamma(T_j)$ for $j = 1, \ldots, J+M$. Notice in particular that $x^*_{J+1} = x$. We classify variables into three groups, $z_1 = (x_J, \ldots, x_1)$, $z_2 = (x_{J+M}, \ldots, x_{J+2})$ and x_{J+1}. We calculate the Hessian with respect to (z_2, z_1) prior to x_{J+1}. Then we can apply Proposition 5.1 and obtain

$$\det \mathrm{Hess}_{x^*_{J+M},\ldots,x^*_1}\Big(\sum_{j=1}^{J+M+1} S_j\Big) = \det \mathrm{Hess}_{x^*_{J+1}}\Big(\sum_{j=1}^{J+M+1} S_j\Big)^\# \times \det \mathrm{Hess}_{z^*_2,z^*_1}\Big(\sum_{j=1}^{J+M+1} S_j\Big). \tag{6.3.12}$$

Here $(\sum_{j=1}^{J+M+1} S_j)^\#$ is the value of $\sum_{j=1}^{J+M+1} S_j$ evaluated at $x_j = x^*_j = \gamma(T_j)$ for $j \neq J+1$. As a consequence of Theorem 2.2,

$$\Big(\sum_{j=1}^{J+M+1} S_j\Big)^\# = S(u,t,x_{J+M+1},x_{J+1}) + S(t,s,x_{J+1},x_0).$$

Since $x^*_{J+1} = x$,

$$\det \mathrm{Hess}_{x^*_{J+1}} \Big(\sum_{j=1}^{J+M+1} S_j \Big)^{\#} = \partial_x^2 S(u, t, x_{J+M+1}, x) + \partial_x^2 S(t, s, x, x_0).$$

On the other hand, $\partial_{x_k} \partial_{x_l} \sum_{j=1}^{J+M+1} S_j = 0$ if $k \le J$ and $l \ge J+2$. Thus the second factor on the right-hand side of (6.3.12) decomposes into

$$\det \mathrm{Hess}_{z^*_2, z^*_1} \Big(\sum_{j=1}^{J+M+1} S_j \Big) = \det \mathrm{Hess}_{z^*_2} \Big(\sum_{j=J+2}^{J+M+1} S_j \Big) \times \det \mathrm{Hess}_{z^*_1} \Big(\sum_{j=1}^{J+1} S_j \Big).$$

By (5.1.8)

$$\det \mathrm{Hess}_{(x^*_{J+M+1}, \dots, x^*_1)} \Big(\sum_{j=1}^{J+M+1} S_j \Big) = \frac{u-s}{\tau_1 \dots \tau_{J+M+1}} D(\Delta_u; u, s, x_{J+M+1}, x_0),$$

$$\det \mathrm{Hess}_{z^*_2} \Big(\sum_{j=J+2}^{J+M+1} S_j \Big) = \frac{u-t}{\tau_{J+2} \dots \tau_{J+M+1}} D(\Delta_\epsilon; u, t, x_{J+M+1}, x_{J+1}),$$

$$\det \mathrm{Hess}_{z^*_1} \Big(\sum_{j=1}^{J+1} S_j \Big) = \frac{t-s}{\tau_1 \tau_2 \dots \tau_{J+1}} D(\Delta; t, s, x_{J+1}, x_0).$$

Substitute these in (6.3.12) and multiply $\tau_1 \dots \tau_{J+M+1}$ to both sides. Then

$$\begin{aligned}
&(u-s) D(\Delta_u; u, s, x_{J+M+1}, x_0) \\
&= \big(\partial^2_{x_{J+1}} S(u, t, x_{J+M+1}, x_{J+1}) + \partial^2_{x_{J+1}} S(t, s, x_{J+1}, x_0)\big)\Big|_{x_{J+1}=x} \\
&\quad \times (u-t)(t-s) D(\Delta_\epsilon; u, t, x_{J+M+1}, x_{J+1}) D(\Delta; t, s, x_{J+1}, x_0)\Big|_{x_{J+1}=x}.
\end{aligned}$$

If $|\Delta| \to 0$ and $|\Delta_\epsilon| \to 0$, then owing to Theorem 5.2

$$\begin{aligned}
D(\Delta_u; u, s, x_{J+M+1}, x_0) &\to D(u, s, w, y) \\
D(\Delta_\epsilon; u, t, x_{J+M+1}, x) &\to D(u, t, w, x) \\
D(\Delta; t, s, x, x_0) &\to D(t, s, x, y).
\end{aligned}$$

Thus

$$\begin{aligned}
&(u-s) D(u, s, w, y) \\
&= (\partial_x^2 S(u, t, w, x) + \partial_x^2 S(t, s, x, y))(u-t) D(u, t, w, x)(t-s) D(t, s, x, y).
\end{aligned}$$

If $|u - t|$ is sufficiently small, then Theorem 5.4 reads

$$D(u, t, \gamma(u), x) = 1 + O(|u - t|^2)$$

and Proposition 2.8 reads

$$\partial_x^2 S(u, t, \gamma(u), x) = (u - t)^{-1} + O(|u - t|).$$

$$\begin{aligned}&(u - t)D(u, t, \gamma(u), x)(\partial_x^2 S(u, t, \gamma(u), x) + \partial_x^2 S(t, s, x, y))\Big|_{x=\gamma(t)} \\ &= 1 + (u - t)\partial_x^2 S(t, s, x, y) + O(|u - t|^2).\end{aligned}$$

Therefore,

$$\begin{aligned}&(u - s)D(u, s, \gamma(u), y) \\ &= (1 + (u - t)\partial_x^2 S(t, s, x, y) + O(|u - t|^2))(t - s)D(t, s, x, y).\end{aligned}$$

In other words,

$$\begin{aligned}&q(u, s, \gamma(u), y) \\ &= q(t, s, \gamma(t), y)(1 + (u - t)\partial_x^2 S(t, s, x, y) + O(|u - t|^2).\end{aligned}$$

Since this means

$$\begin{aligned}&(u - t)^{-1}(q(u, s, \gamma(u), y) - q(t, s, \gamma(t), y)) \\ &= \left(\partial_x^2 S(t, s, x, y)\right) q(t, s, \gamma(t), y) + O(|u - t|),\end{aligned}$$

it is clear that

$$\frac{d}{dt} q(t, s, \gamma(t), y) = \partial_x^2 S(t, s, x, y)\Big|_{x=\gamma(t)} q(t, s, \gamma(t), y). \tag{6.3.13}$$

We have proved (6.3.11). Theorem 6.6 has been proved. □

Theorem 4.12 has been proved.

6.4 Fundamental Solution of the Schrödinger Equation

Theorem 4.13 states that the Feynman path integral is the fundamental solution of the Schrödinger equation. We prove it in this section.

We write $H(t)$ for the Hamiltonian operator

$$H(t) = H(t, x, -i\nu^{-1}\partial_x) = \frac{1}{2}(-i\nu^{-1}\partial_x)^2 + V(t, x). \tag{6.4.1}$$

We assume that $\delta_0^2\kappa_2 < 1$ and $0 < t - s \le \delta_0$ in this section. The following operator was defined in Definition 4.3:

$$K(\nu, t, s)f(x) = \int_{\mathbf{R}} K(\nu, t, s, x, y)f(y)dy. \tag{6.4.2}$$

By Corollary 4.1,

$$K(\nu, t, s, x, y) = E(\nu, t, s, x, y) + R(\nu, t, s, x, y),$$

where

$$E(\nu, t, s, x, y) = \left(\frac{-i\nu}{2\pi(t-s)}\right)^{1/2} e^{i\nu S(t,s,x,y)}, \tag{6.4.3}$$

$$R(\nu, t, s, x, y) = \left(\frac{-i\nu}{2\pi(t-s)}\right)^{1/2} e^{i\nu S(t,s,x,y)} \tag{6.4.4}$$

$$\times \Big(D(t, s, x, y)^{-1/2} - 1 + \nu^{-1}D(t, s, x, y)^{-1/2}r(\nu, t, s, x, y)\Big).$$

Using this, we divide the operator $K(\nu, t, s)$ into two parts:

$$K(\nu, t, s)f(x) = E(\nu, t, s)f(y) + R(\nu, t, s)f(x), \tag{6.4.5}$$

where

$$E(\nu, t, s)f(y) = \int_{\mathbf{R}} E(\nu, t, s, x, y)f(y)\,dy, \tag{6.4.6}$$

$$R(\nu, t, s)f(x) = \int_{\mathbf{R}} R(\nu, t, s, x, y)f(y)\,dy. \tag{6.4.7}$$

It is clear from Lemma 6.1 that $E(\nu, t, s)$ and $R(\nu, t, s)$ are extended to be bounded operators on $L^2(\mathbf{R})$.

Lemma 6.3 *The kernel function $E(\nu, t, s, x, y)$ has the following property:*

$$(i\nu^{-1}\partial_s + H(s))E(\nu, t, s, x, y) = -\frac{i(t-s)}{2\nu}\left(\frac{\nu}{2\pi i(t-s)}\right)^{1/2} e^{i\nu S(t,s,x,y)}\partial_y^2\phi(t, s, x, y).$$

Proof Using Propositon 6.5, we can easily prove the lemma. □

Lemma 6.4 *Suppose $0 < t - s \le \delta_0$ and $f \in C_0^\infty(\mathbf{R})$. Then for any x*

$$\partial_s E(\nu, t, s)f(x) = E(\nu, t, s)(i\nu H(s)f)(x) + \frac{(t-s)i}{2}F(\nu, t, s)f(x) \tag{6.4.8}$$

where

$$F(\nu,t,s)f(x) = \left(\frac{\nu}{2\pi i(t-s)}\right)^{1/2} \int_{\mathbf{R}} \partial_y^2 \phi(t,s,x,y) f(y), dy. \tag{6.4.9}$$

Proof Since $f \in C_0^\infty(\mathbf{R})$, we can differentiate (6.4.6) under the integral sign. Using Lemma 6.3, we obtain

$$\begin{aligned} i\nu^{-1}\partial_s E(\nu,t,s)f(x) &= \int_{\mathbf{R}} (i\nu^{-1}\partial_s E(\nu,t,s,x,y)) f(y)\,dy \\ &= -\int_{\mathbf{R}} \Big(\frac{1}{2}(i\nu^{-1}\partial_y)^2 + V(s,y)\Big) E(\nu,t,s,x,y) f(y)\,dy \\ &\quad + \nu^{-1}(t-s)\int_{\mathbf{R}} F(\nu,t,s,x,y) f(y)\,dy. \end{aligned}$$

Integrating by parts, we prove (6.4.8). Lemma 6.4 has been proved. □

Lemma 6.5 *If $f \in L^2(\mathbf{R})$, then $E(\nu,t,s)f$ and $F(\nu,t,s)f$ are $L^2(\mathbf{R})$-valued strongly continuous functions of (t,s) if $0 < t-s \le \delta_0$. Moreover, if $(1+|x|^2)f \in L^2(\mathbf{R})$ and $\partial_x^2 f \in L^2(\mathbf{R})$, then $E(\nu,t,s)H(s)f$ is a strongly continuous function of (t,s) in the same set.*

Proof Suppose that $f \in C_0^\infty$. Then by Proposition 6.3, $E(\nu,t,s)f(x)$ is a strong continuous function of (t,s) in the set $0 < t-s \le \delta_0$.

If more generally $f \in L^2(\mathbf{R})$, by Lemma 6.1, there exists a constant $A > 0$ such that for any t, s with $0 < t-s \le \delta_0$

$$\|E(\nu,t,s)f\| \le A\|f\|. \tag{6.4.10}$$

This means that the maps $E(\nu,t,s) : L^2(\mathbf{R}) \to L^2(\mathbf{R})$ are equicontinuous. Therefore, we can use the Ascoli–Arzelà principle: For any $\epsilon > 0$ there exists a function $g \in C_0^\infty$ such that $\|f-g\|_{L^2} \le \epsilon$.

$$\begin{aligned} &\|E(\nu,t_n,s_n)f(x) - E(\nu,t,s)f(x)\| \\ &\le \|E(\nu,t_n,s_n)(f-g)\| + \|E(\nu,t_n,s_n)g - E(\nu,t,s)g\| + \|E(\nu,t,s)(g-f)\| \\ &\le 2A\|f-g\| + \|E(\nu,t_n,s_n)g - E(\nu,t,s)g\|. \end{aligned}$$

Hence

$$\limsup_{n\to\infty} \|E(\nu,t_n,s_n)f - E(\nu,t,s)f\| \le 2A\epsilon.$$

This proves $s - \lim_{n\to\infty} \|E(\nu,t_n,s_n)f - E(\nu,t,s)f\| = 0$. Strong continuity of $E(\nu,t,s)f$ is proved.

Since $K(\nu,t,s)f$ for $f \in L^2(\mathbf{R})$ is strongly continuous by Theorem 6.5, $F(\nu,t,s)f$ is also strongly continuous.

Suppose that $(1+|x|^2)f \in L^2(\mathbf{R})$ and $\partial^2 f \in L^2(\mathbf{R})$. Then $H(s)f \in L^2(\mathbf{R})$. Let $(t_0, s_0) \in \mathbf{R}^2$ be so that $0 < t_0 - s_0 \le \delta_0$. Then

$$\begin{aligned}&\|E(\nu,t,s)H(s)f - E(\nu,t_0,s_0)H(s_0)f\|\\&\le \|E(\nu,t,s)(H(s)f - H(s_0))f\| + \|E(\nu,t,s)H(s_0)f - E(\nu,t_0,s_0)H(s_0)f\|.\end{aligned}$$

By (6.4.10)

$$\|E(\nu,t,s)(H(s)f - H(s_0))f\| \le A\|(H(s)f - H(s_0))f\| = A\|(V(s,x) - V(s_0,x))f\|.$$

By Assumption 2.1 on the potential $V(s,x)$ that for each x

$$\lim_{s\to s_0}(V(s,x) - V(s_0,x))f(x) = 0$$

and

$$|(V(s,x) - V(s_0,x))f(x)| \le v_0(1+|x|^2)|f(x)|.$$

Since $(1+|x|^2)|f(x)| \in L^2(\mathbf{R})$ by assumption on f, by Lebesgue's bounded convergence theorem

$$\lim_{s\to s_0}\|(V(s,x) - V(s_0,x))f\| = 0.$$

On the other hand, $H(s_0)f$ is a fixed element of $L^2(\mathbf{R})$. Therefore, strong continuity of $E(\nu,t,s)$ implies that

$$\lim_{(t,s)\to(t_0,s_0)}\|E(\nu,t,s)H(s_0)f - E(\nu,t_0,s_0)H(s_0)f\| = 0.$$

Consequently,

$$\lim_{(t,s)\to(t_0,s_0)}\|E(\nu,t,s)(H(s)f - H(s_0))f\| = 0.$$

Strong continuity of $E(\nu,t,s)H(s)f$ has been proved. □

Theorem 6.7 *Suppose that $f \in L^2(\mathbf{R})$ has the property:*

$$\|\partial_x^2 f\| + \|(1+|x|^2)f\| < \infty. \tag{6.4.11}$$

Then

$$s - \lim_{t-s\to+0} i\nu^{-1}(t-s)^{-1}(K(\nu,t,s)f - f) = H(s)f. \tag{6.4.12}$$

Proof It follows from (4.2.12) in Corollary 4.1 and Lemma 6.1 that for any α, β there exists a constant $C_{\alpha\beta} > 0$ such that

$$\left|\partial_x^\alpha \partial_y^\beta \left(D(t,s,x,y)^{-1/2} - 1 + \nu^{-1} D(t,s,x,y)^{-1/2} r(\nu,t,s,x,y)\right)\right| \le C_{\alpha\beta}(t-s)^2. \tag{6.4.13}$$

Thus by Lemma 6.1 we have

$$\|R(\nu,t,s)f\| \le C(t-s)^2 \|f\|$$

with some constant $C > 0$. This means that

$$s - \lim_{|t-s|\to 0} \|R(\nu,t,s)f\| = 0 \tag{6.4.14}$$

and

$$s - \lim_{|t-s|\to 0} (t-s)^{-1} i\nu^{-1} \|R(\nu,t,s)f\| = 0. \tag{6.4.15}$$

It follows from (6.4.14) and (6.2.30) in Theorem 6.5 that

$$s - \lim_{|t-s|\to 0} E(\nu,t,s)f = f. \tag{6.4.16}$$

Suppose $t > s$ and $f \in C_0^\infty(\mathbf{R})$. Then applying Lemma 6.3, we have

$$\begin{aligned} & i\nu^{-1}\partial_s E(\nu,t,s)f(x) \\ &= \int_{\mathbf{R}} (i\nu^{-1}\partial_s E(\nu,t,s,x,y)) f(y)\,dy \\ &= -\int_{\mathbf{R}} \big(H(s,y,\partial_y)E(\nu,t,s,x,y)\big) f(y)\,dy + \nu^{-1}(t-s)\int_{\mathbf{R}} F(\nu,t,s,x,y) f(y)\,dy, \end{aligned} \tag{6.4.17}$$

where

$$F(\nu,t,s,x,y) = \frac{1}{2}\left(\frac{-i\nu}{2\pi(t-s)}\right)^{1/2} e^{i\nu S(t,s,x,y)} \partial_x^2 \phi(t,s,x,y).$$

We set

$$F(\nu,t,s)f(x) = \int_{\mathbf{R}} F(\nu,t,s,x,y) f(y)\,dy.$$

We can apply Lemma 6.1 and obtain, with some constant $C > 0$,

$$\|F(\nu,t,s)f\| \le C\|f\|.$$

Since $f \in C_0^\infty(\mathbf{R})$, integration by parts gives

$$\int_{\mathbf{R}} \big(-H(s,y,\partial_y)E(\nu,t,s,x,y)\big)f(y)\,dy = -E(\nu,t,s)(H(s)f)(x).$$

Thus

$$i\nu^{-1}\partial_s E(\nu,t,s)f(x) = -E(\nu,t,s)(H(s)f)(x) + \nu^{-1}(t-s)F(\nu,t,s)f(x).$$

For any such t' as $s < t' < t$

$$\begin{aligned} & i\nu^{-1}\Big(E(\nu,t,t')f(x) - E(\nu,t,s)f(x)\Big) \qquad (6.4.18) \\ & = i\nu^{-1}\int_s^{t'} \partial_\sigma E(\nu,t,\sigma)f(x)\,d\sigma \\ & = -\int_s^{t'} (E(\nu,t,\sigma)H(\sigma)f(x) + \nu^{-1}(t-\sigma)F(\nu,t,\sigma)f(x))\,d\sigma. \end{aligned}$$

Since by Lemma 6.5, $E(\nu,t,\sigma)H(\sigma)f$ and $F(\nu,t,\sigma)f$ are L^2-valued strongly continuous functions of (t,σ), we can regard the integrals as Riemannian integrals with values in $L^2(\mathbf{R})$. Thus we can regard the above equality as an equality in $L^2(\mathbf{R})$.

Let $t' \to t$. Then $E(\nu,t,t')f \to f$ because of (6.4.16). Thus

$$i\nu^{-1}(f - E(\nu,t,s)f) = -\int_s^t (E(\nu,t,\sigma)H(\sigma)f + \nu^{-1}(t-\sigma)F(\nu,t,\sigma)f)\,d\sigma. \tag{6.4.19}$$

We proved (6.4.19) for $f \in C_0^\infty(\mathbf{R})$. Now we suppose $f \in L^2(\mathbf{R})$ with the property (6.4.11). Then we can find a sequence of functions $f_n(x) \in C_0^\infty(\mathbf{R})$ such that

$$\|\partial_x^2(f - f_n)\| + \|(1+|x|^2)(f - f_n)\| \to 0, \quad \text{as}\, n \to \infty.$$

We can replace f in (6.4.19) with f_n for each n, and for each n

$$\begin{aligned} & \|H(\sigma)f - H(\sigma)f_n\| \le C\|\partial_x^2(f - f_n)\| + \|(1+|x|^2)(f - f_n)\|, \\ & \|F(\nu,t,\sigma)f - F(\nu,t,\sigma)f_n)\| \le C\|f - f_n\| \end{aligned}$$

with some constant $C > 0$ independent of n. Therefore, letting $n \to \infty$, (6.4.19) regarded as an equality in $L^2(\mathbf{R})$ holds for f satisfying the assumption in Theorem 6.7.

Suppose f satisfies the assumption in Theorem 6.7. Since there exists a constant $C > 0$ such that

$$\left\| \int_s^t (t-\sigma)F(\nu,t,\sigma)f\,d\sigma \right\| \le \nu^{-1}C(t-s)^2\|f\| \tag{6.4.20}$$

we have

$$s-\lim_{t\to s+0}(t-s)^{-1}i\nu^{-1}\Big(E(\nu,t,s)f-f\Big)=s-\lim_{t\to s}(t-s)^{-1}\int_s^t E(\nu,t,\sigma)H(\sigma)f\,ds$$
$$=H(s)f.$$

Equation (6.4.12) follows from this and (6.4.15). This proves the theorem. □

We shall make preparations to prove that $K(\nu,t,s)f(x)$ is the solution of the Schrödinger equation.

Lemma 6.6 *Let $m\geq 0$ be a constant and $a(\nu,t,s.x,y)$ be a function such that for any α,β there exists a constant $C_{\alpha,\beta}>0$ with*

$$|\partial_x^\alpha\partial_y^\beta a(\nu,t,s.x,y)|\leq C_{\alpha,\beta}(1+|x|+|y|)^m. \tag{6.4.21}$$

Define an integral transform

$$F(\nu,t,s)f(x)=\left(\frac{\nu}{2\pi i(t-s)}\right)^{1/2}\int_{\mathbf{R}}a(\nu,t,s.x,y)e^{i\nu S(t,s,x,y)}f(y)\,dy \tag{6.4.22}$$

for $f\in C_0^\infty(\mathbf{R})$. Then

$$\begin{aligned}&\partial_xF(\nu,t,s)f(x)\\&=F(\nu,t,s)(\partial_xf)(x)+\left(\frac{\nu}{2\pi i(t-s)}\right)^{1/2}\int_{\mathbf{R}}\partial_xa(\nu,t,s.x,y)e^{i\nu S(t,s,x,y)}f(y)\,dy\\&+i\nu(t-s)B(\nu,t,s)f(x),\end{aligned} \tag{6.4.23}$$

where

$$B(\nu,t,s)f(x)=\left(\frac{\nu}{2\pi i(t-s)}\right)^{1/2}\int_{\mathbf{R}}b(t,s.x,y)a(t,s,x,y)e^{i\nu S(t,s,x,y)}f(y)\,dy,$$

with

$$b(t,s.x,y)=(\partial_x\phi(t,s,x,y)+\partial_y\phi(t,s,x,y)).$$

Proof

$$\begin{aligned}&\partial_xF(\nu,t,s)f(x)\\&=\left(\frac{\nu}{2\pi i(t-s)}\right)^{1/2}\int_{\mathbf{R}}\partial_xa(\nu,t,s.x,y)e^{i\nu S(t,s,x,y)}f(y)\,dy\\&+\left(\frac{\nu}{2\pi i(t-s)}\right)^{1/2}\int_{\mathbf{R}}(i\nu)\partial_xS(t,s,x,y)a(\nu,t,s.x,y)e^{i\nu S(t,s,x,y)}f(y)\,dy.\end{aligned}$$

Replace $\partial_x S(t,s,x,y)$ with the right-hand side of

$$\partial_x S(t,s,x,y) = -\partial_y S(t,s,x,y) + (t-s)(\partial_x \phi(t,s,x,y) + \partial_y \phi(t,s,x,y)),$$

Then

$$\begin{aligned}
&\partial_x F(\nu,t,s)f(x)\\
&= \left(\frac{\nu}{2\pi i(t-s)}\right)^{1/2} \int_{\mathbf{R}} \partial_x a(\nu,t,s.x,y) e^{i\nu S(t,s,x,y)} f(y)\, dy\\
&+ \left(\frac{\nu}{2\pi i(t-s)}\right)^{1/2} \int_{\mathbf{R}} a(\nu,t,s.x,y)(-\partial_y e^{i\nu S(t,s,x,y)}) f(y)\, dy\\
&+ i\nu(t-s)\left(\frac{\nu}{2\pi i(t-s)}\right)^{1/2} \int_{\mathbf{R}} a(\nu,t,s.x,y)(\partial_x \phi(t,s,x,y) + \partial_y \phi(t,s,x,y)) e^{i\nu S(t,s,x,y)} f(y)\, dy.
\end{aligned}$$

Integration by parts proves (6.4.23).

Lemma 6.7 *Suppose* $|t-s| \le \delta_0$ *and* $f(x) \in C_0^\infty(\mathbf{R})$. *Then* $(1+|x|^2)K(\nu,t,s)f \in L^2(\mathbf{R})$ *and* $\partial_x^2 K(\nu,t,s)f(x) \in L^2(\mathbf{R})$. $(t,s) \to \partial_x^2 K(\nu,t,s)f$ *is a strongly continuous* $L^2(\mathbf{R})$*-valued function in the set* $0 < t-s \le \delta_0$. *The correspondence* $(t,s) \to H(t)K(\nu,t,s)f$ *is an* $L^2(\mathbf{R})$*-valued strongly continuous function if* $0 < t-s \le \delta_0$.

Proof The fact $(1+|x|^2)K(\nu,t,s)f \in L^2(\mathbf{R})$ follows from Lemma 6.2.

By the previous lemma,

$$\partial_x K(\nu,t,s)f(x) = K(\nu,t,s)\partial_x f(x) + G(\nu,t,s)f(x),$$

where

$$G(\nu,t,s)f(x) = \left(\frac{-i\nu}{2\pi(t-s)}\right)^{1/2} \int_{\mathbf{R}} g(\nu,t,s,x,y) e^{i\nu S(t,s,x,y)} f(y)\, dy,$$

with

$$g(\nu,t,s,x,y) = \partial_x k(\nu,t,s,x,y) + (i\nu(t-s))b(t,s,x,y)k(\nu,t,s,x,y).$$

Differentiating once again,

$$\begin{aligned}
&\partial_x^2 K(\nu,t,s)f(x) \qquad (6.4.24)\\
&= K(\nu,t,s)(\partial_x^2 f)(x) + G(\nu,t,s)(\partial_x f)(x) + \partial_x G(\nu,t,s)f(x)\\
&= K(\nu,t,s)(\partial_x^2 f)(x) + 2G(\nu,t,s)(\partial_x f)(x) + P(\nu,t,s)f(x),
\end{aligned}$$

where

$$P(\nu, t, s) f(x) = + \left(\frac{-i\nu}{2\pi(t-s)}\right)^{1/2} \int_{\mathbf{R}} p(\nu, t, s, x, y) e^{i\nu S(t,s,x,y)} f(y)\, dy \tag{6.4.25}$$

with

$$p(\nu, t, s, x, y) = \partial_x g(\nu, t, s, x, y) + i\nu(t-s) b(t, s, , x, y) g(\nu, t, s, x, y). \tag{6.4.26}$$

Since for any α, β there exists a positive constant C such that

$$|\partial_x^\alpha \partial_y^\beta b(t, s, x, y)| \le C(1 + |x| + |y|)^{\max\{0, 1-|\alpha|-|\beta|\}}, \tag{6.4.27}$$

for any α, β there exists $C > 0$ such that

$$|\partial_x^\alpha \partial_y^\beta g(\nu, t, s, x, y)| \le C(1 + |x| + |y|)$$
$$|\partial_x^\alpha \partial_y^\beta p(\nu, t, s, x, y)| \le C(1 + |x| + |y|)^2.$$

Suppose $f(y) = 0$ if $|y| \ge R > 1$. Then by Lemma 6.2 $K(\nu, t, s)(\partial_x^2 f)(x) \in L^2(\mathbf{R})$, $G(\nu, t, s)(\partial_x f)(x) \in L^2(\mathbf{R})$ and $P(\nu, t, s) f(x) \in L^2(\mathbf{R})$. Therefore, $\partial_x^2 K(\nu, t, s) f(x) \in L^2(\mathbf{R})$. Moreover, $(t, s) \to \partial_x^2 K(\nu, t, s) f$ is a strongly continuous $L^2(\mathbf{R})$-valued function in the set $0 < t - s \le \delta_0$.

Next we shall prove strong continuity of $V(t)K(\nu, t, s) f$. Notice that at any fixed x, $t \to V(t)K(\nu, t, s) f(x)$ is continuous. On the other hand, it follows from Assumption 2.1 and Corollary 6.1 with $n = 3$ that

$$|V(t)K(\nu, t, s) f(x)| \le v_0(1 + |x|^2)|K(\nu, t, s) f(x)| \le C(1 + |x|^2) q(x),$$

where

$$q(x) = \begin{cases} (1 + |x|^2) \max_{|\beta| \le 2} \|\partial_y^\beta f\|_C, & \text{if } |x| \le BR, \\ (1 + |x|^2) \left(\frac{\delta_0}{\nu}\right)^{3-1/2} (1 - \kappa_2 \delta_0^2)^{-3} BR|x|^{-3} \max_{|\beta| \le 3} \|\partial_y^\beta f\|_C & \text{if } |x| \ge BR. \end{cases} \tag{6.4.28}$$

It is clear that $q(x) \in L^2(\mathbf{R})$. Therefore by Lebesgue's bounded convergence theorem the correspondence $t \to V(t)K(\nu, t, s) f$ is strongly continuous. It turns out from this and strong continuity of $\partial_x^2 K(\nu, t, s) f$ that $H(t)K(\nu, t, s) f$ is strongly continuous. □

Now we can prove the main theorem, Theorem 4.13, that is:

Theorem 6.8 *Suppose* $0 < t-s \leq \delta_0$ *and* $f \in C_0^\infty(\mathbf{R})$. *Then* $K(\nu, t, s)f$ *is strongly differentiable in* $L^2(\mathbf{R})$ *with respect to* t *and*

$$i\nu^{-1}\frac{d}{dt}K(\nu, t, s)f = H(t)K(\nu, t, s)f, \tag{6.4.29}$$

i.e., $K(\nu, t, s)f(x)$ *is the solution of the Scrödinger equation and the initial condition*

$$i\nu^{-1}\partial_t K(\nu, t, s)f(x) = H(t)K(\nu, t, s)f(x), \tag{6.4.30}$$
$$s - \lim_{|t-s|\to 0} K(\nu, t, s)f = f. \tag{6.4.31}$$

Moreover, for any x, $\lim_{t\to s} K(\nu, t, s)f(x) = f(x)$.

Proof (6.4.31) was already proved in Theorem 6.5.

We shall prove (6.4.29). We set $K(\nu, t, s)f(x) = g(x)$.

$$\epsilon^{-1}(K(\nu, t+\epsilon, s)f(x) - K(\nu, t, s)f(x)) = \epsilon^{-1}(K(\nu, t+\epsilon, t)g(x) - g(x)). \tag{6.4.32}$$

Proposition 6.3 implies that $(1+|x|^2)g(x) \in L^2(\mathbf{R})$. As a consequence of Lemma 6.7, $\partial_x^2 g(x) \in L^2(\mathbf{R})$. We apply Theorem 6.7 to obtain a strong derivative from the right:

$$\begin{aligned} i\nu^{-1}\frac{d^+}{dt}K(\nu, t, s)f &= s - \lim_{\epsilon\to+0} i\nu^{-1}\epsilon^{-1}(K(\nu, t+\epsilon, s)f(x) - K(\nu, t, s)f(x)) \\ &= H(t)K(\nu, t, s)f(x). \end{aligned} \tag{6.4.33}$$

We now prove that $H(t)K(\nu, t, s)f$ is strongly continuous in t if $f \in C_0^\infty$ and $0 < t-s < \delta_0$. Let $t_1, t_2, \ldots$ such that $\lim_{n\to\infty} t_n = t$. Then $\partial_x^2 K(\nu, t_n, s)f(x) \to H(t)K(\nu, t, s)f(x)$ at each x. We will prove $s - \lim_{n\to\infty} \partial_x^2 K(\nu, t_n, s)f = \partial_x^2 K(\nu, t, s)f$.

It follows from Lemma 6.7 and Lebesgue's dominated convergence theorem that

$$\lim_{t'\to t} \|H(t')K(\nu, t', s)f - H(t)K(\nu, t, s)f\| = 0,$$

i.e., $H(t)K(\nu, t, s)f$ is strongly continuous with respect to t.

Theorem 6.8 follows from the next lemma (Dini's theorem). □

Lemma 6.8 (Dini's theorem) *Let* X *be a Banach space equipped with norm* $\|z\|$ *and let* $f : (a, b) \ni t \to f(t) \in X$ *be a strongly continuous function which is strongly differentiable from the right. Suppose that its derivative from the right* $\frac{d^+}{dt}f(t)$ *is strongly continuous. Then* $f(t)$ *is strongly differentiable in* (a, b) *and*

$$\frac{d}{dt}f(t) = \frac{d^+}{dt}f(t) \quad \textit{for } t \in (a, b).$$

Proof Let $c \in (a, b)$ be an arbitrary point. We set

$$g(t) = f(c) + \int_c^t \frac{d^+}{dt} f(s)\, ds. \tag{6.4.34}$$

and $h(t) = f(t) - g(t)$.

Since $g(t)$ is differentiable and $\frac{d}{dt}g(t) = \frac{d^+}{dt}f(s)$, $h(t)$ is strongly differentiable from the right and for any $t \in (a, b)$

$$\frac{d^+}{dt}h(t) = 0.$$

We shall prove that $h(t) = 0$ for any $t \in [c, b)$. Let us choose $\epsilon > 0$ arbitrarily small. Since $h(c) = 0$ and $\frac{d^+}{dt}h(c) = 0$, there exists some $\delta > 0$ such that $\|h(t)\| \le \epsilon(t - c)$ for any $t \in [c, c + \delta)$. Thus if we set

$$\gamma = \sup\{t \in [c, b); \|h(s)\| \le \epsilon(s - c) \quad \text{for any } s \in [c, t)\}, \tag{6.4.35}$$

then $\gamma > c$. We shall prove that $\gamma = b$. If $\gamma < b$, then $\|h(\gamma)\| \le \epsilon(s - c)$ by virtue of continuity of $h(t)$ and $(t - c)$. Since $\frac{d^+}{dt}h(\gamma) = 0$, there exists $\delta_1 > 0$ such that $\|h(s) - h(\gamma)\| \le \epsilon(s - \gamma)$ for any $s \in [\gamma, \gamma + \delta_1)$. It follows from this that $\|h(s)\| \le \|h(s) - h(\gamma)\| + \|h(\gamma)\| \le \epsilon(s - \gamma) + \epsilon(\gamma - c) \le \epsilon(s - c)$. This contradicts the definition (6.4.35) of γ. Thus $\gamma = b$ by contradiction.

Thus $\|h(t)\| \le \epsilon(t - c)$ for any $t \in [c, b)$. Since $\epsilon > 0$ is arbitrarily small, $\|h(t)\| = 0$ in $[c, b)$. We can also choose $c \in (a, b)$ arbitrarily close to a. So we have proved that $\|h(t)\| = 0$ in (a, b). As a consequence $f(t) = g(t)$ in (a, b). We have proved the lemma. □

We give the proof of Corollary 4.3.

Corollary 6.2 *Suppose that $f \in C_0^\infty(\mathbf{R})$. Then $K(\nu, t, s)f(x)$ is differentiable with respect to t and twice continuously differentiable with respect to x. It is the classical solution of the Schrödinger equation and initial condition.*

Proof If $f \in C_0^\infty(\mathbf{R})$, then it is clear that $K(\nu, t, s)f(x)$ is differentiable with respect to t and twice continuously differentiable with respect to x. Therefore the corollary follows from Theorem 6.8. □

6.5 Jacobi's Differential Operator and Morette–Van Vleck Determinant

We shall prove in this section Theorems 4.14 and 4.15.

6.5.1 Nuclear Operator and Infinite-Dimensional Determinant

We recall, without proof, well-known facts concerning the infinite dimensional-determinant. See [43, 59] for detailed discussions.

Let $\mathbf{X}$ and $\mathbf{Y}$ be separable Hilbert spaces.

Proposition 6.6 *Let $T : \mathbf{X} \to \mathbf{Y}$ be a complete continuous linear operator. Let $\{\mu_k\}$ be a non-increasing sequence of positive numbers such that $\{\mu_k^2\}$ is the system of all eigenvalues, counted repeatedly following multiplicity, of the operator T^*T. Then $\{\mu_k\}$ is called the singular values of T. Let x_k be the eigenvector of T^*T such that $T^*Tx_k = \mu_k^2 x_k$. For non-zero μ_k set $x'_k = \mu_k^{-1}Tx_k$. Then we have the canonical expression of T:*

$$Tx = \sum_{k=1}^{\infty} \mu_k(x, x_k)x'_k. \tag{6.5.1}$$

Notice that $\{x'_k\}$ is an ortho-normal system of $\mathbf{Y}$. We define $|T| : \mathbf{X} \to \mathbf{X}$ by $|T|x = \sum_{k=1}^{\infty} \mu_k(x, x_k)x_k$ and $|T^*| : \mathbf{Y} \to \mathbf{Y}$ by $|T^*|y = \sum_{k=1}^{\infty} \mu_k(y, x'_k)x'_k$.

Definition 6.1 Let $T : \mathbf{X} \to \mathbf{Y}$ be a complete continuous linear operator. Define

$$\|T\|_2 = \left(\sum_{k=1}^{\infty} \|Tx_k\|^2\right)^{1/2}, \tag{6.5.2}$$

where $\{x_k\}$ is a complete orthogonal family in $\mathbf{X}$. If the series (6.5.2) does not converge, we set $\|T\|_2 = \infty$. $\|T\|_2$ is called the Schmidt norm. If $\|T\|_2 < \infty$, T is called a Schmidt (or Hilbert–Schmidt) operator. The totality of Schmidt operators is called the Schmidt class. $\|T\|_2$ is independent of choices of complete orthogonal system $\{x_k\}$.

Proposition 6.7 *1. $\|T\| \le \|T\|_2$, where $\|T\|$ denotes the operator norm.*
2. The Schmidt class is a Banach space equipped with the norm $\|T\|_2$.
3. $\|T\|_2 = \|T^*\|_2$, where T^* is the adjoint of T.
4. Let $\mathbf{Z}$ be also a separable Hilbert space. If $T : \mathbf{X} \to \mathbf{Y}$ is a Schmidt operator and $B : \mathbf{Y} \to \mathbf{Z}$ is a bounded operator with norm $\|B\|$, then $BT : \mathbf{X} \to \mathbf{Z}$ is a Schmidt operator and $\|BT\|_2 \le \|T\|_2\|B\|$. If $B : \mathbf{X} \to \mathbf{Y}$ is a bounded operator with norm $\|B\|$ and $T : \mathbf{Y} \to \mathbf{Z}$ is a Schmidt operator, then $TB : \mathbf{X} \to \mathbf{Z}$ is a Schmidt operator and $\|TB\|_2 \le \|B\|\|T\|_2$.

5. *Let $T : \mathbf{X} \to \mathbf{Y}$ be a completely continuous operator and $\{\mu_k\}$ be its singular values. Then $\|T\|_2 = \left(\sum_{k=1}^{\infty} \mu_k^2\right)^{1/2}$.*

Definition 6.2 [43], [59]. Suppose $T : \mathbf{X} \to \mathbf{Y}$ is a completely continuous linear operator and $\{\mu_k : k = 1, 2, \ldots\}$ is the system of its singular values. Then the trace norm of T is defined by

$$\|T\|_1 = \sum_{k=1}^{\infty} \mu_k. \tag{6.5.3}$$

If $\|T\|_1 < \infty$, then T is called a nuclear operator or an operator of trace class.

Proposition 6.8 *1. The set of all nuclear operators forms a Banach space equipped with the norm $\|T\|_1$.*
2. *Let T be a nuclear operator of $\mathbf{X}$. Then its trace norm is defined by*

$$\|T\|_1 = \sup \sum_{n=1}^{\infty} |(Tx_n, y_n)|.$$

Here sup *is taken among all pairs of the complete orthogonal system $\{x_n\} \subset X$ and $\{y_n\} \subset \mathbf{Y}$.*

3.
$$\|T\|_2 \leq \|T\|_1.$$

4.
$$\|T\|_1 = \|T^*\|_1 = \||T|\|_1 = \||T|^{1/2}\|_2^2.$$

5. *If $T : \mathbf{X} \to \mathbf{Y}$ and $S : \mathbf{Y} \to \mathbf{Z}$ are Schmidt operators, then $ST : \mathbf{X} \to \mathbf{Z}$ is a nuclear operator and*
$$\|ST\|_1 \leq \|S\|_2 \|T\|_2.$$

6. *If $T : \mathbf{X} \to \mathbf{Y}$ is a nuclear operator and $S : \mathbf{Y} \to \mathbf{Z}$ is a bounded operator, then $ST : \mathbf{X} \to \mathbf{Y}$ is a nuclear operator and $\|ST\|_1 \leq \|S\|_1 \|T\|$. If $T : \mathbf{X} \to \mathbf{Y}$ is a bounded operator and $S : \mathbf{Y} \to \mathbf{Z}$ is a nuclear operator, then $ST : \mathbf{X} \to \mathbf{Y}$ is a nuclear operator and $\|ST\|_1 \leq \|S\| \|T\|_1$.*

Definition 6.3 Let $\mathbf{X} = \mathbf{Y}$ and $T : \mathbf{X} \to \mathbf{X}$ be a trace class operator. For any complete ortho-normal system $\{x_k\}$

$$\operatorname{tr} T = \sum_{k=1}^{\infty} (Tx_k, x_k)$$

is called the trace of T. Tr T is independent of choice of the system $\{x_k\}$.

Proposition 6.9 *Let $S : \mathbf{X} \to \mathbf{Y}$ and $T : \mathbf{Y} \to \mathbf{X}$ be Schmidt operators. Then*

$$\operatorname{tr} ST = \operatorname{tr} TS.$$

Definition 6.4 Suppose T is a nuclear operator of $\mathbf{X}$. Then for $z \in \mathbf{C}$ the determinant of the operator $(I + zT)$ is defined by

$$\det(I + zT) = \prod_{j=1}^{\infty}(1 + z\lambda_j),$$

where λ_j, $j = 1, 2, 3 \ldots$ is the all eigenvalues of T numbered including multiplicity. This infinite product converges.

Proposition 6.10 *(Cf. [59] Theorem 3.4.) Suppose T is a nuclear operator on $\mathbf{X}$ and T_n, $n = 1, 2, \ldots$, is a sequence of nuclear operators on $\mathbf{X}$ such that*

$$\lim_{n\to\infty} \|T_n - T\|_1 = 0.$$

Then

$$\lim_{n\to\infty} \det(I + T_n) = \det(I + T).$$

Proposition 6.11 *(Cf. [59] Theorem 2.16.) Let A and B be bounded linear operators in a Hilbert space $\mathbf{X}$. Suppose B is a positive definite and nuclear operator. Suppose $\{A_n\}_{n=1}^{\infty}$ is a sequence of linear operators that satisfy*

$$|A|,\ |A^*| \le B, \quad |A_n^*|,\ |A_n| \le B, \quad n = 1, 2, \ldots.$$

If $A_n \to A$ in the weak topology of linear operators, then

$$\lim_{n\to\infty} \|A_n - A\|_1 = 0.$$

6.5.2 The Second Variation of the Action and the Jacobi Operator

We choose $\mathbf{X} = \mathscr{H} = H_0^1(s, s')$. The scalar product of two $\varphi, \psi \in \mathscr{H}$ is, as in Sect. 2.2.3,

$$(\varphi, \psi)_{\mathscr{H}} = \int_s^{s'} \frac{d}{dt}\varphi(t)\frac{d}{dt}\psi(t)\, dt. \tag{6.5.4}$$

Here we considered $\varphi, \psi \in \mathscr{H}$ as functions in $L^2(s, s')$. If we explicitly write this identification map

$$\iota : \mathscr{H} \ni \varphi \longrightarrow \iota\varphi \in L^2(s, s'), \tag{6.5.5}$$

then $\varphi, \psi \in \mathscr{H}$ are identified with $\iota\varphi, \iota\psi \in L^2(s, s')$.

By this notation, the inner product above is written as

$$(\varphi, \psi)_{\mathscr{H}} = \int_s^{s'} \frac{d}{dt}(\iota\varphi(t))\frac{d}{dt}(\iota\psi(t))\,dt = \left(\frac{d}{dt}\iota\varphi, \frac{d}{dt}\iota\psi\right)_{L^2(s,s')}. \tag{6.5.6}$$

Here and hereafter $(\ ,\ )_{L^2(s,s')}$ denotes the inner product in $L^2(s, s')$ and $\|a\|_{L^2(s',s)}$ denotes the norm in $L^2(s, s')$. The linear map ι is continuous owing to the Poincaré inequality (2.2.8) of Proposition 2.2

$$\|\iota\varphi\|_{L^2(s,s')} \leq \frac{|s'-s|}{\pi}\|\varphi\|_{\mathscr{H}}. \tag{6.5.7}$$

Let ι^* be the adjoint map of ι:

$$\iota^* : L^2(s, s') \ni f \longrightarrow \iota^* f \in \mathscr{H}$$

that is, for any $g \in L^2(s, s')$

$$(\iota^* g, \varphi)_{\mathscr{H}} = (g, \iota\varphi)_{L^2(s,s')}. \tag{6.5.8}$$

Let $\omega = \dfrac{\pi}{s'-s}$ and

$$e_n(t) = \sqrt{\frac{2}{(s'-s)}}\sin n\omega(t-s').$$

Then $\{e_n\}_{n=1}^{\infty}$ is an orthogonal basis of $L^2(s, s')$. Moreover, $e_n(t) \in C^1[s, s']$, $\frac{d}{dt}e_n(t) \in L^2(s, s')$ and $f(s) = f(s') = 0$. Therefore, there exists $\varphi_n \in \mathscr{H}$ such that $\iota\varphi_n = (n\omega)^{-1}e_n$, i.e.,

$$\iota\varphi_n(t) = (n\omega)^{-1}e_n(t) = (n\omega)^{-1}\sqrt{\frac{2}{(s'-s)}}\sin n\omega(t-s').$$

Then $\{\varphi_n\}_{n=1}^{\infty}$ is an orthogonal basis of $\mathscr{H}$. It is clear that for $n = 1, 2, \ldots$

$$\iota^* e_n = (n\omega)^{-1}\varphi_n, \quad \iota\iota^* e_n = (n\omega)^{-2}e_n, \quad \iota^*\iota\varphi_n = (n\omega)^{-2}\varphi_n. \tag{6.5.9}$$

Proposition 6.12 *$\iota : \mathscr{H} \to L^2(s, s')$ and $\iota^* : L^2(s, s') \to \mathscr{H}$ are Schmidt operators. $\iota\iota^*$ is a nuclear operator in $L^2(s, s')$ and $\iota^*\iota$ is a nuclear operator in $\mathscr{H}$.*

Proof It turns out from (6.5.9) that the system of singular values of ι is $\{(n\omega)^{-1}\}$.□

The following fact is well-known.

Proposition 6.13 *Let $g(t, s)$ be the Green function defined in Sect. 2.2.3. Then for $f \in L^2(s, s')$, $\iota\iota^* f = Gf$, that is*

$$\iota\iota^* f(t) = \int_s^{s'} g(t,\tau) f(\tau) d\tau.$$

It is clear that $Gf \in H^2([s,s'])$ *and*

$$\frac{d^2}{dt^2} Gf(t) = -f(t), \quad Gf(s) = Gf(s') = 0.$$

Proposition 6.14 *Suppose that* $p(t)$ *is a differentiable function and it is uniformly bounded, together with its derivative. Then for any* $\varphi \in \mathscr{H}$, $p(t)\iota\varphi \in L^2(s,s')$ *belongs to the range of the operator* ι *and* $\iota^{-1} p(t)\iota\varphi$ *is well-defined.* $\iota^{-1} p(t)\iota : \mathscr{H} \ni \varphi \to \iota^{-1} p(t)\iota\varphi \in \mathscr{H}$ *is a bounded linear operator in* $\mathscr{H}$.

Proof Let $\varphi \in \mathscr{H}$. Then it is clear that $f(t) = p(t)\iota\varphi$ belongs to the Sobolev space $H^1(s,s')$ and $f(s) = 0 = f(s')$. This means that $p(t)\iota\varphi \in \iota\mathscr{H}$, i.e. it is the image by ι of some $g \in \mathscr{H}$. We write $g = \iota^{-1} p(t)\iota\varphi$. Then by (6.5.7)

$$\begin{aligned}
\|g\|_{\mathscr{H}} &= \left\| \frac{d}{dt}\iota g \right\|_{L^2(s,s')} = \left\| \frac{d}{dt}(p(t)\iota\varphi(t)) \right\|_{L^2(s,s')} \\
&\le \left\| \frac{d}{dt} p \right\|_{L^\infty(s,s')} \|\iota\varphi\|_{L^2(s,s)} + \|p\|_{L^\infty(s,s')} \left\| \frac{d}{dt}\iota\varphi \right\|_{L^2(s,s')} \\
&\le \left(\frac{|s'-s|}{\pi} \left\| \frac{d}{dt} p \right\|_{L^\infty(s,s')} + \|p\|_{L^\infty(s,s')} \right) \|\varphi\|_{\mathscr{H}}.
\end{aligned}$$

This proves the proposition. □

Now we discuss the second variation of the action at the classical path $\gamma^*(t)$. Since $\gamma^*(t)$ is the stationary point of $S(\gamma)$,

$$\delta S(\gamma^*) = 0.$$

Let $\varphi \in \mathscr{H}$. Then the second variation $\delta^2 S(\gamma^*)(\varphi)$ is the quadratic form

$$\begin{aligned}
\delta^2 S(\gamma^*)(\varphi) &= \frac{d^2}{d\epsilon^2} S(\gamma^* + \epsilon\varphi)\Big|_{\epsilon=0} \qquad (6.5.10)\\
&= (\varphi,\varphi)_{\mathscr{H}} - (\partial_x^2 V(t,\gamma^*(t))\iota\varphi(t), \iota\varphi)_{L^2(\mathbf{R})} \\
&= ((I - \iota^*\partial_x^2 V(t,\gamma^*(t))\iota)\varphi, \varphi)_{\mathscr{H}}.
\end{aligned}$$

We have the following bilinear form on $\mathscr{H} \times \mathscr{H}$ associated with the above quadratic form.

$$\delta^2 S(\gamma^*)(\varphi,\psi) = (\varphi,\psi)_{\mathscr{H}} - (\iota^*\partial_x^2 V(t,\gamma^*(t))\iota\varphi(t), \psi)_{\mathscr{H}}. \qquad (6.5.11)$$

Notice that $\iota^*\partial_x^2 V(t,\gamma^*(t))\iota : \mathscr{H} \to \mathscr{H}$ is a nuclear operator, because $\partial_x^2 V(t,\gamma^*(t))\iota$ is an operator of Schmidt class by Propositions 6.12 and 6.14.

If $\varphi = \iota^* g \in \mathcal{H}$ with some $g \in L^2(s, s')$, then we have from (6.5.11) and (6.5.6) that

$$\begin{aligned}\delta^2 S(\gamma^*)(\iota^* g, \psi) &= (\iota^* g, \psi)_{\mathcal{H}} - (\partial_x^2 V(t, \gamma^*(t))\iota\iota^* g, \iota\psi)_{L^2(s,s')} \\ &= \left(\frac{d}{dt}\iota\iota^* g, \frac{d}{dt}\iota\psi\right)_{L^2(s,s')} - (\partial_x^2 V(t, \gamma^*(t))\iota\iota^* g, \iota\psi)_{L^2(s,s')}.\end{aligned}$$

The integration by parts and Propositon 6.13 give

$$\left(\frac{d}{dt}\iota\iota^* g, \frac{d}{dt}\iota\psi\right)_{L^2(s,s')} = -\left(\frac{d^2}{dt^2}\iota\iota^* g(t), \iota\psi\right)_{L^2(s,s')} = (g, \iota\psi)_{L^2(s,s')}.$$

Thus

$$\delta^2 S(\gamma^*)(\iota^* g, \psi) = \left(J\iota\iota^* g, \iota\psi\right)_{L^2(s,s')}, \tag{6.5.12}$$

where J is the following differential operator called Jacobi's differential operator:

$$Jf(t) = -\frac{d^2}{dt^2} f(t) - \partial_x^2 V(t, \gamma^*(t)) f(t), \qquad \text{for } f \in H^2(s, s'). \tag{6.5.13}$$

It is natural to consider this with Dirichlet boundary condition

$$f(s) = f(s') = 0. \tag{6.5.14}$$

The right-hand side of (6.5.12) is a continuous linear function of $\iota\psi$ in the topology of $L^2(\mathbf{R})$. By continuity it can be extended as a bounded linear form of $f \in L^2(s, s')$:

$$L^2(s, s') \ni f \to (J\iota^*\iota g, f)_{L^2(\mathbf{R})}.$$

Proposition 6.15 *$\iota^*\partial_x^2 V(t, \gamma^*)\iota$, $\partial_x^2 V(t, \gamma^*)\iota\iota^*$ and $\iota\iota^*\partial_x^2 V(t, \gamma^*)$ are nuclear operators.*

Proof By Propositon 6.7, $\partial_x^2 V(t, \gamma^*)\iota$ and $\iota^*\partial_x^2 V(t, \gamma^*)$ are operators of Schmidt class. The proposition follows from this and Proposition 6.8. □

6.5.3 $D(s', s, x, y)$ and Infinite-Dimensional Determinant I

The aim of this subsection is to prove the following theorem.

Theorem 6.9 *If $|s' - s| \leq \delta_0$, then*

$$D(s', s, x, y) = \det(I - \iota^*\partial_x^2 V(t, \gamma^*(t))\iota), \tag{6.5.15}$$

where $\gamma^*(t)$, $t \in [s, s']$ is the classical path such that $\gamma^*(s) = y$ and $\gamma^*(s') = x$.

In view of (6.5.11) we may write

$$D(s', s, x, y) = \det \delta^2 S(\gamma^*(t)). \tag{6.5.16}$$

In order to prove this theorem we approximate $\iota^* \partial_x^2 V(t, \gamma^*(t))\iota$ by its restriction to finite-dimensional subspaces of $\mathscr{H}$. Let Δ be the division of the interval $[s, s']$ into $J+1$ subintervals of equal length. Let $T_k = s + \dfrac{k(s'-s)}{J+1}$ for $k = 0, 1, \dots, J+1$. Then

$$\Delta : s = T_0 < T_1 < \cdots < T_J < T_{J=1} = s',$$

with $|\Delta| = (J+1)^{-1}(s'-s)$.

Let $(x_J, x_{J-1}, \dots, x_1)$ be an arbitrary point of $\mathbf{R}^J$. Let $\gamma_\Delta(t)$ be the piecewise classical path such that $\gamma_\Delta(T_k) = x_k$, for $k = 0, 1, 2, \dots, J+1$. Here and hereafter we define $x_0 = y$ and $x_{J+1} = x$. It is clear that $\gamma_\Delta - \gamma^* \in \mathscr{H}$.

Let $x_k^* = \gamma^*(T_k)$ for $k = 0, 1, 2, \dots$. Define functions ψ_j, by

$$\psi_j = \partial_{x_j} \gamma_\Delta.$$

That means

$$\iota\psi_j(t) = \partial_{x_j} \iota\gamma_\Delta(t)\big|_{x_k = x_k^*, 1 \le k \le J}. \tag{6.5.17}$$

It is clear that $\iota\psi_j(t) = 0$ for $t \notin (T_{j-1}, T_{j+1})$, $\iota\psi_j(T_j) = 1$ and that $\iota\psi_j(t)$ is the solution of the Jacobi equation for $t \in (T_{j-1}, T_j) \cup (T_j, T_{j+1})$,

$$\left(-\frac{d^2}{dt^2} - \partial_x^2 V(t, \gamma^*(t)) \right) u(t) = 0. \tag{6.5.18}$$

In view of this we introduce the following system of elements of $\mathscr{H}$. Let $\Delta^{(n)}$ be the division of $[s, s']$ into 2^{n+1} subintervals of equal length $2^{-(n+1)}(s'-s)$:

$$\Delta^{(n)} : T_0^{(n)} < T_1^{(n)} < T_2^{(n)} < \cdots < T_{2k}^{(n)} < T_{2k+1}^{(n)} < \cdots < T_{2^{(n+1)}}^{(n)} = s', \tag{6.5.19}$$

where $T_j^{(n)} = s' + 2^{-(n+1)} j (s'-s)$ for $j = 0, 1, \dots, 2^{(n+1)}$. We define the following function $\varphi_{n,k} \in \mathscr{H}$, $k = 1, 2, \dots, 2^n$, by

$$\iota\varphi_{n,k}(t) = \begin{cases} 0 & \text{if } t \notin (T_{2k-2}^{(n)}, T_{2k}^{(n)}), \\ 2^{-n/2-1}(s'-s)^{1/2} & \text{if } t = T_{2k-1}^{(n)}, \\ \text{solution of (6.5.18)} & \text{if } t \in (T_{2k-2}^{(n)}, T_{2k-1}^{(n)}) \cup (T_{2k-1}^{(n)}, T_{2k}^{(n)}). \end{cases} \tag{6.5.20}$$

Proposition 6.16 *The vector space spanned by* $\{\varphi_{n,k}; n = 0, 1, \dots, k = 1, 2, \dots, 2^n\}$ *is dense in* $\mathscr{H}$. *Moreover, for any* $\varphi \in \mathscr{H}$ *there exists a unique sequence* $\{c_{n,k}\} \subset \mathbf{R}$ *such that*

$$\varphi = s - \lim_{N\to\infty} \sum_{n\le N} \sum_{k=1}^{2^n} c_{n,k}\varphi_{n,k}, \tag{6.5.21}$$

where $s - \lim$ *means the strong convergence in* $\mathscr{H}$.

Proof We use Theorem 2.20 in Kato's book [43] to prove this proposition.

First we prove the uniqueness of the sequence $\{c_{n,k}\}$. Suppose that (6.5.21) is true for $\varphi = 0$. Then by Sobolev's inequality in Proposition 2.2, the series $\sum_n \sum_k c_{n,k}\iota\varphi_{n,k}(t)$ converges uniformly in $t \in [s, s']$. Hence for any $t \in [s, s]$

$$0 = \sum_{n=0}^{\infty} \sum_{k=1}^{2^n} c_{n,k}\iota\varphi_{n,k}(t). \tag{6.5.22}$$

We prove $c_{n,k} = 0$ for all $n = 0, 1, 2, \ldots,$ and $k = 1, 2, \ldots, 2^n$ by induction on n. We begin with proving $c_{0,1} = 0$. Evaluate both sides of (6.5.22) at the point $2^{-1}(s + s') = T_1^{(0)}$. By definition, $\iota\varphi_{n,k}(T_1^{(0)}) = 0$ for $n = 1, 2, \ldots$ and $k = 1, 2, \ldots, 2^n$. Therefore, $0 = c_{0,1}\iota\varphi_{0,1}(T_1^{(0)})$. Thus $c_{0,1} = 0$, because $\iota\varphi_{0,1}(T_1^{(0)}) \neq 0$ by definition.

Suppose that $c_{n,k} = 0$ is true for $0 \le n \le m - 1$ and $1 \le k \le 2^n$. Then (6.5.22) means that for any $t \in [s, s']$

$$0 = \sum_{n=m}^{\infty} \sum_{k=1}^{2^n} c_{n,k}\iota\varphi_{n,k}(t). \tag{6.5.23}$$

Let $j = 1, \ldots, 2^m$ be arbitrary and consider this equality at at $t = T_{2j-1}^{(m)}$. Then we obtain $0 = c_{m,j}\iota\varphi_{m,j}(T_{2j-1}^{(m)})$, because $\iota\varphi_{n,k}(T_{2j-1}^{(m)}) = 0$ for $n > m$ and $n = m, k = j$. As $\iota\varphi_{m,j}(T_{2j-1}^{(m)}) \neq 0$, we obtain $c_{m,j} = 0$. Since j was arbitrary, $c_{m,k} = 0$ for any $k = 1, 2, \ldots, 2^m$. Thus by induction on n, we have proved that $c_{n,k} = 0$ for any n and k. The expression (6.5.21) is unique.

Now we introduce a good approximation $\{\varphi_{n,k}^0\} \subset \mathscr{H}$ of the system $\{\varphi_{n,k}\}$ as follows:

$$\iota\varphi_{n,k}^0(t) = \begin{cases} 0 & \text{for } t \neq (T_{2k-2}^{(n)}, T_{2k}^{(n)}), \\ 2^{n/2}(s'-s)^{-1/2}(t - T_{2k-2}^{(n)}) & \text{if } t \in [T_{2k-2}^{(n)}, T_{2k-1}^{(n)}], \\ 2^{n/2}(s'-s)^{-1/2}(T_{2k}^{(n)} - t) & \text{if } t \in [T_{2k-1}^{(n)}, T_{2k}^{(n)}]. \end{cases} \tag{6.5.24}$$

Notice that $\varphi_{n,k}^{(0)}(t)$ is a piecewise linear function and $\iota\varphi_{n,k}^{(0)}(t) = \iota\varphi_{n,k}(t)$ for $t \notin (T_{2k-2}^{(n)}, T_{2k-1}^{(n)}) \cup (T_{2k-1}^{(n)}, T_{2k}^{(n)})$. It's derivative

$$\frac{d}{dt}\iota\varphi_{n,k}^{0}(t) = \begin{cases} 0 & \text{for } t \neq [T_{2k-2}^{(n)}, T_{2k}^{(n)}], \\ 2^{n/2}(s'-s)^{-1/2} & \text{if } t \in [T_{2k-2}^{(n)}, T_{2k-1}^{(n)}], \\ -2^{n/2}(s'-s)^{-1/2} & \text{if } t \in [T_{2k-1}^{(n)}, T_{2k}^{(n)}] \end{cases}$$

is the system of Haar's functions, which is the complete ortho-normal system of the orthogonal complement of the constant function 1 of $L^2(s, s')$. Thus $\{\varphi_{n,k}^{(0)}(t)\}$ is a complete ortho-normal system of $\mathscr{H}$. Let $r_{n,k} = \|\varphi_{n,k} - \varphi_{n,k}^{(0)}\|_{\mathscr{H}}$. We claim that for $k = 1, 2, \ldots, 2^n$ and $n = 0, 1, 2, \ldots,$

$$r_{n,k} \leq 2C_2(s'-s)^2\pi^{-2}2^{-2n}, \tag{6.5.25}$$

where $C_2 = \frac{v_2}{2}\left(1 - \frac{\delta_0^2 v_2}{8}\right)^{-1}$ as is given in Proposition 2.5. We assume for the moment that the claim is true and continue the proof of the proposition. By the claim

$$r^2 = \sum_{n=0}^{\infty}\sum_{k=1}^{2^n} r_{n,k}^2 \leq C_2^2\pi^{-4}(s'-s)^4\sum_{n=0}^{\infty} 2^{-3n+2} < \infty. \tag{6.5.26}$$

It follows from this and Theorem 2.20 of [43] that Proposition 6.16 is true.

Now we prove the claim. In order to simplify notation we write $\eta(t) = \iota\varphi_{n,k}(t) - \iota\varphi_{n,k}^{(0)}$, $a = T_{2k-2}^{(n)}$, $b = T_{2k-1}^{(n)}$, $c = T_{2k}^{(n)}$ and $I = [a, c]$ for the moment. Then

$$r_{n,k}^2 = \int_I \left|\frac{d}{dt}\eta(t)\right|^2 dt.$$

Since η is the solution of Jacobi's equation

$$-\frac{d^2}{dt^2}\eta(t) - \partial_x^2 V(t, \gamma^*(t))\eta(t) = \partial_x^2 V(t, \gamma^*(t))\iota\varphi_{n,k}^{(0)}(t) \quad \text{in } (a, b) \cup (b, c) \tag{6.5.27}$$

and boundary conditions

$$\eta(a) = \eta(b) = \eta(c) = 0,$$

we obtain the energy equality

$$\int_I \left|\frac{d}{dt}\eta(t)\right|^2 dt = \int_I \partial_x^2 V(t, \gamma^*(t))\eta(t)^2\, dt + \int_I \partial_x^2 V(t, \gamma^*(t))\iota\varphi_{n,k}^{(0)}(t)\eta(t)\, dt. \tag{6.5.28}$$

It is clear by assumption on the potential $V(t, x)$ that

$$\left|\int_I \partial_x^2 V(t, \gamma^*(t))\eta(t)^2\, dt\right| \leq v_2 \int_I \eta(t)^2\, dt,$$

and

$$\left|\int_I \partial_x^2 V(t,\gamma^*(t))\iota\varphi_{n,k}^{(0)}(t)\eta(t)\,dt\right| \le v_2\left(\int_I |\iota\varphi_{n,k}^{(0)}(t)|^2\,dt\right)^{1/2}\left(\int_I |\eta(t)|^2\,dt\right)^{1/2}.$$

Apply these inequalities, the Poincaré inequality and the fact $\|\varphi_{n,k}^0\|_{\mathscr{H}} = 1$ to energy equality (6.5.28). Then

$$\int_I \left|\frac{d}{dt}\eta(t)\right|^2 dt \le v_2(c-a)^2\pi^{-2}\left(\int_I \left|\frac{d}{dt}\eta(t)\right|^2 dt + \left(\int_I \left|\frac{d}{dt}\eta(t)\right|^2 dt\right)^{1/2}\right).$$

It follows from this that

$$r_{n,k} \le (1 - v_2(c-a)^2\pi^{-2})^{-1} \le v_2(c-a)^2\pi^{-2}.$$

Since $v_2(c-a)^2\pi^{-2} < \dfrac{v_2\delta^2}{8} < 1$ and $c - a = 2^{-n}(s'-s)$, it turns out that

$$r_{n,k} \le 2C_2(s'-s)^2\pi^{-2}2^{-2n}.$$

The claim is now proved. The proof of Proposition 6.16 is now complete. □

Let n be a non-negative integer and $\mathscr{H}_n$ be the supspace of $\mathscr{H}$ spanned by functions $\{\varphi_{j,k};\ j \le n, k = 1, 2, \ldots, 2^j\}$. Let P_n be the orthogonal projection onto $\mathscr{H}_n$.

Proposition 6.17

$$\lim_{n\to\infty} P_n = I, \tag{6.5.29}$$

in the strong topology of operators in $\mathscr{H}$. *Moreover,*

$$\lim_{n\to\infty} \det(I - P_n(\iota^*\partial_x^2 V(t,\gamma^*(t))\iota)P_n) = \det(I - (\iota^*\partial_x^2 V(t,\gamma^*(t))\iota)). \tag{6.5.30}$$

Proof (6.5.29) is clear because of Proposition 6.16. Hence

$$\lim_{n\to\infty} (P_n(\iota^*\partial_x^2 V(t,\gamma^*(t))\iota)P_n = \iota^*\partial_x^2 V(t,\gamma^*(t))\iota$$

in the weak operator topology.

The operator $\iota^*\partial_x^2 V(t,\gamma^*(t))\iota$ is a symmetric operator in $\mathscr{H}$ and

$$|(\iota^*\partial_x^2 V(t,\gamma^*(t))\iota\varphi,\varphi)_{\mathscr{H}}| \le v_2(\iota^*\iota\varphi,\varphi)_{\mathscr{H}} \quad \text{for } \varphi \in \mathscr{H}.$$

It follows from these and the minimax principle that

$$|\iota^*\partial_x^2 V(t,\gamma^*(t))\iota| \le v_2\iota^*\iota, \quad |P_n\iota^*\partial_x^2 V(t,\gamma^*(t))\iota P_n| \le v_2\iota^*\iota.$$

Since $\iota^*\iota$ is nuclear, by Propositions 6.10 and 6.11, equation (6.5.30) is true. □

Let $\Delta^{(n)}$ be the division of $[s, s']$ as (6.5.19). In order to simplify notations we will use Δ for $\Delta^{(n)}$, J for $2^{n+1}-1$, T_j for $T_j^{(n)}$ and τ_j for $T_j - T_{j-1}$. We denote an arbitrary piecewise classical path associated with Δ by γ_Δ and set $x_j = \gamma_\Delta(T_j)$ for $j = 0, 1, 2, \ldots, J+1$. $x_0 = y$ and $x_{J+1} = x$ are fixed end points of γ_Δ as usual. Let γ^* be the classical path and $x_j^* = \gamma^*(T_j)$.

We define $\psi_j \in \mathscr{H}$ by

$$\iota\psi_j(t) = \partial_{x_j}\gamma_{\Delta^{(n)}}(t)\Big|_{x_j=x_j^*}. \tag{6.5.31}$$

As we mentioned earlier, $\iota\psi_j(t) = 0$ for $t \notin (T_{j-1}, T_{j+1})$, $\iota\psi_j(T_j) = 1$ and that $\iota\psi_j(t)$ is the solution of the Jacobi equation

$$\left(-\frac{d^2}{dt^2} - \partial_x^2 V(t, \gamma^*(t))\right)u(t) = 0, \quad \text{for } t \in (T_{j-1}, T_j) \cup (T_j, T_{j+1}). \tag{6.5.32}$$

This means that $\psi_j = P_n\psi_j$, i.e. $\{\psi_j\} \subset \mathscr{H}_n$. On the other hand, the number of ψ_j's is $J - 1 = dim\mathscr{H}_n$ and they are linearly independent. Therefore, $\{\psi_j\}$ is another basis of the subspace $\mathscr{H}_n = P_n\mathscr{H}$.

Let $A_\Delta = (a_{jk})$ be the $J \times J$ matrix expression of $P_n\iota^*\partial_x^2 V(t, \gamma^*)\iota P_n$ with respect to $\{\psi_j\}$. Then

$$P_n(I - \iota^*\partial_x^2 V(t, \gamma^*)\iota)P_n\psi_j = \sum_k^J a_{jk}\psi_k. \tag{6.5.33}$$

Then for any $1 \le j, k \le J$

$$\begin{aligned}\partial_{x_j}\partial_{x_l} S(x_{J+1}, x_J^*, x_{J-1}^*, \ldots, x_1^*, x_0) &= ((I - \iota^*\partial_x^2 V(t, \gamma^*(t))\iota)\psi_j, \psi_l)_{\mathscr{H}} \\ = (P_n(I - \iota^*\partial_x^2 V(t, \gamma^*(t))\iota)P_n\psi_j, \psi_l)_{\mathscr{H}} &= \sum_k a_{jk}b_{kl}.\end{aligned}$$

Here $b_{jk} = (\psi_j, \psi_k)_{\mathscr{H}}$, i.e. the matrix $B_\Delta = (b_{kl})$ is the Gram matrix of $\{\psi_j\}$. Thus we have expression

$$\mathrm{Hess}_{(x^*)} S(x_{J+1}, \ldots, x_0) = A_\Delta B_\Delta. \tag{6.5.34}$$

By notation in Sect. 4.2,

$$\begin{aligned}D(\Delta^{(n)}, s', s, x, y) &= \det H_{\Delta^{(n)}}^{-1} \det \mathrm{Hess}_{(x^*)} S(x_{J+1}, \ldots, x_0) \\ &= \det H_{\Delta^{(n)}}^{-1} \det B_{\Delta^{(n)}} \det A_{\Delta^{(n)}}.\end{aligned} \tag{6.5.35}$$

We wish to prove that $\det H_\Delta^{-1} \det B_\Delta \to 1$ as $n \to \infty$.

Proposition 6.18 *Let h_{jk} be the (j,k)-entry of matrix $H_{\Delta^{(n)}}$. Then there exists a constant $C > 0$ independent of n, j and k such that*

$$|h_{jk} - b_{jk}| \begin{cases} \leq C|\Delta^{(n)}|^3 & \text{if } |j-k| \leq 1, \\ = 0 & \text{if } |j-k| > 1. \end{cases} \tag{6.5.36}$$

Proof It is clear that $h_{j,k} = 0 = b_{j,k}$ in the case $|j-k| > 1$. Thus we have only to prove (6.5.36) in the case $|j-k| \leq 1$. Let $\psi_j^0 \in \mathscr{H}$ be such that $\iota\psi_j^0(t)$ is the piecewise linear function given by

$$\iota\psi_j^0(t) = \begin{cases} 0 & \text{for } t \notin [T_{j-1}, T_{j+1}], \\ \tau_j^{-1}(t - T_{j-1}) & \text{for } t \in [T_{j-1}, T_j], \\ \tau_{j+1}^{-1}(T_{j+1} - t) & \text{for } t \in [T_j, T_{j+1}]. \end{cases} \tag{6.5.37}$$

Then by (4.1.9), $H_{\Delta^{(n)}}$ coincides with the Gram matrix of the family of elements $\{\psi_j^0\} \subset \mathscr{H}$, i.e. for $1 \leq j, k \leq J$,

$$h_{jk} = (\psi_j^0, \psi_k^0)_{\mathscr{H}}. \tag{6.5.38}$$

We shall show that $H_{\Delta^{(n)}}$ and $B_{\Delta^{(n)}}$ are close. Set $\eta_j(t) = \psi_j - \psi_j^0$.

$$\begin{aligned} b_{j,j+1} - h_{j,j+1} &= (\psi_j^0 + \eta_j, \psi_{j+1}^0 + \eta_{j+1})_{\mathscr{H}} - (\psi_j^0, \psi_{j+1}^0)_{\mathscr{H}} \\ &= (\psi_j^0, \eta_{j+1})_{\mathscr{H}} + (\eta_j, \psi_{j+1}^0)_{\mathscr{H}} + (\eta_j, \eta_{j+1})_{\mathscr{H}}. \end{aligned}$$

Since $\dfrac{d}{dt}\iota\psi_j^0(t) = -\tau_{j+1}^{-1}$ for $t \in (T_j, T_{j+1})$ and $\iota\eta_{j+1}(T_j) = \iota\eta_{j+1}(T_{j+1}) = 0$,

$$(\psi_j^0, \eta_{j+1})_{\mathscr{H}} = -\tau_{j+1}^{-1} \int_{T_j}^{T_{j+1}} \frac{d}{dt}\iota\eta_{j+1}(t)\, dt = 0.$$

By the same reasoning, $(\eta_j, \psi_{j+1}^0)_{\mathscr{H}} = 0$. Thus

$$b_{j,j+1} - h_{j,j+1} = (\eta_j, \eta_{j+1})_{\mathscr{H}}.$$

Similarly,

$$\begin{aligned} b_{j,j} - h_{j,j} &= \|\eta_j\|_{\mathscr{H}}^2. \\ b_{j+1.j} - h_{j+1,j} &= (\eta_{j+1}, \eta_j)_{\mathscr{H}}. \end{aligned}$$

We look for the bound of $r_j = \|\eta_j\|_{\mathscr{H}}$. It is easy to see that $\eta_j(t)$ satisfies

$$-\frac{d^2}{dt^2}\eta_j(t)-\partial_x^2 V(t,\gamma^*(t))\eta_j(t)=\partial_x^2 V(t,\gamma^*(t))\iota\psi_j^0(t),\ \text{in } (T_{j-1},T_j)\cup(T_j,T_{j+1}),$$

and $\eta_j(T_{j-1})=\eta_j(T_j)=\eta_j(T_{j+1})=0$. Mimicking the proof of (6.5.25), we obtain, via the energy inequality,

$$\begin{aligned}\|\eta_j\|_{\mathscr{H}} &= \left(\int_{T_{j-1}}^{T_{j+1}}\left|\frac{d}{dt}\eta(t)\right|^2 dt\right)^{1/2} \qquad (6.5.39)\\ &\le C_2\left(\frac{\tau_j+\tau_{j+1}}{\pi}\right)^2\left(\int_{T_{j-1}}^{T_{j+1}}\left|\frac{d}{dt}\iota\psi_j^0(t)\right|^2\right)^{1/2}\\ &\le C_2\left(\frac{\tau_j+\tau_{j+1}}{\pi}\right)^2(\tau_j^{-1}+\tau_{j+1}^{-1})^{1/2}\le C_2\pi^{-2}2^{1-3n/2}(s'-s)^{3/2}.\end{aligned}$$

Here $C_2=2^{-1}v_2(1-8^{-1}\delta_0^2 v_2)^{-1}$. It follows from this that

$$|b_{j,k}-h_{j,k}|\le C_2^2\pi^{-4}2^{2-3n}(s'-s)^3$$

if $|j-k|\le 1$. □

Proposition 6.19 *Let $G=(g_{jk})$ be the matrix $H_{\Delta^{(n)}}^{-1}$. Then:*

1. *There exists a constant $C>0$ such that absolute value of eigenvalues of $G_{\Delta^{(n)}}(B_{\Delta^{(n)}}-H_{\Delta^{(n)}})$ are at most $C(s'-s)|\delta|^4$.*
2. *If $|s'-s|<\delta$, then there exists a constant $C>0$ independent of n such that*

$$\exp(-C2^{3-2n}\delta^4)\le(\det H_{\Delta^{(n)}})^{-1}\det B_{\Delta^{(n)}}\le\exp(C2^{2-2n}\delta^4). \qquad (6.5.40)$$

Proof As $H_{\Delta^{(n)}}^{-1}B_{\Delta^{(n)}}=I+G_\Delta(B_\Delta-H_\Delta)$, we set $Q=G_{\Delta^{(n)}}(B_{\Delta^{(n)}}-H_{\Delta^{(n)}})$. Let q_{jk} be its j,k-element. Then with $J=2^{(n+1)}-1$

$$\begin{aligned}q_{m,n}&=\sum_{j=1}^{J}g_{mj}(b_{j,n}-h_{j,n})\\ &=g_{m,n-1}(b_{n-1,n}-h_{n-1,n})+g_{m,n}(b_{n,n}-h_{n,n})+g_{m,n+1}(b_{n+1,n}-h_{n+1,n}).\end{aligned}$$

Since equality (4.1.14) implies that $|g_{jk}|\le(s'-s)$, we have from Proposition 6.18

$$|q_{m,n}|\le C2^{2-3n}(s'-s)^4,\quad \text{for any } m,n.$$

It follows from this that

$$\sup_m\sum_n|q_{m,n}|\le C2^{2-2n}(s'-s)^4 \quad\text{and}\quad \sup_n\sum_m|q_{m,n}|\le C2^{2-2n}(s'-s)^4.$$

By Schur's inequality the operator norm of Q is bounded:

$$\|Q\| \le C2^{2-2n}(s'-s)^4.$$

This implies that all the eigenvalues λ_j, $j = 1, 2, \ldots, 2^n$ of Q have bounds:

$$|\lambda_j| \le C2^{2-2n}(s'-s)^4.$$

This implies that

$$(1-C2^{2-2n}\delta^4)^{2^n} \le \prod_{j=1}^{2^n}(1-|\lambda_j|) \le \det(I-Q) \le \prod_{j=1}^{2^n}(1+|\lambda_j|) \le (1+C2^{2-2n}\delta^4)^{2^n}.$$

Since $e^{-2x} \le (1-x)$ for $0 \le x < 2^{-1}$,

$$\exp(-C2^{3-2n}\delta^4) \le \det(I-Q) \le \exp(C2^{2-2n}\delta^4)$$

if $C2^{3-2n}\delta^4 < 2^{-1}$. As $I - Q = H_{\Delta^{(n)}}^{-1} B_{\Delta^{(n)}}$, we have proved (6.5.40):

$$\exp(-C2^{3-2n}\delta^4) \le (\det H_{\Delta^{(n)}})^{-1} \det B_{\Delta^{(n)}} \le \exp(C2^{2-2n}\delta^4).$$

□

Theorem 6.10

$$D(s', s, x, y) = \det(I - \iota^* \partial_x^2 V(t, \gamma^*)\iota).$$

Proof Let $\Delta^{(n)}$ be the division of $[s, s']$ into 2^{n+1} subintervals of equal length as above. We use $J = 2^{n+1} - 1$, $\{\psi_j \in \mathcal{H} : j = 1, 2, \ldots, J\}$, P_n as above. By Proposition 6.17 we have

$$\det(I - \iota^* \partial_x^2 V(t, \gamma^*)\iota) = \lim_{n\to\infty} \det P_n(I - \iota^* \partial_x^2 V(t, \gamma^*)\iota P_n) = \lim_{n\to\infty} \det A_{\Delta^{(n)}}, \tag{6.5.41}$$

where A_n is the matrix expression of $P_n(I - \iota^* \partial_x^2 V(t, \gamma^*)\iota)P_n$ with respect to the system $\{\psi_{n,k}\}$. Let B_n be the Gram matrix of $\{\psi_j : j = 1, 2, \ldots, J\}$ in $\{\mathcal{H}\}$. Then it follows from Proposition 6.19 and (6.5.35) that

$$\begin{aligned} D(s', s, x, y) &= \lim_{n\to\infty} D(\Delta^{(n)}, s', s, x, y) \\ &= \lim_{n\to\infty} (\det H_{\Delta^{(n)}})^{-1} \det B_{\Delta^{(n)}} \det A_{\Delta^{(n)}} \\ &= \lim_{n\to\infty} \det A_{\Delta^{(n)}}. \end{aligned}$$

This and (6.5.41) prove the theorem. □

6.5.4 $D(s', s, x, y)$ and Infinite-dimensional Determinant II

We discuss the following operator T in $L^2(s, s')$:

$$Tf(t) = \int_s^{s'} g(t,\tau)\partial_x^2 V(\tau, \gamma^*(\tau)) f(\tau)\, d\tau = G\partial_x^2 V(*, \gamma^*(*)) f(t), \tag{6.5.42}$$

where $g(t, \tau)$ is the integral kernel of Green operator $G = \iota\iota^*$.

We wish to prove

Theorem 6.11 *T is a nuclear operator. And we have*

$$\det(I - T) = D(s', s, x, y). \tag{6.5.43}$$

Proof We use notations in the previous subsection. The subspace in $\mathcal{H}$ spanned by $\{\varphi_{n,k} : k = 1, 2, \ldots, J = 2^{(n+1)} - 1\}$ is denoted by $\mathcal{H}_n$. Since ι is a one-to-one linear mapping, its restriction ι_n to the finite-dimensional space $\mathcal{H}_n$ is an linear isomorphism (not isometry) to the image of $\mathcal{H}_n$ by ι_n, which we denote by $X_n \subset L^2(s', s)$.

We discussed the linear transform $P_n\iota^*\partial_x^2 V(*, \gamma^*(*))\iota P_n$ on $\mathcal{H}_n$. We write this in the following form:

$$P_n\iota^*\partial_x^2 V(*, \gamma^*(*))\iota_n = \iota_n^{-1}\iota_n P_n\iota^*\partial_x^2 V(*, \gamma^*(*))\iota_n.$$

This means that two linear mappings:

$$P_n(I - \iota^*\partial_x^2 V(*, \gamma^*(*))\iota)P_n : \mathcal{H}_n \to \mathcal{H}_n$$

and

$$(I - \iota_n P_n\iota^*\partial_x^2 V(*, \gamma^*(*))) : X_n \to X_n$$

are similar. Therefore,

$$\det(I - P_n\iota^*\partial_x^2 V(*, \gamma^*(*))\iota_n P_n) = \det(I - \iota_n P_n\iota^*\partial_x^2 V(*, \gamma^*(*))). \tag{6.5.44}$$

Let Q_n be the orthogonal projection in $L^2(s, s')$ to X_n. Then

$$\det(I - \iota_n P_n\iota^*\partial_x^2 V(*, \gamma^*(*))) = \det(I - Q_n\iota_n P_n\iota^*\partial_x^2 V(*, \gamma^*(*))Q_n). \tag{6.5.45}$$

We claim that $\iota_n P_n\iota^* \to \iota\iota^*$ in the trace norm of operators in $L^2(s', s)$. Since $\iota_n P_n = \iota P_n$, $\iota_n P_n\iota^* = \iota P_n\iota^*$ is a non-negative symmetric operator and $\iota_n P_n\iota^* = \iota P_n\iota^* \le \iota\iota^*$. Thus $|\iota_n P_n\iota^*| \le |\iota\iota^*|$. It is clear that $\iota_n P_n\iota^*$ converges to $\iota\iota^*$ weakly. It follows from Proposition 6.11 that $\iota_n P_n\iota^*$ converges to $\iota\iota^*$ in the trace norm of operators in $L^2(s', s)$. Applying Proposition 6.10 to (6.5.45),

$$\lim_{n\to\infty} \det(I - Q_n\iota_n P_n\iota^*\partial_x^2 V(*, \gamma^*(*))Q_n) = \det(I - \iota\iota^*\partial_x^2 V(*, \gamma^*(*))).$$

Let $n \to \infty$ in (6.5.44) and using (6.5.41), we obtain

$$D(s, s', x, y) = \det(I - \iota^* \partial_x^2 V(*, \gamma^*(*))\iota) = \det(I - \iota\iota^* \partial_x^2 V(*, \gamma^*(*))).$$

The theorem has been proved. □

Theorem 6.12

$$D(s, s', x, y) = \det(I - \partial_x^2 V(*, \gamma^*(*))G) = \det JG.$$

Here J is the equation of the Jacobi field at the stationary point γ^* of the action functional defined on $\mathscr{H} \cap H^2(s, s')$:

$$Jf(t) = -\frac{d^2}{dt^2} f(t) - \partial_x^2 V(t, \gamma^*(t)) f(t) \quad \textit{for } f(s') = 0 = f(s).$$

Proof Since $\iota_n P_n \iota^* \partial_x^2 V(*, \gamma^*(*))$ is a linear transformation on a finite-dimensional space, its determinant is equal to the determinant of its transpose:

$$\det(I - Q_n \iota_n P_n \iota^* \partial_x^2 V(*, \gamma^*(*))Q_n) = \det(I - Q_n \partial_x^2 V(*, \gamma^*(*))\iota P_n \iota_n^*). \quad (6.5.46)$$

Taking the limit of this as $n \to \infty$, we have the first equality of the theorem:

$$D(s, s', x, y) = \det(I - \partial_x^2 V(*, \gamma^*(*))G).$$

The second equality is clear, because

$$-\frac{d^2}{dt^2} G f(t) = f(t).$$

the theorem has been proved. □

6.6 The Second Term of the Semi-classical Asymptotic Formula

Suppose $F(\gamma) \equiv 1$ and $|s' - s| \leq \delta_0$. Then the Feynman path integral is the fundamental solution of the Schrödinger equation. In this case the second term of semi-classical asymptotics is known from Birkhoff [11], cf. also [49]. We shall prove, in this section, the same formula by a method different from that of Birkhoff. Our proof is based on a one step sharper stationary phase method. This enables one to obtain an estimate of the remainder.

Let $[s, s'] \ni t \to \gamma^*(t) \in \mathbf{R}$ be the unique classical path starting from y at time s and arriving at x at time s'. We recall notations used in Sects. 4.2 and 5.1. Let Δ be an arbitrary division of $[s, s']$ as before:

$$\Delta : s = T_0 < T_1 < \cdots < T_{J+1} = s'. \tag{6.6.1}$$

Theorem 6.13 *Suppose $F(\gamma) \equiv 1$ and $|s' - s| \leq \delta_0$. Then we have the equality:*

$$\begin{aligned}
&k(\Delta; \nu, s', s, x_{J+1}, x_0) \qquad (6.6.2)\\
&= D_{x_J^*,\ldots,x_1^*}(S_{J+1,1}; x_{J+1}, x_0)^{-1/2}\Big(1 + \nu^{-1} i (s' - s) p(\Delta; s', s, x_{J+1}, x_0)\Big)\\
&+ \nu^{-1}(s' - s)^2 |\Delta| q(\Delta; s', s, \nu, x_{J+1}, x_0) + \nu^{-2}(s' - s)^2 r(\Delta; \nu, s', s, x_{J+1}, x_0),
\end{aligned}$$

where $x_0 = y$, $x_{J+1} = x$ and

$$\begin{aligned}
&(s' - s) p(\Delta; s', s, x_{J+1}, x_0) \qquad (6.6.3)\\
&= \sum_{k=1}^{J} \frac{(T_k - s)}{2(T_{k+1} - s)} \tau_{k+1}\\
&\times \left\{ D_{x_{k-1}^*,\ldots,x_1^*}\Big(\sum_{j=1}^{k} S_j; x_k, x_0\Big)^{1/2} \partial_{x_k}^2 \Big(D_{x_{k-1}^*,\ldots,x_1^*}\Big(\sum_{j=1}^{k} S_j; x_k, x_0\Big)^{-1/2}\Big) \right\}\Big|_{x_k = \gamma^*(T_k)},
\end{aligned}$$

and functions $q(\Delta; \nu, s', s, x_{J+1}, x_0)$ and $r(\Delta; \nu, s', s, x_{J+1}, x_0)$ satisfies the following estimates: For any $K \geq 0$ there exists a constant $C_K > 0$ independent of Δ such that inequalities

$$|\partial_{x_{J+1}}^{\alpha_{J+1}} \partial_{x_0}^{\alpha_0} q(\Delta; s', s, \nu, x_{J+1}, x_0)| \leq C_K, \tag{6.6.4}$$

$$|\partial_{x_{J+1}}^{\alpha_{J+1}} \partial_{x_0}^{\alpha_0} r(\Delta; s', s, \nu, x_{J+1}, x_0)| \leq C_K \tag{6.6.5}$$

hold if $|\alpha_0| \leq K$ and $|\alpha_{J+1}| \leq K$.

Proof Given in Theorem 8.7 of Sect. 8.4.3, Chap. 8. □

Making use of the above theorem, we can prove:

Theorem 6.14 *If $F(\gamma) \equiv 1$ and $|s' - s| \leq \delta_0$, then*

$$\begin{aligned}
&k(\nu, s', s, x, y) \qquad (6.6.6)\\
&= D(s', s, x, y)^{-1/2}\big(1 + \nu^{-1} i (s' - s) p(s', s, x, y)\big) + \nu^{-2}(s' - s)^2 r(\nu, s', s, x, y).
\end{aligned}$$

Here

$$(s' - s) p(s', s, x, y) = \frac{1}{2} \int_s^{s'} D(t, s, \gamma^*(t), y)^{1/2} \partial_z^2 \Big(D(t, s, z, y)^{-1/2} \Big)\Big|_{z = \gamma^*(t)} dt. \tag{6.6.7}$$

Moreover, the following estimate holds: For any integer $K \geq 0$ there exists a constant $C_K > 0$ such that

$$\left|\partial_x^\alpha \partial_y^\beta r(\nu, s', s, x, y)\right| \leq C_K \tag{6.6.8}$$

if $|\alpha| \leq K$ and $|\beta| \leq K$.

Proof The key point of the proof is the following proposition.

Proposition 6.20 *Define*

$$f(t, s, x, y) = D(t, s, \gamma^*(t), y)^{1/2} \partial_z^2 (D(t, s, z, y)^{-1/2})\Big|_{z=\gamma^*(t)}.$$

Then for any α, β there exists a constant $C_{\alpha,\beta} > 0$ such that

$$\begin{aligned} &\sup_{(x,y)\in\mathbf{R}^2} \left|\partial_y^\beta \partial_x^\alpha (s' - s) p(\Delta; s', s, x, y) - 2^{-1} \sum_{k=1}^{J} \tau_{k+1} \partial_x^\alpha \partial_y^\beta f(T_k, s, x, y)\right| \\ &\leq C_{\alpha,\beta}(|\Delta||s' - s|^2 + |\Delta|^{1/2}(s' - s)^{1/2}). \end{aligned} \tag{6.6.9}$$

We will prove this proposition later. Now we assume, for the moment, that this proposition is true and continue the proof of Theorem 6.14.

Since for any α and β $\partial_x^\alpha \partial_y^\beta f(t, s, x, y)$ is a continuous function of t,

$$\lim_{|\Delta|\to 0} \sum_{k=1}^{J} \tau_{k+1} \partial_x^\alpha \partial_y^\beta f(T_k, s, x, y) = \int_s^{s'} \partial_x^\alpha \partial_y^\beta f(t, s, x, y)\, dt.$$

It follows from this and (6.6.9) that

$$\lim_{|\Delta|\to 0} (s' - s) \partial_{x_{J+1}}^\alpha \partial_{x_0}^\beta p(\Delta; s', s, x_{J+1}, x_0) = \frac{1}{2} \int_s^{s'} \partial_x^\alpha \partial_y^\beta f(t, s, x, y)\, dt. \tag{6.6.10}$$

It is clear from (6.6.4) that

$$\lim_{|\Delta|\to 0} \nu^{-1}(s' - s)^2 |\Delta| \partial_x^\alpha \partial_y^\beta q(\Delta; \nu, s', s, x_{J+1}, x_0) = 0. \tag{6.6.11}$$

On the other hand, by Theorem 4.6

$$\lim_{|\Delta|\to 0} \partial_x^\alpha \partial_y^\beta k(\Delta; \nu, s', s, x_{J+1}, x_0) = \partial_x^\alpha \partial_y^\beta k(\nu, s', s, x_{J+1}, x_0) \tag{6.6.12}$$

It turns out from this, (6.6.10) and (6.6.4) that $\partial_x^\alpha \partial_y^\beta r(\Delta; \nu, s', s, x_{J+1}, x_0)$ converges to the limit as $|\Delta| \to 0$. Set

$$\lim_{|\Delta|\to 0} \partial_x^\alpha \partial_y^\beta r(\Delta; \nu, s', s, x_{J+1}, x_0) = r(\nu, s', s, x, y). \tag{6.6.13}$$

Then we obtain (6.6.6) and (6.6.8). This proves Theorem 6.14 up to the proof of Proposition 6.20. □

We prove Proposition 6.20.

Proof Set

$$\begin{aligned} & g_k(x_k, x_0) \\ & = \Big(D_{x^*_{k-1},\dots,x^*_1}\Big(\sum_{j=1}^{k} S_j; x_k, x_0\Big)^{1/2} \partial^2_{x_k}\Big(D_{x^*_{k-1},\dots,x^*_1}\Big(\sum_{j=1}^{k} S_j; x_k, x_0\Big)^{-1/2}\Big)\Big). \end{aligned} \tag{6.6.14}$$

Then

$$(s'-s)p(\Delta; s', s, x_{J+1}, x_0) = \sum_{k=1}^{J} \frac{(T_k - s)}{2(T_{k+1}-s)} \tau_{k+1} g_k(\gamma^*(T_k), x_0). \tag{6.6.15}$$

We denote by Δ_k the division of $[s, T_k]$:

$$\Delta_k : s = T_0 < T_1 < \cdots < T_k.$$

Then by (5.1.10), $D_{x^*_{k-1},\dots,x^*_1}(\sum_{j=1}^{k} S_j; x_k, x_0) = D(\Delta_k, x_k, x_0)$. By Theorem 5.2, for any $K \ge 0$ there exist $\rho_K > 0$ and $C_K > 0$ independent of k such that for any $(x_k, x_0) \in \mathbf{R}^2$

$$\Big|\partial^\alpha_{x_k}\partial^\beta_{x_0}\Big(D_{x^*_{k-1},\dots,x^*_1}\Big(\sum_{j=1}^{k} S_j; x_k, x_0\Big) - D(T_k, s, x_k, y)\Big)\Big| \le C_K|\Delta|(T_k - s),$$

as far as $|\alpha| \le K$, $|\beta| \le K$ and $|\Delta| \le \rho_K$. By Lemma 4.3, there exists a constant $C > 0$ independent of k and Δ such that

$$0 < C \le D(T_k, s, x_k, y) \quad \text{and} \quad 0 < C < D_{x^*_{k-1},\dots,x^*_1}\Big(\sum_{j=1}^{k} S_j; x_k, x_0\Big).$$

Thus for any $K \ge 0$ there exist $\rho_K > 0$ and $C_K > 0$ independent of k such that if $|\alpha| \le K$, $|\beta| \le K$ and $|\Delta| \le \rho_K$,

$$\begin{aligned} \sup_{(x_k, x_0)\in\mathbf{R}^2} & \Big|\partial^\alpha_{x_k}\partial^\beta_{x_0} g_k(x_k, x_0) \\ & - \partial^\alpha_{x_k}\partial^\beta_{x_0} D(T_k, s, x_k, y)^{1/2}\partial^2_{x_k}(D(T_k, s, x_k, y)^{-1/2})\Big)\Big| \\ \le & \; C_K|\Delta|(T_k - s). \end{aligned} \tag{6.6.16}$$

Hence

$$\left|\partial_x^\alpha \partial_y^\beta \sum_{k=1}^J \frac{(T_k - s)}{2(T_{k+1} - s)} \tau_{k+1}(f(T_k, s, x, y) - g_k(\gamma^*(T_k), y))\right|$$
$$\leq \sum_{k=1}^J C_K \frac{(T_k - s)^2}{2(T_{k+1} - s)} \tau_{k+1}|\Delta| \leq \frac{1}{2} C_K |(s' - s)^2| \Delta|.$$

On the other hand

$$\frac{(T_k - s)}{2(T_{k+1} - s)} \tau_{k+1} = \frac{1}{2}\left(\tau_{k+1} - \frac{\tau_{k+1}^2}{T_{k+1} - s}\right). \tag{6.6.17}$$

Since $|\tau_{k+1}(T_{k+1} - s)^{-1}| \leq 1$, $|\tau_{k+1}(T_{k+1} - s)^{-1}| \leq |\tau_{k+1}(T_{k+1} - s)^{-1}|^{1/2}$. Thus

$$\sum_{k=1}^J \frac{\tau_{k+1}^2}{T_{k+1} - s} \leq \sum_{k=1}^J \tau_{k+1}\left(\frac{\tau_{k+1}}{T_{k+1} - s}\right)^{1/2} \leq \sum_{k=1}^J \tau_{k+1}^{1/2} \int_{T_k}^{T_{k+1}} (t - s)^{-1/2} dt$$
$$\leq |\Delta|^{1/2} \int_{T_1}^{T_{J+1}} (t - s)^{-1/2} dt \leq 2|\Delta|^{1/2}(s' - s)^{1/2}.$$

It follows from this that for any α and β there exists a constant $C_{\alpha,\beta} > 0$ with the property

$$\sup_{(x,y)\in\mathbf{R}^2} \left|\partial_x^\alpha \partial_y^\beta \Big(\sum_{k=1}^J \frac{(T_k - s)}{2(T_{k+1} - s)} \tau_{k+1} f(T_k, s, x, y) - 2^{-1} \sum_{k=1}^J \tau_{k+1} f(T_k, s, x, y)\Big)\right|$$
$$\leq C_{\alpha,\beta}|\Delta|^{1/2}|s' - s|^{1/2}.$$

Therefore, for any α and β there exists a constant $C_{\alpha,\beta} > 0$ such that for any

$$\sup_{(x,y)\in\mathbf{R}^2} \left|\partial_{x_0}^\beta \partial_{x_{J+1}}^\alpha \sum_{k=1}^J \frac{(T_k - s)}{2(T_{k+1} - s)} \tau_{k+1} g(T_k, \gamma^*(T_k), x_0)\right.$$
$$\left. - \partial_{x_0}^\beta \partial_{x_{J+1}}^\alpha \sum_{k=1}^J \tau_{k+1} 2^{-1} f_k(T_k, s, \gamma^*(T_k), y)\right|$$
$$\leq C_{\alpha,\beta}\big(|\Delta||s' - s|^2 + |\Delta|^{1/2}(s' - s)^{1/2}\big). \tag{6.6.18}$$

This and (6.6.15) prove Proposition 6.20.

The proof of Theorem 6.14 has been completed. □

Now we compare Theorem 6.14 to Birkhoff's formula in [11].

We write t for s'. Define

$$v_0(t,s,x,y) = \left(\frac{\nu}{2\pi i(t-s)}\right)^{1/2} D(t,s,x,y)^{-1/2},$$

$$v_1(t,s,x,y) = -\left(\frac{\nu}{2\pi i(t-s)}\right)^{1/2} D(t,s,x,y)^{-1/2}(t-s)p(t,s,x,y).$$

Then by Theorem 6.14

$$\int_{\Omega_{x,y}} e^{i\nu S(\gamma)}\mathscr{D}\gamma = e^{i\nu S(t,s,x,y)}(v_0(t,s,x,y) + (i\nu)^{-1}v_1(t,s,x,y) + \mathscr{O}(\nu^{-2})).$$

We proved in Theorem 6.6 that $v_0(t,s,x,y)$ satisfies the first transport equation:

$$\frac{\delta}{\delta t}v_0(t,s,x,y) + \frac{1}{2}\partial_x^2 S(t,s,x,y)v_0(t,s,x,y) = 0, \tag{6.6.19}$$

where the linear first-order differential operator

$$\frac{\delta}{\delta t} = \partial_t + \partial_x S(t,s,x,y)\partial_x \tag{6.6.20}$$

is the operator that expresses differentiation along the classical path $\gamma^*_{x,y}(t)$ starting from y at time s and arriving at x at time t.

Theorem 6.15 *$v_1(t,s.x.y)$ satisfies the second transport equation:*

$$\frac{\delta}{\delta t}v_1(t,s,x,y) + \frac{1}{2}\partial_x^2 S(t,s,x,y)v_1(t,s,x,y) + \frac{1}{2}\partial_x^2 v_0(t,s,x,y) = 0,$$

and the initial condition

$$v_1(s,s,x,y) = 0.$$

The function $v_1(t,s,x,y)$ coincides with the second term of Birkhoff's asymptotic expansion formula [11, 49].

Proof We can write

$$v_1(t,s,x,y) = -v_0(t,s,x,y)(t-s)p(t,s,x,y).$$

Differentiate $v_1(t,s,x,y)$ along the classical path $\gamma^*_{x,y}(t)$. Then

$$\frac{\delta}{\delta t}v_1(t,s,x,y)$$
$$= -\left(\frac{\delta}{\delta t}v_0(t,s,x,y)(t-s)p(t,s,x,y) + v_0(t,s,x,y)\frac{\delta}{\delta t}(t-s)p(t,s,x,y)\right).$$

It follows from the definition of $p(t,s,x,y)$ in Theorem 6.14 that

$$\begin{aligned}\frac{\delta}{\delta t}(t-s)p(t,s,x,y) &= 2^{-1}D(t,s,\gamma^*(t),y)^{1/2}\partial_z^2 D(t,s,z,y)\Big|_{z=\gamma^*_{x,y}(t)}^{-1/2} \\ &= 2^{-1}D(t,s,x,y)^{1/2}\partial_x^2 D(t,s,x,y)^{-1/2}.\end{aligned}$$

Therefore,

$$\begin{aligned}v_0(t,s,x,y)\frac{\delta}{\delta t}(t-s)p(t,s,x,y) &= \left(\frac{\nu}{2\pi i(t-s)}\right)^{1/2} 2^{-1}\partial_x^2 D(t,s,x,y)^{-1/2} \\ &= 2^{-1}\partial_x^2 v_0(t,s,x,y).\end{aligned}$$

Using the first transport equation (6.6.19), we have

$$\begin{aligned}&\frac{\delta}{\delta t}v_1(t,s,x,y) \\ &= -\left(-\frac{1}{2}\partial_x^2 S(t,s,0,x,y)v_0(t,s,x,y)(t-s)p(t,s,x,y)+\frac{1}{2}\partial_x^2 v_0(t,s,x,y)\right) \\ &= -\frac{1}{2}\left(\partial_x^2 S(t,s,x,y)v_1(t,s,x,y)+\partial_x^2 v_0(t,s,x,y)\right).\end{aligned}$$

This proves that $v_1(t,s,x,y)$ satisfies the second transport equation of Birkhoff.

The initial condition $\lim_{t\to 0} v_1(t,s,x,y) = 0$ is clear. In other words, $v_1(t,s,x,y)$ and the second term of Birkhoff's formula satisfies the same differential equation and the same initial condition. Therefore, they coincide with each other. Theorem 6.15 has been proved. □

Remark 6.2 The above theorem means that we have proved Birkhoff's formula up to the second term by completely different method from that of Birkhoff [11] In the case $F(\gamma) \not\equiv 1$ let $k(F)$ be as (4.2.32) in Theorem 4.10. Then it seems true that the following formula,

$$\begin{aligned}&\left|k(F;\nu,t,s,x,y) - D(t,s,x,y)^{-1/2}\left(F(\gamma^*)+\nu^{-1}i(t-s)p(t,s,\nu,x,y)\right)\right| \\ &\le C\nu^{-2}|t-s|^2,\end{aligned}$$

holds even in this case if

$$(t-s)p(t,s,\nu,x,y) = \lim_{|\Delta|\to 0} p(\Delta,\nu,x,y)$$

exists. Here $p(\Delta, \nu, x, y)$ is given by (5.3.10) in Theorem 5.5. The second term of semi-classical asymptotic behavior of Feynman path integrals with integrand $F(\gamma)$ was discussed in this way by [31]. See also [29].

Remark 6.3 So far, the length of the interval $|t - s|$ is very small. If we wish to discuss the case of larger interval of time, knowledge on the calculus of variations in the large, for instance [52], will play an essential rôle. See also [49].

Part II
Supplement–Some Results of Real Analysis

Abstract Main results stated in Chap. 4 are proved in Chaps. 5 and 6 . However, results of L_2-boundedness theorem and stationary phase method over a space of large dimension are cited but their proof are not given there. Their proofs are given here, L_2-boundedness theorem is proved in Chap. 9 and the stationary phase method over a space of large dimension is given in Chaps. 7 and 8 in Part II.

Chapter 7
Kumano-go–Taniguchi Theorem

Abstract The proof of stationary phase method for oscillatory integrals over a space of large dimension is divided into two steps. As the first step we shall prove a theorem of Kumano-go-Taniguchi's type in this chapter.

7.1 Statement of Kumano-go–Taniguchi Theorem

Let T be a positive constant. We discuss oscillatory integrals of the following form in this chapter.

$$\begin{aligned} & I(\{\tau_j\}; S, a_\lambda, \nu)(x_{J+1}, x_0) \qquad (7.1.1) \\ & = \prod_{j=1}^{J} \left(\frac{-i\nu}{2\pi\tau_j}\right)^{\frac{1}{2}} \int_{\mathbf{R}^J} e^{i\nu S(x_{J+1},\dots,x_0)} a_\lambda(x_{J+1},\dots,x_0) \prod_{j=1}^{J} dx_j. \end{aligned}$$

Here $\{\tau_j\}_{j=1}^{J+1}$ are positive constants satisfying

$$\sum_{j=1}^{J+1} \tau_j = T.$$

The phase function is of the following form

$$S(x_{J+1}, x_J, \dots, x_1, x_0) = \sum_{j=1}^{J+1} \frac{(x_j - x_{j-1})^2}{2\tau_j} + \tau_j \phi_j(x_j, x_{j-1}).$$

And the amplitude function $a_\lambda(x_{J+1}, x_J, \dots, x_1, x_0)$ is a function of $(x_{J+1}, x_J, \dots, x_1, x_0)$ with parameter λ.

The assumption for functions ϕ_j is the following

Assumption 7.1 Every $\phi_j(x, y)$ is a real valued function of $C^\infty(\mathbf{R}^2)$ class. There exists an increasing sequence $0 < \kappa_0 \le \kappa_1 \le \kappa_2 \le \cdots \le \kappa_k \le \dots$ independent of j

D. Fujiwara, *Rigorous Time Slicing Approach to Feynman Path Integrals*, Mathematical Physics Studies, DOI 10.1007/978-4-431-56553-6_7

such that if $\alpha + \beta = n$,

$$|\partial^{\alpha}_{x_j}\partial^{\beta}_{x_{j-1}}\phi_j(x_j, x_{j-1})| \le \kappa_n(1 + |x_j| + |x_{j-1}|)^{(2-n)_+}, \tag{7.1.2}$$

where $(2-n)_+$ means $\max\{0, 2-n\}$,

Remark 7.1 Let Δ be the division of the interval $[a, b]$ as in Sect. 1.1.4. Then $\tau_1, \tau_2, \ldots, \tau_{J+1}$ are positive numbers satisfying $\tau_1 + \tau_2 + \ldots, +\tau_{J+1} = |b-a|$. Let $\phi(T_j, T_{j-1}, x_j, x_{j-1})$ be the function defined from the classical action $S(T_j, T_{j-1}, x_j, x_{j-1})$ by

$$S(T_j, T_{j-1}, x_j, x_{j-1}) = \frac{(x_j - x_{j-1})^2}{2\tau_j} + \tau_j\phi(T_j, T_{j-1}, x_j, x_{j-1})$$

as in (4.1.5). If we set $\phi_j(x_j, x_{j-1}) = \phi(T_j, T_{j-1}, x_j, x_{j-1})$, then the above Assumption 7.1 clearly holds. However, in the present chapter we do not assume that $\phi_j(x_j, x_{j-1}) = \phi(T_j, T_{j-1}, x_j, x_{j-1})$. Results of the present chapter are applicable for $\{\phi_j(x_j, x_{j-1})\}_j$ other than $\{\phi(T_j, T_{j-1}, x_j, x_{j-1})\}_j$.

The following assumption concerns the amplitude function $a_\lambda(x_{J+1}, \ldots, x_0)$.

Assumption 7.2 Function $a_\lambda(x_{J+1}, \ldots, x_0)$ with parameter λ is infinitely differentiable with respect to $(x_{J+1}, \ldots, x_0)$. There exist non-negative numbers m, λ and a sequence $0 < A_0 \le A_1 \le A_2 \le A_3 \le \ldots$ such that

$$\left|\prod_{j=0}^{J+1}\partial^{\alpha_j}_{x_j} a_\lambda(x_{J+1}, x_J, \ldots, x_1, x_0)\right| \le A_K(1 + |\lambda| + \sum_{j=0}^{J+1}|x_j|)^m, \tag{7.1.3}$$

if $\alpha_j \le K$ for all $j = 0, 1, 2, \ldots, J+1$.

The following Proposition is proved in just the same way as Proposition 4.2.

Proposition 7.1 *For any fixed (x_{J+1}, x_0) there exists a positive constant δ such that if $T \le \delta$ then there exists one and only one stationary point $(x_J^*, \ldots, x_1^*)$ of the phase function $S(x_{J+1}, \ldots, x_0)$ regarded as a function of $x_1, x_2, \ldots, x_J$.*

We write simply x for $(x_1, \ldots, x_J)$ and regard x_0, x_{J+1} as parameters.

The main result of this chapter is the following

Theorem 7.3 (Kumano-go–Taniguchi) *Assume that the phase function S satisfies Assumption 7.1 and the amplitude satisfies Assumption 7.2. Then there exists a positive constant δ depending only on κ_2 such that if $T \le \delta_0$, the oscillatory integral (7.1.1) can be expressed as:*

$$\begin{aligned} &I(\{\tau_j\}; S, a_\lambda, \nu)(x_{J+1}, x_0) \\ &= \left(\frac{-\nu i}{2\pi T}\right)^{1/2} e^{i\nu S^*(x_{J+1}, x_0)} b_\lambda(\{\tau_j\}; \nu, x_{J+1}, x_0). \end{aligned} \tag{7.1.4}$$

Here $S^*(x_{J+1}, x_0) = S(x_{J+1}, x_J^*, \dots, x_1^*, x_0)$.

Moreover, for any non-negative integer n there exists a positive constant C_n *and a non-negative integer* $K(n)$ *such that for any* (x_0, x_{J+1})

$$\left|\partial_{x_0}^{\alpha_0}\partial_{x_{J+1}}^{\alpha_{J+1}} b_\lambda(\{\tau_j\}; \nu, x_{J+1}, x_0)\right| \le C_n^{J+1} A_{K(n)}(1 + |\lambda| + |x_0| + |x_{J+1}|)^m, \quad (7.1.5)$$

if $|\alpha_0|, |\alpha_{J+1}| \le n$. *Here* $K(n)$, C_n *are depending only on* $\{\kappa_n\}_{n\ge 2}$ *and independent of* $\{\tau_j\}$, J, λ, ν *and of the amplitude* a_λ.

Since differentiation with respect to λ commutes with integration with respect to $(x_J, \dots, x_1)$ in (7.1.1), we have the following corollary to Theorem 7.3.

Corollary 7.1 *Suppose* δ *be as in Theorem 7.3 and* $T \le \delta$. *Let* $\mathscr{L}$ *be any differential operator with respect to the parameter* λ. *If* $|\alpha_0| \le n$, $|\alpha_{J+1}| \le n$, *then with* C_n *and* $K(n)$ *as in Theorem 7.3*

$$\begin{aligned}(1 + |\lambda| + |x_0| + |x_{J+1}|)^{-m}&\left|\mathscr{L}\partial_{x_0}^{\alpha_0}\partial_{x_{J+1}}^{\alpha_{J+1}} b_\lambda(\{\tau_j\}; \nu, x_{J+1}, x_0)\right| \\ \le C_n^{J+1} \sup_{(x_J,\dots,x_1)\in\mathbf{R}^J} &(1 + |\lambda| + \sum_{j=0}^{J+1}|x_j|)^{-m}\left|\mathscr{L}\prod_{j=0}^{J+1}\partial_{x_j}^{\alpha_j} a_\lambda(x_{J+1}, x_J, \dots, x_1, x_0)\right|.\end{aligned}$$

In order to prove this theorem we use the well known formula

$$\lim_{\varepsilon\to+0}(\frac{\nu}{2\pi})\int_{\mathbf{R}} e^{-\frac{\varepsilon}{2}\xi^2 - i\nu(\frac{1}{2}\tau\xi^2 + (x-y)\xi)}d\xi = \left(\frac{-i\nu}{2\pi\tau}\right)^{\frac{1}{2}} e^{i\nu\frac{(x-y)^2}{2\tau}}.$$

Hence, we can write

$$\begin{aligned}&I(\{\tau_j\}; S, a_\lambda, \nu)(x_{J+1}, x_0) \quad (7.1.6)\\ &= \left(\frac{\nu}{2\pi}\right)^{J+1}\int_{\mathbf{R}^{2J+1}} e^{-i\nu\Phi(x_{J+1},\xi_{J+1},x_J,\xi_J,\dots,\xi_1,x_0)} a_\lambda(x_{J+1}, x_J, \dots, x_1, x_0)\, d\xi_{J+1}\prod_{j=1}^{J} d\xi_j dx_j\end{aligned}$$

with

$$\Phi(x_{J+1}, \xi_{J+1}, x_J, \xi_J, \dots, x_1, \xi_1, x_0) = \sum_{j=1}^{J+1}\tau_j(\frac{1}{2}|\xi_j|^2 - \phi_j(x_j, x_{j-1})) + (x_j - x_{j-1})\xi_j.$$

In view of this, we discuss in this section the following oscillatory integral

$$\begin{aligned}&I(\Phi, a_\lambda, \nu)(x_{J+1}, x_0) \quad (7.1.7)\\ &= \left(\frac{\nu}{2\pi}\right)^{J+1}\int_{\mathbf{R}^{2J+1}} e^{-i\nu\Phi(x_{J+1},\xi_{J+1},x_J,\xi_J,\dots,x_1,\xi_1,x_0)}\end{aligned}$$

$$\times\, a_\lambda(x_{J+1},\xi_{J+1},\dots,x_1,\xi_1,x_0)\,d\xi_{J+1}\prod_{j=1}^{J} d\xi_j dx_j.$$

The phase function is supposed to assume the following assumptions:

Assumption 7.4 The phase function $\Phi(x_{J+1},\xi_{J+1},\dots,x_1,\xi_1,x_0)$ is of the form

$$\Phi(x_{J+1},\xi_J,\dots,x_1,\xi_1,x_0)=\sum_{j=1}^{J+1}\tau_j\phi_j(x_j,\xi_j,x_{j-1})+(x_j-x_{j-1})\xi_j. \tag{7.1.8}$$

The function ϕ_j is a real valued C^∞ function for $j=1,2,\dots,J+1$. There exists a positive nondecreasing sequence $\kappa_0\le\kappa_1\le\kappa_2\le\dots$ such that if $n=|\alpha|+|\beta|+|\gamma|$

$$|\partial_{x_j}^\alpha\partial_{\xi_j}^\beta\partial_{x_{j-1}}^\gamma\phi_j(x_j,\xi_j,x_{j-1})|\le\kappa_n(1+|x_j|+|\xi_j|+|x_{j-1}|)^{(2-n)_+}, \tag{7.1.9}$$

for $j=1,2,\dots,J+1$

The conditions for amplitude function is the following: Let m be a non-negative constant. We call the amplitude $a_\lambda(x_{J+1},\xi_{J+1},\dots,x_1,\xi_1,x_0)$ is of order m, if it satisfies the following assumption:

Assumption 7.5 For any non-negative integer $K\ge 0$ there exists a positive constant A_K such that if $|\alpha_j|\le K$ for $0\le j\le J+1$ and $|\beta_j|\le K$ for $1\le j\le J+1$, then

$$\left|\partial_{x_0}^{\alpha_0}\left(\prod_{j=1}^{J+1}\partial_{x_j}^{\alpha_j}\partial_{\xi_j}^{\beta_j}\right)a_\lambda(x_{J+1},\xi_{J+1},x_J,\xi_J,\dots,x_1,\xi_1,x_0)\right| \tag{7.1.10}$$

$$\le A_K(1+|\lambda|+\sum_{j=0}^{J+1}|x_j|+\sum_{j=1}^{J+1}|\xi_j|)^m.$$

We write

$$I(\Phi,a_\lambda,\nu)(x_{J+1},x_0)=\left(\frac{\nu}{2\pi}\right)\int_{\mathbf{R}}e^{-i\nu(x_{J+1}-x_0)\xi_{J+1}}K(x_{J+1},\xi_{J+1},x_0)d\xi_{J+1}, \tag{7.1.11}$$

Here

$$K(x_{J+1},\xi_{J+1},x_0) \tag{7.1.12}$$

$$=\left(\frac{\nu}{2\pi}\right)^J\int_{\mathbf{R}^{2J}}e^{-i\nu\Phi}a_\lambda(x_{J+1},\xi_{J+1},x_J,\dots,x_1,\xi_1,x_0)\prod_{j=1}^{J}d\xi_j dx_j,$$

and

$$\Phi = \tau_{J+1}\phi_{J+1}(x_{J+1}, \xi_{J+1}, x_J) - (x_J - x_0)\xi_{J+1} + \sum_{j=1}^{J}\{\tau_j\phi_j(x_j, \xi_j, x_{j-1}) + (x_j - x_{j-1})\xi_j\}. \tag{7.1.13}$$

Let κ_2 be as in (7.1.9). We fix a positive constant δ_1 such that

$$6\kappa_2\delta_1 < 1. \tag{7.1.14}$$

Proposition 7.2 *If $T \le \delta_1$, then for any fixed $(x_{J+1}, \xi_{J+1}, x_0) \in \mathbf{R}^3$ there exists one and only one stationary point $(x_1^*, \dots, x_J^*, \xi_1^*, \dots, \xi_J^*)$ of*

$$(x_1, x_2, \dots, x_J, \xi_1, \dots, \xi_J) \to \Phi(x_{J+1}, \xi_{J+1}, \dots, x_1, \xi_1, x_0).$$

Theorem 7.3 will be proved using the following theorem.

Theorem 7.6 (Kumano-go–Taniguchi) *Suppose that Assumptions 7.4 and 7.5 hold with some m. We assume further $T \le \delta_1$. We define*

$$\Phi^*(x_{J+1}, \xi_{J+1}, x_0) = \Phi(x_{J+1}, \xi_{J+1}, x_J^*, \xi_J^*, \dots, x_1^*, \xi_1^*, x_0)$$

and

$$\tilde{b}(x_{J+1}, \xi_{J+1}, x_0) = e^{i\nu\Phi^*(x_{J+1},\xi_{J+1},x_0)}K(x_{J+1}, \xi_{J+1}, x_0). \tag{7.1.15}$$

Then for any non-negative integer k there exist a positive constant C_k and a natural number $K(k)$ such that as far as $|\alpha| + |\beta| + |\gamma| \le k$

$$|\partial_{x_{J+1}}^{\alpha}\partial_{\xi_{J+1}}^{\beta}\partial_{x_0}^{\gamma}\tilde{b}(x_{J+1}, \xi_{J+1}, x_0)| \le C_k^{J+1}A_{K(k)}(1 + |\lambda| + |x_{J+1}| + |\xi_{J+1}| + |x_0|)^m. \tag{7.1.16}$$

Constants $K(k)$ and C_k are independent of J, λ, $\nu \ge 1$ and of $\{A_K\}_{K=0}^{\infty}$.

Proof of this theorem will be given in Sect. 7.4. Before the proof we make preparation of the proof in Sect. 7.2 and in Sect. 7.3.

Remark 7.2 This theorem first appeared in [45] in a slightly different form. Discussions here is due to [26] and [33].

Remark 7.3 The following facts are simple but useful consequence of Assumption 7.4.

$$\begin{aligned}
&\partial_{\xi_j}\partial_{\xi_k}\Phi = 0 \quad \text{if } j \ne k,\\
&\partial_{x_j}\partial_{\xi_k}\Phi = 0 \quad \text{if } j < k-1 \text{ or } j > k,\\
&\partial_{x_j}\partial_{x_k}\Phi = 0 \quad \text{if } |k-j| \ge 2.
\end{aligned}$$

7.2 The Stationary Point of the Phase Function

We prove Proposition 7.2. Let $\{\mathbf{e}_j; j = 1, 2, \ldots, 2J\}$ be the standard basis of $\mathbf{R}^{2J}$, i.e. $\mathbf{e}_j$ is the vector whose j-th component equals 1 and all the other components equal 0. In this section we fix x_{J+1}, ξ_{J+1} and x_0. We write (x, ξ) for $(x_1, \ldots, x_J, \xi_1, \ldots, \xi_J)$,i.e., $(x, \xi) = \sum_{j=1}^{J} x_j \mathbf{e}_j + \sum_{j=1}^{J} \xi_j \mathbf{e}_{J+j}$. We abbreviate $\sum_{j=1}^{J} |x_j|$ to $|x|$ and $\sum_{j=1}^{J} |\xi_j|$ to $|\xi|$. We use two norms of a vector $(x, \xi) \in \mathbf{R}^{2J}$:

$$\|(x, \xi)\|_{l^1} = |x| + |\xi| = |x_1| + |x_2| + \ldots + |x_J| + |\xi_1| + |\xi_2| + \ldots + |\xi_J|$$

and

$$\|(x, \xi)\|_{l^\infty} = \max\{|x_1|, |x_2|, \ldots, |x_J|, |\xi_1|, |\xi_2|, \ldots, |\xi_J|\}.$$

If A is a linear mapping from $\mathbf{R}^{2J}$ to itself. The symbol $\|A\|_{l^p \to l^q}$ stands for its operator norm from the sourse space $(\mathbf{R}^{2J}, \| \ \|_{l^p})$ to the target space $(\mathbf{R}^{2J}, \| \ \|_{l^q})$, where p and q equals either 1 or ∞.

Let Φ be the function defined by (7.1.13). We define for $j = 1, 2, \ldots, J$

$$z_j(x_j, \xi_j, x_{j-1}) = \partial_{\xi_j} \Phi \quad \text{and} \quad \zeta_j(x_{j+1}, \xi_{j+1}, x_j, \xi_j, x_{j-1}) = \partial_{x_j} \Phi. \tag{7.2.1}$$

We define also the map

$$\mathscr{F} : \mathbf{R}^{2J} \ni (x, \xi) \longrightarrow (z, \zeta) \in \mathbf{R}^{2J}. \tag{7.2.2}$$

We wish to prove that $\mathscr{F}$ is a diffeomorphism of $\mathbf{R}^{2J}$. Calculation shows that

$$z_j = x_j - x_{j-1} + \tau_j \partial_{\xi_j} \phi_j(x_j, \xi_j, x_{j-1}) \tag{7.2.3}$$
$$\zeta_j = -\xi_{j+1} + \xi_j + \tau_{j+1} \partial_{x_j} \phi_{j+1}(x_{j+1}, \xi_{j+1}, x_j) + \tau_j \partial_{x_j} \phi_j(x_j, \xi_j, x_{j-1}). \tag{7.2.4}$$

We decompose $\mathscr{F}$ into the sum of a linear part Δ and a nonlinear mapping $\mathscr{K}$,i.e.,

$$(z, \zeta) = \mathscr{F}(x, \xi) = \Delta(x, \xi) + \mathscr{K}(x, \xi). \tag{7.2.5}$$

Although the letter Δ is the same letter as is used to express the name of division (1.2.4) of time interval, no confusion will be caused. When we use Δ to express the name of division of the interval in this and the next section, we shall explicitly mention so.

The linear map Δ is the mapping $\mathbf{R}^{2J} \ni (x, \xi) \to (y, \eta) \in \mathbf{R}^{2J}$ given by

$$y = \sum_{j=1}^{J} y_j \mathbf{e}_j = x_1 \mathbf{e}_1 + \sum_{j=2}^{J} (x_j - x_{j-1}) \mathbf{e}_j,$$

$$\eta = \sum_{j=1}^{J} \eta_j \mathbf{e}_{J+j} = \sum_{j=1}^{J-1} (\xi_j - \xi_{j+1}) \mathbf{e}_{J+j} + \xi_J \mathbf{e}_{2J}.$$

The map $\mathcal{K} : \mathbf{R}^{2J} \ni (x, \xi) \to (y, \eta) \in \mathbf{R}^{2J}$ is given by

$$y = \sum_{j=1}^{J} y_j \mathbf{e}_j = -x_0 \mathbf{e}_1 + \sum_{j=1}^{J} \tau_j \partial_{\xi_j} \phi_j \mathbf{e}_j, \tag{7.2.6}$$

$$\eta = \sum_{j=1}^{J} \eta_j \mathbf{e}_j = -\xi_{J+1} \mathbf{e}_{2J} + \sum_{j=1}^{J} (\tau_{j+1} \partial_{x_j} \phi_{j+1} + \tau_j \partial_{x_j} \phi_j) \mathbf{e}_{j+J}. \tag{7.2.7}$$

Here and hereafter we use abbreviation $\partial_{\xi_1} \phi_1$ for $\partial_{\xi_1} \phi_1(x_1, \xi_1, x_0)$, etc..

We consider differential map $D\mathcal{F}_{(x,\xi)}$ of $\mathcal{F}$ at a point (x, ξ). We write by $D\mathcal{F}_{(x,\xi)}(x', \xi')$ for the image of a vector (x', ξ') by the differential map $D\mathcal{F}_{(x,\xi)}$. In accordance with (7.2.5) we have decomposition

$$D\mathcal{F}_{(x,\xi)}(x', \xi') = \Delta(x', \xi') + D\mathcal{K}_{(x,\xi)}(x', \xi'), \tag{7.2.8}$$

where $D\mathcal{K}_{(x,\xi)}(x', \xi')$ is the image of (x', ξ') by the differential map $D\mathcal{K}_{(x,\xi)}$ of $\mathcal{K}$ at (x, ξ).

We express a vector in $\mathbf{R}^{2J}$ as a column vector. And we can give matrix expression of linear maps Δ and $D\mathcal{K}_{(x,\xi)}$. The subspace spanned by $\{\mathbf{e}_j; j = 1, 2, \dots, J\}$ and the subspace spanned by $\{\mathbf{e}_{j+J}; j = 1, 2, \dots, J\}$ are invariant subspaces of Δ and Δ splits into two parts. Its matrix expression has block expression:

$$\Delta = \begin{pmatrix} \Delta_1 & 0 \\ 0 & {}^t\Delta_1 \end{pmatrix},$$

where Δ_1 and ${}^t\Delta_1$ are two $J \times J$ square matrix defined by

$$\Delta_1 = \begin{pmatrix} 1 & 0 & 0 & \dots\dots \\ -1 & 1 & 0 & \dots\dots \\ \dots & \dots & \dots & \dots\dots \\ \dots & \dots & \dots & \dots\dots \\ \dots & \dots\dots & -1 & 1 \end{pmatrix} \quad \text{and} \quad {}^t\Delta_1 = \begin{pmatrix} 1 & -1 & 0 & \dots\dots \\ 0 & 1 & -1 & 0 \dots \\ \dots & \dots & \dots & \dots\dots \\ \dots & \dots & \dots & \dots\dots \\ \dots & \dots\dots & 0 & 1 \end{pmatrix}.$$

The matrix corresponding to $D\mathcal{K}_{(x,\xi)}$ has also block decomposition:

$$D\mathcal{K}_{(x,\xi)} = \begin{pmatrix} A & B \\ C & {}^tA \end{pmatrix}, \tag{7.2.9}$$

where A, B and C are $J \times J$ square matrix with the following entries:

$$A = \begin{pmatrix} \tau_1 \partial_{x_1} \partial_{\xi_1} \phi_1 & 0 & 0 & 0 & \dots & 0 \\ \tau_2 \partial_{x_1} \partial_{\xi_2} \phi_2 & \tau_2 \partial_{x_2} \partial_{\xi_2} \phi_2 & 0 & 0 & \dots & 0 \\ 0 & \tau_3 \partial_{x_2} \partial_{\xi_3} \phi_3 & \tau_3 \partial_{x_3} \partial_{\xi_3} \phi_3 & 0 & \dots & \dots \\ 0 & 0 & \tau_4 \partial_{x_3} \partial_{\xi_4} \phi_4 & \tau_4 \partial_{x_4} \partial_{\xi_4} \phi_4 & \dots & \dots \\ 0 & \dots & \dots & \dots & \dots & 0 \\ 0 & \dots & \dots & \dots & \dots & \tau_J \partial_{x_J} \partial_{\xi_J} \phi_J \end{pmatrix}$$

$$B = \begin{pmatrix} \tau_1 \partial_{\xi_1}{}^2 \phi_1 & 0 & 0 & 0 \dots & 0 \\ 0 & \tau_2 \partial_{\xi_2}{}^2 \phi_2 & 0 & 0 \dots & 0 \\ 0 & 0 & \tau_3 \partial_{\xi_3}{}^2 \phi_3 & 0 \dots & \dots \\ \dots & \dots & \dots & \dots \dots & \dots \\ 0 & 0 & 0 & \dots \dots & \tau_J \partial_{\xi_J}{}^2 \phi_J \end{pmatrix}$$

$$C = \begin{pmatrix} \tau_1 \partial_{x_1}{}^2 \phi_1 + \tau_2 \partial_{x_1}{}^2 \phi_2 & \tau_2 \partial_{x_2} \partial_{x_1} \phi_2 & 0 & 0 \dots \\ \tau_2 \partial_{x_1} \partial_{x_2} \phi_2 & \tau_2 \partial_{x_2}{}^2 \phi_2 + \tau_3 \partial_{x_2}{}^2 \phi_3 & \tau_3 \partial_{x_3} \partial_{x_2} \phi_3 & 0 \dots \\ 0 & \dots & \dots & \dots \dots \\ \dots & \dots & \dots & \dots \dots \\ 0 & 0 & \dots & \dots \dots \end{pmatrix}.$$

The linear map Δ is non-singular:

$$\Delta^{-1} = \begin{pmatrix} \Delta_1^{-1} & 0 \\ 0 & {}^t\Delta_1^{-1} \end{pmatrix},$$

where

$$\Delta_1^{-1} = \begin{pmatrix} 1 & 0 & 0 & \dots & \\ 1 & 1 & 0 & 0 & \dots \\ 1 & 1 & 1 & 0 & \dots \\ \dots & & & & \\ 1 & 1 & \dots & 1 & 1 \end{pmatrix}.$$

The next remark is clear.

Remark 7.4 The following inequality holds for any $(x, \xi) \in \mathbf{R}^{2J}$:

$$\|\Delta^{-1}(x, \xi)\|_{l^\infty} \le \max\{|x|, |\xi|\} \le \|(x, \xi)\|_{l^1}.$$

Equations (7.2.5) and (7.2.8) are written as

$$\Delta^{-1}(z, \zeta) = (x, \xi) + \Delta^{-1}\mathscr{K}(x, \xi), \tag{7.2.10}$$

$$\Delta^{-1} D\mathscr{F}_{(x,\xi)}[(x', \xi')] = (I + \Delta^{-1} D\mathscr{K}_{(x,\xi)})(x', \xi'). \tag{7.2.11}$$

Hereafter we shall often identify linear mappings Δ and $D\mathcal{K}_{(x,\xi)}$, e.t.c.with their corresponding matrix.

Lemma 7.1

$$\|\Delta^{-1}D\mathcal{K}_{(x,\xi)}\|_{l^\infty\to l^\infty} \le 6\kappa_2 T.$$

Proof Let $(x',\xi')\in\mathbf{R}^{2J}$, $A=\|(x',\xi')\|_{l^\infty}$ and $(z,\zeta)=D\mathcal{K}_{(x,\xi)}[(x',\xi')]$. Then the definition of $D\mathcal{K}_{(x,\xi)}$ and Assumption 7.4 give

$$\begin{aligned}|z| &\le A\sum_{j=2}^{J}|\tau_j\partial_{x_{j-1}}\partial_{\xi_j}\phi_j| + A\sum_{j=1}^{J}(|\tau_j\partial_{x_j}\partial_{\xi_j}\phi_j| + |\tau_j{\partial_{\xi_j}}^2\phi_j|)\\ &\le A\sum_{j=1}^{J}3\kappa_2\tau_j \le 3\kappa_2 TA.\end{aligned}$$

Similarly,

$$\begin{aligned}|\zeta| &\le A\sum_{j=1}^{J-1}|\tau_{j+1}\partial_{x_j}\partial_{x_{j+1}}\phi_{j+1}| + A\sum_{j=1}^{J}|\tau_j{\partial_{x_j}}^2\phi_j + \tau_{j+1}{\partial_{x_j}}^2\phi_{j+1}| + A\sum_{j=1}^{J-1}|\tau_{j+1}\partial_{x_{j+1}}\partial_{x_j}\phi_{j+1}|\\ &\quad + A\sum_{j=1}^{J-1}|\tau_{j+1}\partial_{x_j}\partial_{\xi_{j+1}}\phi_{j+1}| + A\sum_{j=1}^{J}|\tau_j\partial_{x_j}\partial_{\xi_j}\phi_j|\\ &\le A\sum_{j=1}^{J-1}\kappa_2\tau_{j+1} + A\sum_{j=1}^{J}\kappa_2(\tau_j+\tau_{j+1}) + A\sum_{j=1}^{J-1}\kappa_2\tau_{j+1} + A\sum_{j=1}^{J-1}\kappa_2\tau_{j+1} + A\sum_{j=1}^{J}\kappa_2\tau_j\\ &\le 6\kappa_2 TA.\end{aligned}$$

It follows from these and Remark 7.4 that

$$\|\Delta^{-1}(z,\zeta)\|_{l^\infty} \le 6\kappa_2 TA.$$

This proves Lemma. □

Corollary 7.2 *Let δ_1 be so small as $6\kappa_2\delta_1 < 1$. If $T\le\delta_1$, then $(I\pm\Delta^{-1}D\mathcal{K}_{(x,\xi)})$ is a non singular matrix. We have*

$$\begin{aligned}&\|(I\pm\Delta^{-1}D\mathcal{K}_{(x,\xi)})^{-1}\|_{l^\infty\to l^\infty} \le (1-6\kappa_2 T)^{-1},\\ &\|((I\pm\Delta^{-1}D\mathcal{K}_{(x,\xi)})^{-1} - I)\|_{l^\infty\to l^\infty} \le 6\kappa_2 T(1-6\kappa_2 T)^{-1}.\\ &\|((I\pm\Delta^{-1}D\mathcal{K}_{(x,\xi)})^{-1} - (I\mp\Delta^{-1}D\mathcal{K}_{(x,\xi)}))\|_{l^\infty\to l^\infty}\\ &\qquad \le (6\kappa_2 T)^2(1-6\kappa_2 T)^{-1}.\end{aligned} \tag{7.2.12}$$

Lemma 7.2 *For any* $(x,\xi), (x',\xi') \in \mathbf{R}^{2J}$

$$\|\Delta^{-1}\mathcal{K}(x,\xi) - \Delta^{-1}\mathcal{K}(x',\xi')\|_{l^\infty} \le 6\kappa_2 T\|(x,\xi) - (x',\xi'))\|_{l^\infty}.$$

Proof Since Δ^{-1} is a linear map we have

$$\Delta^{-1}\mathcal{K}(x,\xi) - \Delta^{-1}\mathcal{K}(x',\xi') = \int_0^1 \Delta^{-1}D\mathcal{K}_{(x(t),\xi(t))}[(x,\xi) - (x',\xi')]dt, \tag{7.2.13}$$

where $(x(t),\xi(t)) = t(x,\xi) + (1-t)(x',\xi')$. It follows from this and Lemma 7.1 that

$$\|\Delta^{-1}\mathcal{K}(x,\xi) - \Delta^{-1}\mathcal{K}(x',\xi')\|_{l^\infty} \le 6\kappa_2 T\|(x,\xi) - (x',\xi')\|_{l^\infty}.$$

Lemma has been proved. □

We can now prove Proposition 7.2.

Proposition 7.3 *Let* δ_1 *be so small that* $6\kappa_2\delta_1 < 1$. *If* $T \le \delta_1$, *then the map* $\mathcal{F}$: $(x,\xi) \to (z,\zeta)$ *is a global diffeomorphism from* $\mathbf{R}^{2J}$ *to itself. In particular, there exists one and only one stationary point* (x^*,ξ^*) *of* Φ. *The stationary point* (x^*,ξ^*) *satisfies the following equation:*

$$\Delta(x^*,\xi^*) + \mathcal{K}(x^*,\xi^*) = 0. \tag{7.2.14}$$

Let $J(\mathcal{F})$ *be the jacobian determinant of* $\mathcal{F}$. *Then*

$$(1 - 6\kappa_2 T)^{2J} \le |J(\mathcal{F})| \le (1 + 6\kappa_2 T)^{2J}. \tag{7.2.15}$$

Proof Recall that $6\kappa_2\delta_1 < 1$. If $T \le \delta_1$,the mapping $\Delta^{-1}\mathcal{K}$ is a strict contraction mapping in $(\mathbf{R}^{2J}, l^\infty)$ because of Lemma 7.2 and the definition of δ_1.

Once an arbitrary point $(z,\zeta) \in \mathbf{R}^{2J}$ is fixed, the map

$$(x,\xi) \to \Delta^{-1}(z,\zeta) - \Delta^{-1}\mathcal{K}(x,\xi)$$

is also a contraction mapping. Therefore, it follows from the contraction mapping theorem that this map has one and only one fixed point (x,ξ). The fixed point (x,ξ) satisfies the equation

$$\Delta^{-1}(z,\zeta) - \Delta^{-1}\mathcal{K}(x,\xi) = (x,\xi).$$

This and (7.2.5) prove that

$$(z,\zeta) = \mathcal{F}(x,\xi).$$

Thus we have proved that $\mathcal{F}$ is a one to one and onto mapping from $\mathbf{R}^{2J}$ to itself.

We have

$$J(\mathscr{F}) = \det \Delta \det \left(I + \Delta^{-1} D\mathscr{K}_{(x,\xi)}\right) = \det \left(I + \Delta^{-1} D\mathscr{K}_{(x,\xi)}\right),$$

because $\det \Delta^{-1} = 1$. And every eigenvalue λ of $\Delta^{-1} D\mathscr{K}_{(x,\xi)}$ satisfies $|\lambda| \le 6\kappa_2 T$ because of Lemma 7.1. Therefore,

$$(1 - 6\kappa_2 T)^{2J} \le |\det \left(I + \Delta^{-1} D\mathscr{K}_{(x,\xi)}\right)| \le (1 + 6\kappa_2 T)^{2J}.$$

We have proved (7.2.15).

Assumption $T < \delta_1$ and (7.2.15) implies that

$$0 < (1 - 6\kappa_2 \delta_1)^{2J} \le |J(\mathscr{F})|$$

and hence the inverse map $\mathscr{F}^{-1}$ is infinitely differentiable. Therefore, $\mathscr{F}$ is a diffeomorphism of $\mathbf{R}^{2J}$.

The stationary point (x^*, ξ^*) is nothing but $\mathscr{F}^{-1}(0, 0)$. It exists uniquely, because $\mathscr{F}$ is a one to one and onto mapping. □

Lemma 7.3 *Assume that* $T \le \delta_1$. *Then for any* $(x, \xi) \in \mathbf{R}^{2J}$

$$\begin{aligned}(1 + 6\kappa_2 T)^{-1} \|\Delta^{-1}\mathscr{F}(x, \xi)\|_{l^\infty} &\le \|(x - x^*, \xi - \xi^*)\|_{l^\infty} \\ &\le (1 - 6\kappa_2 T)^{-1} \|\Delta^{-1}\mathscr{F}(x, \xi)\|_{l^\infty}.\end{aligned}$$

And

$$\|(x - x^*, \xi - \xi^*) - \Delta^{-1}\mathscr{F}(x, \xi)\|_{l^\infty} \le 6\kappa_2 T (1 - 6\kappa_2 T)^{-1} \|\Delta^{-1}\mathscr{F}(x, \xi)\|_{l^\infty}.$$

Proof We set $(z, \zeta) = \mathscr{F}(x, \xi)$. Then we have, as (7.2.5),

$$\Delta^{-1}(z, \zeta) = (x, \xi) + \Delta^{-1}\mathscr{K}(x, \xi).$$

Subtracting

$$(0, 0) = (x^*, \xi^*) + \Delta^{-1}\mathscr{K}(x^*, \xi^*)$$

from this, we have

$$\Delta^{-1}(z, \zeta) = (x - x^*, \xi - \xi^*) + \Delta^{-1}\mathscr{K}(x, \xi) - \Delta^{-1}\mathscr{K}(x^*, \xi^*).$$

Using Lemma 7.2, we have

$$\|\Delta^{-1}(z, \zeta) - (x - x^*, \xi - \xi^*)\|_{l^\infty} \le 6\kappa_2 T \|(x - x^*, \xi - \xi^*)\|_{l^\infty}.$$

The first estimate of the Lemma follows from here. Replacing $\|(x - x^*, \xi - \xi^*)\|_{l^\infty}$ on the right hand side by the result of the first estimate, we have the second estimate.

Corollary 7.3 *Assume that $T \le \delta_1$. Then the following inequalities holds:*

$$\|(x^*, \xi^*) - \Delta^{-1}(x_0\mathbf{e}_1 + \xi_{J+1}\mathbf{e}_{2J})\|_{l^\infty} \le CT(1 + |x_{J+1}| + |\xi_{J+1}| + |x_0|),$$

with $C = (6\kappa_2 + 2\kappa_1)(1 - 6\kappa_2\delta_1)^{-1}$. In other words, for $j = 1, 2, \ldots, J$,

$$\begin{aligned} |x_j^* - x_0| &\le CT(1 + |x_{J+1}| + |\xi_{J+1}| + |x_0|),\\ |\xi_j^* - \xi_{J+1}| &\le CT(1 + |x_{J+1}| + |\xi_{J+1}| + |x_0|). \end{aligned}$$

We have also

$$\|(x^*, \xi^*)\|_{l^\infty} \le (1 - 6\kappa_2 T)^{-1}(1 + 2\kappa_1 T)(1 + |x_{J+1}| + |\xi_{J+1}| + |x_0|).$$

Proof Set $(x, \xi) = (0, 0)$ in Lemma 7.3. Then

$$\|(x^*, \xi^*) + \Delta^{-1}\mathscr{F}(0,0)\|_{l^\infty} \le 6\kappa_2 T(1 - 6\kappa_2 T)^{-1}\|\Delta^{-1}\mathscr{F}(0,0)\|_{l^\infty}.$$

Calculation and Remark 7.4 show

$$\begin{aligned} \|\Delta^{-1}(\mathscr{F}(0,0) + x_0\mathbf{e}_1 + \xi_{J+1}\mathbf{e}_{2J})\|_{l^\infty} &\le \|\mathscr{F}(0,0) + x_0\mathbf{e}_1 + \xi_{J+1}\mathbf{e}_{2J}\|_{l^1}\\ &\le 3\kappa_1 T(1 + |x_{J+1}| + |\xi_{J+1}| + |x_0|). \end{aligned}$$

This implies on one hand

$$\|\Delta^{-1}\mathscr{F}(0,0)\|_{l^\infty} \le (1 + 3\kappa_1 T)(1 + |x_{J+1}| + |\xi_{J+1}| + |x_0|),$$

and on the other hand

$$\begin{aligned} &\|(x^*, \xi^*) - \Delta^{-1}(x_0\mathbf{e}_1 + \xi_{J+1}\mathbf{e}_{2J})\|_{l^\infty}\\ &\le 6\kappa_2 T(1 - 6\kappa_2 T)^{-1}\|\Delta^{-1}\mathscr{F}(0,0)\|_{l^\infty}\\ &\quad + 2\kappa_1 T(1 + |x_{J+1}| + |\xi_{J+1}| + |x_0|). \end{aligned}$$

Combining these, we have

$$\begin{aligned} &\|(x^*, \xi^*) - \Delta^{-1}(x_0\mathbf{e}_1 + \xi_{J+1}\mathbf{e}_{2J})\|_{l^\infty}\\ &\le T(6\kappa_2(1 - 6\kappa_2\delta_1)^{-1}(1 + 3\kappa_1\delta_1) + 3\kappa_1)(1 + |x_{J+1}| + |\xi_{J+1}| + |x_0|)\\ &= T(6\kappa_2 + 2\kappa_1)(1 - 6\kappa_2\delta_1)^{-1}(1 + |x_{J+1}| + |\xi_{J+1}| + |x_0|). \end{aligned}$$

Similarly, we have

$$\|(x^*,\xi^*)\|_{l^\infty} \le (1+2\kappa_1 T)(1-6\kappa_2\delta_1)^{-1}(1+|x_{J+1}+|\xi_{J+1}|+|x_0|).$$

We have proved Corollary. □

7.3 Parameter Dependence of the Stationary Point

In the previous section parameters $(x_0, x_{J+1}, \xi_{J+1})$ are fixed. In this section we consider them as variables and study derivatives of (x^*,ξ^*) with respect to parameters.

It follows from definition (7.2.6) that for any (x,ξ) and x_0, ξ_{J+1}, x_{J+1},

$$\begin{aligned}
&\partial_{x_0}\mathcal{K}(x,\xi) = -(1+\tau_1\partial_{x_0}\partial_{\xi_1}\phi_1(x_1,\xi,x_0))\mathbf{e}_1 + \tau_1\partial_{x_0}\partial_{x_1}\phi_1(x_1,\xi_1,x_0)\mathbf{e}_{J+1},\\
&\partial_{x_{J+1}}\mathcal{K}(x,\xi) = \tau_{J+1}\partial_{x_{J+1}}\partial_{x_J}\phi_{J+1}(x_{J+1},\xi_{J+1},x_J)\mathbf{e}_{2J},\\
&\partial_{\xi_{J+1}}\mathcal{K}(x,\xi) = (-1+\tau_{J+1}\partial_{\xi_{J+1}}\partial_{x_J}\phi(x_{J+1},\xi_{J+1},x_J))\mathbf{e}_{2J}.
\end{aligned} \tag{7.3.1}$$

We abbreviate the linear map $(z,\eta)\to D\mathcal{K}_{(x^*,\xi^*)}[(z,\eta)]$ to $W(x^*,\xi^*)(z,\eta)$.

Lemma 7.4 *If $T\le\delta_1$, then there exists a positive constant C depending only on κ_2 and δ_1 such that*

$$\begin{aligned}
&\|\partial_{x_0}(x^*,\xi^*) + \Delta^{-1}\partial_{x_0}\mathcal{K}(x^*,\xi^*) + \Delta^{-1}D\mathcal{K}_{(x^*,\xi^*)}[\Delta^{-1}\mathbf{e}_1]\|_\infty \le C(\kappa_2 T)^2,\\
&\|\partial_{x_{J+1}}(x^*,\xi^*) + \Delta^{-1}\partial_{x_{J+1}}\mathcal{K}(x^*,\xi^*)\|_\infty \le C(\kappa_2 T)^2,\\
&\|\partial_{\xi_{J+1}}(x^*,\xi^*) + \Delta^{-1}\partial_{\xi_{J+1}}\mathcal{K}(x^*,\xi^*) + \Delta^{-1}D\mathcal{K}_{(x^*,\xi^*)}[\Delta^{-1}\mathbf{e}_{2J}]\|_\infty\\
&\quad\le C(\kappa_2 T)^2,
\end{aligned} \tag{7.3.2}$$

where we may choose $C = 42(1-6\kappa_2 T)^{-1}$.

Proof By (7.2.14) in Proposition 7.3, (x^*,ξ^*) satisfies

$$(x^*,\xi^*) + \Delta^{-1}\mathcal{K}(x^*,\xi^*) = 0.$$

Differentiating this, we have

$$\partial_{\xi_{J+1}}(x^*,\xi^*) + \Delta^{-1}D\mathcal{K}_{(x^*,\xi^*)}[\partial_{\xi_{J+1}}(x^*,\xi^*)] + \Delta^{-1}(\partial_{\xi_{J+1}}\mathcal{K})(x^*,\xi^*) = 0. \tag{7.3.3}$$

Here and hereafter we write $(\partial_{\xi_{J+1}}\mathcal{K})(x^*,\xi^*)$ for $\left(\partial_{\xi_{J+1}}\mathcal{K}(x,\xi)\right)\big|_{(x,\xi)=(x^*,\xi^*)}$.

By (7.3.1),

$$\|\Delta^{-1}(\partial_{\xi_{J+1}}\mathcal{K})(x^*,\xi^*)\|_{l^\infty} \le (1+\tau_{J+1}\kappa_2).$$

It follows from this, (7.3.3) and (7.2.12) in Corollary 7.2 that

$$\|\big((\partial_{\xi_{J+1}}(x^*,\xi^*) + (I - \Delta^{-1}D\mathcal{K}_{(x^*,\xi^*)})\big)\Delta^{-1}(\partial_{\xi_{J+1}}\mathcal{K})(x^*,\xi^*)\|_{l^\infty}$$
$$\le (6\kappa_2 T)^2(1-6\kappa_2 T)^{-1}(1+\tau_{J+1}\kappa_2). \tag{7.3.4}$$

Since (7.3.1) implies that

$$\|\Delta^{-1}(\partial_{\xi_{J+1}}\mathcal{K})(x^*,\xi^*) + \mathbf{e}_{2J})\|_{l^\infty} \le \kappa_2 T,$$

we obtain by Lemma 7.1

$$\|\Delta^{-1}D\mathcal{K}_{(x^*,\xi^*)}((\Delta^{-1}\partial_{\xi_{J+1}}\mathcal{K})(x^*,\xi^*) + \Delta^{-1}\mathbf{e}_{2J})\|_{l^\infty} \le 6(\kappa_2 T)^2.$$

Combining this with (7.3.4), we obtain (7.3.2).

Similarly, other inequalities of Lemma can be proved.

□

Corollary 7.4 *If $T \le \delta_1$, then there exists a positive constant C which may depends on κ_2, δ_1 but not on J such that for $k = 1, 2, \ldots, J$*

$$|\partial_{\xi_{J+1}}x_k^* + \sum_{j=1}^{k}\tau_j(\partial_{\xi_j}^2\phi_j)(x_j^*,\xi_j^*,x_{j-1}^*)| \le 42(1-6\kappa_2 T)^{-1}(\kappa_2 T)^2,$$

where $x_0^* = x_0$.

Proof Corollary follows from that

$$\begin{aligned}\Delta\partial_{\xi_{J+1}}\mathcal{K}(x^*,\xi^*) &+ \Delta^{-1}D\mathcal{K}_{(x^*,\xi^*)}(\Delta^{-1}\mathbf{e}_{2J}) \\ &= \sum_{j=1}^{J}(-1+\tau_{J+1}\partial_{x_J}\partial_{\xi_{J+1}}\phi_{J+1}(x_{J+1},\xi_{J+1},x_J^*))\mathbf{e}_{j+J} \\ &+ \sum_{k=1}^{J}\Big(\sum_{j=1}^{k}\tau_j(\partial_{\xi_j}^2\phi_j)(x_j^*,\xi_j^*,x_{j-1}^*)\Big)\mathbf{e}_k.\end{aligned} \tag{7.3.5}$$

□

Corollary 7.5 *If $T \le \delta_1$, then there exists a positive constant C depending only on κ_2 and δ_1 such that for $j = 1, 2, \ldots, J$*

$$\begin{aligned}|\partial_{x_0}x_j^* - 1| \le C\kappa_2 T, &\quad |\partial_{x_0}\xi_j^*| \le C\kappa_2 T, \\ |\partial_{x_{J+1}}x_j^*| \le C\kappa_2 T, &\quad |\partial_{x_{J+1}}\xi_j^*| \le C\kappa_2 T, \\ |\partial_{\xi_{J+1}}x_j^*| \le C\kappa_2 T, &\quad |\partial_{\xi_{J+1}}\xi_j^* - 1| \le C\kappa_2 T.\end{aligned} \tag{7.3.6}$$

Here we may choose $C = 7\kappa_2(1-6\kappa_2 T)^{-1}$.

Proof It follows from (7.3.5) that

$$\|\Delta\partial_{\xi_{J+1}}\mathscr{K}(x^*,\xi^*)+\Delta^{-1}D\mathscr{K}_{(x^*,\xi^*)}(\Delta^{-1}\mathbf{e}_{2J})+\sum_{j=1}^{J}\mathbf{e}_{J+j}\|_{l^\infty}$$
$$\le \kappa_2 T+\tau_{J+1}\kappa_2.$$

(7.3.6) follows from this and (7.3.1).

Other inequalities are proved in the similar manner. □

We write $(\partial_{x_0}\mathscr{K})(x^*,\xi^*)$ for $\left(\partial_{x_0}\mathscr{K}(x,\xi)\right)\big|_{(x,\xi)=(x^*,\xi^*)}$, e.t.c.. Moreover we often use abbreviation $(\partial_{x_0}\mathscr{K})^*$ for $(\partial_{x_0}\mathscr{K})(x^*,\xi^*)$, e.t.c..

Proposition 7.4 *Assume $T\le\delta_1$. For any integers $n\ge 1$ there exists a positive constant C which may depend on δ_1 and on κ_l for $(2\le l\le n+1)$ but is independent of J such that for $j=1,2,\dots,J$*

$$|\partial_{x_0}^\alpha\partial_{x_{J+1}}^\beta\partial_{\xi_{J+1}}^\gamma(x_j^*-x_0)|\le CT,\quad |\partial_{x_0}^\alpha\partial_{x_{J+1}}^\beta\partial_{\xi_{J+1}}^\gamma(\xi_j^*-\xi_{J+1})|\le CT,$$

as far as $1\le\alpha+\beta+\gamma\le n$.

Proof We set $(y,\eta)=\mathscr{K}(x,\xi)$ and $(y^*,\eta^*)=\mathscr{K}(x^*,\xi^*)$. Then

$$\begin{aligned} y_1^* &= -x_0+\tau_1(\partial_{\xi_1}\phi_1)(x_1^*,\xi_1^*,x_0), \\ y_j^* &= \tau_j(\partial_{\xi_j}\phi_j)(x_j^*,\xi_j^*,x_{j-1}^*) \qquad \text{for } j=2,3,\dots,J, \end{aligned} \tag{7.3.7}$$

and for $j=1,2,3,\dots,J-1$

$$\begin{aligned} \eta_j^* &= \tau_j(\partial_{x_j}\phi_j)(x_j^*,\xi_j^*,x_{j-1}^*)+\tau_{j+1}(\partial_{x_j}\phi_{j+1})(x_{j+1}^*,\xi_{j+1}^*,x_j^*), \\ \eta_J^* &= -\xi_{J+1}+\tau_J(\partial_{x_J}\phi_J)(x_J^*,\xi_J^*,x_{J-1}^*)+\tau_{J+1}(\partial_{x_J}\phi_{J+1})(x_{J+1},\xi_{J+1},x_J^*). \end{aligned}$$

Here x_0^* stands for x_0. We shall prove the proposition by induction on $\alpha+\beta+\gamma$.

The case $\alpha+\beta+\gamma=1$ is proved in Corollary 7.5.

Assume that $n\ge 2$ and Proposition is true in the case $\alpha+\beta+\gamma\le n-1$. Let (α,β,γ) be such that $\alpha+\beta+\gamma=n$. We can easily see that for $j=1,2,3,\dots,J$

$$\begin{aligned} &\partial_{x_0}^\alpha\partial_{x_{J+1}}^\beta\partial_{\xi_{J+1}}^\gamma((\partial_{\xi_j}\phi_j)(x_j^*,\xi_j^*,x_{j-1}^*)) \\ &= (\partial_{x_{j-1}}\partial_{\xi_j}\phi_j)^*\cdot\partial_{x_0}^\alpha\partial_{x_{J+1}}^\beta\partial_{\xi_{J+1}}^\gamma x_{j-1}^*+(\partial_{x_j}\partial_{\xi_j}\phi_j)^*\cdot\partial_{x_0}^\alpha\partial_{x_{J+1}}^\beta\partial_{\xi_{J+1}}^\gamma x_j^* \\ &\quad +(\partial_{\xi_j}^2\phi_j)^*\cdot\partial_{x_0}^\alpha\partial_{x_{J+1}}^\beta\partial_{\xi_{J+1}}^\gamma\xi_j^* \\ &\quad +R. \end{aligned}$$

The remainder term R is a polynomial of derivatives of $\partial_{\xi_j}\phi_j(x_j,\xi_j,x_{j-1})$ by (x_j,ξ_j,x_{j-1}) evaluated at $(x_j^*,\xi_j^*,x_{j-1}^*)$ of order at most n and such derivatives of x_j^*, x_{j-1}^*, ξ_j^* as $\partial_{x_0}^{\alpha'}\partial_{x_{J+1}}^{\beta'}\partial_{\xi_{J+1}}^{\gamma'}x_j^*$, $\partial_{x_0}^{\alpha'}\partial_{x_{J+1}}^{\beta'}\partial_{\xi_{J+1}}^{\gamma'}x_{j-1}^*$ and $\partial_{x_0}^{\alpha'}\partial_{x_{J+1}}^{\beta'}\partial_{\xi_{J+1}}^{\gamma'}\xi_j^*$ with $\alpha'+\beta'+\gamma'\le n-1$. Therefore, by induction hypothesis,

$$|R|\le C$$

with some positive constant C which may depend on κ_l, $(2\le l\le n+1)$ and δ_1 but is independent of J. It follows from this and similar estimates that for $j=1,2,3,\dots,J$

$$\begin{aligned}
&\partial_{x_0}^{\alpha}\partial_{x_{J+1}}^{\beta}\partial_{\xi_{J+1}}^{\gamma}(y_j^*) \qquad (7.3.8)\\
&=\tau_j\Big((\partial_{x_{j-1}}\partial_{\xi_j}\phi_j)^*\partial_{x_0}^{\alpha}\partial_{x_{J+1}}^{\beta}\partial_{\xi_{J+1}}^{\gamma}(x_{j-1}^*)+(\partial_{x_j}\partial_{\xi_j}\phi_j)^*\partial_{x_0}^{\alpha}\partial_{x_{J+1}}^{\beta}\partial_{\xi_{J+1}}^{\gamma}(x_j^*)\\
&\quad+(\partial_{\xi_j}^2\phi_j)^*\partial_{x_0}^{\alpha}\partial_{x_{J+1}}^{\beta}\partial_{\xi_{J+1}}^{\gamma}(\xi_j^*)\Big)\\
&\quad+\tau_j p_j,
\end{aligned}$$

and

$$\begin{aligned}
&\partial_{x_0}^{\alpha}\partial_{x_{J+1}}^{\beta}\partial_{\xi_{J+1}}^{\gamma}(\eta_j^*) \qquad (7.3.9)\\
&=\tau_j\Big((\partial_{x_{j-1}}\partial_{x_j}\phi_j)^*\partial_{x_0}^{\alpha}\partial_{x_{J+1}}^{\beta}\partial_{\xi_{J+1}}^{\gamma}(x_{j-1}^*)+(\partial_{x_j}^2\phi_j)^*\partial_{x_0}^{\alpha}\partial_{x_{J+1}}^{\beta}\partial_{\xi_{J+1}}^{\gamma}(x_j^*)\\
&\quad+(\partial_{\xi_j}\partial_{x_j}\phi_j)^*\partial_{x_0}^{\alpha}\partial_{x_{J+1}}^{\beta}\partial_{\xi_{J+1}}^{\gamma}(\xi_j^*)\Big)\\
&\quad+\tau_{j+1}\Big((\partial_{x_j}^2\phi_{j+1})^*\partial_{x_0}^{\alpha}\partial_{x_{J+1}}^{\beta}\partial_{\xi_{J+1}}^{\gamma}(x_j^*)+(\partial_{x_{j+1}}\partial_{x_j}\phi_{j+1})^*\partial_{x_0}^{\alpha}\partial_{x_{J+1}}^{\beta}\partial_{\xi_{J+1}}^{\gamma}(x_{j+1}^*)\\
&\quad+(\partial_{\xi_{j+1}}\partial_{x_j}\phi_{j+1})^*\partial_{x_0}^{\alpha}\partial_{x_{J+1}}^{\beta}\partial_{\xi_{J+1}}^{\gamma}(\xi_{j+1}^*)\Big)\\
&\quad+\tau_j q_{j1}+\tau_{j+1}q_{j2}.
\end{aligned}$$

Here we write ξ_{J+1}^* for ξ_{J+1} and we have $|p_j|\le C$, $|q_{j1}|\le C$ and $|q_{j2}|\le C$. We define the vector $(r,\rho)=(r_1,r_2,\dots,r_J,\rho_1,\rho_2,\dots,\rho_J)\in\mathbf{R}^{2J}$ by the equation

$$r_j=\tau_j p_j,\quad \rho_j=\tau_j q_{j1}+\tau_{j+1}q_{j2}\quad \text{for } j=1,2,\dots,J.$$

Then (7.3.8) and (7.3.9) imply that

$$\Delta\partial_{x_0}^{\alpha}\partial_{x_{J+1}}^{\beta}\partial_{\xi_{J+1}}^{\gamma}(x^*,\xi^*)+D\mathscr{K}_{(x^*,\xi^*)}[\partial_{x_0}^{\alpha}\partial_{x_{J+1}}^{\beta}\partial_{\xi_{J+1}}^{\gamma}(x^*,\xi^*)]=-(r,\rho).$$

Thus

$$\partial_{x_0}^{\alpha}\partial_{x_{J+1}}^{\beta}\partial_{\xi_{J+1}}^{\gamma}(x^*,\xi^*)=-(I+\Delta^{-1}D\mathscr{K}_{(x^*,\xi^*)})^{-1}\Delta^{-1}(r,\rho).$$

Since $\|\Delta^{-1}(r,\rho)\|_{l^\infty}\le\|(r,\rho)\|_{l^1}\le 3CT$, we have

$$\|\partial_{x_0}^{\alpha}\partial_{x_{J+1}}^{\beta}\partial_{\xi_{J+1}}^{\gamma}(x^*,\xi^*)\|_{l^\infty} \leq 3TC(1-6\kappa_2 T)^{-1}.$$

We have proved Proposition in the case $\alpha+\beta+\gamma = n \geq 2$, because

$$\partial_{x_0}^{\alpha}\partial_{x_{J+1}}^{\beta}\partial_{\xi_{J+1}}^{\gamma}x_0 = 0, \quad \partial_{x_0}^{\alpha}\partial_{x_{J+1}}^{\beta}\partial_{\xi_{J+1}}^{\gamma}\eta_{J+1} = 0.$$

Induction process has been completed. We have proved Proposition. □

The set of C^∞ functions vanishing at (x^*,ξ^*) is an ideal of the ring of C^∞ functions on $\mathbf{R}^{2J}$. The set of functions $\{x_j - x_j^*, \xi_j - \xi_j^*; \ j = 1, 2, \ldots, J\}$ is a basis of the ideal. Let $(z,\zeta) = \mathscr{F}(x,\xi)$. If $T \leq \delta_1$, then the set $\{z_j, \zeta_j : j = 1, 2, \ldots, J\}$ is also a basis because of Lemma 7.3. We shall discuss the matrix expression of change of basis from the first basis to the second.

Subtracting

$$(0,0) = \Delta(x^*,\xi^*) + \mathscr{K}(x^*,\xi^*)$$

from (7.2.5), we have

$$(z,\zeta) = \Delta(x-x^*,\xi-\xi^*) + \mathscr{K}(x,\xi) - \mathscr{K}(x^*,\xi^*).$$

Just as we did in the proof of Lemma 7.2 we have

$$(z,\zeta) = \Delta(x-x^*,\xi-\xi^*) + \int_0^1 D\mathscr{K}_{(x(t),\xi(t))}[(x-x^*,\xi-\xi^*)]dt,$$

here $(x(t),\xi(t)) = t(x,\xi) + (1-t)(x^*,\xi^*)$ for $0 \leq t \leq 1$. We write L for the matrix expression of $\int_0^1 D\mathscr{K}_{(x(t),\xi(t))}dt$. Then

$$P = \Delta + L$$

is the matrix expression for the change of basis of the ideal,i.e.,

$$(z,\zeta) = P\ \ (x-x^*,\xi-\xi^*).$$

It is clear that the matrix L is decomposed into blocks:

$$L = \begin{pmatrix} \tilde{A} & \tilde{B} \\ \tilde{C} & {}^t\tilde{A} \end{pmatrix}. \tag{7.3.10}$$

Here $\tilde{A}$, $\tilde{B}$ and $\tilde{C}$ are $J \times J$ square matrices with the following entries:

$$\tilde{A} = \begin{pmatrix} a_{11} & 0 & 0 & 0 & \dots & 0 \\ a_{21} & a_{22} & 0 & 0 & \dots & 0 \\ 0 & a_{32} & a_{33} & 0 & \dots & 0 \\ 0 & 0 & a_{43} & a_{44} & \dots & 0 \\ \dots & \dots & \dots & \dots & \dots & \\ \dots & \dots & \dots & \dots & \dots & a_{JJ} \end{pmatrix} \qquad \tilde{B} = \begin{pmatrix} b_{11} & 0 & 0 & 0 & \dots & 0 \\ 0 & b_{21} & 0 & 0 & \dots & 0 \\ 0 & 0 & b_{33} & 0 & \dots & \dots \\ \dots & \dots & \dots & \dots & \dots & \dots \\ 0 & 0 & 0 & \dots & \dots & b_{JJ} \end{pmatrix}$$

and

$$\tilde{C} = \begin{pmatrix} c_{11} & c_{12} & 0 & 0 & 0 & \dots & \dots \\ c_{21} & c_{22} & c_{23} & 0 & 0 & \dots & \dots \\ 0 & c_{32} & c_{33} & c_{34} & 0 & \dots & \dots \\ \dots & \dots & \dots & \dots & \dots & \dots & \dots \\ 0 & 0 & \dots & \dots & \dots & \dots & c_{JJ} \end{pmatrix},$$

where the non-vanishing elements of $\tilde{A}$, $\tilde{B}$ and $\tilde{C}$ are

$$a_{jj} = \tau_j \int_0^1 (\partial_{x_j}\partial_{\xi_j}\phi_j)(x_j(t), \xi_j(t), x_{j-1}(t))\,dt \quad \text{for } j = 1, 2, \dots, J,$$

$$a_{jj-1} = \tau_j \int_0^1 (\partial_{x_{j-1}}\partial_{\xi_j}\phi_j)(x_j(t), \xi_j(t), x_{j-1}(t))\,dt \quad \text{for } j = 2, \dots, J,$$

$$b_{jj} = \tau_j \int_0^1 (\partial_{\xi_j}^2\phi_j)(x_j(t), \xi_j(t), x_{j-1}(t))\,dt \quad \text{for } j = 1, 2, \dots, J,$$

$$c_{jj-1} = \tau_j \int_0^1 (\partial_{x_{j-1}}\partial_{x_j}\phi_j)(x_j(t), \xi_j(t), x_{j-1}(t))\,dt \quad \text{for } j = 2, 3, \dots, J,$$

$$c_{jj} = \tau_j \int_0^1 (\partial_{x_j}^2\phi_j)(x_j(t), \xi_j(t), x_{j-1}(t))\,dt + \tau_{j+1} \int_0^1 (\partial_{x_j}^2\phi_{j+1})(x_{j+1}(t), \xi_{j+1}(t), x_j(t))\,dt \quad \text{for } j = 1, 2, \dots, J,$$

$$c_{jj+1} = \tau_{j+1} \int_0^1 (\partial_{x_{j+1}}\partial_{x_j}\phi_{j+1})(x_{j+1}(t), \xi_{j+1}(t), x_j(t))\,dt \quad \text{for } j = 1, 2, \dots, J-1.$$

Other elements of the above matrices are all 0.

Proposition 7.5 *If $T \leq \delta_1$, then we have*

1. $\|\Delta^{-1}L\|_{l^\infty \to l^\infty} \leq 6\kappa_2 T$.
2. *If $T \leq \delta_1$, then $(I + \Delta^{-1}L)$ and P are nonsingular matrices. Moreover, $\|(I + \Delta^{-1}L)^{-1}\|_{l^\infty \to l^\infty} \leq (1 - 6\kappa_2\delta_1)^{-1}$.*

Proof The first assertion of the proposition follows from Lemma 7.1.
If $T \leq \delta_1$, then $\|\Delta^{-1}L\|_{l^\infty \to l^\infty} < 6\kappa_2\delta_1 < 1$. Therefore $(I + \Delta^{-1}L)$ is invertible and $\|(I + \Delta^{-1}L)^{-1}\|_{l^\infty \to l^\infty} < (1 - 6\kappa_2\delta_1)^{-1}$. □

Every entry of L is a function of (x, ξ) and x_0, x_{J+1}, ξ_{J+1}. By virtue of Remark 7.3, we can obtain a good bound for derivatives of L.

Proposition 7.6 *Let $(\alpha, \beta) = (\alpha_1, \alpha_2, \ldots, \alpha_J, \beta_1, \beta_2, \ldots, \beta_J)$ be a multi-index and $\alpha_0, \alpha_{J+1}, \beta_{J+1}$ be non-negative integers. Let $K > 0$ be a constant. Suppose that $\alpha_j \leq K$ for $0 \leq j \leq J+1$ and $\beta_j \leq K$ for $1 \leq j \leq J+1$ and that $\partial_{x_0}^{\alpha_0} \partial_{x_{J+1}}^{\alpha_{J+1}} \partial_{\xi_{J+1}}^{\beta_{J+1}} \partial_x^\alpha \partial_\xi^\beta L$ does not identically vanish. Then there exists a positive constant c_K depending only on κ_n with $2 \leq n \leq 6K+2$ and δ_1 but not on J such that one of the following cases occurs:*

Case 1 *If $\beta_k \neq 0$ with some k $(1 \leq k \leq J)$, then $\alpha_j = 0$ for $j \neq k-1, k$ and $\beta_j = 0$ for $j \neq k$ and*

$$\|\partial_{x_0}^{\alpha_0} \partial_{x_{J+1}}^{\alpha_{J+1}} \partial_{\xi_{J+1}}^{\beta_{J+1}} \partial_{x_{k-1}}^{\alpha_{k-1}} \partial_{x_k}^{\alpha_k} \partial_{\xi_k}^{\beta_k} L\|_{l^\infty \to l^1} \leq c_K \tau_k. \tag{7.3.11}$$

Case 2 *If $\beta_j = 0$ for all $j = 1, 2, \ldots, J$ and there exists some k $(1 \leq k \leq J)$ such that $\alpha_j = 0$ for all $j < k$ and $\alpha_k > 0$, then $\alpha_j = 0$ for all $j > k+1$ and*

$$\|\partial_{x_0}^{\alpha_0} \partial_{x_{J+1}}^{\alpha_{J+1}} \partial_{\xi_{J+1}}^{\beta_{J+1}} \partial_{x_k}^{\alpha_k} \partial_{x_{k+1}}^{\alpha_{k+1}} L\|_{l^\infty \to l^1} \leq c_K (\tau_k + \tau_{k+1}). \tag{7.3.12}$$

Case 3 *If $\alpha_j = 0$ and $\beta_j = 0$ for all $j = 1, 2, \ldots, J$, then*

$$\|\partial_{x_0}^{\alpha_0} \partial_{x_{J+1}}^{\alpha_{J+1}} \partial_{\xi_{J+1}}^{\beta_{J+1}} L\|_{l^\infty \to l^1} \leq c_K T. \tag{7.3.13}$$

Proof Assume that $\partial_{x_0}^{\alpha_0} \partial_{x_{J+1}}^{\alpha_{J+1}} \partial_{\xi_{J+1}}^{\beta_{J+1}} \partial_x^\alpha \partial_\xi^\beta L$ does not identically vanish.

Case 1. If $\beta_k \neq 0$ with some $1 \leq k \leq J$, then it follows from Remark 7.3 that $\alpha_j = 0$ for $j \neq k-1, k$ and $\beta_j = 0$ for $j \neq k$. We know that $|\partial_{x_{k-1}}^{\alpha_{k-1}} \partial_{x_k}^{\alpha_k} \partial_{\xi_k}^{\beta_k} a_{jl}| = 0$ if $j \neq k$ and that

$$\begin{aligned} &\partial_{x_{k-1}}^{\alpha_{k-1}} \partial_{x_k}^{\alpha_k} \partial_{\xi_k}^{\beta_k} a_{kk-1} \\ &= \tau_k \int_0^1 t^{\alpha_{k-1}+\alpha_k+\beta_k} (\partial_{x_{k-1}}^{\alpha_{k-1}+1} \partial_{x_k}^{\alpha_k} \partial_{\xi_k}^{\beta_k+1} \phi_k)(x_k(t), \xi_k(t), x_{k-1}(t))) \, dt. \end{aligned}$$

Hence

$$\begin{aligned} &\partial_{x_0}^{\alpha_0} \partial_{x_{J+1}}^{\alpha_{J+1}} \partial_{\xi_{J+1}}^{\beta_{J+1}} \partial_{x_{k-1}}^{\alpha_{k-1}} \partial_{x_k}^{\alpha_k} \partial_{\xi_k}^{\beta_k} a_{kk-1} \\ &= \tau_k \int_0^1 t^{\alpha_{k-1}+\alpha_k+\beta_k} \partial_{x_0}^{\alpha_0} \partial_{x_{J+1}}^{\alpha_{J+1}} \partial_{\xi_{J+1}}^{\beta_{J+1}} ((\partial_{x_{k-1}}^{\alpha_{k-1}+1} \partial_{x_k}^{\alpha_k} \partial_{\xi_k}^{\beta_k+1} \phi_k)(x_k(t), \xi_k(t), x_{k-1}(t))) \, dt. \end{aligned}$$

As a result of Proposition 7.4 and definition of $(x(t), \xi(t))$ there exists a positive constant c depending on κ_n, $2 \leq n \leq 3K+1$ and on δ_1 such that for $j = 1, 2, \ldots, J$

$$|\partial_{x_0}^{\alpha'_0} \partial_{x_{J+1}}^{\alpha'_{J+1}} \partial_{\xi_{J+1}}^{\beta'_{J+1}} x_j(t)| \leq c \quad \text{and} \quad |\partial_{x_0}^{\alpha'_0} \partial_{x_{J+1}}^{\alpha'_{J+1}} \partial_{\xi_{J+1}}^{\beta'_{J+1}} \xi_j(t)| \leq c,$$

as far as $\alpha'_0, \alpha'_{J+1}, \beta'_{J+1} \leq K$. Thus

$$|\partial_{x_0}^{\alpha_0}\partial_{x_{J+1}}^{\alpha_{J+1}}\partial_{\xi_{J+1}}^{\beta_{J+1}}((\partial_{x_{k-1}}^{\alpha_{k-1}+1}\partial_{x_k}^{\alpha_k}\partial_{\xi_k}^{\beta_k+1}\phi_k)(x_k(t),\xi_k(t),x_{k-1}(t)))| \le c$$

with another constant $c > 0$ depending only on κ_n $(2 \le n \le 6K+2)$ and on δ_1. Therefore, there exists a positive constant c such that

$$|\partial_{x_0}^{\alpha_0}\partial_{x_{J+1}}^{\alpha_{J+1}}\partial_{\xi_{J+1}}^{\beta_{J+1}}\partial_{x_{k-1}}^{\alpha_{k-1}}\partial_{x_k}^{\alpha_k}\partial_{\xi_k}^{\beta_k} a_{k\,k-1}| \le c\tau_k,$$

here c depends only on κ_n, $2 \le n \le 6K+2$ and δ_1.

Similarly,

$$|\partial_{x_0}^{\alpha_0}\partial_{x_{J+1}}^{\alpha_{J+1}}\partial_{\xi_{J+1}}^{\beta_{J+1}}\partial_{x_{k-1}}^{\alpha_{k-1}}\partial_{x_k}^{\alpha_k}\partial_{\xi_k}^{\beta_k} a_{k\,k}| \le c\tau_k.$$

It is clear that $\partial_{\xi_k} b_{jj} = 0$ for $j \ne k$. In the same way as above we can prove that there exists a positive constant c such that

$$|\partial_{x_0}^{\alpha_0}\partial_{x_{J+1}}^{\alpha_{J+1}}\partial_{\xi_{J+1}}^{\beta_{J+1}}\partial_{x_{k-1}}^{\alpha_{k-1}}\partial_{x_k}^{\alpha_k}\partial_{\xi_k}^{\beta_k} b_{k\,k}| \le c\tau_k.$$

Similarly $\partial_{\xi_k} c_{jl} = 0$ unless (j,l) equals one of $(k-1,k-1)$, $(k-1,k)$, $(k,k-1)$ and (k,k). We can prove as above

$$|\partial_{x_0}^{\alpha_0}\partial_{x_{J+1}}^{\alpha_{J+1}}\partial_{\xi_{J+1}}^{\beta_{J+1}}\partial_{x_{k-1}}^{\alpha_{k-1}}\partial_{x_k}^{\alpha_k}\partial_{\xi_k}^{\beta_k} c_{k-1\,k-1}| \le c\tau_k, \quad |\partial_{x_0}^{\alpha_0}\partial_{x_{J+1}}^{\alpha_{J+1}}\partial_{\xi_{J+1}}^{\beta_{J+1}}\partial_{x_{k-1}}^{\alpha_{k-1}}\partial_{x_k}^{\alpha_k}\partial_{\xi_k}^{\beta_k} c_{k-1\,k}| \le c\tau_k,$$
$$|\partial_{x_0}^{\alpha_0}\partial_{x_{J+1}}^{\alpha_{J+1}}\partial_{\xi_{J+1}}^{\beta_{J+1}}\partial_{x_{k-1}}^{\alpha_{k-1}}\partial_{x_k}^{\alpha_k}\partial_{\xi_k}^{\beta_k} c_{k\,k-1}| \le c\tau_k, \quad |\partial_{x_0}^{\alpha_0}\partial_{x_{J+1}}^{\alpha_{J+1}}\partial_{\xi_{J+1}}^{\beta_{J+1}}\partial_{x_{k-1}}^{\alpha_{k-1}}\partial_{x_k}^{\alpha_k}\partial_{\xi_k}^{\beta_k} c_{k\,k}| \le c\tau_k.$$

Therefore, there exists a positive constant c depending on κ_n, $2 \le n \le 6K+2$ and on δ_1 such that

$$\begin{aligned}
&\|\partial_{x_0}^{\alpha_0}\partial_{x_{J+1}}^{\alpha_{J+1}}\partial_{\xi_{J+1}}^{\beta_{J+1}}\partial_x^{\alpha}\partial_\xi^{\beta} L\|_{l^\infty\to l^1}\\
&\le 2|\partial_{x_0}^{\alpha_0}\partial_{x_{J+1}}^{\alpha_{J+1}}\partial_{\xi_{J+1}}^{\beta_{J+1}}\partial_{x_{k-1}}^{\alpha_{k-1}}\partial_{x_k}^{\alpha_k}\partial_{\xi_k}^{\beta_k} a_{k\,k-1}| + 2|\partial_{x_0}^{\alpha_0}\partial_{x_{J+1}}^{\alpha_{J+1}}\partial_{\xi_{J+1}}^{\beta_{J+1}}\partial_{x_{k-1}}^{\alpha_{k-1}}\partial_{x_k}^{\alpha_k}\partial_{\xi_k}^{\beta_k} a_{k\,k}|\\
&+ |\partial_{x_0}^{\alpha_0}\partial_{x_{J+1}}^{\alpha_{J+1}}\partial_{\xi_{J+1}}^{\beta_{J+1}}\partial_{x_{k-1}}^{\alpha_{k-1}}\partial_{x_k}^{\alpha_k}\partial_{\xi_k}^{\beta_k} b_{k\,k}|\\
&+ |\partial_{x_0}^{\alpha_0}\partial_{x_{J+1}}^{\alpha_{J+1}}\partial_{\xi_{J+1}}^{\beta_{J+1}}\partial_{x_{k-1}}^{\alpha_{k-1}}\partial_{x_k}^{\alpha_k}\partial_{\xi_k}^{\beta_k} c_{k-1\,k-1}| + |\partial_{x_0}^{\alpha_0}\partial_{x_{J+1}}^{\alpha_{J+1}}\partial_{\xi_{J+1}}^{\beta_{J+1}}\partial_{x_{k-1}}^{\alpha_{k-1}}\partial_{x_k}^{\alpha_k}\partial_{\xi_k}^{\beta_k} c_{k-1\,k}|\\
&+ |\partial_{x_0}^{\alpha_0}\partial_{x_{J+1}}^{\alpha_{J+1}}\partial_{\xi_{J+1}}^{\beta_{J+1}}\partial_{x_{k-1}}^{\alpha_{k-1}}\partial_{x_k}^{\alpha_k}\partial_{\xi_k}^{\beta_k} c_{k\,k-1}| + |\partial_{x_0}^{\alpha_0}\partial_{x_{J+1}}^{\alpha_{J+1}}\partial_{\xi_{J+1}}^{\beta_{J+1}}\partial_{x_{k-1}}^{\alpha_{k-1}}\partial_{x_k}^{\alpha_k}\partial_{\xi_k}^{\beta_k} c_{k\,k}|\\
&\le c\tau_k.
\end{aligned}$$

Case 2 Next we shall discuss the case $\partial_{x_0}^{\alpha_0}\partial_{x_{J+1}}^{\alpha_{J+1}}\partial_{\xi_{J+1}}^{\beta_{J+1}}\partial_x^{\alpha}\partial_\xi^{\beta} L$ does not identically vanish and $\beta_j = 0$ for all $1 \le j \le J$ but $\alpha \ne 0$. Let k, $1 \le k \le J$ be such that $\alpha_k > 0$ but $\alpha_j = 0$ for all j with $1 \le j < k$. Then it follows from Remark 7.3 that $\alpha_j = 0$ for all $j > k+1$. Furthermore, $\partial_{x_k}^{\alpha_k} a_{jl} = 0$ unless $j = k, k+1$ and discussions similar to the discussions above give

$$|\partial_{x_0}^{\alpha_0}\partial_{x_{J+1}}^{\alpha_{J+1}}\partial_{\xi_{J+1}}^{\beta_{J+1}}\partial_{x_k}^{\alpha_k}\partial_{x_{k+1}}^{\alpha_{k+1}}a_{kk-1}| \le \begin{cases} 0 & \text{when } \alpha_{k+1} \neq 0 \\ c\tau_k & \text{when } \alpha_{k+1} = 0 \end{cases},$$

$$|\partial_{x_0}^{\alpha_0}\partial_{x_{J+1}}^{\alpha_{J+1}}\partial_{\xi_{J+1}}^{\beta_{J+1}}\partial_{x_k}^{\alpha_k}\partial_{x_{k+1}}^{\alpha_{k+1}}a_{kk}| \le \begin{cases} 0 & \text{when } \alpha_{k+1} \neq 0 \\ c\tau_k & \text{when } \alpha_{k+1} = 0 \end{cases},$$

$$|\partial_{x_0}^{\alpha_0}\partial_{x_{J+1}}^{\alpha_{J+1}}\partial_{\xi_{J+1}}^{\beta_{J+1}}\partial_{x_k}^{\alpha_k}\partial_{x_{k+1}}^{\alpha_{k+1}}a_{k+1k}| \le c\tau_{k+1},$$

$$|\partial_{x_0}^{\alpha_0}\partial_{x_{J+1}}^{\alpha_{J+1}}\partial_{\xi_{J+1}}^{\beta_{J+1}}\partial_{x_k}^{\alpha_k}\partial_{x_{k+1}}^{\alpha_{k+1}}a_{k+1k+1}| \le c\tau_{k+1}.$$

If $j \neq k, k+1$ then $\partial_{x_k}b_{jj} = 0$. And we have estimates

$$|\partial_{x_0}^{\alpha_0}\partial_{x_{J+1}}^{\alpha_{J+1}}\partial_{\xi_{J+1}}^{\beta_{J+1}}\partial_{x_k}^{\alpha_k}\partial_{x_{k+1}}^{\alpha_{k+1}}b_{kk}| \le \begin{cases} 0 & \text{when } \alpha_{k+1} \neq 0 \\ c\tau_k & \text{when } \alpha_{k+1} = 0, \end{cases}$$

$$|\partial_{x_0}^{\alpha_0}\partial_{x_{J+1}}^{\alpha_{J+1}}\partial_{\xi_{J+1}}^{\beta_{J+1}}\partial_{x_k}^{\alpha_k}\partial_{x_{k+1}}^{\alpha_{k+1}}b_{k+1k+1}| \le c\tau_{k+1}.$$

It is clear from Remark 7.3 that $\partial_{x_k}c_{jl} = 0$ unless (j,l) is one of $(k-1,k-1)$, $(k-1,k)$, $(k,k-1)$, (k,k), $(k,k+1)$, $(k+1,k)$ and $(k+1,k+1)$. And the followings hold:

$$|\partial_{x_0}^{\alpha_0}\partial_{x_{J+1}}^{\alpha_{J+1}}\partial_{\xi_{J+1}}^{\beta_{J+1}}\partial_{x_k}^{\alpha_k}\partial_{x_{k+1}}^{\alpha_{k+1}}c_{k-1k-1}| \le \begin{cases} 0 & \text{when } \alpha_{k+1} \neq 0 \\ c\tau_k & \text{when } \alpha_{k+1} = 0, \end{cases}$$

$$|\partial_{x_0}^{\alpha_0}\partial_{x_{J+1}}^{\alpha_{J+1}}\partial_{\xi_{J+1}}^{\beta_{J+1}}\partial_{x_k}^{\alpha_k}\partial_{x_{k+1}}^{\alpha_{k+1}}c_{k-1k}| \le \begin{cases} 0 & \text{when } \alpha_{k+1} \neq 0 \\ c\tau_k & \text{when } \alpha_{k+1} = 0, \end{cases}$$

$$|\partial_{x_0}^{\alpha_0}\partial_{x_{J+1}}^{\alpha_{J+1}}\partial_{\xi_{J+1}}^{\beta_{J+1}}\partial_{x_k}^{\alpha_k}\partial_{x_{k+1}}^{\alpha_{k+1}}c_{kk-1}| \le \begin{cases} 0 & \text{when } \alpha_{k+1} \neq 0 \\ c\tau_k & \text{when } \alpha_{k+1} = 0, \end{cases}$$

$$|\partial_{x_0}^{\alpha_0}\partial_{x_{J+1}}^{\alpha_{J+1}}\partial_{\xi_{J+1}}^{\beta_{J+1}}\partial_{x_k}^{\alpha_k}\partial_{x_{k+1}}^{\alpha_{k+1}}c_{kk}| \le \begin{cases} c\tau_{k+1} & \text{when } \alpha_{k+1} \neq 0 \\ c(\tau_k + \tau_{k+1}) & \text{when } \alpha_{k+1} = 0, \end{cases}$$

$$|\partial_{x_0}^{\alpha_0}\partial_{x_{J+1}}^{\alpha_{J+1}}\partial_{\xi_{J+1}}^{\beta_{J+1}}\partial_{x_k}^{\alpha_k}\partial_{x_{k+1}}^{\alpha_{k+1}}c_{kk+1}| \le c\tau_{k+1},$$

$$|\partial_{x_0}^{\alpha_0}\partial_{x_{J+1}}^{\alpha_{J+1}}\partial_{\xi_{J+1}}^{\beta_{J+1}}\partial_{x_k}^{\alpha_k}\partial_{x_{k+1}}^{\alpha_{k+1}}c_{k+1k}| \le c\tau_{k+1},$$

$$|\partial_{x_0}^{\alpha_0}\partial_{x_{J+1}}^{\alpha_{J+1}}\partial_{\xi_{J+1}}^{\beta_{J+1}}\partial_{x_k}^{\alpha_k}\partial_{x_{k+1}}^{\alpha_{k+1}}c_{k+1k+1}| \le c\tau_{k+1}.$$

Therefore, similar discussion as above proves that there exists a positive constant c depending only on κ_j, $(2 \le j \le 5K+2)$ and on δ_1 such that

$$\left\| \partial_{x_0}^{\alpha_0}\partial_{x_{J+1}}^{\alpha_{J+1}}\partial_{\xi_{J+1}}^{\beta_{J+1}}\partial_{x_k}^{\alpha_k}\partial_{x_{k+1}}^{\alpha_{k+1}} L \right\|_{l^\infty \to l^1} \le c(\tau_k + \tau_{k+1}).$$

Case 3 Finally, we shall discuss the case $\alpha_j = \beta_j = 0$ for $1 \le j \le J$ and $\partial_{x_0}^{\alpha_0} \partial_{x_{J+1}}^{\alpha_{J+1}} \partial_{\xi_{J+1}}^{\beta_{J+1}} L$ does not identically vanish. If $\alpha_0 + \alpha_{J+1} + \beta_{J+1} = 0$, then discussion similar to the proof of Lemma 7.1 proves that

$$\|L\|_{l^\infty \to l^1} \le 6\kappa_2 T.$$

Assume that $\alpha_0 + \alpha_{J+1} + \beta_{J+1} > 0$. Then using Proposition 7.4, we have for any $1 \le k \le J$

$$|\partial_{x_0}^{\alpha_0} \partial_{x_{J+1}}^{\alpha_{J+1}} \partial_{\xi_{J+1}}^{\beta_{J+1}} a_{kk-1}| \le c\tau_k, \quad |\partial_{x_0}^{\alpha_0} \partial_{x_{J+1}}^{\alpha_{J+1}} \partial_{\xi_{J+1}}^{\beta_{J+1}} a_{kk}| \le c\tau_k,$$

$$|\partial_{x_0}^{\alpha_0} \partial_{x_{J+1}}^{\alpha_{J+1}} \partial_{\xi_{J+1}}^{\beta_{J+1}} b_{kk}| \le c\tau_k,$$

$$\begin{aligned}
&|\partial_{x_0}^{\alpha_0} \partial_{x_{J+1}}^{\alpha_{J+1}} \partial_{\xi_{J+1}}^{\beta_{J+1}} c_{k,k-1}| \le c\tau_k, \quad |\partial_{x_0}^{\alpha_0} \partial_{x_{J+1}}^{\alpha_{J+1}} \partial_{\xi_{J+1}}^{\beta_{J+1}} c_{kk}| \le c(\tau_k + \tau_{k+1}),\\
&|\partial_{x_0}^{\alpha_0} \partial_{x_{J+1}}^{\alpha_{J+1}} \partial_{\xi_{J+1}}^{\beta_{J+1}} c_{k,k+1}| \le c\tau_{k+1}.
\end{aligned}$$

Therefore, taking sum of all these from $k = 1$ to J, we can find a constant $c > 0$ depending only on κ_j for $2 \le j \le 3K + 2$ and δ_1 such that

$$\begin{aligned}
&\|\partial_{x_0}^{\alpha_0} \partial_{x_{J+1}}^{\alpha_{J+1}} \partial_{\xi_{J+1}}^{\beta_{J+1}} L\|_{l^\infty \to l^1}\\
&\le \sum_{k=2}^{J} (2|\partial_{x_0}^{\alpha_0} \partial_{x_{J+1}}^{\alpha_{J+1}} \partial_{\xi_{J+1}}^{\beta_{J+1}} a_{kk-1}| + |\partial_{x_0}^{\alpha_0} \partial_{x_{J+1}}^{\alpha_{J+1}} \partial_{\xi_{J+1}}^{\beta_{J+1}} c_{k,k-1}|)\\
&+ \sum_{k=1}^{J} \Big(2|\partial_{x_0}^{\alpha_0} \partial_{x_{J+1}}^{\alpha_{J+1}} \partial_{\xi_{J+1}}^{\beta_{J+1}} a_{kk}| + |\partial_{x_0}^{\alpha_0} \partial_{x_{J+1}}^{\alpha_{J+1}} \partial_{\xi_{J+1}}^{\beta_{J+1}} b_{kk}| + |\partial_{x_0}^{\alpha_0} \partial_{x_{J+1}}^{\alpha_{J+1}} \partial_{\xi_{J+1}}^{\beta_{J+1}} c_{kk}|\Big)\\
&+ \sum_{k=1}^{J-1} |\partial_{x_0}^{\alpha_0} \partial_{x_{J+1}}^{\alpha_{J+1}} \partial_{\xi_{J+1}}^{\beta_{J+1}} c_{k,k+1}|\\
&\le cT.
\end{aligned}$$

Proposition has been proved. □

Assume that $T \le \delta_1$. Then it follows from Proposition 7.5 that $Q = (I + \Delta^{-1}L)^{-1} = P^{-1}\Delta$ exists. This means that

$$(x - x^*, \xi - \xi^*) = Q\Delta^{-1}(z, \zeta). \tag{7.3.14}$$

We shall prove that each entry of matrix Q satisfies Assumption 7.5 in Sect. 7.1.

Lemma 7.5 *Assume that $T \le \delta_1$. Let $Q = (I + \Delta^{-1}L)^{-1}$ and $C_0 = (1 - 6\kappa_2\delta_1)^{-1}$. Then there exists a non-decreasing sequence $C_0 \le C_1 \le C_2 \le \cdots \le C_n \le \ldots$ such that if $|\alpha_j| \le K$ $(j = 0, \ldots, J+1)$ and $|\beta_j| \le K$ $(j = 1, 2, \ldots, J+1)$, then*

$$\left\| \prod_{j=0}^{J+1} \partial_{x_j}^{\alpha_j} \prod_{j=1}^{J+1} \partial_{\xi_j}^{\beta_j} Q \right\|_{l^\infty \to l^\infty} \leq C_K^{|\alpha_0| + \sum_{j=1}^{J+1}(|\alpha_j| + |\beta_j|)}. \tag{7.3.15}$$

We can so choose C_K for $K = 2, 3, \dots$ as

$$C_K = \max\{C_{K-1},\ 2^{6K+2} C_0 c_K T\}, \tag{7.3.16}$$

where c_K is the constant given in Proposition 7.6, which depends on κ_n with $2 \leq n \leq 6K+2$ and on δ_1 but not on J.

Proof We change notation throughout this proof. We write α for $(\alpha_0, \alpha_1, \dots, \alpha_{J+1})$ and β for $(\beta_1, \beta_2, \dots, \beta_{J+1})$. We define $m(\alpha, \beta) = \max\{\alpha_j, \beta_k :\ 0 \leq j \leq J+1, 1 \leq k \leq J+1\}$ for any index (α, β). Let (α', β') be $(\alpha'_0, \dots, \alpha'_{J+1}, \beta'_1, \dots, \beta'_{J+1})$. We say $(\alpha, \beta) > (\alpha', \beta')$ if $\alpha_j \geq \alpha'_j$ and $\beta_j \geq \beta'_j$ for each j and $(\alpha, \beta) \neq (\alpha', \beta')$. This relation ">" is a partial order relation among indices. It is clear that $m(\alpha, \beta) \geq m(\alpha', \beta')$ if $(\alpha, \beta) > (\alpha', \beta')$.

We prove the lemma by induction on $|\alpha| + |\beta| = \sum_{j=0}^{J+1} \alpha_j + \sum_{j=1}^{J+1} \beta_j$.

In the case $|\alpha| + |\beta| = 0$ the statement is true because of Proposition 7.5.

Assume that $n \geq 1$ and Lemma is true in the case $|\alpha| + |\beta| \leq n-1$. Let (α, β) be any index such that $|\alpha| + |\beta| = n$ and $K = m(\alpha, \beta)$. We abbreviate $\prod_{j=0}^{J+1} \partial_{x_j}^{\alpha_j} \prod_{j=1}^{J+1} \partial_{\xi_j}^{\alpha_j}$ to $\partial_x^\alpha \partial_\xi^\beta$. Since $(I + \Delta^{-1} L) Q = I$, $\partial_x^\alpha \partial_\xi^\beta \left((I + \Delta^{-1} L) Q\right) = 0$. By Leibnitz's rule,

$$-(I + \Delta^{-1} L) \partial_x^\alpha \partial_\xi^\beta Q = \sum_{\substack{(\alpha', \beta') \leq (\alpha, \beta) \\ (0,0) \neq (\alpha', \beta')}} \binom{\alpha}{\alpha'} \binom{\beta}{\beta'} \Delta^{-1} \partial_x^{\alpha'} \partial_\xi^{\beta'} L \partial_x^{\alpha - \alpha'} \partial_\xi^{\beta - \beta'} Q.$$

Since $\|(I + \Delta^{-1} L)^{-1}\|_{l^\infty \to l^\infty} \leq C_0$ and $\|\Delta^{-1}\|_{l^1 \to l^\infty} \leq 1$,

$$\begin{aligned}
&\|\partial_x^\alpha \partial_\xi^\beta Q\|_{l^\infty \to l^\infty} \\
&\leq C_0 \sum_{\substack{(\alpha', \beta') \leq (\alpha, \beta) \\ (0,0) \neq (\alpha', \beta')}} \binom{\alpha}{\alpha'} \binom{\beta}{\beta'} \|\Delta^{-1} \partial_x^{\alpha'} \partial_\xi^{\beta'} L\|_{l^\infty \to l^\infty} \|\partial_x^{\alpha - \alpha'} \partial_\xi^{\beta - \beta'} Q\|_{l^\infty \to l^\infty} \\
&\leq C_0 \sum_{\substack{(\alpha', \beta') \leq (\alpha, \beta) \\ (0,0) \neq (\alpha', \beta')}} \binom{\alpha}{\alpha'} \binom{\beta}{\beta'} \|\partial_x^{\alpha'} \partial_\xi^{\beta'} L\|_{l^\infty \to l^1} \|\partial_x^{\alpha - \alpha'} \partial_\xi^{\beta - \beta'} Q\|_{l^\infty \to l^\infty}.
\end{aligned}$$

If $(0,0) \neq (\alpha', \beta') \leq (\alpha, \beta)$, then $m(\alpha', \beta') \leq m(\alpha, \beta) = K$ and $|\alpha - \alpha'| + |\beta - \beta'| \leq n-1$. It follows from this and the induction hypothesis that

$$\|\partial_x^{\alpha - \alpha'} \partial_\xi^{\beta - \beta'} Q\|_{l^\infty \to l^\infty} \leq C_K^{|\alpha - \alpha'| + |\beta - \beta'|} \leq C_K^{n-1},$$

because $C_K \geq C_j > 1$ for $0 \leq j \leq K$. Hence

$$\|\partial_x^\alpha \partial_\xi^\beta Q\|_{l^\infty \to l^\infty} \leq C_0 C_K^{n-1} \sum_{\substack{(\alpha',\beta')\leq(\alpha,\beta)\\(0,0)\neq(\alpha',\beta')}} \binom{\alpha}{\alpha'}\binom{\beta}{\beta'} \|\partial_x^{\alpha'} \partial_\xi^{\beta'} L\|_{l^\infty \to l^1}.$$

Proposition 7.6 shows that all indices (α', β') such as $\partial_x^{\alpha'} \partial_\xi^{\beta'} L \neq 0$ are classified into three classes. Using the classification, we write

$$\begin{aligned}
&\sum_{\substack{(\alpha',\beta')\leq(\alpha,\beta)\\(0,0)\neq(\alpha',\beta')}} \binom{\alpha}{\alpha'}\binom{\beta}{\beta'} \|\partial_x^{\alpha'} \partial_\xi^{\beta'} L\|_{l^\infty \to l^1} \qquad (7.3.17)\\
=&\sum_{k=1}^{J} \sum_{\substack{\alpha_0'\leq\alpha_0\\ \alpha'_{J+1}\leq\alpha_{J+1}\\ \beta'_{J+1}\leq\beta_{J+1}}} \sum_{\substack{0\neq\beta_k'\\ \beta_k'\leq\beta_k}} \sum_{\substack{\alpha_{k-1}'\leq\alpha_{k-1}\\ \alpha_k'\leq\alpha_k}} \binom{\alpha_0}{\alpha_0'}\binom{\alpha_{J+1}}{\alpha'_{J+1}}\binom{\beta_{J+1}}{\beta'_{J+1}}\binom{\alpha_{k-1}}{\alpha'_{k-1}}\binom{\alpha_k}{\alpha'_k}\binom{\beta_k}{\beta'_k}\\
&\times \|\partial_{x_0}^{\alpha_0'} \partial_{x_{J+1}}^{\alpha'_{J+1}} \partial_{\xi_{J+1}}^{\beta'_{J+1}} \partial_{x_{k-1}}^{\alpha'_{k-1}} \partial_{x_k}^{\alpha'_k} \partial_{\xi_k}^{\beta'_k} L\|_{l^\infty \to l^1}\\
&+\sum_{k=1}^{J} \sum_{\substack{\alpha_0'\leq\alpha_0\\ \alpha'_{J+1}\leq\alpha_{J+1}\\ \beta'_{J+1}\leq\beta_{J+1}}} \sum_{\substack{0\neq\alpha_k'\\ \alpha_k'\leq\alpha_k}} \sum_{\alpha'_{k+1}\leq\alpha_{k+1}} \binom{\alpha_0}{\alpha_0'}\binom{\alpha_{J+1}}{\alpha'_{J+1}}\binom{\beta_{J+1}}{\beta'_{J+1}}\binom{\alpha_k}{\alpha'_k}\binom{\alpha_{k+1}}{\alpha'_{k+1}}\\
&\times \|\partial_{x_0}^{\alpha_0'} \partial_{x_{J+1}}^{\alpha'_{J+1}} \partial_{\xi_{J+1}}^{\beta'_{J+1}} \partial_{x_k}^{\alpha'_k} \partial_{x_{k+1}}^{\alpha'_{k+1}} L\|_{l^\infty \to l^1}\\
&+\sum_{\substack{\alpha_0'\leq\alpha_0,\alpha'_{J+1}\leq\alpha_{J+1},\beta'_{J+1}\leq\beta_{J+1}\\ 1\leq\alpha_0+\alpha_{J+1}+\beta_{J+1}}} \binom{\alpha_0}{\alpha_0'}\binom{\alpha_{J+1}}{\alpha'_{J+1}}\binom{\beta_{J+1}}{\beta'_{J+1}} \|\partial_{x_0}^{\alpha_0'} \partial_{x_{J+1}}^{\alpha'_{J+1}} \partial_{\xi_{J+1}}^{\beta'_{J+1}} L\|_{l^\infty \to l^1}.
\end{aligned}$$

Proposition 7.6 says that in the case $\beta_k' \neq 0$ there exists a positive constant c_K depending on κ_j, $(2 \leq j \leq 6K+2)$ and δ_1 but not on J such that

$$\|\partial_{x_0}^{\alpha_0'} \partial_{x_{J+1}}^{\alpha'_{J+1}} \partial_{\xi_{J+1}}^{\beta'_{J+1}} \partial_{x_{k-1}}^{\alpha'_{k-1}} \partial_{x_k}^{\alpha'_k} \partial_{\xi_k}^{\beta'_k} L\|_{l^\infty \to l^1} \leq c_K \tau_k.$$

Thus the first term of the right hand side of (7.3.17) is bounded by

$$\begin{aligned}
&\sum_{k=1}^{J} \sum_{\substack{\alpha_0'\leq\alpha_0\\ \alpha'_{J+1}\leq\alpha_{J+1}\\ \beta'_{J+1}\leq\beta_{J+1}}} \sum_{\substack{0\neq\beta_k'\\ \beta_k'\leq\beta_k}} \sum_{\substack{\alpha_{k-1}'\leq\alpha_{k-1}\\ \alpha_k'\leq\alpha_k}} \binom{\alpha_0}{\alpha_0'}\binom{\alpha_{J+1}}{\alpha'_{J+1}}\binom{\beta_{J+1}}{\beta'_{J+1}}\binom{\alpha_{k-1}}{\alpha'_{k-1}}\binom{\alpha_k}{\alpha'_k}\binom{\beta_k}{\beta'_k} c_K \tau_k\\
&\leq \sum_{k=1}^{J} 2^{6K} c_K \tau_k \leq 2^{6K} c_K T.
\end{aligned}$$

Similarly, it follows from Proposition 7.6 that the second term of the right hand side of (7.3.17) is bounded by

$$\sum_{k=1}^{J} \sum_{\substack{\alpha'_0 \le \alpha_0 \\ \alpha'_{J+1} \le \alpha_{J+1} \\ \beta'_{J+1} \le \beta_{J+1}}} \sum_{\substack{0 \ne \alpha'_k \\ \alpha'_k \le \alpha_k}} \sum_{\alpha'_{k+1} \le \alpha_{k+1}} \binom{\alpha_0}{\alpha'_0} \binom{\alpha_{J+1}}{\alpha'_{J+1}} \binom{\beta_{J+1}}{\beta'_{J+1}} \binom{\alpha_k}{\alpha'_k} \binom{\alpha_{k+1}}{\alpha'_{k+1}} c_K(\tau_k + \tau_{k+1})$$

$$\le \sum_{k=1}^{J} 2^{5K} c_K(\tau_k + \tau_{k+1}) \le 2^{5K+1} c_K T.$$

It follows from Proposition 7.6 that the third term of the right side of (7.3.17) is bounded by

$$\sum_{\substack{\alpha'_0 \le \alpha_0, \alpha'_{J+1} \le \alpha_{J+1}, \beta'_{J+1} \le \beta_{J+1} \\ 1 \le \alpha_0 + \alpha_{J+1} + \beta_{J+1}}} \binom{\alpha_0}{\alpha'_0} \binom{\alpha_{J+1}}{\alpha'_{J+1}} \binom{\beta_{J+1}}{\beta'_{J+1}} \| \partial_{x_0}^{\alpha'_0} \partial_{x_{J+1}}^{\alpha'_{J+1}} \partial_{\xi_{J+1}}^{\beta'_{J+1}} L \|_{l^\infty \to l^1}$$

$$\le \sum_{\substack{\alpha'_0 \le \alpha_0, \alpha'_{J+1} \le \alpha_{J+1}, \beta'_{J+1} \le \beta_{J+1} \\ 1 \le \alpha_0 + \alpha_{J+1} + \beta_{J+1}}} \binom{\alpha_0}{\alpha'_0} \binom{\alpha_{J+1}}{\alpha'_{J+1}} \binom{\beta_{J+1}}{\beta'_{J+1}} c_K T$$

$$\le 2^{3K} c_K T.$$

Taking sum of these, we obtain

$$\sum_{\substack{(\alpha', \beta') \le (\alpha, \beta) \\ (0,0) \ne (\alpha', \beta')}} \binom{\alpha}{\alpha'} \binom{\beta}{\beta'} \| \partial_x^{\alpha'} \partial_\xi^{\beta'} L \|_{l^\infty \to l^1}$$

$$\le 2^{6K} c_K |T + 2^{5K+1} c_K T + 2^{3K} c_K T \le 2^{6K+2} c_K T.$$

Therefore, by the choice of C_K as in (7.3.16) we have

$$\| \partial_x^\alpha \partial_\xi^\beta Q \|_{l^\infty \to l^\infty} \le C_0 2^{6K+2} c_K T C_K^{n-1} \le C_K^n.$$

We have proved the estimate (7.3.15) in the case $|\alpha| + |\beta| = n$. Induction has been completed. □

7.4 Proof of Kumano-go–Taniguchi Theorem

We discuss

$$K(x_{J+1}, \xi_{J+1}, x_0) \tag{7.4.1}$$
$$= \left(\frac{\nu}{2\pi}\right)^J \int_{\mathbf{R}^{2J}} e^{-i\nu\Phi} a_\lambda(x_{J+1}, \xi_{J+1}, x_J, \ldots, x_1, \xi_1, x_0) \prod_{j=1}^{J} d\xi_j dx_j,$$

and

$$\Phi = \tau_{J+1}\phi_{J+1}(x_{J+1}, \xi_{J+1}, x_J) - (x_J - x_0)\xi_{J+1}$$
$$+ \sum_{j=1}^{J} \{\tau_j \phi_j(x_j, \xi_j, x_{j-1}) + (x_j - x_{j-1})\xi_j\}.$$

Here $a_\lambda(x_{J+1}, \xi_{J+1}, x_J, \ldots, x_1, \xi_1, x_0)$ satisfies Assumption 7.5 and $\phi_j(x_j, \xi_j, x_{j-1})$ satisfies (7.1.9). We begin with the next

Proposition 7.7 *If $T \le \delta_1$, then the oscillatory integral (7.1.12) is well defined. Function*

$$\tilde{b}(x_{J+1}, \xi_{J+1}, x_0) = e^{i\nu\Phi^*(x_{J+1}, \xi_{J+1}, x_0)} K(x_{J+1}, \xi_{J+1}, x_0)$$

as in (7.1.15) is also defined.

Proof It follows from Assumptions 7.4, 7.5 and Proposition 7.3 that we can apply Theorem 3.3 to the oscillatory integral (7.1.12). □

We divide the proof of Kumano-go–Taniguchi Theorem 7.6 into four steps. Steps 1° and 2° are devoted to preparation. Inequality (7.1.16) in the case $\alpha + \beta + \gamma = 0$ is proved in Step 3°. The general case is proved in Step 4°.

Step 1°.

We use integration by parts carefully. Let (z, ζ) be as in (7.2.1). Then we define

$$M_j = \frac{1 + i\zeta_j(x_{j+1}, \xi_{j+1}, x_j, \xi_j, x_{j-1})\partial_{x_j}}{1 + \nu\zeta_j(x_{j+1}, \xi_{j+1}, x_j, \xi_j, x_{j-1})^2}, \quad \text{for} \quad 1 \le j \le J,$$

$$N_j = \frac{1 + iz_j(x_j, \xi_j, x_{j-1})\partial_{\xi_j}}{1 + \nu z_j(x_j, \xi_j, x_{j-1})^2}, \quad \text{for} \quad 1 \le j \le J.$$

It is clear that

$$M_j e^{-i\nu\Phi} = e^{-i\nu\Phi}, \quad N_j e^{-i\nu\Phi} = e^{-i\nu\Phi}.$$

Let M_j^* denote the formal adjoint operator of M_j. Then integration by parts gives

$$\begin{aligned} &K(x_{J+1}, \xi_{J+1}, x_0) \\ &= \left(\frac{\nu}{2\pi}\right)^J \int_{\mathbf{R}^{2J}} M_j\left(e^{-i\nu\Phi}\right) a_\lambda(x_{J+1}, \xi_{J+1}, \ldots, x_1, \xi_1, x_0) \prod_{j=1}^J d\xi_j dx_j \\ &= \left(\frac{\nu}{2\pi}\right)^J \int_{\mathbf{R}^{2J}} e^{-i\nu\Phi} M_j^*(a_\lambda(x_{J+1}, \xi_{J+1}, \ldots, x_1, \xi_1, x_0)) \prod_{j=1}^J d\xi_j dx_j. \end{aligned}$$

Similarly, using N_j in place of M_j, we have

$$\begin{aligned} &K(x_{J+1}, \xi_{J+1}, x_0) \\ &= \left(\frac{\nu}{2\pi}\right)^J \int_{\mathbf{R}^{2J}} N_j\left(e^{-i\nu\Phi}\right) a_\lambda(x_{J+1}, \xi_{J+1}, \ldots, x_1, \xi_1, x_0) \prod_{j=1}^J d\xi_j dx_j \\ &= \left(\frac{\nu}{2\pi}\right)^J \int_{\mathbf{R}^{2J}} e^{-i\nu\Phi} N_j^*(a_\lambda(x_{J+1}, \xi_{J+1}, \ldots, x_1, \xi_1, x_0)) \prod_{j=1}^J d\xi_j dx_j. \end{aligned}$$

Let r be such integer as $m + 1 < r \leq m + 2$. Repeating the above discussions, we have

$$K(x_{J+1}, \xi_{J+1}, x_0) = \left(\frac{\nu}{2\pi}\right)^J \int_{R^{2J}} e^{-i\nu\Phi} N_J^{*r} M_J^{*r} \ldots N_1^{*r} M_1^{*r} a_\lambda \prod_{j=1}^J d\xi_j dx_j. \tag{7.4.2}$$

We set for simplicity

$$D_{x_j} = \nu^{-1/2}\partial_{x_j}, \quad D_{\xi_j} = \nu^{-1/2}\partial_{\xi_j}.$$

Then

$$M_j^* = a_{jj}(x_{j+1}, \xi_{j+1}, x_j, \xi_j, x_{j-1}) D_{x_j} + a_{j0}(x_{j+1}, \xi_{j+1}, x_j, \xi_j, x_{j-1}), \tag{7.4.3}$$

$$N_j^* = b_{jj}(x_j, \xi_j, x_{j-1}) D_{\xi_j} + b_{j0}(x_j, \xi_j, x_{j-1}). \tag{7.4.4}$$

Here,

$$\begin{aligned} a_{jj} &= -\frac{i\nu^{1/2}\zeta_j}{1 + \nu\zeta_j^2}, \quad a_{j0} = \frac{1}{1 + \nu\zeta_j^2} - D_{x_j}\{\frac{i\nu^{1/2}\zeta_j}{1 + \nu\zeta_j^2}\}, \\ b_{jj} &= -\frac{i\nu^{1/2}z_j}{1 + \nu z_j^2}, \quad b_{j0} = \frac{1}{1 + \nu z_j^2} - D_{\xi_j}\{\frac{i\nu^{1/2}z_j}{1 + \nu z_j^2}\}. \end{aligned}$$

Since all the derivatives of ζ_j and z_j are bounded functions, for any natural number k there exists a positive constant c_k depending only on κ_n, $(2 \le n \le 5k+2)$, but independent of J and ν, $(\nu \ge 1)$, such that for $j = 1, 2, \ldots, J$

$$|D_{x_{j+1}}^{\alpha_{j+1}} D_{\xi_{j+1}}^{\beta_{j+1}} D_{x_j}^{\alpha_j} D_{\xi_j}^{\beta_j} D_{x_{j-1}}^{\alpha_{j-1}} a_{jj}| \le c_k(1+\nu\zeta_j^2)^{-1/2}, \tag{7.4.5}$$

$$|D_{x_{j+1}}^{\alpha_{j+1}} D_{\xi_{j+1}}^{\beta_{j+1}} D_{x_j}^{\alpha_j} D_{\xi_j}^{\beta_j} D_{x_{j-1}}^{\alpha_{j-1}} a_{j0}| \le c_k(1+\nu\zeta_j^2)^{-1/2},$$

$$|D_{x_j}^{\alpha_j} D_{\xi_j}^{\beta_j} D_{x_{j-1}}^{\alpha_{j-1}} b_{jj}| \le c_k(1+\nu z_j^2)^{-1/2},$$

$$|D_{x_j}^{\alpha_j} D_{\xi_j}^{\beta_j} D_{x_{j-1}}^{\alpha_{j-1}} b_{j0}| \le c_k(1+\nu z_j^2)^{-1/2}, \tag{7.4.6}$$

as far as $\alpha_{j+1} \le k$, $\beta_{j+1} \le k$, $\alpha_j \le k$, $\beta_j \le k$ and $\alpha_{j-1} \le k$.

Step 2°.
We shall look for bound of the integrand of (7.4.2). The key point is that, although there are many functions, each function a_{jj}, a_{j0}, b_{jj} and b_{j0} depends only on at most five variables $(x_{j+1}, \xi_{j+1}, x_j, \xi_j, x_{j-1})$. To see this advantage more clearly, we discuss it with a simpler notation. Let L and K be a natural number. Let $y = (y_1, y_2, \ldots, y_K)$ be the coordinate of a point in $\mathbf{R}^K$. Using the symbol $D_k = \nu^{-1/2}\partial_{y_k}$, we consider the product $P_L(y, D)P_{L-1}(y, D) \cdots P_1(y, D)$ of L linear first order differential operators

$$P_j(y, D) = \sum_{k=1}^{K} p_{jk}(y)D_k + p_{j0}(y), \qquad j = 1, 2, \ldots, L. \tag{7.4.7}$$

We define sets of indices:

$$\begin{aligned} K_1(j) &= \{k;\quad p_{jk}(y) \neq 0 \quad \text{for some} \quad y\}, \\ J_1(k) &= \{j;\quad p_{jk}(y) \neq 0 \quad \text{for some} \quad y\}, \\ K_2(j) &= \{l;\quad D_{y_l} p_{jk}(y) \neq 0 \quad \text{for some} \quad y \quad \text{and some} \quad k \in K_1(j) \cup \{0\}\}, \\ J_2(l) &= \{j;\quad D_{y_l} p_{jk}(y) \neq 0 \quad \text{for some} \quad y \in \mathbf{R}^K \text{and some} \quad k \in K_1(j) \cup \{0\}\}. \end{aligned}$$

If A is a set, we write $\#A$ for the number of elements in the set A. We define

$$\rho_1 = \max_k \#J_1(k), \quad \rho_2 = \max_k \#J_2(k),$$
$$\sigma_1 = \max_j \#K_1(j), \quad \sigma_2 = \max_j \#K_2(j).$$

For any j we set $K_0(j) = \{0\} \cup K_1(j)$, $\prod K_0 = \prod_{j=1}^{L} K_0(j)$ and $\mathbf{Z}_+ = \{0, 1, 2, 3, \ldots\}$. Then we can rewrite the product $P_L(y, D)P_{L-1}(y, D) \ldots P_1(y, D)$ of operators in the following form:

$$
\begin{aligned}
&P_L(y, D)P_{L-1}(y, D) \cdots P_1(y, D) \qquad (7.4.8)\\
&= \sum_{\{k_j\}} \sum_{\{\alpha_{jk_j}\}} \sum_{\beta} C(\{k_j\}, \{\alpha_{jk_j}\}, \beta) \left(\prod_{j=1}^{L} D^{\alpha_{jk_j}} p_{jk_j}(y) \right) D^{\beta}.
\end{aligned}
$$

Here the sequence $\{k_j\}$ ranges over the set $\prod K_0$, $\alpha_{jk_j} = (\alpha_{jk_j1}, \ldots, \alpha_{jk_jK})$ and $\beta = (\beta_1, \beta_2, \ldots, \beta_K)$ are multi-indices which range over the set $\mathbf{Z}_+^K$, and $C(\{k_j\}, \{\alpha_{jk_j}\}, \beta)$ is a non-negative integer D

Proposition 7.8 *1. If $C(\{k_j\}, \{\alpha_{jk_j}\}, \beta) \neq 0$, then we have $\beta_k \leq \rho_1$ for any k with $1 \leq k \leq K$, $|\alpha_{jk_jl}| \leq \rho_1$ and $|\alpha_{jk_j}| = \alpha_{jk_j1} + \alpha_{jk_j1} + \cdots + \alpha_{jk_jK} \leq \rho_1\sigma_2$ for any j, k_j.*
2. We have

$$
\sum_{\{k_j\}} \sum_{\{\alpha_{jk_j}\}} \sum_{\beta} C(\{k_j\}, \{\alpha_{jk_j}\}, \beta) \leq (1 + \sigma_1(\rho_2 + 1))^L. \qquad (7.4.9)
$$

Proof Let $\beta = (\beta_1, \beta_2, \ldots, \beta_K)$ be such that $C(\{k_j\}, \{\alpha_{jk_j}\}, \{\beta\}) \neq 0$ for some $\{k_j\}$ and $\{\alpha_{jk_j}\}$. Assume k-th component $\beta_k \neq 0$. Since every operator $P_j(y, D)$ is of 1-st order, there must be at least β_k operators such that $p_{jk}(y) \neq 0$. This means $|\beta_k| \leq \#J_1(k) = \rho_1$.

Choose any coefficient $p_{jk_j}(y), k_j \in K_1(j)$ in $P_j(y, D)$. Then $D_l p_{jk_j}(y)$ does not vanish identically if and only if $l \in K_2(j)$. That is $\alpha_{jk_jl} \neq 0$ if and only if $l \in K_2(j)$. For fixed l, D_l is contained in $P_n(y, D)$ if and only if $n \in J_1(l)$, i.e., $|\alpha_{jk_jl}| = \#J_1(l) \leq \rho_1$. Thus $|\alpha_{jk_j}| \leq \sigma_2 \cdot \rho_1$. The first part of Proposition has been proved.

Now we prove (7.4.9) by induction on L. For the sake of simplicity we set

$$
B(L) = \sum_{\{k_j\}} \sum_{\{\alpha_{jk_j}\}} \sum_{\beta} C(\{k_j\}, \{\alpha_{jk_j}\}, \beta).
$$

It is clear that

$$
B(1) \leq \#K_1(1) + 1 \leq \sigma_1 + 1.
$$

Thus (7.4.9) for $L = 1$ is true.

Assume that (7.4.9) for $L = n - 1$ is true. Then (7.4.8) with L replaced by $n - 1$ is

$$
\begin{aligned}
&P_{n-1}(y, D)P_{n-2}(y, D) \ldots P_1(y, D) \qquad (7.4.10)\\
&= \sum_{\{k_j\}} \sum_{\{\alpha_{jk_j}\}} \sum_{\beta} C(\{k_j\}, \{\alpha_{jk_j}\}, \beta) \prod_{j=1}^{n-1} (D_y^{\alpha_{jk_j}} p_{jk_j}(y)) D_y^{\beta}.
\end{aligned}
$$

Here $k_j \in K_0(j)$ for $j = 1, 2, \ldots, n-1$. We operate

$$P_n(y, D) = \sum_{l \in K_1(n)} p_{nl}(y) D_l + p_{n0}(y)$$

to the equality (7.4.10) from the left. Since multiplication operator $p_{n0}(y)$ commutes with $D^{\alpha_{jk}} p_{jk}(y)$ and $p_{j0}(y)$, contribution of this term to $B(n)$ is at most $B(n-1)$. On the other hand, the differential operator D_l commutes with every factor of $\prod_{j=1}^{n-1}(D^{\alpha_{jk_j}} p_{jk_j}(y))$ if $j \notin J_2(l)$. It follows from Leibnitz rule that $D_l \prod_{j=1}^{n-1}(D^{\alpha_{jk_j}} p_{jk_j}(y))$ has at most $\#J_2(l) + 1$ terms. Thus contribution of D_l in $P_n(y, D)$ to $B(n)$ is at most $B(n-1) \cdot (\rho_2 + 1)$. The operator $P_n(y, D)$ contains $\#K_1(n)$ such terms as D_l. Therefore, we have

$$B(n) \le B(n-1) + B(n-1)(\rho_2 + 1)\sigma_1 = B(n-1)(1 + (\rho_2 + 1)\sigma_1).$$

This proves (7.4.9) in the case $L = n$. Induction process has been completed. Proposition has been proved. □

Step 3°. Proof of Kumano-go–Tanguchi Theorem in the case $\alpha_{J+1} + \beta_{J+1} + \alpha_0 = 0$.

Lemma 7.6 *Suppose that Assumptions 7.4 and 7.5 hold with some m. We assume further $T \le \delta_1$. Let r be the integral part of $m + 2$. Then*

There exists a constant $C_0 \ge 1$ independent of J, x_0, ξ_{J+1} and x_{J+1} but may depend on $\kappa_j (2 \le j \le 5r + 2)$ such that

$$|\tilde{b}(x_{J+1}, \xi_{J+1}, x_0)| \le C_0^{J+1} A_r (1 + |\lambda| + |x_{J+1}| + |\xi_{J+1}| + |x_0|)^m. \tag{7.4.11}$$

If we choose $K(0) = r$, this proves (7.1.16) for $\alpha_{J+1} + \beta_{J+1} + \alpha_0 = 0$.

Proof Since $\Phi^*(x_{J+1}, \xi_{J+1}, x_0) = \Phi(x_{J+1}, \xi_{J+1}, x_J^*, \xi_J^*, \ldots, x_1^*, \xi_1^*, x_0)$ is well defined, we have only to obtain bound of $K(x_{J+1}, \xi_{J+1}, x_0)$.

Now we can apply Proposition 7.8 to the product $N_J^{*r} M_J^{*r} \ldots N_1^{*r} M_1^{*r}$ of operators. In the present case $L = 2rJ$ and $K = 2J$. Parameters x_{J+1}, ξ_{J+1}, x_0 are fixed and independent variables $y_1, y_2, \ldots, y_{2J}$ are

$$y_{2j-1} = x_j, \quad j = 1, 2, \ldots, J,$$
$$y_{2j} = \xi_j, \quad j = 1, 2, \ldots, J.$$

And for $k = 1, 2, \ldots, J$

$$P_{(2k-2)r+j} = M_k^*, \quad \text{for } 1 \le j \le r,$$
$$P_{(2k-1)r+j} = N_k^*, \quad \text{for } 1 \le j \le r.$$

Each coefficients p_{jk} is nothing but one of $a_{jj}, a_{j0}, b_{jj}, b_{j0}$. It is clear that in this case

$$\rho_1 = r, \quad \rho_2 = 5r, \quad \sigma_1 = 1, \quad \sigma_2 = 5.$$

Thus

$$\rho_1\sigma_2 = 5r, \quad (1+\sigma_1(\rho_2+1)) = 2+5r \quad \text{and}\ \beta_k \le r \text{ for any } k. \tag{7.4.12}$$

In accordance with (7.4.8) we write

$$N_J^{*r} M_J^{*r} \cdots N_1^{*r} M_1^{*r} a_\lambda(y) =$$
$$= \sum_{\{k_j\}} \sum_{\{\alpha_{jk_j}\}} \sum_{\beta} C(\{k_j\}, \{\alpha_{jk_j}\}, \beta) \prod_{j=1}^{L} (D_y^{\alpha_{jk_j}} p_{jk_j}(y)) D_y^\beta a_\lambda(y).$$

Since $\beta_k \le r$ for any $k = 1, 2, \ldots, 2J$ and $\nu \ge 1$, Assumption (7.1.10) gives

$$|D^\beta a(y)| \le A_r \nu^{-|\beta|/2}(1 + |\lambda| + |x_{J+1}| + |\xi_{J+1}| + |x_0| + \sum_{j=1}^{J}(|x_j| + |\xi_j|))^m$$
$$\le A_r(1 + |\lambda| + |x_{J+1}| + |\xi_{J+1}| + |x_0| + \sum_{j=1}^{J}(|x_j| + |\xi_j|))^m.$$

As $|\alpha_{jk_j}| \le \rho_1\sigma_2 = 5r$ for each jk_j, it follows from inequalities from (7.4.5) to (7.4.6) that

$$|\prod_{j=1}^{L}(D_y^{\alpha_{jk_j}} p_{jk_j}(y))| \le \prod_{j=1}^{J} \left(c_{5r}^2(1+\nu\zeta_j^2)^{-1/2}(1+\nu z_j^2)^{-1/2}\right)^r.$$

Here c_r is the constant appeared in estimates from (7.4.5) to (7.4.6). Since $(1 + \sigma_1(\rho_2+1)) = 2+5r$ and $L = 2rJ$, inequality (7.4.9) of Proposition 7.8 gives

$$\sum_{\{k_j\}} \sum_{\{\alpha_{jk_j}\}} \sum_{\beta} C(\{k_j\}, \{\alpha_{jk_j}\}, \beta) \le (1+\sigma_1(\rho_2+1))^L = (2+5r)^{2rJ}. \tag{7.4.13}$$

Therefore, we have

$$|N_J^{*r} M_J^{*r} \cdots N_1^{*r} M_1^{*r} a_\lambda(x_{J+1}.\xi_{J+1}.x_J, \ldots, \xi_1, x_0)| \tag{7.4.14}$$
$$\le c_{5r}^{2rJ}(2+5r)^{2rJ} A_r(1 + |\lambda| + |x_{J+1}| + |\xi_{J+1}| + |x_0| + \sum_{j=1}^{J}(|x_j| + |\xi_j|))^m$$

$$\times \prod_{J=1}^{J} (1 + \nu \zeta_j^2)^{-r/2} (1 + \nu z_j^2)^{-r/2}.$$

Notice that

$$\sum_{j=1}^{J} (|x_j| + |\xi_j|) \le \sum_{j=1}^{J} (|x_j - x_j^*| + |\xi_j - \xi_j^*| + |x_j^*| + |\xi_j^*|).$$

It follows from Lemma 7.3 that

$$\sum_{j=1}^{J} (|x_j - x_j^*| + |\xi_j - \xi_j^*|) \le 2J \|(x - x^*, \xi - \xi^*)\|_{l^\infty}$$
$$\le 2J(1 - 6\kappa_2 T)^{-1} \sum_{j=1}^{J} (|z_j| + |\zeta_j|).$$

Moreover, Corollary 7.3 implies that

$$\sum_{j=1}^{J} |x_j^*| + |\xi_j^*| \le 2J \|(x^*, \xi^*)\|_{l^\infty}$$
$$\le 2J(1 - 6\kappa_2 T)^{-1}(1 + 2\kappa_1 T)(1 + |x_{J+1}| + |\xi_{J+1}| + |x_0|).$$

Thus

$$\begin{aligned}
&(1 + |\lambda| + |x_{J+1}| + |\xi_{J+1}| + |x_0| + \sum_{j=1}^{J} (|x_j| + |\xi_j|)) \\
&\le 2J(1 - 6\kappa_2 T)^{-1} \sum_{j=1}^{J} (|z_j| + |\zeta_j|) \\
&+ (1 + 2J(1 - 6\kappa_2 T)^{-1}(1 + 2\kappa_1 T))(1 + |\lambda| + |x_{J+1}| + |\xi_{J+1}| + |x_0|) \\
&\le (1 + 2J(1 - 6\kappa_2 T)^{-1}(1 + 2\kappa_1 T)) \\
&\qquad \times (1 + |\lambda| + |x_{J+1}| + |\xi_{J+1}| + |x_0| + \sum_{j=1}^{J} (|z_j| + |\zeta_j|) \\
&\le 4J(1 - 6\kappa_2 T)^{-1}(1 + 2\kappa_1 T) \\
&\qquad \times (1 + |\lambda| + |x_{J+1}| + |\xi_{J+1}| + |x_0|)(1 + \sum_{j=1}^{J} (|z_j| + |\zeta_j|)).
\end{aligned}$$

Since

$$(1+\sum_{j=1}^{J}(|z_j|+|\zeta_j|)) \le \prod_{j=1}^{J}\sqrt{2}(1+\nu z_j^2)^{1/2}\sqrt{2}(1+\nu\zeta_j^2)^{1/2},$$

we have

$$\begin{aligned}
&(1+|\lambda|+|x_{J+1}|+|\xi_{J+1}|+|x_0|+\sum_{j=1}^{J}(|x_j|+|\xi_j|))^m \\
&\le (4J)^m 2^{jm}(1-6\kappa_2 T)^{-m}(1+2\kappa_1 T)^m(1+|\lambda|+|x_{J+1}|+|\xi_{J+1}|+|x_0|)^m \\
&\quad\times\prod_{j=1}^{J}(1+\nu|z_j|^2)^{m/2}(1+\nu|\zeta_j|^2)^{m/2} \\
&= J^m 2^{jm} B_0(1+|\lambda|+|x_{J+1}|+|\xi_{J+1}|+|x_0|)^m\prod_{j=1}^{J}(1+\nu|z_j|^2)^{m/2}(1+\nu|\zeta_j|^2)^{m/2},
\end{aligned}$$

where

$$B_0 = 4^m(1-6\kappa_2 T)^{-m}(1+2\kappa_1 T)^m.$$

Therefore,

$$\begin{aligned}
&|N_J^{*r}M_J^{*r}\cdots N_1^{*r}M_1^{*r}a(x_{J+1}.\xi_{J+1}.x_J,\ldots,\xi_1,x_0)| \qquad (7.4.15)\\
&\le c_{5r}^{2rJ}(2+5r)^{2rJ}A_r J^m 2^{jm} B_0 \\
&\times(1+|\lambda|+|x_{J+1}|+|\xi_{J+1}|+|x_0|)^m\prod_{J=1}^{J}(1+\nu|z_j|^2)^{(m-r)/2}(1+\nu|\zeta_j|^2)^{(m-r)/2}.
\end{aligned}$$

It follows from this and (7.4.2) that

$$\begin{aligned}
&|K(x_{J+1},\xi_{J+1},x_0)| \\
&\le c_{5r}^{2rJ}(2+5r)^{2rJ}A_r J^m 2^{Jm} B_0(1+|\lambda|+|x_{J+1}|+|\xi_{J+1}|+|x_0|)^m \\
&\times\left(\frac{\nu}{2\pi}\right)^J\int_{\mathbf{R}^{2J}}\prod_{J=1}^{J}(1+\nu\zeta_j^2)^{(m-r)/2}(1+\nu z_j^2)^{(m-r)/2}\prod_{j=1}^{J}d\xi_j dx_j.
\end{aligned}$$

Changing variables $(x,\xi)\to(z,\zeta)$ and using the estimate (7.2.15) of the Jacobian, we obtain

$$\int_{\mathbf{R}^{2J}}\prod_{J=1}^{J}(1+\nu\zeta_j^2)^{(m-r)/2}(1+\nu z_j^2)^{(m-r)/2}\prod_{j=1}^{J}d\xi_j dx_j$$

$$\le \int_{\mathbf{R}^{2J}} \prod_{J=1}^{J} (1+\nu\zeta_j^2)^{(m-r)/2}(1+\nu z_j^2)^{(m-r)/2}(1-6\kappa_2 T)^{-2J} \prod_{j=1}^{J} dz_j d\zeta_j$$

$$\le \nu^{-J}(1-6\kappa_2 T)^{-2J}\left(\int_{\mathbf{R}}(1+t^2)^{(m-r)/2}dt\right)^{2J}.$$

Therefore,

$$|K(x_{J+1},\xi_{J+1},x_0)| \le B_0 J^m B_1^J A_r(1+|\lambda|+|x_{J+1}|+|\xi_{J+1}|+|x_0|)^m. \quad (7.4.16)$$

Here

$$B_1 = c_{5r}^{2r}(2+5r)^{2r}(1-6\kappa_2 T)^{-2}2^m(2\pi)^{-1}\left(\int_{\mathbf{R}}(1+t^2)^{(m-r)/2}dt\right)^2.$$

Since $J \le e^J$,

$$\begin{aligned}|\tilde{b}(x_{J+1},\xi_{J+1},x_0)| &= |K(x_{J+1},\xi_{J+1},x_0)| \\ &\le B_0(e^m B_1)^J A_r(1+|\lambda|+|x_{J+1}|+|\xi_{J+1}|+|x_0|)^m.\end{aligned}$$

Define

$$C_0 = \max\{B_0, e^m B_1\}. \quad (7.4.17)$$

Then

$$|\tilde{b}(x_{J+1},\xi_{J+1},x_0)| \le C_0^{J+1} A_r(1+|\lambda|+|x_{J+1}|+|\xi_{J+1}|+|x_0|)^m. \quad (7.4.18)$$

If we choose $K(0)=r$, this proves (7.1.16) for $\alpha_{J+1}+\beta_{J+1}+\alpha_0=0$. Therefore, Kumano-go-Taniguchi theorem for $\alpha_{J+1}+\beta_{J+1}+\alpha_0=0$ has been proved. Lemma 7.6 has been proved. □

Remark 7.5 Constants B_0 depends only on m, δ_1 and κ_1,κ_2. But B_1 depends only on m, κ_j for $2 \le j \le 5m+2$.

Step 4°. We prove the estimate (7.1.16) for derivatives of $\tilde{b}(x_{J+1},\xi_{J+1},x_0)$. It is clear that

$$\partial_{x_{J+1}}\tilde{b}(x_{J+1},\xi_{J+1},x_0) \quad (7.4.19)$$

$$= \left(\frac{\nu}{2\pi}\right)^J \int_{\mathbf{R}^{2J}} \partial_{x_{J+1}}\left(e^{-i\nu(\Phi-\Phi^*)}a_\lambda(x_{J+1},\xi_{J+1},x_J,\ldots,x_1,\xi_1,x_0)\right)\prod_{j=1}^{J}dx_j d\xi_j$$

$$= I_1 + I_2.$$

Here,

$$I_1 = -i\nu \left(\frac{\nu}{2\pi}\right)^J \times \int_{\mathbf{R}^{2J}} e^{-i\nu(\Phi-\Phi^*)} \partial_{x_{J+1}}(\Phi - \Phi^*) a_\lambda(x_{J+1}, \xi_{J+1}, x_J, \dots, x_1, \xi_1, x_0) \prod_{j=1}^{J} dx_j d\xi_j, \tag{7.4.20}$$

and

$$I_2 = \left(\frac{\nu}{2\pi}\right)^J \int_{\mathbf{R}^{2J}} e^{-i\nu(\Phi-\Phi^*)} \partial_{x_{J+1}} a_\lambda(x_{J+1}, \xi_{J+1}, x_J, \dots, x_1, \xi_1, x_0) \prod_{j=1}^{J} dx_j d\xi_j.$$

We shall discuss I_1. As (x^*, ξ^*) is the stationary point of Φ, we have

$$\begin{aligned}\partial_{x_{J+1}}(\Phi - \Phi^*) &= \tau_{J+1}\left(\partial_{x_{J+1}}\phi_{J+1}(x_{J+1}, \xi_{J+1}, x_J) - \partial_{x_{J+1}}\phi_{J+1}(x_{J+1}, \xi_{J+1}, x_J)\big|_{x_J = x_J^*}\right) \\ &= \tau_{J+1} p(x_{J+1}, \xi_{J+1}, x_J)(x_J - x_J^*).\end{aligned}$$

Here,

$$p(x_{J+1}, \xi_{J+1}, x_J) = \int_0^1 \partial_{x_J}\partial_{x_{J+1}}\phi_{J+1}(x_{J+1}, \xi_{J+1}, x_J(t))\, dt.$$

with $x_J(t) = tx_J + (1-t)x_J^*$.

Let $q_{jk}(x_{J+1}.\xi_{J+1}, x, \xi, x_0)$ be the (j,k) entry of the matrix Q in Lemma7.5. Here x and ξ are abbreviation of $(x_1, x_2, \dots, x_J)$ and $(\xi_1, \xi_2, \dots, \xi_J)$, respectively. Since $(x - x^*, \xi - \xi^*) = Q\Delta^{-1}(z, \zeta)$,

$$\begin{aligned} x_J - x_J^* &= \sum_{k=1}^{J} q_{Jk}(x_{J+1}.\xi_{J+1}, x, \xi, x_0)(\sum_{j=1}^{k} z_j) + q_{JJ+k}(x_{J+1}.\xi_{J+1}, x, \xi, x_0)(\sum_{j=k}^{J}\zeta_j) \\ &= \sum_{j=1}^{J} z_j \sum_{k=j}^{J} q_{Jk}(x_{J+1}.\xi_{J+1}, x, \xi, x_0) + \sum_{j=1}^{J}\zeta_j \sum_{k=1}^{j} q_{JJ+k}(x_{J+1}.\xi_{J+1}, x, \xi, x_0).\end{aligned}$$

Thus

$$\begin{aligned}&\partial_{x_{J+1}}(\Phi - \Phi^*) \\ &= p(x_{J+1}, \xi_{J+1}, x_J) \\ &\times (\sum_{j=1}^{J} z_j \sum_{k=j}^{J} q_{Jk}(x_{J+1}.\xi_{J+1}, x, \xi, x_0) + \sum_{j=1}^{J}\zeta_j \sum_{k=1}^{j} q_{JJ+k}(x_{J+1}.\xi_{J+1}, x, \xi, x_0)).\end{aligned}$$

Replacing this in (7.4.20) and using definition (7.2.1) of z_k and ζ_k, we have

$$I_1 = \left(\frac{\nu}{2\pi}\right)^J e^{i\nu\Phi^*} \int_{\mathbf{R}^{2J}} \left(\sum_{j=1}^{J} f_j(x_{J+1}.\xi_{J+1}, x, \xi, x_0)\partial_{\xi_j} + \sum_{j=1}^{J} g_j(x_{J+1}.\xi_{J+1}, x, \xi, x_0)\partial_{x_j} \right) (e^{-i\nu\Phi}) \prod_{j=1}^{J} dx_j d\xi_j, \tag{7.4.21}$$

where

$$\begin{aligned} &f_j(x_{J+1}.\xi_{J+1}, x, \xi, x_0) \\ &= p(x_{J+1}, \xi_{J+1}, x_J)a_\lambda(x_{J+1}.\xi_{J+1}, x, \xi, x_0) \sum_{k=j}^{J} q_{Jk}(x_{J+1}.\xi_{J+1}, x, \xi, x_0), \end{aligned} \tag{7.4.22}$$

$$\begin{aligned} &g_j(x_{J+1}.\xi_{J+1}, x, \xi, x_0) \\ &= p(x_{J+1}, \xi_{J+1}, x_J)a_\lambda(x_{J+1}.\xi_{J+1}, x, \xi, x_0) \sum_{k=1}^{j} q_{J\,J+k}(x_{J+1}.\xi_{J+1}, x, \xi, x_0). \end{aligned} \tag{7.4.23}$$

Integration by parts gives

$$I_1 = -\left(\frac{\nu}{2\pi}\right)^J e^{i\nu\Phi^*} \int_{\mathbf{R}^{2J}} \left(\sum_{j=1}^{J} \partial_{\xi_j} f_j(x_{J+1}.\xi_{J+1}, x, \xi, x_0) + \sum_{j=1}^{J} \partial_{x_j} g_j(x_{J+1}.\xi_{J+1}, x, \xi, x_0) \right) e^{-i\nu\Phi} \prod_{j=1}^{J} dx_j d\xi_j.$$

Taking the sum of I_1 and I_2, we have

$$\begin{aligned} &\partial_{x_{J+1}} \tilde{b}(x_{J+1}, \xi_{J+1}, x_0) \\ &= \left(\frac{\nu}{2\pi}\right)^J \int_{\mathbf{R}^{2J}} e^{-i\nu(\Phi-\Phi^*)} h(x_{J+1}.\xi_{J+1}, x, \xi, x_0) \prod_{j=1}^{J} dx_j d\xi_j, \end{aligned} \tag{7.4.24}$$

where

$$\begin{aligned} &h(x_{J+1}.\xi_{J+1}, x, \xi, x_0) \\ &= \partial_{x_{J+1}} a(x_{J+1}, \xi_{J+1}, x, \xi, x_0) \\ &\quad - \sum_{j=1}^{J} \partial_{\xi_j} f_j(x_{J+1}.\xi_{J+1}, x, \xi, x_0) - \sum_{j=1}^{J} \partial_{x_j} g_j(x_{J+1}.\xi_{J+1}, x, \xi, x_0). \end{aligned}$$

We shall show that the function h satisfies Assumption 7.5 with different constants. Let $K \geq 0$ be an integer and $|\alpha_j| \leq K, |\beta_j| \leq K$ for $j = 1, 2, \ldots, J+1$ and $|\alpha_0| \leq K$. Then function p depends only on three variables $(x_{J+1}, \xi_{J+1}, x_0)$ and

$$|\partial_{x_{J+1}}^{\alpha_{J+1}} \partial_{\xi_{J+1}}^{\beta_{J+1}} \partial_{x_J}^{\alpha_J} p(x_{x_{J+1}}, \xi_{J+1}, x_J)| \leq \kappa_{2+3K}.$$

By Lemma 7.5, for any j, k and $K \geq 0$ there exists a constant $c_K > 0$ depending only on $\{\kappa_n\}_{n=2}^{6K+2}$ such that

$$\left| \prod_{j=0}^{J+1} \partial_{x_j}^{\alpha_j} \prod_{j=1}^{J+1} \partial_{\xi_j}^{\beta_j} q_{jk} \right| \leq c_K^{K(2J+3)},$$

where we have replaced the symbol C_K in Lemma 7.5 with c_K in order to avoid confusion of notation.

It follows from Assumption 7.5, Leibnitz' formula and these inequalities that for $(x_{J+1}, \xi_{J+1}, x_J, \xi_J, \ldots, \xi_1, x_0)$

$$\left| \prod_{j=0}^{J+1} \partial_{x_j}^{\alpha_j} \prod_{j=1}^{J+1} \partial_{\xi_j}^{\beta_j} q_{jk} p a_\lambda \right| \leq \kappa_{2+3K} 2^{K(2J+6)} c_K^{K(2J+3)} A_K \left(1 + |\lambda| + \sum_{j=0}^{J+1} |x_j| + \sum_{j=1}^{J+1} |\xi_j|\right)^m,$$

if $|\alpha_j| \leq K, |\beta_j| \leq K$ for $j = 1, \ldots, J+1$ and $|\alpha_0| \leq K$. Thus, for $j = 1, 2, \ldots, J$

$$|\prod_{j=0}^{J+1} \partial_{x_j}^{\alpha_j} \prod_{j=1}^{J+1} \partial_{\xi_j}^{\beta_j} f_j|$$
$$\leq (J - j + 1)\kappa_{2+3K} 2^{K(2J+6)} c_K^{K(2J+3)} A_K (1 + |\lambda| + \sum_{j=0}^{J+1} |x_j| + \sum_{j=1}^{J+1} |\xi_j|)^m,$$

$$|\prod_{j=0}^{J+1} \partial_{x_j}^{\alpha_j} \prod_{j=1}^{J+1} \partial_{\xi_j}^{\beta_j} g_j| \leq j\kappa_{2+3K} 2^{K(2J+6)} c_K^{K(2J+3)} A_K (1 + |\lambda| + \sum_{j=0}^{J+1} |x_j| + \sum_{j=1}^{J+1} |\xi_j|)^m.$$

It follows from these that

$$|\prod_{j=0}^{J+1} \partial_{x_j}^{\alpha_j} \prod_{j=1}^{J+1} \partial_{\xi_j}^{\beta_j} \sum_{j=1}^{J} \partial_{x_j} f_j|$$
$$\leq \frac{1}{2} J(J+1)\kappa_{2+3(K+1)} 2^{(K+1)(2J+6)} c_{K+1}^{(K+1)(2J+6)} A_{K+1} (1 + |\lambda| + \sum_{j=0}^{J+1} |x_j| + \sum_{j=1}^{J+1} |\xi_j|)^m.$$

Similarly,

$$
\begin{aligned}
&|\prod_{j=0}^{J+1}\partial_{x_j}^{\alpha_j}\prod_{j=1}^{J+1}\partial_{\xi_j}^{\beta_j}\sum_{j=1}^{J}\partial_{x_j}g_j| \\
&\le \frac{1}{2}J(J+1)\kappa_{2+3(K+1)}2^{(K+1)(2J+6)}c_{K+1}^{(K+1)(2J+6)}A_{K+1}(1+|\lambda|+\sum_{j=0}^{J+1}|x_j|+\sum_{j=1}^{J+1}|\xi_j|)^m.
\end{aligned}
$$

On the other hand, by Assumption 7.5

$$
|\prod_{j=0}^{J+1}\partial_{x_j}^{\alpha_j}\prod_{j=1}^{J+1}\partial_{\xi_j}^{\beta_j}\partial_{x_{J+1}}a_\lambda| \le A_{K+1}(1+|\lambda|+\sum_{j=0}^{J+1}|x_j|+\sum_{j=1}^{J+1}|\xi_j|)^m.
$$

Consequently,

$$
\begin{aligned}
&|\prod_{j=0}^{J+1}\partial_{x_j}^{\alpha_j}\prod_{j=1}^{J+1}\partial_{\xi_j}^{\beta_j}h| \\
&\le \big(1+J(J+1)\kappa_{2+3(K+1)}2^{(K+1)(2J+6)}c_{K+1}^{(K+1)(2J+6)}\big) \\
&\qquad \times A_{K+1}(1+|\lambda|+\sum_{j=0}^{J+1}|x_j|+\sum_{j=1}^{J+1}|\xi_j|)^m.
\end{aligned}
$$

This means that h satisfies Assumption 7.5 A_K replaced by

$$
\big(1+J(J+1)\kappa_{2+3(K+1)}2^{(K+1)(2J+6)}c_{K+1}^{(K+1)(2J+6)}\big)A_{K+1}
$$

for any K.

Therefore, we can apply Lemma 7.6 to (7.4.24) and obtain that

$$
\begin{aligned}
&|\partial_{x_{J+1}}\tilde{b}(x_{J+1},\xi_{J+1},x_0)| \\
&\le C_0^{J+1}\big(1+J(J+1)\kappa_{2+3(r+1)}2^{(r+1)(2J+6)}c_{r+1}^{(r+1)(2J+6)}\big) \\
&\qquad \times A_{r+1}(1+|\lambda|+|x_{J+1}|+|\xi_{J+1}|+|x_0|)^m.
\end{aligned}
$$

We choose a constant $C_1 \ge 1$ depending only on $\{\kappa_n\}_{n=2}^{6r+8}$ so that

$$
C_1^{J+1} \ge C_0^{J+1}\big(1+J(J+1)\kappa_{2+3(r+1)}2^{(r+1)(2J+6)}c_{r+1}^{(r+1)(2J+6)}\big).
$$

Then

$$
|\partial_{x_{J+1}}\tilde{b}(x_{J+1},\xi_{J+1},x_0)| \le C_1^{J+1}A_{r+1}(1+|\lambda|+|x_{J+1}|+|\xi_{J+1}|+|x_0|)^m.
$$

Since $C_1 \geq 1$ we have proved (7.1.16) in the case $\alpha = 1, \beta = 0, \gamma = 0$ if we set $K(1) = r + 1$.

Repeating similar discussions, we can prove (7.1.16) in general cases.

The aim of Step 4° is attained.

Consequently, we have completed proof of Kumano-go–Taniguchi Theorem 7.6. □

Remark 7.6 Kumano-go–Taniguchi theorem first appeared in [45] in a little different form. The method of proof presented here was given in [33] in the case $m = 0$.

7.5 Proof of Theorem 7.3

As a preparation of the proof of Theorem 7.3 we discuss oscillatory integral (7.1.11). Using the result of Theorem 7.6, we can write

$$\begin{aligned} &I(S, a_\lambda, \nu)(x_{J+1}, x_0)\\ &= \left(\frac{\nu}{2\pi}\right) \int_{\mathbf{R}} e^{-i\nu(x_{J+1}-x_0)\xi_{J+1}} K(x_{J+1}, \xi_{J+1}, x_0) d\xi_{J+1}\\ &= \left(\frac{\nu}{2\pi}\right) \int_{\mathbf{R}} e^{-i\nu((x_{J+1}-x_0)\xi_{J+1}+\Phi^*(x_{J+1},\xi_{J+1},x_0))} \tilde{b}(x_{J+1}, \xi_{J+1}, x_0) d\xi_{J+1}. \end{aligned}$$

We set $\phi^*(x_{J+1}, \xi_{J+1}, x_0) = T^{-1}\Phi^*(x_{J+1}, \xi_{J+1}, x_0)$. Then

$$\begin{aligned} &I(S, a_\lambda, \nu)(x_{J+1}, x_0)\\ &= \left(\frac{\nu}{2\pi}\right) \int_{\mathbf{R}} e^{-i\nu(\xi_{J+1}(x_{J+1}-x_0)+T\phi^*(x_{J+1},\xi_{J+1},x_0))} \tilde{b}(x_{J+1}, \xi_{J+1}, x_0) d\xi_{J+1}. \end{aligned}$$

For the proof of Theorem 7.3, we set, in accordance to Eq. (7.1.6),

$$\begin{aligned} \Phi(x_{J+1}, \xi_{J+1}, x_J, \dots, \xi_1, x_0) &= \tau_{J+1}(\frac{1}{2}|\xi_{J+1}|^2 - \phi_{J+1}(x_{J+1}, x_J)) + (x_0 - x_J)\xi_{J+1}\\ &+ \sum_{j=1}^{J} \tau_j(\frac{1}{2}|\xi_j|^2 - \phi_j(x_j, x_{j-1})) + (x_j - x_{j-1})\xi_j. \end{aligned} \tag{7.5.1}$$

Here $\phi_j(x_j, x_{j-1})$ satisfies Assumption 7.1 with constants $\kappa_n, n = 1, 2, \dots,$.

In this case, $\phi_j(x_j, \xi_j, x_{j-1})$ of Assumption 7.4 equals $|\xi_j|^2/2 - \phi_j(x_j, x_{j-1})$. Thus $\partial_{x_j}\partial_{\xi_j}\phi_j(x_j, \xi_j, x_{j-1}) = \partial_{x_{j-1}}\partial_{\xi_j}\phi_j(x_j, \xi_j, x_{j-1}) = 0$. Using this property, we can strengthen some of conclusions in Sects. 7.2 and 7.3 in this case.

Hessian matrix of Φ is the following $2J \times 2J$ matrix:

$$Hess_{(x,\xi)}\Phi = \begin{pmatrix} -W & \Delta_1 \\ {}^t\Delta_1 & B \end{pmatrix}. \tag{7.5.2}$$

Here $W = W(x) = (w_{jk}(x))_{j,k=1}^J$ is the $J \times J$ matrix with j, k element $w_{jk} = w_{jk}(x)$

$$w_{jk} = \begin{cases} \partial_{x_j}^2(\tau_j\phi_j(x_j, x_{j-1}) + \tau_{j+1}\phi_{j+1}(x_{j+1}, x_j)) & \text{if } k = j, \\ \tau_j \partial_{x_{j-1}}\partial_{x_j}\phi_j(x_j, x_{j-1}) & \text{if } k = j-1, \\ \tau_{j+1}\partial_{x_{j+1}}\partial_{x_j}\phi_{j+1}(x_{j+1}, x_j) & \text{if } k = j+1, \\ 0 & \text{if } |j-k| \geq 2, \end{cases} \tag{7.5.3}$$

and B is the $J \times J$-matrix

$$B = \begin{pmatrix} \tau_1 & 0 & 0\,0\,0 \dots \\ 0 & \tau_2 & 0\,0\,0 \dots \\ 0 & 0 & \tau_3\,0\,0 \dots \\ \dots\dots & & \end{pmatrix}. \tag{7.5.4}$$

We wrote C for the matrix W in (7.2.9). We changed notation to avoid confusion, because we often write various constants by C.

Notice that

$$\begin{pmatrix} I & -{}^t\Delta_1^{-1}B \\ 0 & I \end{pmatrix} \begin{pmatrix} 0 & {}^t\Delta_1^{-1} \\ \Delta_1^{-1} & 0 \end{pmatrix} \begin{pmatrix} -W & \Delta_1 \\ {}^t\Delta_1 & B \end{pmatrix} = \begin{pmatrix} I + {}^t\Delta_1^{-1}B\Delta_1^{-1}W & 0 \\ -{}^t\Delta_1^{-1}W & I \end{pmatrix}.$$

Therefore,

$$|\det Hess_{(x,\xi)}\Phi| = |\det(I + {}^t\Delta_1^{-1}B\Delta_1^{-1}W)|.$$

Since $\|\Delta_1^{-1}W\|_{l^\infty \to l^\infty} \leq 4\kappa_2 T$ and $\|\Delta_1^{-1}B\|_{l^\infty \to l^\infty} \leq T$,

$$\|{}^t\Delta_1^{-1}B\Delta_1^{-1}W\|_{l^\infty \to l^\infty} \leq 4\kappa_2 T^2. \tag{7.5.5}$$

Therefore, if $4\kappa_2 T^2 < 1$, then

$$(1 + 4\kappa_2 T^2)^{2J} \geq |\det Hess_{(x,\xi)}\Phi| \geq (1 - 4\kappa_2 T^2)^{2J}.$$

It turns out that $\Phi(x, \xi)$ satisfies Assumption 3.1. This corresponds to Proposition 7.3.

Now we discuss bound of (x^*, ξ^*) in this case. The map $\mathscr{K}(x, \xi)$ is of the form

$$\mathscr{K}(x,\xi) = -x_0\mathbf{e}_1 + \sum_{j=1}^{J} \tau_j\xi_j\mathbf{e}_j$$
$$- \xi_{J+1}\mathbf{e}_{2J} - \sum_{j=1}^{J}(\tau_{j+1}\partial_{x_j}\phi_{j+1}(x_{j+1}, x_j) + \tau_j\partial_{x_j}\phi_j(x_j, x_{j-1}))\mathbf{e}_{J+j}.$$

The stationary point (x^*, ξ^*) is the solution of

$$\Delta(x^*, \xi^*) + \mathscr{K}(x^*, \xi^*) = 0, \tag{7.5.6}$$

or equivalently,

$$(x^*, \xi^*) + \Delta^{-1}\mathscr{K}(x^*, \xi^*) = 0. \tag{7.5.7}$$

More precisely, for $j = 1, 2, \dots, J$,

$$x_j^* - x_0 + \sum_{k=1}^{j} \tau_k\xi_k^* = 0, \tag{7.5.8}$$

$$\xi_j^* - \xi_{J+1} - \sum_{k=j}^{J}(\tau_{k+1}\partial_{x_k}\phi_{k+1}(x_{k+1}^*, x_k^*) + \tau_k\partial_{x_k}\phi_k(x_k^*, x_{k-1}^*)) = 0, \tag{7.5.9}$$

where $x_{J+1}^* = x_{J+1}$ and $x_0^* = x_0$.

The matrix expression of differential of $\mathscr{K}$ is

$$D\mathscr{K}_{(x,\xi)} = \begin{pmatrix} 0 & B \\ -W & 0 \end{pmatrix},$$

Thus

$$\Delta^{-1}D\mathscr{K}_{(x,\xi)} = \begin{pmatrix} 0 & \Delta_1^{-1}B \\ -{}^t\Delta_1^{-1}W & 0 \end{pmatrix}. \tag{7.5.10}$$

Corresponding to Corollary 7.3, we have

Proposition 7.9 *Suppose* $4\kappa_2\delta^2 < 1$ *and* $T \le \delta$. *Then there exists a constant* $C > 0$ *depending on* κ_1, κ_2 *and* δ *such that*

$$\|(x^*, \xi^*)\|_{l^\infty} \le C(1 + |x_0| + |x_{J+1}| + |\xi_{J+1}|).$$

Here we can choose

$$C = (1 - 4\kappa_2\delta^2)^{-1}(1 + (1 + 4\kappa_2)\delta)(1 + 2\kappa_1\delta). \tag{7.5.11}$$

Moreover,

$$\|(x^*, \xi^*) - \Delta^{-1}(x_0\mathbf{e}_1 + \xi_{J+1}\mathbf{e}_{2J})\|_{l^\infty} \le CT(1 + |x_0| + |x_{J+1} + |\xi_{J+1}|) \tag{7.5.12}$$

with some constant $C > 0$ depending on $\kappa_1, \kappa_2, \delta$.

Proof By Taylor's formula

$$(x^*, \xi^*) = -\Delta^{-1}\mathscr{K}(x^*, \xi^*) = -\Delta^{-1}\mathscr{K}(0, 0) - \Delta^{-1}\widetilde{D\mathscr{K}} \cdot (x^*, \xi^*), \tag{7.5.13}$$

where $\widetilde{D\mathscr{K}} \cdot (x^*, \xi^*)$ is the image of the vector (x^*, ξ^*) by the linear map $\widetilde{D\mathscr{K}}$ of which the matrix expression is

$$\widetilde{D\mathscr{K}} = \begin{pmatrix} 0 & \Delta_1^{-1}B \\ -{}^t\Delta_1^{-1}\widetilde{W} & 0 \end{pmatrix}$$

with

$$\widetilde{W} = \int_0^1 W(tx^*)\,dt.$$

It follows from (7.5.13) that

$$\begin{aligned}(x^*, \xi^*) &= -\Delta^{-1}\mathscr{K}(0, 0) + \Delta^{-1}\widetilde{D\mathscr{K}} \cdot \left(\Delta^{-1}\mathscr{K}(0, 0) + \Delta^{-1}\widetilde{D\mathscr{K}} \cdot (x^*, \xi^*)\right) \\ &= -\Delta^{-1}\mathscr{K}(0, 0) + \Delta^{-1}\widetilde{D\mathscr{K}} \cdot \Delta^{-1}\mathscr{K}(0, 0) + (\Delta^{-1}\widetilde{D\mathscr{K}})^2 \cdot (x^*, \xi^*).\end{aligned}$$

Here

$$(\Delta^{-1}\widetilde{D\mathscr{K}})^2 = \begin{pmatrix} -\Delta_1^{-1}B{}^t\Delta_1^{-1}\widetilde{W} & 0 \\ 0 & -{}^t\Delta_1^{-1}\widetilde{W}\Delta_1^{-1}B \end{pmatrix}.$$

and

$$\|(\Delta^{-1}\widetilde{D\mathscr{K}})^2\|_{l^\infty \to l^\infty} \le 4\kappa_2 T^2. \tag{7.5.14}$$

Suppose $4\kappa_2\delta^2 < 1$ and $T \le \delta$. Then $(I - (\Delta^{-1}\widetilde{D\mathscr{K}})^2)$ is invertible and

$$(x^*, \xi^*) = (I - (\Delta^{-1}\widetilde{D\mathscr{K}})^2)^{-1}\left(-\Delta^{-1}\mathscr{K}(0, 0) + \Delta^{-1}\widetilde{D\mathscr{K}} \cdot \Delta^{-1}\mathscr{K}(0, 0)\right). \tag{7.5.15}$$

On the other hand,

$$\mathscr{K}(0,0) = -x_0\mathbf{e}_1 - \xi_{J+1}\mathbf{e}_{2J} + \mathscr{K}_0(x_{J+1}, x_0),$$

where

$$\begin{aligned}\mathscr{K}_0(x_{J+1}, x_0) = &-(\tau_1\partial_{x_1}\phi_1(0, x_0) + \tau_2\partial_{x_1}\phi_2(0,0))\mathbf{e_{J+1}}\\ &-\sum_{j=2}^{J-1}(\tau_j\partial_{x_j}\phi_j(0,0) + \tau_{j+1}\partial_{x_j}\phi_{j+1}(0,0))\mathbf{e_{J+j}}\\ &-(\tau_J\partial_{x_J}\phi_J(0,0) + \tau_{J+1}\partial_{x_J}\phi_{J+1}(x_{J+1},0))\mathbf{e_{2J}}.\end{aligned}$$

It follows from this and Assumption 7.1 that

$$\|\mathscr{K}_0(x_{J+1}, x_0)\|_{l^1} \leq 2\kappa_1 T(1 + |x_0| + |x_{J+1}|).$$

Hence

$$\|{}^t\Delta^{-1}\mathscr{K}_0(x_{J+1}, x_0)\|_{l^\infty} \leq 2\kappa_1 T(1 + |x_0| + |x_{J+1}|). \tag{7.5.16}$$

Since

$$\Delta^{-1}\mathscr{K}(0,0) = -\Delta^{-1}(x_0\mathbf{e}_1 + \xi_{J+1}\mathbf{e}_{2J}) + {}^t\Delta_1^{-1}\mathscr{K}_0(x_{J+1}, x_0),$$

$$\|\Delta^{-1}\mathscr{K}(0,0)\|_{l^\infty} \leq (|x_0| + |\xi_{J+1}| + 2\kappa_1 T(1 + |x_0| + |x_{J+1}|)).$$

We have $\|\Delta^{-1}\widetilde{D\mathscr{K}}\|_{l^\infty\to l^\infty} \leq \max\{1, 4\kappa_2\}T$, because $\|\Delta_1^{-1}B\|_{l^\infty\to l^\infty} \leq T$ and $\|{}^t\Delta_1^{-1}\widetilde{W}\|_{l^\infty\to l^\infty} \leq 4\kappa_2 T$. Thus

$$\begin{aligned}&\|\Delta^{-1}\widetilde{D\mathscr{K}}\cdot\Delta^{-1}\mathscr{K}(0,0)\|_{l^\infty}\\ &\leq (1 + 4\kappa_2)T(|x_0| + |\xi_{J+1}| + 2\kappa_1 T(1 + |x_0| + |x_{J+1}|)).\end{aligned} \tag{7.5.17}$$

Therefore,

$$\begin{aligned}&\|(x^*, \xi^*)\|_{l^\infty}\\ &\leq (1 - 4\kappa_2 T^2)^{-1}(1 + (1 + 4\kappa_2)T)(|x_0| + |\xi_{J+1}| + 2\kappa_1 T(1 + |x_0| + |x_{J+1}|))\\ &\leq C(\kappa_1, \kappa_2)(1 + |x_0| + |x_{J+1}| + |\xi_{J+1}|),\end{aligned} \tag{7.5.18}$$

where

$$C(\kappa_1, \kappa_2) = (1 - 4\kappa_2\delta^2)^{-1}(1 + (1 + 4\kappa_2)\delta)(1 + 2\kappa_1\delta). \tag{7.5.19}$$

By (7.5.13)

$$
\begin{aligned}
&(x^*, \xi^*) - \Delta^{-1}(x_0\mathbf{e}_1 + \xi_{J+1}\mathbf{e}_{2J}) \\
&= -\Delta^{-1}\mathscr{K}_0 + \Delta^{-1}\widetilde{D\mathscr{K}} \cdot \Delta^{-1}\mathscr{K}(0,0) + (\Delta^{-1}\widetilde{D\mathscr{K}})^2 \cdot (x^*, \xi^*).
\end{aligned} \tag{7.5.20}
$$

It follows from this, (7.5.16), (7.5.17) and (7.5.18) that

$$
\|(x^*, \xi^*) - \Delta^{-1}(x_0\mathbf{e}_1 + \xi_{J+1}\mathbf{e}_{2J})\|_{l^\infty} \le CT(1 + |x_0| + |x_{J+1} + |\xi_{J+1}|) \tag{7.5.21}
$$

with some constant $C > 0$ depending on κ_1, κ_2. □

Corresponding to Proposition 7.4 and its corollary, we have

Proposition 7.10 *Suppose that $T < \delta$ with $4\kappa_2\delta < 1$. Then the following inequalities hold:*

$$
\begin{aligned}
&\|\partial_{x_0}(x^*, \xi^*)\|_{l^\infty} \le (1 - 4\kappa_2 T^2)^{-1}(1 + T)(1 + \tau_1\kappa_2), && (7.5.22)\\
&\|\partial_{x_0}(x^*, \xi^*) - (1 + \tau_1\partial^2_{x_0}\phi_1(x^*, x_0))(I - \Delta_1^{-1}B)\Delta_1^{-1}\mathbf{e}_1\|_{l^\infty} \le 4\kappa_2 T^2, && (7.5.23)\\
&\|\partial_{x_{J+1}}(x^*, \xi^*)\|_{l^\infty} \le (1 - 4\kappa_2 T^2)^{-1}\kappa_2\tau_{J+1}(1 + 4\kappa_2 T), && (7.5.24)\\
&\|\partial_{x_{J+1}}(x^*, \xi^*) - \tau_{J+1}\partial_{x_{J+1}}\partial_{x_J}\phi_{J+1}(x_{J+1}, x_J^*)'{}^t\Delta_1^{-1}\mathbf{e}_{2J}\|_{l^\infty} \le 4\kappa_2 T^2, && (7.5.25)\\
&\|(\partial_{\xi_{J+1}}(x^*, \xi^*)\|_{l^\infty} \le (1 - 4\kappa_2 T^2)^{-1}(1 + 4\kappa_2 T), && (7.5.26)\\
&\|(\partial_{\xi_{J+1}}(x^*, \xi^*) - (I - \Delta^{-1}D\mathscr{K}_{(x^*,\xi^*)})'{}^t\Delta_1^{-1}\mathbf{e}_{2J}\|_{l^\infty} \\
&\qquad \le 4\kappa_2 T^2(1 - 4\kappa_2 T^2)^{-1}(1 + 4\kappa_2 T), && (7.5.27)\\
&\|\partial_{\xi_{J+1}} x^*\| \le (1 - 4\kappa_2 T^2)^{-1}T, && (7.5.28)\\
&|\partial_{\xi_{J+1}} x_J^* + (\tau_1 + \cdots + \tau_J)| \le 4\kappa_2(1 - 4\kappa_2 T^2)^{-1}T^3. && (7.5.29)
\end{aligned}
$$

Proof Let $\partial_{\xi_{J+1}}(x^*, \xi^*) = (y, \eta)$. Differentiating (7.5.7), we have

$$
(y, \eta) = -\Delta^{-1}(\partial_{\xi_{J+1}}\mathscr{K})|_{x=x^*,\xi=\xi^*} - \Delta^{-1}D\mathscr{K}_{(x^*,\xi^*)} \cdot (y, \eta).
$$

Since $\partial_{\xi_{J+1}}\mathscr{K}|_{x=x^*,\xi=\xi^*} = -\mathbf{e}_{2J}$,

$$
(y, \eta) = \Delta^{-1}\mathbf{e}_{2J} - \Delta^{-1}D\mathscr{K}_{(x^*,\xi^*)} \cdot (y, \eta).
$$

It follows from this that

$$
(y, \eta) = \Delta^{-1}\mathbf{e}_{2J} - \Delta^{-1}D\mathscr{K}_{(x^*,\xi^*)} \cdot \left(\Delta^{-1}\mathbf{e}_{2J} - \Delta^{-1}D\mathscr{K}_{(x^*,\xi^*)} \cdot (y, \eta)\right). \tag{7.5.30}
$$

Notice that

$$
\Delta^{-1}D\mathscr{K}_{(x^*,\xi^*)} = \begin{pmatrix} 0 & \Delta_1^{-1}B \\ -{}^t\Delta_1^{-1}W(x^*) & 0 \end{pmatrix}. \tag{7.5.31}
$$

Therefore,

$$\|\Delta^{-1}D\mathscr{K}_{(x^*,\xi^*)}\|_{l^\infty\to l^\infty} \le \max\{1, 4\kappa_2\}T, \tag{7.5.32}$$

and

$$\left(\Delta D\mathscr{K}_{(x^*,\xi^*)}\right)^2 = \begin{pmatrix} -\Delta_1^{-1}B^t\Delta_1^{-1}W(x^*) & 0 \\ 0 & -{}^t\Delta_1^{-1}W(x^*)\Delta_1^{-1}B \end{pmatrix}. \tag{7.5.33}$$

This implies that

$$\|\left(\Delta^{-1}D\mathscr{K}_{(x^*,\xi^*)}\right)^2\|_{l^\infty\to l^\infty} \le 4\kappa_2 T^2. \tag{7.5.34}$$

Consequently, $I - \left(\Delta^{-1}D\mathscr{K}_{(x^*,\xi^*)}\right)^2$ is invertible and

$$(y,\eta) = (I - (\Delta^{-1}D\mathscr{K}_{(x^*,\xi^*)})^2)^{-1}(I - \Delta^{-1}D\mathscr{K}_{(x^*,\xi^*)})\Delta^{-1}\mathbf{e}_{2J} \tag{7.5.35}$$

if $4\kappa_2 T^2 < 1$. Thus

$$\|(y,\eta)\|_{l^\infty} \le (1 - 4\kappa_2 T^2)^{-1}(1 + 4\kappa_2 T).$$

This proves (7.5.27).

Moreover, looking at the subspace span by $\mathbf{e}_1, \dots, \mathbf{e}_J$ of Eq. (7.5.30), we have

$$y = -\Delta_1^{-1}B\,{}^t\Delta_1^{-1}\mathbf{e}_J + {}^t\Delta_1^{-1}W\Delta_1^{-1}By. \tag{7.5.36}$$

This implies that

$$y = -(I - {}^t\Delta_1^{-1}W\Delta_1^{-1}B)^{-1}(\Delta^{-1}B\,{}^t\Delta_1^{-1}\mathbf{e}_J)$$

and

$$\|y\|_{l^\infty} \le (1 - 4\kappa_2 T^2)^{-1}T.$$

This proves (7.5.29). Using this, (7.5.34) and (7.5.36) again, we have

$$\|y + \Delta^{-1}B^t\Delta_1^{-1}\mathbf{e}_J\|_{l^\infty} \le 4\kappa_2(1 - 4\kappa_2 T^2)^{-1}T^3.$$

This implies

$$|y_J + (\tau_1 + \cdots + \tau_J)| \le 4\kappa_2(1 - 4\kappa_2 T^2)^{-1}T^3.$$

(7.5.29) has been proved.

Now we prove (7.5.22) and (7.5.23). Let $(y,\eta) = \partial_{x_0}(x^*,\xi^*)$. Then

$$(y,\eta) = -\Delta^{-1}\partial_{x_0}\mathscr{K}\big|_{(x,\xi)=(x^*,\xi^*)} - \Delta^{-1}D\mathscr{K}_{(x^*,\xi^*)}[(y,\eta)].$$

Thus by iteration

$$(y,\eta) = -(I + \Delta^{-1} D\mathscr{K}_{(x^*,\xi^*)})[\Delta^{-1}\partial_{x_0}\mathscr{K}\big|_{(x^*,\xi^*)}] + \left(\Delta^{-1} D\mathscr{K}_{(x^*,\xi^*)}\right)^2 [(y,\eta)]. \tag{7.5.37}$$

It follows from this, (7.5.31), (7.5.33) and

$$\Delta^{-1}\partial_{x_0}\mathscr{K}\big|_{(x,\xi)} = (1 + \tau_1\phi_1(x_1^*, x_0))\Delta^{-1}\mathbf{e}_1$$

that

$$(y,\eta) = (1 + \tau_1\phi_1(x_1^*, x_0))(I + \Delta_1^{-1}B)\Delta^{-1}\mathbf{e}_1) + \left(\Delta^{-1} D\mathscr{K}_{(x^*,\xi^*)}\right)^2 [(y,\eta)].$$

Therefore, we have

$$(y,\eta) = (1 + \tau_1\phi_1(x_1^*, x_0)) \left(I - \left(\Delta^{-1} D\mathscr{K}_{(x^*,\xi^*)}\right)^2\right)^{-1} [(I + \Delta_1^{-1}B)\Delta^{-1}\mathbf{e}_1)]. \tag{7.5.38}$$

Inequalities (7.5.22) and (7.5.23) follow from this.

Next we prove (7.5.24) and (7.5.25). Let $(y,\eta) = \partial_{x_{J+1}}(x^*, \xi^*)$. Then

$$(y,\eta) = \tau_{J+1}\partial_{x_{J+1}}\partial_{x_J}\phi_{J+1}(x_{J+1}, x_J^*)\Delta^{-1}\mathbf{e}_{2J} - \Delta^{-1} D\mathscr{K}_{(x^*,\xi^*)}[(y,\eta)].$$

By iteration

$$\begin{aligned}(y,\eta) =& \tau_{J+1}\partial_{x_{J+1}}\partial_{x_J}\phi_{J+1}(x_{J+1}, x_J^*)(I - \Delta^{-1} D\mathscr{K}_{(x^*,\xi^*)})\Delta^{-1}\mathbf{e}_{2J} \\ &+ \left(\Delta^{-1} D\mathscr{K}_{(x^*,\xi^*)}\right)^2 [(y,\eta)].\end{aligned}$$

Therefore,

$$\begin{aligned}&(y,\eta) \\ &= \tau_{J+1}\partial_{x_{J+1}}\partial_{x_J}\phi_{J+1}(x_{J+1}, x_J^*) \left(I - \left(\Delta^{-1} D\mathscr{K}_{(x^*,\xi^*)}\right)^2\right)^{-1} (I - \Delta_1^{-1}B)\Delta_1^{-1}\mathbf{e}_{2J}.\end{aligned} \tag{7.5.39}$$

(7.5.24) and (7.5.25) follow from (7.5.39). □

Remark 7.7 We set $Q = I + R$ in Lemma 7.5. Then R satisfies $R = \Delta^{-1}L - \Delta^{-1}LR$. By iteration we have $(I - (\Delta^{-1}L)^2)R = \Delta^{-1}L - (\Delta^{-1}L)^2$. If $A = 0$ in (7.2.9), then similar reasoning as in the proof of Proposition 7.10 is valid.

The stationary value $\Phi^*(x_{J+1}, \xi_{J+1}, x_0) = \Phi(x_{J+1}, \xi_{J+1}, x^*, \xi^*, x_0)$ is defined. We set $\phi^*(x_{J+1}, \xi_{J+1}, x_0) = T^{-1}\Phi(x_{J+1}, \xi_{J+1}, x^*, \xi^*, x_0)$.

Theorem 7.7 *Assume that* $8\kappa_2\delta^2 < 1$. *Then there exists an increasing sequence* $0 < \kappa_0^* \le \kappa_1^* \le \kappa_2^* \le \cdots \le \kappa_k^* \le \ldots$ *such that if* $\alpha_{J+1} + \alpha_0 + \beta_{J+1} = n$,

$$|\partial_{x_0}^{\alpha_0}\partial_{\xi_{J+1}}^{\beta_{J+1}}\partial_{x_{J+1}}^{\alpha_{J+1}}\phi^*(x_{J+1},\xi_{J+1},x_0)| \le \kappa_n^*(1+|x_{J+1}|+|\xi_{J+1}|+|x_0|)^{(2-n)_+}, \tag{7.5.40}$$

where $(2-n)_+$ *means* $\max\{0, 2-n\}$. *There exists a positive constant* C *depending only on* κ_2, δ_1 *such that*

$$|\partial_{\xi_{J+1}}^2\phi^*(x_{J+1},\xi_{J+1},x_0) - 1| \le CT^2, \tag{7.5.41}$$

We may choose $C = 4\kappa_2(1-4\kappa_2\delta^2)^{-1}$.

Proof First we prove (7.5.40) for $n = 0$. Recall (7.5.1). Since (7.5.6) holds, $(x_j^* - x_{j-1}^*)\xi_j^* = -\tau_j\xi_j^{*2}$ for $j = 1, 2, \ldots, J$. Therefore,

$$\begin{aligned}&\Phi(x_{J+1},\xi_{J+1},x^*,\xi^*,x_0)\\ &= \tau_{J+1}(\frac{1}{2}|\xi_{J+1}|^2 - \phi_{J+1}(x_{J+1},x_J^*)) + (x_0 - x_J^*)\xi_{J+1} - \sum_{j=1}^{J}\tau_j(\frac{1}{2}|\xi_j^*|^2 + \phi_j(x_j^*,x_{j-1}^*)),\end{aligned}$$

where $x_0^* = x_0$. Hence

$$\begin{aligned}&|\Phi(x_{J+1},\xi_{J+1},x^*,\xi^*,x_0)| \\ &\le (\frac{\tau_{J+1}}{2}|\xi_{J+1}|^2 + \tau_{J+1}\kappa_0(1+|x_{J+1}|^2+|x_J^*|^2) + |x_0 - x_J^*||\xi_{J+1}| \\ &\quad + \frac{1}{2}T\|\xi^*\|_{l^\infty}^2 + \kappa_0 T(1+2\|x^*\|_{l^\infty}).\end{aligned} \tag{7.5.42}$$

By (7.5.18) and (7.5.8) there exists a constant $C > 0$ depending on κ_1, κ_2 such that

$$|x_0 - x_J^*| \le T\|\xi^*\|_{l^\infty} \le CT(1+|x_0|+|x_{J+1}|+|\xi_{J+1}|). \tag{7.5.43}$$

It follows from these and (7.5.18) that

$$|\Phi(x_{J+1},\xi_{J+1},x^*,\xi^*,x_0)| \le CT(1+|x_0|^2+|x_{J+1}|^2+|\xi_{J+1}|^2)$$

with another constant $C > 0$ depending only on $\kappa_0, \kappa_1, \kappa_2$. (7.5.40) has been proved.

Let us discuss (7.5.40) with $n = 1$. Since (x^*, ξ^*) is the stationary point,

$$\begin{aligned}\partial_{x_0}\Phi(x_{J+1},\xi_{J+1},x^*,\xi^*,x_0) &= \partial_{x_0}\Phi(x_{J+1},\xi_{J+1},x,\xi,x_0)\Big|_{x=x^*,\xi=\xi^*}\\ &= \xi_{J+1} - \xi_1^* + \tau_1(\partial_{x_0}\phi_1)(x_1^*,x_0).\end{aligned}$$

By (7.5.9),

$$
\begin{aligned}
|\xi_j^* - \xi_{J+1}| &\le 4T\kappa_1(1 + |x_{J+1}| + \|x^*\|_{l^\infty}) \\
&\le \kappa_1 T(1 + |x_{J+1}| + C(1 + |x_0| + |\xi_{J+1}|)).
\end{aligned}
$$

Thus there exists a constant $C > 0$ depending on κ_1, κ_2 such that

$$
|\partial_{x_0} \Phi(x_{J+1}, \xi_{J+1}, x^*, \xi^*, x_0)| \le CT(1 + |x_0| + |x_{J+1}| + |\xi_{J+1}|).
$$

Next we discuss $\partial_{\xi_{J+1}} \Phi(x_{J+1}, \xi_{J+1}, x^*, \xi^*, x_0)$.

$$
\begin{aligned}
\partial_{\xi_{J+1}} \Phi(x_{J+1}, \xi_{J+1}, x^*, \xi^*, x_0) &= \partial_{\xi_{J+1}} \Phi(x_{J+1}, \xi_{J+1}, x, \xi, x_0)\Big|_{x=x^*, \xi=\xi^*} \\
&= \tau_{J+1}\xi_{J+1} + x_0 - x_J^*.
\end{aligned}
$$

Using (7.5.43), we obtain

$$
|\partial_{\xi_{J+1}} \Phi(x_{J+1}, \xi_{J+1}, x^*, \xi^*, x_0)| \le T|\xi_{J+1}| + CT(1 + |x_0| + |\xi_{J+1}|).
$$

Finally,

$$
\begin{aligned}
&|\partial_{x_{J+1}} \Phi(x_{J+1}, \xi_{J+1}, x^*, \xi^*, x_0)| \le \tau_{J+1} |\partial_{x_{J+1}} \phi_{J+1}(x_{J+1}, x_J^*)| \\
&\le T\kappa_1(1 + |x_{J+1}| + |x_J^*|) \le T\kappa_1(1 + |x_{J+1}| + C(|x_0| + T|\xi_{J+1}| + |x_0|).
\end{aligned}
$$

(7.5.40) for $n = 1$ has been proved.

We shall prove (7.5.41).

$$
\begin{aligned}
\partial_{\xi_{J+1}}^2 \Phi(x_{J+1}, \xi_{J+1}, x^*, \xi^*, x_0)| &= \partial_{\xi_{J+1}}(\tau_{J+1}\xi_{J+1} + x_0 - x_J^*) \\
&= \tau_{J+1} - \partial_{\xi_{J+1}} x_J^*.
\end{aligned}
\tag{7.5.44}
$$

Apply (7.5.29) to this. Then

$$
|\partial_{\xi_{J+1}}^2 \Phi(x_{J+1}, \xi_{J+1}, x^*, \xi^*, x_0) - T| \le 4\kappa_2(1 - 4\kappa_2 T^2)^{-1} T^3.
$$

This proves (7.5.41) with $C = 4\kappa_2(1 - 4\kappa_2\delta^2)^{-1}$.

The rest of Theorem is proved in the same manner. □

Now we give proof of the main Theorem 7.3. It concerns with the oscillatory integral of the following form:

$$
\begin{aligned}
&I(\{\tau_j\}; S, a_\lambda, \nu)(x_{J+1}, x_0) \\
&= \prod_{j=1}^{J+1} \left(\frac{-i\nu}{2\pi\tau_j} \right)^{\frac{1}{2}} \int_{\mathbf{R}^J} e^{i\nu S(x_{J+1}, \ldots, x_0)} a_\lambda(x_{J+1}, x_J, \ldots, x_1, x_0) \prod_{j=1}^{J} dx_j,
\end{aligned}
$$

where

$$S(x_{J+1}, x_J, \dots, x_1, x_0) = \sum_{j=1}^{J+1} \frac{(x_j - x_{j-1})^2}{2\tau_j} + \tau_j \phi_j(x_j, x_{j-1}).$$

As we discussed in Sect. 7.1 we can write

$$\begin{aligned} &I(\{\tau_j\}; S, a_\lambda, \nu)(x_{J+1}, x_0) \\ &= \left(\frac{\nu}{2\pi}\right) \int_{\mathbf{R}} e^{-i\nu(x_{J+1}-x_0)\xi_{J+1}} K(x_{J+1}, \xi_{J+1}, x_0) d\xi_{J+1}, \end{aligned}$$

where

$$K(x_{J+1}, \xi_{J+1}, x_0) = \left(\frac{\nu}{2\pi}\right)^J \int_{\mathbf{R}^{2J}} e^{-i\nu\Phi} a_\lambda(x_{J+1}, x_J, \dots, x_1, x_0) \prod_{j=1}^{J} d\xi_j dx_j,$$

and

$$\begin{aligned} \Phi = &\tau_{J+1}(\frac{1}{2}|\xi_{J+1}|^2 - \phi_{J+1}(x_{J+1}, x_J)) - (x_J - x_0)\xi_{J+1} \\ &+ \sum_{j=1}^{J} \{\tau_j(\frac{1}{2}|\xi_j|^2 - \phi_j(x_j, x_{j-1})) + (x_j - x_{j-1})\xi_j\}. \end{aligned}$$

Let

$$\phi_j(x_j, \xi_j, x_{j-1}) = \frac{1}{2}|\xi_j|^2 - \phi_j(x_j, x_{j-1}). \tag{7.5.45}$$

Function $\phi_j(x_j, \xi, x_{j-1})$ satisfies Assumption 7.4, because the function $\phi_j(x_j, x_{j-1})$ satisfies Assumption 7.1. As $a(x_{J+1}, x_J, \dots, x_1, x_0)$ satisfies Assumption 7.2, it obviously satisfies Assumption 7.5. Thus we can apply Theorem 7.6 and obtain

$$\begin{aligned} &I(\{\tau_j\}; S, a_\lambda, \nu)(x_{J+1}, x_0) \qquad (7.5.46) \\ &= \left(\frac{\nu}{2\pi}\right) \int_{\mathbf{R}} e^{-i\nu(\xi_{J+1}(x_{J+1}-x_0)+\Phi^*(x_{J+1},\xi_{J+1},x_0))} \tilde{b}(x_{J+1}, \xi_{J+1}, x_0) d\xi_{J+1}. \end{aligned}$$

The amplitude function $\tilde{b}(x_{J+1}, \xi_{J+1}, x_0)$ satisfies (7.1.16) and the phase function $\Phi^*(x_{J+1}, \xi_{J+1}, x_0)$ has the property stated in Theorem 7.7.

Integral (7.5.46) is an oscillatory integral with respect to ξ_{J+1}. (7.5.41) in Theorem 7.7 shows that

$$|\partial^2_{\xi_{J+1}} \Phi^*(x_{J+1}, \xi_{J+1}, x_0) - T| \le CT^2, \qquad C = 4\kappa_2(1 - 4\kappa_2 T^2)^{-1}.$$

Let δ be a positive constant such that

$$\delta C \le 1/2 \quad \text{and}\, \delta \le \delta_1. \tag{7.5.47}$$

If $T \le \delta$, then

$$2^{-1} \le |T^{-1}\partial^2_{\xi_{J+1}}\Phi^*(x_{J+1}, \xi_{J+1}, x_0)| \le 2.$$

This implies that the Hessian is not degenerate and we can apply the stationary phase method for one variable ξ_{J+1} to (7.5.46).

Let ξ^*_{J+1} be the stationary point of $(x_{J+1} - x_0)\xi_{J+1} - \Phi^*(x_{J+1}, \xi_{J+1}, x_0)$ with respect to ξ_{J+1}, i.e.,

$$\partial_{\xi_{J+1}}((x_{J+1} - x_0)\xi_{J+1} - \Phi^*(x_{J+1}, \xi_{J+1}, x_0)) = 0.$$

Then we obtain

$$\begin{aligned} &I(\{\tau_j\}; S, a_\lambda, \nu)(x_{J+1}, x_0) \\ &= \left(\frac{\nu}{2\pi i \partial^2_{\xi_{J+1}}\Phi^*(x_{J+1}, \xi^*_{J+1}, x_0)}\right)^{1/2} e^{-i\nu((x_{J+1}-x_0)\xi^*_{J+1} - \Phi^*(x_{J+1}, \xi^*_{J+1}, x_0))} \\ &\quad \times b_1(\nu, x_{J+1}, x_0) \\ &= \left(\frac{-i\nu}{2\pi T}\right)^{\frac{1}{2}} e^{-i\nu((x_{J+1}-x_0)\xi^*_{J+1} - \Phi^*(x_{J+1}, \xi^*_{J+1}, x_0))} b_\lambda(\{\tau_j\}; \nu, x_{J+1}, x_0), \end{aligned} \tag{7.5.48}$$

where

$$b_\lambda(\{\tau_j\}; \nu, x_{J+1}, x_0) = T^{1/2}(\partial^2_{\xi_{J+1}}\Phi^*(x_{J+1}, \xi_{J+1}, x_0))^{-1/2} b_1(\nu, x_{J+1}, x_0).$$

By Corollary 3.6, for any non-negative integer k there exist a positive constant $C'_k \ge 1$ and a natural number k' such that if $|\alpha_0| \le k$ and $|\alpha_{J+1}| \le k$,

$$\begin{aligned} &(1 + |\lambda| + |x_{J+1}| + |x_0|)^{-m} |\partial^{\alpha_{J+1}}_{x_{J+1}} \partial^{\alpha_0}_{x_0} b_1(\nu, x_{J+1}, x_0)| \\ &\le C'_k \max_{\substack{|\beta| \le k' \\ \alpha \le \alpha_{J+1}, \gamma \le \alpha_0}} \sup_{\xi_{J+1} \in \mathbf{R}} (1 + |\lambda| + |x_{J+1}| + |\xi_{J+1}| + |x_0|)^{-m} \\ &\qquad\qquad \times |\partial^\alpha_{x_{J+1}} \partial^\gamma_{x_0} \partial^\beta_{\xi_{J+1}} \tilde{b}(x_{J+1}, \xi_{J+1}, x_0)|. \end{aligned}$$

Combining this with the estimate (7.1.16) in Kumano-go-Taniguchi Theorem 7.6 and setting $n = 2k + k'$, we have

$$\begin{aligned} |\partial^{\alpha_{J+1}}_{x_{J+1}} \partial^{\alpha_0}_{x_0} b_1(\nu, x_{J+1}, x_0)| &\le C'_k C_n^{J+1} A_{K(n)} (1 + |\lambda| + |x_0| + |x_{J+1}|)^m \\ &\le (C'_k C_n)^{J+1} A_{K(n)} (1 + |\lambda| + |x_0| + |x_{J+1}|)^m. \end{aligned}$$

Here $K(n)$ depends only on k and C_n is independent of J.

Since the function

$$T^{1/2}(\partial^2_{\xi_{J+1}}\Phi^*(x_{J+1},\xi_{J+1},x_0))^{-1/2}$$

together with its derivatives of all order are bounded functions of (x_{J+1},ξ_{J+1},x_0). There is another constant C_k such that

$$|\partial^{\alpha_{J+1}}_{x_{J+1}}\partial^{\alpha_0}_{x_0}b_\lambda(\{\tau_j\};\nu,x_{J+1},x_0)|\le(C_k)^{J+1}A_{K(n)}(1+|\lambda|+|x_0|+|x_{J+1}|)^m. \tag{7.5.49}$$

This has proved (7.1.5), because n depends only on k.

We claim the following

Proposition 7.11 *If $T\le\delta$, then*

$$(x_{J+1}-x_0)\xi^*_{J+1}-\Phi^*(x_{J+1},\xi^*_{J+1},x_0)=-S^*(x_{J+1},x_0).$$

Here $S^*(x_{J+1},x_0)$ stands for the stationary value of $S(x_{J+1},x_J,x_{J-1},\dots,x_1,x_0)$ regarded as a function of $(x_J,x_{J-1},\dots,x_1)$.

Proof Note that

$$\Phi^*(x_{J+1},\xi_{J+1},x_0)=\Phi(x_{J+1},\xi_{J+1},x,\xi,x_0)|_{x=x^*,\xi=\xi^*}.$$

Here $(x^*,\xi^*)=(x^*_J,\xi^*_J,\dots,x^*_1,\xi^*_1)$ is the solution of equations

$$\begin{aligned}
&0=x^*_1-x_0+\tau_1\xi^*_1\\
&0=x^*_j-x^*_{j-1}+\tau_j\xi^*_j,\qquad \text{for } j=2,3,\dots,J\\
&0=\xi^*_j-\xi^*_{j+1}+\tau_j\partial_{x_j}\phi_j(x^*_j,x^*_{j-1})+\tau_{j+1}\partial_{x_j}\phi_{j+1}(x^*_{j+1},x^*_j)\qquad \text{for } j=1,2,\dots,J-1\\
&0=\xi^*_J-\xi^*_{J+1}+\tau_J\partial_{x_J}\phi_J(x^*_J,x^*_{J-1})+\tau_{J+1}\partial_{x_J}\phi_{J+1}(x_{J+1},x^*_J).
\end{aligned}$$

In addition ξ^*_{J+1} satisfies

$$0=x_{J+1}-x^*_J+\tau_{J+1}\xi^*_{J+1}.$$

We can eliminate $\xi^*_1,\xi^*_1,\dots,\xi^*_{J+1}$ from this system of equations and obtain

$$0=-\frac{x^*_1-x_0}{\tau_1}+\frac{x^*_2-x^*_1}{\tau_2}+\tau_1\partial_{x_1}\phi_j(x^*_1,x_0)+\tau_2\partial_{x_1}\phi_2(x^*_2,x^*_1),$$

For $j=2,3,\dots,J-1$

$$0=-\frac{x^*_j-x^*_{j-1}}{\tau_j}+\frac{x^*_{j+1}-x^*_j}{\tau_{j+1}}+\tau_j\partial_{x_j}\phi_j(x^*_j,x^*_{j-1})+\tau_{j+1}\partial_{x_j}\phi_{j+1}(x^*_{j+1},x^*_j),$$

$$0=-\frac{x^*_J-x^*_{J-1}}{\tau_J}+\frac{x_{J+1}-x^*_J}{\tau_{J+1}}+\tau_J\partial_{x_J}\phi_J(x^*_J,x^*_{J-1})+\tau_{J+1}\partial_{x_J}\phi_{J+1}(x_{J+1},x^*_J).$$

This system of equations means that $(x_1^*, x_2^*, \ldots, x_J^*)$ is the stationary point of the function

$$S(x_{J+1}, x_J, \ldots, x_1, x_0) = \sum_{j=1}^{J+1} \frac{(x_j - x_{j-1})^2}{2\tau_j} + \tau_j \phi_j(x_j, x_{j-1}).$$

As to the stationary value we have

$$\begin{aligned}
&(x_{J+1} - x_0)\xi_{J+1}^* - \Phi^*(x_{J+1}, \xi_{J+1}^*, x_0) \\
&= -\frac{(x_{J+1} - x_J)^2}{2\tau_{J+1}} - \tau_{J+1}\phi_{J+1}(x_{J+1}, x_J^*) - \sum_{j=1}^{J} \Big(\frac{(x_j^* - x_{j-1}^*)^2}{2\tau_j} + \tau_j \phi_j(x_j^*, x_{j-1}^*) \Big) \\
&= -S(x_{J+1}, x_J^*, \ldots, x_1^*, x_0).
\end{aligned}$$

The proposition is proved. □

As the result of this proposition and (7.5.48) we have

$$I(\{\tau_j\}; S, a, \nu)(x_{J+1}, x_0) = \Big(\frac{\nu}{2\pi i T}\Big)^{1/2} e^{i\nu S^*(x_{J+1}, x_0)} b_\lambda(\{\tau_j\}; \nu, x_{J+1}, x_0).$$

As was proved by (7.5.49), Theorem 7.3 has been proved completely. □

The number δ in (7.5.47) may be very small compared with δ_0. In the following we shall prove that we can make $\delta = \delta_0$ in Theorem 7.3.

We introduce a notational convention. We define $T_j = \sum_{k=1}^{j} \tau_k$ for $j = 1, 2, \ldots$ $J+1$ and $T_0 = 0$, $T_{J+1} = T$. We obtain the division

$$\Delta : 0 = T_0 < T_1 < T_2 < \cdots < T_J < T_{J+1} = T$$

of the interval $[0, T]$. Let j, l be two integers such that $1 \le j < l \le J+1$. Then we set $T(l, j) = \tau_l + \cdots + \tau_j$ and

$$S_{l,j}(x_l, \ldots, x_{j-1}) = \sum_{k=j}^{l} S_k(x_k, x_{k-1}).$$

We understand that $S_{j,j}(x_j, x_{j-1}) = S_j(x_j, x_{j-1})$. Notice that $S_{J+1,1}(x_{J+1}, x_J, \ldots, x_1, x_0) = S(x_{J+1}, x_J, \ldots, x_1, x_0)$. Let $S_{l,j}^*(x_l, x_{j-1})$ be the stationary value of $S_{l,j}(x_l, \ldots, x_{j-1})$ at the stationary point of the function

$$(x_{l-1}, \ldots, x_j) \to S_l(x_l, x_{l-1}) + S_{l-1}(x_{l-1}, x_{l-2}) + \cdots + S_j(x_j, x_{j-1}) = S_{l,j}(x_l, \ldots, x_{j-1}).$$

Proposition 7.12 *Theorem 7.3 holds for δ_0.*

Proof First let δ be as in (7.5.47). We choose the subsequence $T_{j_1} < T_{j_2} < \cdots < T_{j_s} < T_{j_{s+1}} = T_{J+1}$ in the following way. Let j_1 be the smallest number $j \le J+1$ such that $T_j - T_0 \ge \delta$, let j_2 be the smallest $j \le J+1$ such that $T_j - T_{j_1} \ge \delta$, ..., j_k be the smallest number $j \le J+1$ such that $T_j - T_{j_{k-1}} \ge \delta$, for $k = 1, 2, \dots, s$ and $T_{j_{s+1}} = T_{J+1}$. Then $s \le \delta^{-1} T$, which is independent of J.

If $j_1 - 1 > 0$, then $0 < T_{j_1-1} - T_0 < \delta$. We can apply Theorem 7.3 to the integration with respect to variables $x_1, x_2, \dots, x_{j_1-2}$. As the result, we have

$$\prod_{j=1}^{j_1-1} \left(\frac{-i\nu}{2\pi\tau_j}\right)^{\frac{1}{2}} \int_{\mathbf{R}^{j_1-2}} e^{i\nu S_{j_1-1,1}(x_{j_1-1}, x_{j_1-2}, \dots, x_1, x_0)} \tag{7.5.50}$$

$$\times\ a_\lambda(x_{J+1}, x_J, \dots, x_{j_1}, x_{j_1-1}, x_{j_1-2}, \dots, x_1, x_0) \prod_{j=1}^{j_1-2} dx_j$$

$$= \left(\frac{-i\nu}{2\pi(T_{j_1-1} - T_0)}\right)^{\frac{1}{2}} e^{i\nu S^*_{j_1-1,1}(x_{j_1-1}, x_0)} a_{\lambda, j_1}(x_{J+1}, s_J, \dots, x_{j_1}, x_{j_1-1}, x_0)$$

with some $a_{\lambda, j_1}(x_{J+1}, s_J, \dots, x_{j_1}, x_{j_1-1}, x_0)$. Let $\mathscr{L} = \prod_{j=j_1}^{J+1} \partial_{x_j}^{\beta_j}$. Then if $|\alpha_0| \le n$ and $|\alpha_{j_1-1}| \le n$, then by Corollary 7.1 we have

$$\left| \left(1 + |\lambda| + |x_0| + \sum_{j=j_1-1}^{J+1} |x_j|\right)^{-m} |\partial_{x_0}^{\alpha_0} \partial_{x_{j_1-1}}^{\alpha_{j_1-1}} \mathscr{L} a_{\lambda, j_1}(x_{J+1}, \dots, x_{j_1-1}, x_0) \right|$$

$$\le C_n^{j_1-1} \max_{\substack{|\beta_j| \le K(n),\\ 0 \le j \le, j_1-1}} \sup_{(x_{j_1-1}, \dots, x_0) \in \mathbf{R}^{j_1-2}} \left(1 + |\lambda| + \sum_{j=0}^{J+1} |x_j|\right)^{-m}$$

$$\times \left| \mathscr{L} \partial_{x_0}^{\alpha_0} \partial_{x_{j_1-1}}^{\alpha_{j_1-1}} \prod_{j=1}^{j_1-1} \partial_{x_{j_1-1}}^{\alpha_{j_1-1}} a_\lambda(x_{J+1}, x_J, \dots, x_1, x_0) \right|.$$

Next we replace (7.5.50) in (7.1.1) and integrate (7.1.1) with respect to variables $(x_{j_2-2}, \dots, x_{j_1+1})$ and apply Theorem 7.3. Afterward we integrate with respect to $(x_{j_3-2}, \dots, x_{j_2+1})$, ... In this way, we continue this operation in succession. Then finally, we have

$$I(\{\tau_j\}; S, a_\lambda, \nu)(x_{J+1}, x_0) \tag{7.5.51}$$

$$= \left(\frac{-i\nu}{2\pi(T_{J+1} - T_{j_s})}\right)^{\frac{1}{2}} \prod_{r=1}^{s} \left(\frac{-i\nu}{2\pi(T_{j_r-1} - T_{j_{r-1}})}\right)^{\frac{1}{2}} \prod_{r=1}^{s} \left(\frac{-i\nu}{2\pi\tau_{j_r}}\right)^{\frac{1}{2}}$$

$$\times \int_{\mathbf{R}^{2s}} e^{i\nu S_{j_s, j_{s-1}, \dots, j_1}(x_{J+1}, x_{j_s}, x_{j_s-1}, \dots, x_{j_1}, x_{j_1-1}, x_0)}$$

$$\times\ a_{\lambda,j_s,j_{s-1},\dots,j_1}(x_{J+1},x_{j_s},x_{j_s-1},\dots,x_{j_1},x_{j_1-1},x_0)\prod_{r=1}^{s}dx_{j_r}dx_{j_r-1}.$$

Here

$$\begin{aligned}&S_{j_s,j_{s-1},\dots,j_1}(x_{J+1},x_{j_s},x_{j_s-1},\dots,x_{j_1},x_{j_1-1},x_0) \\ &= S^*_{J+1,j_s+1}(x_{J+1},x_{j_s})+\sum_{r=1}^{s}\left(S_{j_r}(x_{j_r},x_{j_r-1})+S^*_{j_r-1,j_{r-1}+1}(x_{j_r-1},x_{j_{r-1}})\right),\end{aligned} \tag{7.5.52}$$

and it follows from repeated use of Corollary 7.1 that for any integer $n\geq 0$ if $|\beta_0|\leq n$, $|\beta_{J+1}|\leq n$ and $|\beta_{j_k-1}|\leq n$, $|\beta_{j_k}|\leq n$ for $1\leq k\leq s$, then

$$\begin{aligned}&\left|\partial_{x_0}^{\beta_0}\partial_{x_{J+1}}^{\beta_{J+1}}\prod_{k=1}^{s}\partial_{x_{j_k-1}}^{\beta_{j_k-1}}\partial_{x_{j_k}}^{\beta_{j_k}}a_{\lambda,j_s,j_{s-1},\dots,j_1}(x_{J+1},x_{j_s},x_{j_s-1},\dots,x_{j_1},x_{j_1-1},x_0)\right| \\ &\leq C_n^{J-2s+1}A_{K(n)}\Big(1+|\lambda|+|x_{J+1}|+|x_0|+\sum_{r=1}^{s}(|x_{j_r}|+|x_{j_r-1}|)\Big)^m.\end{aligned} \tag{7.5.53}$$

We apply the stationary phase method Theorem 3.9 to (7.5.51). Then

$$I(\{\tau_j\};S,a,\nu)(x_{J+1},x_0)=\left(\frac{\nu}{2\pi iT}\right)^{1/2}e^{i\nu S^*(x_{J+1},x_0)}b_\lambda(\{\tau_j\};\nu,x_{J+1},x_0).$$

For any n there exists a constant $k(n)$ and a constant C'_n such that if $|\alpha_0|\leq n$ and $|\alpha_{J+1}|\leq n$, then

$$\begin{aligned}&(1+|\lambda|+|x_{J+1}|+|x_0|)^{-m}|\partial_{x_{J+1}}^{\alpha_{J+1}}\partial_{x_0}^{\alpha_0}b_\lambda(\{\tau_j\};\nu,x_{J+1},x_0)| \\ &\leq C'_n\max\sup\big(1+|x_{J+1}|+|x_0|+\sum_{r=1}^{s}(|x_{j_r}+|x_{j_r-1}|)\big)^{-m} \\ &\times\left|\partial_{x_{J+1}}^{\beta_{J+1}}\partial_{x_0}^{\beta_0}\prod_{r=1}^{s}\partial_{x_{j_r}}^{\beta_{j_r}}\partial_{x_{j_r-1}}^{\beta_{j_r-1}}a_{\lambda,j_s,j_{s-1},\dots,j_1}(x_{J+1},x_{j_s},x_{j_s-1},\dots,x_{j_1},x_{j_1-1},x_0)\right|.\end{aligned} \tag{7.5.54}$$

Here sup is taken with respect to $(x_{J+1},x_{j_s},x_{j_s-1},\dots,x_{j_1},x_{j_1-1},x_0)\in\mathbf{R}^{2s+2}$ and max is taken among $\beta_{j_r},\beta_{j_r-1}$ for $1\leq r\leq s$ and β_{J+1},β_0 satisfying $|\beta_{j_r}|,|\beta_{j_r-1}|\leq k(n)$, $|\beta_{J+1}|\leq k(n)$ and $|\beta_0|\leq k(n)$. Apply (7.5.53) to the right-hand side of (7.5.54). Then we have

$$\begin{aligned}&(1+|\lambda|+|x_{J+1}|+|x_0|)^{-m}|\partial_{x_{J+1}}^{\alpha_{J+1}}\partial_{x_0}^{\alpha_0}b_\lambda(\{\tau_j\};\nu,x_{J+1},x_0)| \\ &\leq C'_nC_{k(n)}^{J-2s+1}A_{K(k(n))} \\ &\leq B_n^{J+1}A_{M(n)},\end{aligned} \tag{7.5.55}$$

if we define $M(n) = K(k(n))$ and choose B_n so that $B_n \geq C_{k(n)}$ and $C'_n B_n^{-2s} \leq 1$. Since s has the upper bound $\delta^{-1}T$ independent of J, we can choose $k(n)$, C'_n and B_n, $M(n)$ independent of J. It follows from this and (7.5.55) that if we replace $K(n)$ by $M(n)$ and C_n by other constant B_n, then Theorem 7.3 with δ replaced by δ_1 is valid. We have proved that Theorem 7.3 for δ_0 holds. □

Chapter 8
Stationary Phase Method for Oscillatory Integrals over a Space of Large Dimension

Abstract We shall prove in this chapter the stationary phase method for oscillatory integrals over a space of large dimension. An estimate of the remainder term is given, which is independent of the dimension. This theorem enables us to discuss the time slicing approximation of Feynman path integrals when the dimension of the space goes to ∞. This was the central tool of our discussions in Sect. 5.4 of Chap. 5.

8.1 Assumption for Phase Function

We shall discuss oscillatory integrals of the following form in this chapter:

$$\begin{aligned} & I(\{\tau_j\}; S, a_\lambda, \nu)(x_{J+1}, x_0) \\ &= \prod_{j=1}^{J} \left(\frac{-i\nu}{2\pi\tau_j}\right)^{\frac{1}{2}} \int_{\mathbf{R}^J} e^{i\nu S(x_{J+1},\dots,x_0)} a_\lambda(x_{J+1}, \dots, x_0) \prod_{j=1}^{J} dx_j. \end{aligned} \tag{8.1.1}$$

Here $\{\tau_j\}_{j=1}^{J+1}$ are positive constants with the property:

$$\sum_{j=1}^{J+1} \tau_j = T. \tag{8.1.2}$$

We write simply x for $(x_1, \dots, x_J)$ and regard x_0, x_{J+1} as parameters.

The phase function $S(x_{J+1}, \dots, x_0)$ is assumed to satisfy the following assumption.

Assumption 8.1 $S(x_{J+1}, \dots, x_0)$ is of the form

$$S(x_{J+1}, \dots, x_0) = \sum_{j=1}^{J+1} S_j(x_j, x_{j-1}), \tag{8.1.3}$$

where for each $j = 1, 2, \dots, J+1$

D. Fujiwara, *Rigorous Time Slicing Approach to Feynman Path Integrals*, Mathematical Physics Studies, DOI 10.1007/978-4-431-56553-6_8

$$S_j(x_j, x_{j-1}) = \frac{(x_j - x_{j-1})^2}{2\tau_j} + \tau_j \phi_j(x_j, x_{j-1}).$$

There exists an increasing sequence of positive constants $\kappa_0 \le \kappa_1 \le \kappa_2 \le \dots$ independent of j such that for any $(x_j, x_{j-1}) \in \mathbf{R}^2$

$$|\partial_{x_j}^{\alpha} \partial_{x_{j-1}}^{\beta} \phi_j(x_j, x_{j-1})| \le \kappa_k (1 + |x_j| + |x_{j-1}|)^{(2-k)_+}, \tag{8.1.4}$$

if $k = |\alpha| + |\beta|$, where $(2-k)_+ = max\{0, 2-k\}$.

Remark 8.1 The action function of the piecewise classical path discussed in Chap. 2 satisfies the above assumption. However, we do not assume that $S(x_{J+1}, \dots, x_0)$ is the action of a piecewise classical path in this chapter.

We assume that the amplitude function $a_\lambda(x_{J+1}, \dots, x_0)$ fulfills Assumption 7.2.

8.1.1 Stationary Point of the Phase Function

In this subsection we shall discuss properties of the stationary point and stationary value of the phase function. One will find many discussions which are common with those in Sect. 4.1. Let $\{\mathbf{e}_j;\ j = 1, 2, \dots, J\}$ be the standard basis of $\mathbf{R}^J$, i.e., $\mathbf{e}_j$ is the vector whose j-th component equals 1 and all the other components equal 0. The gradient vector of $S(x_{J+1}, \dots, x_0)$ is

$$\sum_{j=1}^{J} \partial_{x_j} S(x_{J+1}, \dots, x_0)\mathbf{e}_j = H_0 x + \mathscr{K}(x) - z.$$

Here H_0 is the matrix

$$H_0 = \begin{pmatrix} \frac{1}{\tau_1} + \frac{1}{\tau_2} & -\frac{1}{\tau_2} & 0 & 0 & \cdots & 0 & 0 \\ -\frac{1}{\tau_2} & \frac{1}{\tau_2} + \frac{1}{\tau_3} & -\frac{1}{\tau_3} & 0 & \cdots & 0 & 0 \\ \cdots & \cdots & \cdots & \cdots & \cdots & \cdots & \cdots \\ \cdots & \cdots & \cdots & \cdots & \cdots & 0 & -\frac{1}{\tau_J} \\ 0 & 0 & 0 & \cdots & 0 & -\frac{1}{\tau_{J-1}} & \frac{1}{\tau_J} + \frac{1}{\tau_{J+1}} \end{pmatrix}, \tag{8.1.5}$$

$\mathscr{K}$ is a map $\mathbf{R}^J \ni x \to \mathscr{K}(x) \in \mathbf{R}^J$ defined by

$$\mathscr{K}(x) = \sum_{j=1}^{J} (\tau_j \partial_{x_j} \phi_j(x_j, x_{j-1}) + \tau_{j+1} \partial_{x_j} \phi_{j+1}(x_{j+1}, x_j))\mathbf{e}_j, \tag{8.1.6}$$

and z is the vector

$$z = \frac{x_0}{\tau_1}\mathbf{e}_1 + \frac{x_{J+1}}{\tau_{J+1}}\mathbf{e}_J.$$

The stationary point x^* is the solution of the system of equations

$$0 = H_0 x^* + \mathscr{K}(x^*) - z. \tag{8.1.7}$$

H_0 is invertible, because

$$\det H_0 = \frac{T}{\tau_1 \tau_2 \cdots \tau_{J+1}} > 0.$$

We write G for H_0^{-1}. Then equality (8.1.7) is equivalent to

$$x^* + G\mathscr{K}(x^*) - Gz = 0. \tag{8.1.8}$$

Let $H(x) = (h_{jk}(x_{J+1}, x, x_0))_{jk=1}^{J}$ be the Hessian matrix of $S(x_{J+1}, x_J, \dots, x_1, x_0)$:

$$h_{jk}(x_{J+1}, x, x_0) = \partial_{x_j} \partial_{x_k} S(x_{J+1}, x_J, \dots, x_1.x_0), \quad j, k = 1, 2, \dots, J.$$

We often write $h_{jk}(x)$ for $h_{jk}(x_{J+1}, x, x_0)$, omitting x_{J+1} and x_0. Matrix $H(x)$ is a function of x with parameters x_{J+1}, x_0. Let $W(x)$ be the matrix expression of the differential $D\mathscr{K}(x)$. Then

$$H(x) = H_0 + W(x) = H_0(I + GW(x)). \tag{8.1.9}$$

The (j, k) entry w_{jk} of $W(x)$ is given by

$$w_{jk} = \begin{cases} \partial_{x_j}^2 (\tau_j \phi_j(x_j, x_{j-1}) + \tau_{j+1}\phi_{j+1}(x_{j+1}, x_j)) & \text{if } k = j, \\ \tau_j \partial_{x_{j-1}} \partial_{x_j} \phi_j(x_j, x_{j-1}) & \text{if } k = j - 1, \\ \tau_{j+1} \partial_{x_{j+1}} \partial_{x_j} \phi_{j+1}(x_{j+1}, x_j) & \text{if } k = j + 1, \\ 0 & \text{if } |j - k| \geq 2. \end{cases}$$

Let g_{jk} be (j, k) entry of the matrix $G = H_0^{-1}$. Then by (4.1.14),

$$g_{kl} = \begin{cases} T^{-1}(\tau_1 + \cdots + \tau_k)(\tau_{l+1} + \cdots + \tau_{J+1}) & \text{if } 1 \leq k \leq l \leq J, \\ T^{-1}(\tau_1 + \cdots + \tau_l)(\tau_{k+1} + \cdots + \tau_{J+1}) & \text{if } 1 \leq l \leq k \leq J. \end{cases}$$

Definition 8.1 For any vector $\xi = (\xi_1, \xi_2, \dots, \xi_J) \in \mathbf{R}^J$ we define the following two norms:

$$\|\xi\|_{l^\infty} = \max(|\xi_1|, \dots, |\xi_J|), \quad \|\xi\|_{l^1} = \sum_{j=1}^{J} |\xi_j|. \tag{8.1.10}$$

Remark 8.2 It follows from definition (8.1.6) of $\mathscr{K}(x)$ and (8.1.4) of Assumption 8.1 that

$$\|\mathscr{K}(x)\|_{l^1} \leq 4\kappa_1 T(1+\max\{|x_0|, |x_{J+1}|, \|x\|_{l^\infty}\}). \tag{8.1.11}$$

For any $\xi \in \mathbf{R}^J$, let $(G\xi)_j$ be the j-th component of the vector $G\xi$. Then:

Lemma 8.1 *The following inequalities hold for any* $\xi \in \mathbf{R}^J$.

$$\begin{aligned}
\|W(x)\xi\|_{l^1} &\leq 4\kappa_2 T\|\xi\|_{l^\infty}, \\
|(G\xi)_j| &\leq \min\{\tau_1+\cdots+\tau_j, \tau_{j+1}+\cdots+\tau_{J+1}\}\|\xi\|_{l^1}, \\
\|G\xi\|_{l^\infty} &\leq \frac{T}{4}\|\xi\|_{l^1}, \\
\|GW(x)\xi\|_{l^\infty} &\leq \kappa_2 T^2\|\xi\|_{l^\infty}.
\end{aligned}$$

The proof of the lemma is omitted, because it is the same as that of of Lemma 4.1. Let $\delta_0 > 0$ be so small that $\kappa_2\delta_0^2 < 2^{-1}$. We always assume that $T \leq \delta_0$.

Lemma 8.2 *If* $T \leq \delta_0$, *then for any* $(x_0, x_1, \ldots, x_J, x_{J+1}) \in \mathbf{R}^{J+2}$

$$\begin{aligned}
(1-\kappa_2 T^2)^J &\leq |\det(I+GW(x))| \leq (1+\kappa_2 T^2)^J, \\
0 < (1-\kappa_2 T^2)^J \frac{T}{\tau_1\tau_2\cdots\tau_{J+1}} &\leq |\det(H_0+W(x))|.
\end{aligned}$$

The proof of this lemma is omitted, because it is similar to that of Lemma 4.2. The next theorem is a simple consequence of Assumption 8.1 and Lemma 8.2.

Theorem 8.2 *Suppose that* $S(x_{J+1}, x_J, \ldots, x_1, x_0)$ *fulfills Assumption 8.1 and that* $T \leq \delta_0$. *Then the phase* $S(x_{J+1}, x_J, \ldots, x_1, x_0)$ *satisfies (3.2.7) and (3.2.8) of Assumption 3.1.*

Applying Lemma 3.1, we have the following corollaries.

Corollary 8.1 *Under the assumption of Theorem 8.2 there exists a unique stationary point* $x^* = (x_1^*, \ldots, x_J^*)$ *of the phase function* $S(x_{J+1}, x_J, \ldots, x_1, x_0)$ *for any* (x_{J+1}, x_0).

Corollary 8.2 *Under the assumption of Theorem 8.2 the oscillatory integral (8.1.1) is well-defined if* $a_\lambda(x_{J+1}, \ldots, x_0)$ *fulfills Assumption 7.2.*

We write $S(\widehat{x_{J+1}, x_0})$ for the stationary value $S(x_{J+1}, x_J^*, \ldots, x_1^*, x_0)$. We shall look for bounds of derivatives of with respect to (x_{J+1}, x_0) and prove that $S(\widehat{x_{J+1}, x_0})$ has the property described in Theorem 2.3. The rest of this section is devoted to that purpose. If $S_j(x_j, x_{j-1})$ is the classical action $S(T_j, T_{j-1}, x_j, x_{j-1})$ defined by (2.3.2) corresponding to the potential $V(t, x)$ on the time interval $[T_{j-1}, T_j]$, then $S(\widehat{x_{J+1}, x_0})$ is also the action, i.e. $S(\widehat{x_{J+1}, x_0}) = S(T_{J+1}, T_0, x_{J+1}, x_0)$. Hence we know such bounds by Theorem 2.3. However, as $S_j(x_j, x_{j-1})$ is not necessarily the classical action in the present chapter, we have to prove $S(\widehat{x_{J+1}, x_0})$ has the property described in Theorem 2.3.

We start from (8.1.7) and study $\partial_{x_0}^{\alpha}\partial_{x_{J+1}}^{\beta}x^*$. Afterward, we will obtain bounds of $\partial_{x_0}^{\alpha}\partial_{x_{J+1}}^{\beta}S(\widehat{x_{J+1}, x_0})$.

Notice that

$$\begin{aligned}\partial_{x_0}\mathscr{K}(x) &= \tau_1\partial_{x_0}\partial_{x_1}\phi_1(x_1, x_0)\mathbf{e}_1,\\ \partial_{x_{J+1}}\mathscr{K}(x) &= \tau_{J+1}\partial_{x_{J+1}}\partial_{x_J}\phi_{J+1}(x_{J+1}, x_J)\mathbf{e}_J.\end{aligned} \tag{8.1.12}$$

Since

$$Gz = T^{-1}\sum_{j=1}^{J}((\tau_{j+1} + \cdots + \tau_{J+1})x_0 + (\tau_1 + \cdots + \tau_j)x_{J+1})\mathbf{e}_j, \tag{8.1.13}$$

$$\|Gz\|_{l^\infty} \le \max\{|x_0|, |x_{J+1}|\}. \tag{8.1.14}$$

Notice that

$$Gz = \sum_{j=1}^{J}(x_0 + T^{-1}(\tau_1 + \cdots + \tau_j)(x_{J+1} - x_0))\mathbf{e}_j.$$

We can regard Gz as a good approximation of x^* if T is small.

Proposition 8.1 *If $T \le \delta_0$, then*

$$\|x^* - Gz\|_{l^\infty} \le \kappa_1 T^2(1 - \kappa_2\delta_0^2)^{-1}(1 + |x_0| + |x_{J+1}|). \tag{8.1.15}$$

Proof Since the differential $D\mathscr{K}(x)$ of $\mathscr{K}(x)$ at x is $W(x)$, we have

$$\mathscr{K}(x^*) = \mathscr{K}(Gz) + \int_0^1 W(tx^* + (1-t)Gz)(x^* - Gz)\,dt.$$

By (8.1.8) and Lemma 8.1

$$\begin{aligned}\|x^* - Gz\|_{l^\infty} &\le \|G\mathscr{K}(Gz)\|_{l^\infty} + \int_0^1 \|GW(tx^* + (1-t)Gz)(x^* - Gz)\|_{l^\infty}\,dt\\ &\le \frac{T}{4}\|\mathscr{K}(Gz)\|_{l^1} + \kappa_2 T^2\|x^* - Gz\|_{l^\infty}.\end{aligned}$$

It turns out from this, (8.1.14) and (8.1.11) that

$$\begin{aligned}\|x^* - Gz\|_{l^\infty} &\le 4^{-1}(1 - \kappa_2 T^2)^{-1}T\|\mathscr{K}(Gz)\|_{l^1}\\ &\le (1 - \kappa_2\delta_0^2)^{-1}T^2\kappa_1(1 + |x_0| + |x_{J+1}|).\end{aligned}$$

□

Corollary 8.3 *If $T \le \delta_0$, then*

$$\begin{aligned}\|x^*\|_{l^\infty} &\le (1 + (1-\kappa_2\delta_0^2)^{-1}T^2\kappa_1)(1 + |x_0| + |x_{J+1}|)\\ &\le (1 + 2\kappa_1 T^2)(1 + |x_0| + |x_{J+1}|).\end{aligned}$$

Proof Notice that $(1-\kappa_2\delta_0^2)^{-1} \le 2$. The corollary follows from Proposition 8.1 and (8.1.14). □

We define $y = x^* - Gz$. Then (8.1.8) means that

$$y = -G\mathscr{K}(Gz + y). \tag{8.1.16}$$

Differentiate (8.1.16) with respect x_0 and use $\partial_{x_0} z = \tau_1^{-1}\mathbf{e}_1$. Then

$$\partial_{x_0} y = -GW(x^*)\partial_{x_0} y - \tau_1^{-1}GW(x^*)G\mathbf{e}_1 - G(\partial_{x_0}\mathscr{K})(x^*).$$

This and similar discussion prove

$$(I + GW(x^*))\partial_{x_0} y = -\tau_1^{-1}GW(x^*)G\mathbf{e}_1 - G\partial_{x_0}\mathscr{K}(x^*), \tag{8.1.17}$$
$$(I + GW(x^*))\partial_{x_{J+1}} y = -\tau_{J+1}^{-1}GW(x^*)G\mathbf{e}_J - G\partial_{x_{J+1}}\mathscr{K}(x^*). \tag{8.1.18}$$

Proposition 8.2 *Suppose $T \le \delta_0$. Then*

$$\begin{aligned}\|\partial_{x_0} x^* - \partial_{x_0} Gz\|_{l^\infty} &\le 2\kappa_2(1-\kappa_2\delta_0^2)^{-1}T^2,\\ \|\partial_{x_{J+1}} x^* - \partial_{x_{J+1}} Gz\|_{l^\infty} &\le 2\kappa_2(1-\kappa_2\delta_0^2)^{-1}T^2.\end{aligned}$$

Proof Since

$$\|G\mathbf{e}_1\|_{l^\infty} = \Big\| \sum_{j=1}^{J} T^{-1}\tau_1(\tau_{j+1} + \cdots + \tau_{J+1})\mathbf{e}_j \Big\|_{l^\infty} \le \tau_1,$$

It follows from this and (8.1.12) that

$$\|\tau_1^{-1}GW(x^*)G\mathbf{e}_1\|_{l^\infty} \le \kappa_2 T^2 \quad \text{and} \quad \|G(\partial_{x_0}\mathscr{K})(x^*)\|_{l^\infty} \le \kappa_2\tau_1^2.$$

Therefore,

$$\|(I + GW(x^*))\partial_{x_0} y\|_{l^\infty} \le 2\kappa_2 T^2.$$

Consequently,

$$\|\partial_{x_0} y\|_{l^\infty} \le (1-\kappa_2\delta_0^2)^{-1}2\kappa_2 T^2.$$

This proves the first inequality of the proposition. The second one can be proved similarly. □

The following corollaries are simple consequences of Proposition 8.2.

Corollary 8.4

$$\left\| \partial_{x_0} x^* - \sum_{j=1}^{J} \frac{\tau_{j+1} + \tau_{j+2} + \cdots + \tau_{J+1}}{T} \mathbf{e}_j \right\|_{l^\infty} \leq 2\kappa_2 T^2 (1 - \kappa_2 \delta_0^2)^{-1},$$

$$\left\| \partial_{x_{J+1}} x^* - \sum_{j=1}^{J} \frac{\tau_1 + \tau_2 + \cdots + \tau_j}{T} \mathbf{e}_j \right\|_{l^\infty} \leq 2\kappa_2 T^2 (1 - \kappa_2 \delta_0^2)^{-1}.$$

Corollary 8.5 *Suppose $T \leq \delta_0$. Then*

$$\|\partial_{x_0} x^*\|_{l^\infty} \leq 1 + 2\kappa_2 T^2 (1 - \kappa_2 \delta_0^2)^{-1}, \qquad \|\partial_{x_{J+1}} x^*\|_{l^\infty} \leq 1 + 2\kappa_2 T^2 (1 - \kappa_2 \delta_0^2)^{-1}.$$

Corollary 8.6 *Suppose $T \leq \delta_0$. Then there exists a positive constant C independent of J such that*

$$\|\partial_{x_0} \mathscr{K}(x^*)\|_{l^1} \leq 8(1 - \kappa_2 \delta_0^2)^{-1} \kappa_2 T, \quad \|\partial_{x_{J+1}} \mathscr{K}(x^*)\|_{l^1} \leq 8(1 - \kappa_2 \delta_0^2)^{-1} \kappa_2 T.$$

Proof It follows from Lemma 8.1, Corollary 8.5 and (8.1.12) that

$$\begin{aligned} \|\partial_{x_0} \mathscr{K}(x^*)\|_{l^1} &\leq \|W(x^*) \partial_{x_0} x^*\|_{l^1} + \|(\partial_{x_0} \mathscr{K})(x^*)\|_{l^1} \\ &\leq 4\kappa_2 T (1 + 2\kappa_2 T^2 (1 - \kappa_2 \delta_0^2)^{-1}) + \kappa_2 \tau_1 \leq 8\kappa_2 T (1 - \kappa_2 \delta_0^2)^{-1}. \end{aligned}$$

We have proved the first inequality. The second one is proved in a similar manner. □

Proposition 8.3 *For any $n \geq 2$ there exists a positive constant C_n depending on $\kappa_2, \ldots, \kappa_{n+1}$ such that if $\alpha + \beta = n$,*

$$\|\partial_{x_0}^{\alpha} \partial_{x_{J+1}}^{\beta} x^*\|_{l^\infty} \leq C_n T^2. \tag{8.1.19}$$

Proof If $\alpha + \beta \geq 2$, then $\partial_{x_0}^{\alpha} \partial_{x_{J+1}}^{\beta} Gz = 0$. Therefore, it is sufficient for us to prove that for any $n \geq 1$ there exists a positive constant C depending on $\kappa_2, \ldots, \kappa_{n+1}$ such that if $\alpha + \beta = n$,

$$\|\partial_{x_0}^{\alpha} \partial_{x_{J+1}}^{\beta} (x^* - Gz)\|_{l^\infty} \leq C_n T^2. \tag{8.1.20}$$

We prove (8.1.20) by induction on n. In the case $n = 1$ (8.1.20) is true by Proposition 8.2.

Assuming that (8.1.20) is true for $1 \leq |\alpha| + |\beta| < n$, we shall prove (8.1.19) in the case $2 \leq |\alpha| + |\beta| = n$. We begin with the following remark: Define $C = \max\{1 + 2\kappa_2 T^2(1 - \kappa_2\delta^2)^{-1}, C_2, \ldots, C_{n-1}\}$. Then by the induction hypothesis and Corollary 8.5, there exists a positive constant C depending on $\kappa_2, \ldots, \kappa_n$ such that

$$\|\partial_{x_0}^{\alpha}\partial_{x_{J+1}}^{\beta} x^*\|_{l^\infty} \leq C_{n-1} \quad \text{if } 1 \leq |\alpha| + |\beta| < n. \tag{8.1.21}$$

Let α, β be arbitrary pair that satisfies $|\alpha| + |\beta| = n$.

We shall abbreviate $\partial_{x_0}^{k}\partial_{x_{J+1}}^{l} x^*(x_{J+1}, x_0)$ to $x^{*(k,l)}$ and $\partial_{x_0}^{k}\partial_{x_{J+1}}^{l} x_j^*(x_{J+1}, x_0)$ to $x_j^{*(k,l)}$. We shall sometimes write x_0^* for x_0 and x_{J+1}^* for x_{J+1} as notational convention. We prove (8.1.20) by differentiating (8.1.8). We shall look for the bound of $\partial_{x_0}^{\alpha}\partial_{x_{J+1}}^{\beta}\mathscr{K}(x^*)$. By the chain rule, for $j = 1, \ldots, J+1$,

$$\begin{aligned} &\partial_{x_0}^{\alpha}\partial_{x_{J+1}}^{\beta}(\partial_{x_j}\phi_j)(x_j^*, x_{j-1}^*) \\ &= (\partial_{x_{j-1}}\partial_{x_j}\phi_j)(x_j^*, x_{j-1}^*)x_{j-1}^{*(\alpha,\beta)} + (\partial_{x_j}^2\phi_j)(x_j^*, x_{j-1}^*)x_j^{*(\alpha,\beta)} + r_{j,1}, \end{aligned} \tag{8.1.22}$$

where $r_{j,1}$ is of the form

$$\begin{aligned} r_{j,1} = \sum_{2\leq k+l\leq n} (\partial_{x_j}^{k}\partial_{x_{j-1}}^{l}\partial_{x_j}\phi_j)(x_j^*, x_{j-1}^*) {\sum}' C(k, l, \{(\alpha_i', \beta_i')\}_{i=1}^{k}, \{(\alpha_i'', \beta_i'')\}_{i=1}^{l}) \\ \times x_j^{*(\alpha_1',\beta_1')} \cdots x_j^{*(\alpha_k',\beta_k')} x_{j-1}^{*(\alpha_1'',\beta_1'')} \cdots x_{j-1}^{*(\alpha_l'',\beta_l'')}. \end{aligned}$$

The symbol ${\sum}'$ means the summation over the set of all partitions of the pair (α, β) to $k + l$ non-zero pairs $\{(\alpha_i', \beta_i')\}_{i=1}^{k}$ and $\{(\alpha_i'', \beta_i'')\}_{i=1}^{l}$ that satisfy

$$\alpha_i' + \beta_i' \neq 0 \quad \text{for } i = 1, \ldots, k, \quad \alpha_i'' + \beta_i'' \neq 0 \quad \text{for } i = 1, \ldots, l, \tag{8.1.23}$$

$$(\alpha_1', \beta_1') + \cdots + (\alpha_k', \beta_k') + (\alpha_1'', \beta_1'') + \cdots + (\alpha_l'', \beta_l'') = (\alpha, \beta), \tag{8.1.24}$$

and $C(k, l, \{(\alpha_i', \beta_i')\}_{i=1}^{k}, \{(\alpha_i'', \beta_i'')\}_{i=1}^{l})$ is a combinatorial non-negative constant depending only on k, l, $\{(\alpha_i', \beta_i')\}_{i=1}^{k}$ and $\{(\alpha_i'', \beta_i'')\}_{i=1}^{l}$.

Since (8.1.23) and (8.1.24) imply that $\alpha_i' + \beta_i' < n$ and $\alpha_i'' + \beta_i'' < n$, we can apply (8.1.21) and obtain

$$|x_j^{*(\alpha_i',\beta_i')}| \leq C_{n-1} \quad \text{and} \quad |x_{j-1}^{*(\alpha_i'',\beta_i'')}| \leq C_{n-1}. \tag{8.1.25}$$

As $C(k, l, \{(\alpha_i', \beta_i')\}_{i=1}^{k}, \{(\alpha_i'', \beta_i'')\}_{i=1}^{l})$ depends on k, l, partition $\{(\alpha_i', \beta_i')\}_{i=1}^{k}$ and $\{(\alpha_i'', \beta_i'')\}_{i=1}^{l}$ but not on j, there exists a positive constant C which depends on $\kappa_2, \ldots, \kappa_{n+1}$ but not on j such that

$$|r_{j,1}| \leq C. \tag{8.1.26}$$

Similarly, we have

$$\begin{aligned}&\partial_{x_0}^{\alpha}\partial_{x_{J+1}}^{\beta}(\partial_{x_{j-1}}\phi_j)(x_j^*, x_{j-1}^*) \\ &= (\partial_{x_{j-1}}^2\phi_j)(x_j^*, x_{j-1}^*)\partial_{x_0}^{\alpha}\partial_{x_{J+1}}^{\beta}x_{j-1}^* + (\partial_{x_{j-1}}\partial_{x_j}\phi_j)(x_j^*, x_{j-1}^*)\partial_{x_0}^{\alpha}\partial_{x_{J+1}}^{\beta}x_j^* + r_{j,2},\end{aligned} \tag{8.1.27}$$

and there exists a constant $C > 0$ depending on $\kappa_2, \ldots, \kappa_{n+1}$ and independent of J such that

$$|r_{j,2}| \le C. \tag{8.1.28}$$

It turns out from these that the j-th component of $\partial_{x_0}^{\alpha}\partial_{x_{J+1}}^{\beta}\mathscr{K}(x^*)$ equals

$$\begin{aligned}&\partial_{x_0}^{\alpha}\partial_{x_{J+1}}^{\beta}(\tau_j\partial_{x_j}\phi_j(x_j^*, x_{j-1}^*) + \tau_{j+1}\partial_{x_j}\phi_{j+1}(x_{j+1}^*, x_j^*)) \\ &= \tau_j(\partial_{x_{j-1}}\partial_{x_j}\phi_j)^* x_{j-1}^{*(\alpha,\beta)} + (\tau_j(\partial_{x_j}^2\phi_j)^* + \tau_{j+1}(\partial_{x_j}^2\phi_{j+1})^*)x_j^{*(\alpha,\beta)} \\ &+ \tau_{j+1}(\partial_{x_{j+1}}\partial_{x_j}\phi_{j+1})^* x_{j+1}^{*(\alpha,\beta)} \\ &+ \tau_j r_{j,1} + \tau_{j+1} r_{j+1,2} \\ &= w_{j,j-1}^* x_{j-1}^{*(\alpha,\beta)} + w_{j,j}^* x_j^{*(\alpha,\beta)} + w_{j,j+1}^* x_{j+1}^{*(\alpha,\beta)} + \tau_j r_{j,1} + \tau_{j+1} r_{j+1,2}.\end{aligned} \tag{8.1.29}$$

Here, for short we denoted $(\partial_{x_{j-1}}\partial_{x_j}\phi_j)(x_j^*, x_{j-1}^*)$ by $(\partial_{x_{j-1}}\partial_{x_j}\phi_j)^*$, $w_{j,j-1}(x^*)$ by $w_{j,j-1}^*$ and so on. Define a vector u by

$$u = \sum_{j=1}^{N}(\tau_j r_{j,1} + \tau_{j+1} r_{j+1,2})\mathbf{e}_j. \tag{8.1.30}$$

Then it turns out from (8.1.8) that

$$0 = (I + GW(x^*))x_j^{*(\alpha,\beta)} + Gu, \tag{8.1.31}$$

because the condition $|\alpha| + |\beta| \ge 2$ implies $\partial_{x_0}^{\alpha}\partial_{x_{J+1}}^{\beta}z = 0$. Therefore,

$$\|x^{*(\alpha,\beta)}\|_{l^\infty} \le (1 - \kappa_2 T^2)^{-1}\|Gu\|_{l^\infty} \le (1 - \kappa_2 T^2)^{-1}4^{-1}T\|u\|_{l^1}. \tag{8.1.32}$$

On the other hand, by (8.1.26) and (8.1.28),

$$\|u\|_{l^1} \le \sum_{j=1}^{J}(\tau_j|r_{j,1}| + \tau_{j+1}|r_{j+1,2}|) \le 2CT. \tag{8.1.33}$$

Therefore, we have proved that there exists a constant $C > 0$ independent of J but depending on $\kappa_2, \ldots, \kappa_{n+1}$ such that

$$\|x^{*(\alpha,\beta)}\|_{l^\infty} \leq (1 - \kappa_2 T^2)^{-1} C T^2. \tag{8.1.34}$$

The induction is now complete. We have proved (8.1.19). □

Corollary 8.7 *Let $x_0^* = x_0$ and $x_{J+1}^* = x_{J+1}$ as notational convention. For any α, β with $|\alpha| + |\beta| = n \geq 1$ and $j = 1, \ldots, J+1$ there exists a constant $C > 0$ independent of J but dependent on $\kappa_2, \ldots, \kappa_{n+1}$ such that for $0 \leq j \leq J+1$,*

$$|\partial_{x_0}^\alpha \partial_{x_{J+1}}^\beta (\partial_{x_j} \phi_j)(x_j^*, x_{j-1}^*)| \leq C, \quad |\partial_{x_0}^\alpha \partial_{x_{J+1}}^\beta (\partial_{x_{j-1}} \phi_j)(x_j^*, x_{j-1}^*)| \leq C,$$

and

$$\|\partial_{x_0}^\alpha \partial_{x_{J+1}}^\beta \mathscr{K}(x^*)\|_{l^1} \leq CT.$$

Proof Let $|\alpha| + |\beta| = n$. We have (8.1.22) and (8.1.26). Thus applying Proposition 8.3, we obtain

$$\begin{aligned}
&|\partial_{x_0}^\alpha \partial_{x_{J+1}}^\beta (\partial_{x_j} \phi_j)(x_j^*, x_{j-1}^*)| \\
&\leq |(\partial_{x_{j-1}} \partial_{x_j} \phi_j)(x_j^*, x_{j-1}^*)||x_{j-1}^{*(\alpha,\beta)}| + |(\partial_{x_j}^2 \phi_j)(x_j^*, x_{j-1}^*)||x_j^{*(\alpha,\beta)}| + |r_{j,1}| \\
&\leq \kappa_2 C + \kappa_2 C + C.
\end{aligned}$$

Therefore the first inequality is proved. The second inequality is proved similarly.

It follows from the first two inequalities of the corollary and (8.1.6) that

$$\begin{aligned}
&\|\partial_{x_0}^\alpha \partial_{x_{J+1}}^\beta \mathscr{K}(x^*)\|_{l^1} \\
&\leq \sum_{j=1}^{J+1} (\tau_j |\partial_{x_0}^\alpha \partial_{x_{J+1}}^\beta (\partial_{x_j} \phi_j)(x_j^*, x_{j-1}^*)| + \tau_{j+1} |\partial_{x_0}^\alpha \partial_{x_{J+1}}^\beta (\partial_{x_j} \phi_{j+1})(x_{j+1}^*, x_j^*)|) \\
&\leq 2CT.
\end{aligned}$$

This proves the last inequality. □

8.1.2 Stationary Value of the Phase Function

We shall abbreviate the stationary value of $S(x_{J+1}, x_J, \ldots, x_1, x_0)$ to $S(\widehat{x_{J+1}, x_0})$, i.e. $S(\widehat{x_{J+1}, x_0}) = S(x_{J+1}, x_J^*, \ldots, x_1^*, x_0)$.

Proposition 8.4 *Suppose that $T \leq \delta_0$. Then $S(\widehat{x_{J+1}, x_0})$ is of the form:*

$$S(\widehat{x_{J+1}, x_0}) = \frac{|x_{J+1} - x_0|^2}{2T} + T\phi_{J+1,0}(x_{J+1}, x_0). \tag{8.1.35}$$

There exist positive constants κ_0^* and κ_1^* depending on κ_0, κ_1 and κ_2 such that

$$|\phi_{J+1,0}(x_{J+1}, x_0)| \le \kappa_0^*(1+|x_0|+|x_{J+1}|)^2 \tag{8.1.36}$$
$$|\partial_{x_0}^{\alpha}\partial_{x_{J+1}}^{\beta}\phi_{J+1,0}(x_{J+1}, x_0)| \le \kappa_1^*(1+|x_0|+|x_{J+1}|), \quad \text{if } |\alpha|+|\beta|=1. \tag{8.1.37}$$

Moreover, for any positive integer $k \ge 2$ there exists a positive constant κ_k^* depending on $\kappa_2.\kappa_3, \dots, \kappa_k$ such that if $|\alpha|+|\beta|=k$

$$|\partial_{x_0}^{\alpha}\partial_{x_{J+1}}^{\beta}\phi_{J+1,0}(x_{J+1}, x_0)| \le \kappa_k^*. \tag{8.1.38}$$

In particular, in the case $k=2$, $\kappa_2^* = 10\kappa_2(1-\kappa_2\delta_0^2)^{-1}$ is enough.

Remark 8.3 If $S(x_{J+1}, \dots, x_0)$ is the action of a piecewise classical path such as (3.1.6) in Sect. 3.1, then it is clear that we can choose $\kappa_k^* = \kappa_k$.

Proof Let $u = -\mathscr{K}(x^*)$. Then $x^* = G(z+u)$. Replace x^* by $G(z+u)$ in the following equality:

$$S(\widehat{x_{J+1}, x_0}) = \frac{1}{2}(H_0 x^*, x^*) - \frac{x_1^* x_0}{\tau_1} - \frac{x_J^* x_{J+1}}{\tau_{J+1}} + \frac{x_0^2}{2\tau_1} + \frac{x_{J+1}^2}{2\tau_{J+1}} + \sum_{j=1}^{J+1}\tau_j\phi_j(x_j^*, x_{j-1}^*),$$

where $x_0^* = x_0$ and $x_{J+1}^* = x_{J+1}$. Then it turns out that

$$S(\widehat{x_{J+1}, x_0}) = -\frac{1}{2}(z, Gz) + \frac{x_0^2}{2\tau_1} + \frac{x_{J+1}^2}{2\tau_{J+1}} + \frac{1}{2}(Gu, u) + \sum_{j=1}^{J+1}\tau_j\phi_j(x_j^*, x_{j-1}^*).$$

Calculation shows that

$$-\frac{1}{2}(z, Gz) + \frac{x_0^2}{2\tau_1} + \frac{x_{J+1}^2}{2\tau_{J+1}} = \frac{|x_{J+1}-x_0|^2}{2T}.$$

Define

$$\phi_{J+1,0}(x_{J+1}, x_0) = \frac{1}{2T}(Gu, u) + T^{-1}\sum_{j=1}^{J+1}\tau_j\phi_j(x_j^*, x_{j-1}^*).$$

Then we have (8.1.35).

We show (8.1.36). By (8.1.11)

$$|(u, Gu)| \le \|u\|_{l^1}\|Gu\|_{l^\infty} \le \frac{T}{4}\|u\|_{l^1}^2 \le 4T^3\kappa_1^2(1+\|x^*\|_{l^\infty}+|x_{J+1}|+|x_0|)^2.$$

On the other hand

$$\Big|\sum_{j=1}^{J+1}\tau_j\phi_j(x_j^*,x_{j-1}^*)\Big|\le 2T\kappa_0(1+\|x^*\|_{l^\infty}+|x_{J+1}|+|x_0|)^2.$$

Therefore, taking the sum of these,

$$|\phi_{J+1,0}(x_{J+1},x_0)|\le(2T^2\kappa_1^2+2\kappa_0)(1+\|x^*\|_{l^\infty}+|x_{J+1}|+|x_0|)^2.$$

Inequality (8.1.36) follows from this and Corollary 8.3.

We shall discuss (8.1.38). Since $S(\widehat{x_{J+1},x_0})$ is the critical value, we have

$$\partial_{x_0}S(\widehat{x_{J+1},x_0})=\partial_{x_0}S(x_{J+1},x_J,\dots,x_1,x_0)\big|_{x=x^*}=\frac{(x_0-x_1^*)}{\tau_1}+\tau_1\partial_{x_0}\phi_1(x_1,x_0)|_{x_1=x_1^*}.$$

It turns out from (8.1.8) that

$$\frac{(x_0-x_1^*)}{\tau_1}=\frac{1}{T}(x_0-x_{J+1})-\frac{1}{\tau_1}\sum_{j=1}^{J}g_{1j}u_j.$$

Since $T\partial_{x_0}\phi_{J+1,0}(x_{J+1},x_0)=\partial_{x_0}S(\widehat{x_{J+1},x_0})-\dfrac{1}{T}(x_0-x_{J+1})$,

$$T\partial_{x_0}\phi_{J+1,0}(x_{J+1},x_0)=-\frac{1}{\tau_1}\sum_{j=1}^{J}g_{1j}u_j+\tau_1\partial_{x_0}\phi_1(x_1^*,x_0). \tag{8.1.39}$$

As $g_{1j}\le\tau_1$ for $j=1,2,\dots,J$, we have, by (8.1.11), that

$$\begin{aligned}|T\partial_{x_0}\phi_{J+1,0}(x_{J+1},x_0)|&\le\|u\|_{l^1}+\tau_1|\partial_{x_0}\phi_1(x_1^*,x_0)|\\&\le 4T\kappa_1(1+\|x^*\|_{l^\infty}+|x_{J+1}|+|x_0|)+\tau_1\kappa_1(1+\|x^*\|_{l^\infty}+|x_0|).\end{aligned}$$

(8.1.37) for $|\alpha|=1$, $|\beta|=0$ follows from this and Corollary 8.3. Similarly, (8.1.37) for $|\alpha|=0$, $|\beta|=1$ is proved.

Differentiating (8.1.39) once more, we obtain

$$\begin{aligned}&T\partial_{x_0}^2\phi_{J+1,0}(x_{J+1},x_0)\\&=-\frac{1}{\tau_1}\sum_{j=1}^{J}g_{1j}\partial_{x_0}u_j+\tau_1(\partial_{x_1}\partial_{x_0}\phi_1)(x_1^*,x_0)\partial_{x_0}x_1^*+\tau_1(\partial_{x_0}^2\phi_1)(x_1^*,x_0).\end{aligned}$$

It turns out from Corollaries 8.5 and 8.6 that

$$
\begin{aligned}
&T|\partial_{x_0}^2 \phi_{J+1,0}(x_{J+1}, x_0)| \le \|\partial_{x_0} u\|_{l^1} + \tau_1(\kappa_2|\partial_{x_0} x_1^*| + \kappa_2) \\
&\le 8\kappa_2 T(1-2\kappa_2\delta_0^2)^{-1} + \kappa_2 T(2 + 2\kappa_2 T^2(1-\kappa_2\delta_0^2)^{-1}) \le 10\kappa_2 T(1-2\kappa_2\delta_0^2)^{-1}.
\end{aligned}
$$

(8.1.38) for $\alpha = 2, \beta = 0$ has been proved. Similarly, we can prove (8.1.38) for other cases of $k = 2$.

Suppose $n \ge 2$. Let α, β be such that $|\alpha| + |\beta| = n$. Differentiating (8.1.39), we obtain

$$
T\partial_{x_0}^{\alpha+1}\partial_{x_{J+1}}^{\beta}\phi_{J+1,0}(x_{J+1}, x_0) = -\frac{1}{\tau_1}\sum_{j=1}^{J} g_{1j}\partial_{x_0}^{\alpha}\partial_{x_{J+1}}^{\beta} u_j + \tau_1\partial_{x_0}^{\alpha}\partial_{x_{J+1}}^{\beta}(\partial_{x_0}\phi_1)(x_1^*, x_0).
$$

By Corollary 8.7, there exists a constant $C > 0$ depending on $\kappa_2, \dots, \kappa_{n+1}$ such that

$$
\|\partial_{x_0}^{\alpha}\partial_{x_{J+1}}^{\beta} u\|_{l^1} \le CT, \quad \text{and } |\partial_{x_0}^{\alpha}\partial_{x_{J+1}}^{\beta}(\partial_{x_0}\phi_1)(x_1^*, x_0)| \le C.
$$

Therefore,

$$
|\partial_{x_0}^{\alpha+1}\partial_{x_{J+1}}^{\beta}\phi_{J+1,0}(x_{J+1}, x_0)| \le 2C. \tag{8.1.40}
$$

The proposition is proved. □

8.2 Statement of Theorem

We always assume that $2\kappa_2\delta_0^2 < 1$ and $T \le \delta_0$. We discuss

$$
\begin{aligned}
&I(\{\tau_j\}; S, a_\lambda, \nu)(x_{J+1}, x_0) \qquad (8.2.1)\\
&= \prod_{j=1}^{J+1}\left(\frac{-i\nu}{2\pi\tau_j}\right)^{\frac{1}{2}} \int_{\mathbf{R}^J} e^{i\nu S(x_{J+1}, x_J, \dots, x_1, x_0)} a_\lambda(x_{J+1}, x_J, \dots, x_1, x_0) \prod_{j=1}^{J} dx_j.
\end{aligned}
$$

The phase function is supposed to fulfill Assumption 8.1.

Before stating assumptions for the amplitude, we introduce a notational convention. Let j, l be two integers such that $1 \le j < l \le J+1$. Then we set $T(l, j) = \tau_l + \cdots + \tau_j$ and

$$
S_{l,j}(x_l, \dots, x_{j-1}) = \sum_{k=j}^{l} S_k(x_k, x_{k-1}).
$$

We understand that $S_{j,j}(x_j, x_{j-1}) = S_j(x_j, x_{j-1})$. Note that $S_{J+1,1}(x_{J+1}, x_J, \ldots, x_1, x_0) = S(x_{J+1}, x_J, \ldots, x_1, x_0)$. Since $T(l, j) \le T \le \delta_0$, it follows from Corollary 8.1 that for any fixed $(x_l, x_{j-1}) \in \mathbf{R}^2$, the stationary point $(x^*_{l-1}, x^*_{l-2}, \ldots, x^*_j)$ of the function

$$(x_{l-1}, \ldots, x_j) \to S_l(x_l, x_{l-1}) + S_{l-1}(x_{l-1}, x_{l-2}) + \cdots + S_j(x_j, x_{j-1}) = S_{l,j}(x_l, \ldots, x_{j-1})$$

exists uniquely. We write $x^*_k(x_l, x_{j-1})$ for x^*_k if we wish to express its dependence on (x_l, x_{j-1}). Let $S^*_{l,j}(x_l, x_{j-1})$ be the stationary value of $S_{l,j}(x_l, \ldots, x_{j-1})$ at the stationary point, i.e.

$$S^*_{l,j}(x_l, x_{j-1}) = S_l(x_l, x^*_{l-1}) + \sum_{k=j+1}^{l-1} S_k(x^*_k(x_l, x_{j-1}), x^*_{k-1}(x_l, x_{j-1})) + S_j(x^*_j, x_{j-1}). \tag{8.2.2}$$

For any function $f(x_{J+1}, x_J, \ldots, x_l, \ldots, x_k, \ldots, x_0)$ of $(x_0, \ldots, x_k, \ldots, x_l, \ldots, x_J, x_{J+1})$ replace x_j by $x^*_j(x_l, x_k)$, $k < j < l$. Then we define a new function $f(x_{J+1}, \ldots, \widehat{x_l, x_k}, \ldots, x_0)$ of $(x_0, \ldots, x_k, x_l, \ldots, x_J, x_{J+1})$, i.e.,

$$f(x_{J+1}, \ldots, \widehat{x_l, x_k}, \ldots, x_0) = f(x_{J+1}, x_J, \ldots, x_l, \ldots, x_k, \ldots, x_0)\big|_{x_j = x^*_j(x_l, x_k), fork<j<l}.$$

If $k + 1 = l$, then we define

$$f(x_{J+1}, \ldots, \widehat{x_{k+1}, x_k}, \ldots, x_0) = f(x_{J+1}, x_J, \ldots, x_{k+1}, x_k, \ldots, x_0).$$

By definition, if $0 \le j < l \le J + 1$

$$S^*_{l,j}(x_l, x_{j-1}) = S_{l,j}(\widehat{x_l, x_{j-1}}).$$

The assumption for the amplitude $a_\lambda(x_{J+1}, \ldots, x_0)$ is the following:

Assumption 8.3 Let m be a non-negative number. The amplitude function $a_\lambda = a_\lambda(x_{J+1}, x_J, \ldots, x_1, x_0)$ may depend on a parameter λ as well as $(x_{J+1}, x_J, \ldots, x_1, x_0)$. For any integer $K \ge 0$ there exist constants $A_K > 0$ and $X_K \ge 1$ such that

1.
$$\left|\left(\prod_{j=0}^{J+1} \partial_{x_j}^{\alpha_j}\right) a_\lambda(x_{J+1}, x_J, \ldots, x_1, x_0)\right| \tag{8.2.3}$$
$$\le A_K X_K^{J+2}(1 + |\lambda| + |x_{J+1}| + |x_J| + \cdots + |x_1| + |x_0|)^m.$$

   if $|\alpha_j| \le K$ for all $j = 0, 1, \ldots, J + 1$.
2. Let $0 = j_0 < j_1 - 1 < j_1 < j_2 - 1 < \cdots < j_n < j_{n+1} - 1 = J + 1$, $1 \le n \le J/2$ be any sequence of indices. Let $\{\alpha_j\}$ be a sequence of indices which satisfies $|\alpha_j| \le K$. Then for any $(x_0, x_{j_1-1}, x_{j_1}, \ldots, x_{j_{n-1}}, x_{j_n-1}, x_{j_n}, x_{J+1}) \in \mathbf{R}^{2n+2}$

$$\Big|\partial_{x_0}^{\alpha_0}\partial_{x_{J+1}}^{\alpha_{J+1}}\Big(\prod_{u=1}^{n}\partial_{x_{j_u-1}}^{\alpha_{j_u-1}}\partial_{x_{j_u}}^{\alpha_{j_u}}\Big)a_\lambda(x_{J+1},\widehat{x_{j_n}},x_{j_n-1},\widehat{x_{j_{n-1}}},\dots,\widehat{x_{j_1-1},x_0})\Big|$$
$$\le A_K X_K^{2n+2}(1+|\lambda|+|x_{J+1}|+|x_{j_n}|+|x_{j_n-1}|+|x_{j_{n-1}}|+\cdots+|x_{j_1-1}|+|x_0|)^m.$$

3. Let $0=j_0<j_1-1<j_1<j_2-1<\cdots<j_n=J+1$, $1\le n\le (J+1)/2$ be any sequence of indices and let $\{\alpha_j\}$ be multi-indices satisfying $|\alpha_j|\le K$. Then for any $(x_0,x_{j_1-1},x_{j_1},\dots,x_{j_{n-1}},x_J,x_{J+1})\in\mathbf{R}^{2n+2}$

$$\Big|\partial_{x_0}^{\alpha_0}\Big(\prod_{u=1}^{n}\partial_{x_{j_u-1}}^{\alpha_{j_u-1}}\partial_{x_{j_u}}^{\alpha_{j_u}}\Big)a_\lambda(x_{J+1},\widehat{x_J,x_{j_n}},x_{j_n-1},\widehat{x_{j_{n-1}}},\dots,\widehat{x_{j_1-1},x_0})\Big|$$
$$\le A_K X_K^{2n+1}(1+|\lambda|+|x_{J+1}|+|x_J|+|x_{j_n}|+|x_{j_{n-1}}|+\cdots+|x_{j_1}|+|x_{j_1-1}|+|x_0|)^m.$$

Theorem 8.4 (Stationary phase method over a space of large dimension) *Suppose that the phase function fulfills Assumption 8.1 and the amplitude function fulfills Assumption 8.3. Let δ_0 be the positive constant for which the Kumano-go–Taniguchi theorem, Theorem 7.3, holds with κ_2^* instead of κ_2. Then the stationary point x^* of the phase function is unique and there holds the following equality:*

$$\begin{aligned}&I(\{\tau_j\},S,a_\lambda,\nu)(x_{J+1},x_0)\\&=\Big(\frac{-\nu i}{2\pi T}\Big)^{1/2}\exp\{i\nu S(\widehat{x_{J+1},x_0})\}\\&\times D(\{\tau_j\};x_{J+1},x_0)^{-1/2}(a_\lambda(\widehat{x_{J+1},x_0})+\nu^{-1}r_\lambda(\{\tau_j\};\nu,x_{J+1},x_0)).\end{aligned}\tag{8.2.4}$$

Here $D(\{\tau_j\};x_{J+1},x_0)=\det(I+W(x^*)H_0^{-1})$. The remainder $r_\lambda(\{\tau_j\};\nu,x_{J+1},x_0)$ satisfies the following estimates. For any non-negative integer k there exists a positive constant C_k and a natural number $M(k)$ such that for any $(x_{J+1},x_0)\in\mathbf{R}^2$

$$\begin{aligned}&\big|\partial_{x_0}^{\alpha_0}\partial_{x_{J+1}}^{\alpha_{J+1}}r_\lambda(\{\tau_j\};\nu,x_{J+1},x_0)\big|\\&\le A_{M(k)}(1+|\lambda|+|x_{J+1}|+|x_0|)^m\nu\Big(\prod_{j=1}^{J+1}(1+C_kX_{M(k)}^2\nu^{-1}\tau_j)-1\Big),\end{aligned}\tag{8.2.5}$$

if $|\alpha_0|,|\alpha_{J+1}|\le k$. Here $M(k)$ and C_k do not depend on $\{\tau_j\}$, J, $a_\lambda(x_{J+1},\dots,x_0)$, (x_{J+1},x_0) and ν.

If $z_k>0$ for any $k=1,2,\dots,J+1$, then it is clear that

$$\Big(\prod_{j=1}^{J+1}(1+z_k)-1\Big)\le e^{\sum_{j=1}^{J+1}z_k}\sum_{j=1}^{J+1}z_k.$$

Thus

Corollary 8.8 *For any non-negative integer k there exists a positive constant C_k and a natural number $M(k)$ such that for any (x_{J+1},x_0)*

$$
\begin{aligned}
&\left|\partial_{x_0}^{\alpha_0}\partial_{x_{J+1}}^{\alpha_{J+1}} r_\lambda(\{\tau_j\};\nu,x_{J+1},x_0)\right| \\
&\quad\le C_k A_{M(k)} X_{M(K)}^2 T e^{\nu^{-1}C_k X_{M(K)}^2 T}(1+|\lambda|+|x_{J+1}|+|x_0|)^m.
\end{aligned}
\tag{8.2.6}
$$

if $|\alpha_0|, |\alpha_{J+1}| \le k$. *Here* $M(k)$ *and* C_k *do not depend on* $\{\tau_j\}$, J, $a_\lambda(x_{J+1},\dots,x_0)$, (x_{J+1},x_0) *and* ν.

Corollary 8.9 *Suppose that the same assumption as in the theorem above. We can write*

$$
\begin{aligned}
&I(\{\tau_j\}, S, a_\lambda, \nu)(x_{J+1},x_0) \\
&= \left(\frac{-\nu i}{2\pi T}\right)^{1/2} \exp\{i\nu S(\widehat{x_{J+1},x_0})\} D(\{\tau_j\};x_{J+1},x_0)^{-1/2}\tilde a(\{\tau_j\};\nu,x_{J+1},x_0).
\end{aligned}
\tag{8.2.7}
$$

For any non-negative integer k *there exists a positive constant* C_k *and a natural number* $M(k)$ *such that for any* (x_{J+1},x_0) *we have*

$$
\left|\partial_{x_0}^{\alpha_0}\partial_{x_{J+1}}^{\alpha_{J+1}} \tilde a_\lambda(\{\tau_j\};\nu,x_{J+1},x_0)\right| \le A_{M(k)}(1+|\lambda|+|x_{J+1}|+|x_0|)^m e^{\nu^{-1}C_k X_{M(k)}^2 T},
$$

if $|\alpha_0|, |\alpha_{J+1}| \le k$. *Here* $M(k)$ *and* C_k *do not depend on* $\{\tau_j\}$, J, $a_\lambda(x_{J+1},\dots,x_0)$, (x_{J+1},x_0) *and* ν.

Proof of the theorem is divided into several steps. In the next subsection we discuss the simplest case $J = 1$. Afterward we discuss general case.

8.3 Proof of Stationary Phase Method in the Case of J = 1

In this section the case of $J = 1$ is proved. Suppose $T \le \delta_0$ and $\tau_1 + \tau_2 = T$.

The oscillatory integral we discuss is the following integral with parameter λ:

$$
\begin{aligned}
&I(\{\tau_1,\tau_2\}; S, a_\lambda, \nu)(x_2,x_0) \\
&= \left(\frac{\nu}{2\pi i\tau_2}\right)^{1/2}\left(\frac{\nu}{2\pi i\tau_1}\right)^{1/2}\int_{\mathbf{R}} e^{i\nu(S_2(x_2,x_1)+S_1(x_1,x_0))} a_\lambda(x_2,x_1,x_0)\,dx_1,
\end{aligned}
\tag{8.3.1}
$$

where the phase function $S(x_2,x_1,x_0) = S_1(x_1,x_0) + S_2(x_2,x_1)$ fulfills Assumption 8.1 and the amplitude a_λ satisfies Assumption 8.3 with $J = 1$.

Let $p \ge 0$. In accordance with Sect. 3.2, we define norms for functions $f_\lambda(x_2,x_1,x_0)$ and $g_\lambda(x_2,x_0)$;

$$
\begin{aligned}
\|f_\lambda\|_{(p)} &= \sup_{(x_2,x_1,x_0)\in\mathbf{R}^3} (1+|x_2|+|x_1|+|x_0|+|\lambda|)^{-p}|f_\lambda(x_2,x_1,x_0)| \\
|||g_\lambda|||_{(p)} &= \sup_{(x_2,x_0)\in\mathbf{R}^2} (1+|x_2|+|x_0|+|\lambda|)^{-p}|g_\lambda(x_2,x_0)|.
\end{aligned}
$$

As we discussed in the previous section there exists one and only one critical point $x_1^* = x_1^*(x_2, x_0)$ of $S_2(x_2, x_1) + S_1(x_1, x_0)$, which is a smooth function of (x_2, x_0). Let

$$D_{x_1^*}(S_2 + S_1; x_2, x_0) = \left(\frac{\tau_2\tau_1}{\tau_2 + \tau_1}\right) \det \mathrm{Hess}_{x_1^*}(S_2 + S_1),$$

where $\mathrm{Hess}_{x_1^*}(S_2 + S_1)$ means the Hessian of $S_2 + S_1$ with respect to x_1 at its critical point x_1^*.

Theorem 8.5 *Let $S_{2,1}^*(x_2, x_0) = S_2(x_2, x_1^*) + S_1(x_1^*, x_0)$. Then*

$$\begin{aligned} &I(\{\tau_1, \tau_2\}; S, a_\lambda, \nu)(x_2, x_0) \qquad (8.3.2)\\ &= \left(\frac{\nu}{2\pi i(\tau_2 + \tau_1)}\right)^{1/2} e^{i\nu S_{2,1}^*(x_2,x_0)} D_{x_1^*}(S_2 + S_1; x_2, x_0)^{-1/2}\\ &\times \Bigg(a_\lambda(x_2, x_1^*, x_0) - \frac{\tau_2\tau_1}{2\nu i(\tau_2 + \tau_1)} D_{x_1^*}(S_2 + S_1; x_2, x_0)^{-1} (\partial_{x_1}^2 a_\lambda)(x_2, x_1^*, x_0)\\ &\quad + \frac{(\tau_2\tau_1)^2}{\nu(\tau_2 + \tau_1)} b_{1,\lambda}(x_2, x_0) + \left(\frac{\tau_2\tau_1}{\nu(\tau_2 + \tau_1)}\right)^2 b_{2,\lambda}(\nu, x_2, x_0)\Bigg). \end{aligned}$$

$b_{1,\lambda}(x_2, x_0)$ is of the form

$$b_{1,\lambda}(x_2, x_0) = a_\lambda(x_2, x_1^*, x_0)c_0(x_2, x_0) + \partial_{x_1} a_\lambda(x_2, x_1^*, x_0)c_1(x_2, x_0), \qquad (8.3.3)$$

where $c_0(x_2, x_0)$ and $c_1(x_2, x_0)$ are algebraic functions of $\partial_{x_1}^k (S_2(x_2, x_1^*) + S_1(x_1^*, x_0))$ with $k = 2, 3$. Moreover, for any $K \geq 0$ there exist a constant $C_K > 0$ and an integer $M(K) \geq 0$ such that if $|\alpha| \leq K$ and $|\beta| \leq K$, then

$$|||\partial_{x_2}^\alpha \partial_{x_0}^\beta b_{1,\lambda}|||_{(m)} \leq C_K \max_{\alpha' \leq \alpha, \beta' \leq \beta, \gamma \leq M(K)} \|\partial_{x_2}^{\alpha'} \partial_{x_0}^{\beta'} \partial_{x_1}^\gamma a_\lambda\|_{(m)}, \qquad (8.3.4)$$

$$|||\partial_{x_2}^\alpha \partial_{x_0}^\beta b_{2,\lambda}|||_{(m)} \leq C_K \max_{\alpha' \leq \alpha, \beta' \leq \beta, \gamma \leq M(K)} \|\partial_{x_2}^{\alpha'} \partial_{x_0}^{\beta'} \partial_{x_1}^\gamma a_\lambda\|_{(m)}. \qquad (8.3.5)$$

Proof Let $\sigma = \dfrac{\tau_2\tau_1}{\tau_2 + \tau_1}$ and $\mu = \nu\sigma^{-1}$. Then

$$\nu(S_2(x_2, x_1) + S_1(x_1, x_0)) = \mu\Phi(x_2, x_1, x_0)$$

with

$$\Phi(x_2, x_1, x_0) = \frac{1}{2}\left(x_1 - \frac{\tau_2 x_0 + \tau_1 x_2}{\tau_2 + \tau_1}\right)^2 + \tau_2\tau_1\phi_{2,1}(x_2, x_1, x_0) + \frac{\sigma(x_2 - x_0)^2}{2(\tau_1 + \tau_2)},$$

where

$$\phi_{2,1}(x_2, x_1, x_0) = \frac{\tau_2\phi_2(x_2, x_1) + \tau_1\phi_1(x_1, x_0)}{\tau_2 + \tau_1}. \tag{8.3.6}$$

We write

$$\begin{aligned} &I(\{\tau_1, \tau_2\}; S, a_\lambda, \nu)(x, z) \\ &= \left(\frac{\nu}{2\pi i(\tau_1 + \tau_2)}\right)^{1/2} \left(\frac{\mu}{2\pi i}\right)^{1/2} \int_{\mathbf{R}} e^{i\mu\Phi(x_2,x_1,x_0)} a_\lambda(x_2, x_1, x_0)dx_1. \end{aligned} \tag{8.3.7}$$

The stationary point x_1^* of $S_2(x_2, x_1) + S_1(x_1, x_0)$ is also the stationary point of $\Phi(x_2, x_1, x_0)$, and

$$\partial_{x_1}^2 \Phi(x_2, x_1, x_0) = 1 + \tau_1\tau_2\partial_{x_1}^2 \phi_{2,1}(x_2, x_1, x_0) \geq 1 - 4^{-1}\delta_0^2\kappa_2 > 0$$

under our assumption. This means that $\Phi(x_2, x_1, x_0)$ satisfies Assumption 3.1. Thus we regard the right-hand side of (8.3.7) as an oscillatory integral with parameter μ. We apply Theorem 3.7 to this and obtain

$$\begin{aligned} &\left(\frac{\mu}{2\pi i}\right)^{1/2} \int_{\mathbf{R}} e^{i\mu\Phi(x_2,x_1,x_0)} a_\lambda(x_2, x_1, x_0)dx_1 \\ &= H_0^{-1/2} e^{i\mu\Phi(x_2,x_1^*,x_0)} \Big(a_\lambda(x_2, x_1^*, x_0) \\ &\qquad + (i\mu)^{-1}R_1(a_\lambda)(x_2, x_0) + (i\mu)^{-2}R_2(a_\lambda)(\nu, x_2, x_0)\Big). \end{aligned} \tag{8.3.8}$$

Here

$$H_0 = \partial_{x_1}^2 \Phi(x_2, x_1, x_0)|_{x_1=x_1^*} = 1 + \tau_1\tau_2\partial_{x_1}^2 \phi_{2,1}(x_2, x_1, x_0)|_{x_1=x_1^*} = D(S_2 + S_1; x_2, x_0). \tag{8.3.9}$$

Define $b_{1,\lambda}$ and $b_{2,\lambda}$ by

$$R_1(a_\lambda) = -\frac{1}{2}H_0^{-1}\partial_{x_1}^2 a_\lambda(x_2, x_1, x_0)|_{x_1=x_1^*} - i\tau_1\tau_2 b_{1,\lambda} \quad \text{and} \quad b_{2,\lambda} = H_0^{1/2}R_2(a_\lambda).$$

Then equality (8.3.2) holds.

It follows from the definition of $b_{2,\lambda}$, (8.1.4) and (3.3.28) that

$$|||\partial_{x_2}^{\alpha_2}\partial_{x_0}^{\alpha_0} b_{2,\lambda}|||_{(m)} \leq C \max_{\alpha_2' \leq \alpha_2, \alpha_0' \leq \alpha_0, |\alpha_1| \leq 2K+m+6} \|\partial_{x_2}^{\alpha_2'}\partial_{x_0}^{\alpha_0'}\partial_{x_1}^{\alpha_1} a_\lambda\|_{(m)} \tag{8.3.10}$$

as far as $\alpha_2 \leq K$ and $\alpha_0 \leq K$. Therefore, (8.3.5) is proved.

We prove (8.3.4). Let $\zeta(x_2, x_1, x_0) = \partial_{x_1}\Phi(x_2, x_1, x_0)$. Since x_1^* is the stationary point of $\Phi(x_2, x_1, x_0)$,

$$\zeta(x_2, x_1, x_0) = x_1 - x_1^* + \tau_2\tau_1(\partial_{x_1}\phi_{2,1}(x_2, x_1, x_0) - \partial_{x_1}\phi_{2,1}(x_2, x_1, x_0)|_{x_1=x_1^*})$$
$$= (x_1 - x_1^*)(1 + \tau_2\tau_1 u(x_2, x_1, x_0)),$$

where

$$u(x_2, x_1, x_0) = \int_0^1 \partial_{x_1}^2 \phi_{2,1}(x_2, tx_1 + (1-t)x_1^*, x_0)\, dt.$$

As Assumption 8.1 implies

$$\tau_2\tau_1|u(x_2, x_1, x_0)| \le 4^{-1}T^2\kappa_2 < 1,$$

$q(x_2, x_1, x_0) = (1 + \tau_1\tau_2 u(x_2, x_1, x_0))^{-1}$ is defined for all (x_2, x_1, x_0). We have

$$x_1 - x_1^* = q(x_2, x_1, x_0)\zeta(x_2, x_1, x_0),$$

which corresponds to (3.2.43) of Lemma 3.6.

Let z be the variable such that the change of coordinates $x_1 \to z = z(x_2, x_1, x_0)$ fulfills Morse's lemma, i.e.

$$\Phi(x_2, x_1, x_0) = \Phi(x_2, x_1^*, x_0) + \frac{1}{2}z^2,$$

and $J(x_2, x_1, x_0) = dz(x_2, x_1, x_0)/dx_1$ be the Jacobian of the change of coordinates. We divide $R_1(a_\lambda)$ of (8.3.8) into three parts following (3.3.27) in Theorem 3.7.

$$R_1(a_\lambda) = f_1 + f_2 + f_3, \tag{8.3.11}$$
$$f_1(x_2, x_0) = -\frac{1}{2}(a_\lambda(x_2, x_1^*, x_0)q(x_2, x_1^*, x_0)\partial_z^2 J(x_2, x_1(z), x_0)^{-1}|_{z=0},$$
$$f_2(x_2, x_0) = \frac{1}{2}\partial_{x_1} a_\lambda(x_2, x_1, x_0)|_{x_1=x_1^*} q(x_2, x_1^*, x_0)^2 \partial_{x_1}^3 \Phi(x_2, x_1, x_0)|_{x_1=x_1^*},$$
$$f_3(x_2, x_0) = -\frac{1}{2}\partial_{x_1}^2 a_\lambda(x_2, x_1, x_0)|_{x_1=x_1^*} q(x_2, x_1^*, x_0).$$

Since $u(x_2, x_1^*, x_0) = \partial_{x_1}^2 \phi_{2,1}(x_2, x_1, x_0)|_{x_1=x_1^*}$,

$$q(x_2, x_1^*, x_0) = H_0^{-1}.$$

This means

$$f_3(x_2, x_0) = -\frac{1}{2}H_0^{-1}\partial_{x_1}^2 a_\lambda(x_2, x_1, x_0)|_{x_1=x_1^*}.$$

Hence by definition

$$-i\tau_1\tau_2 b_{1,\lambda}(x_2, x_0) = f_1(x_2, x_0) + f_2(x_2, x_0).$$

Therefore, it is sufficient for the proof of (8.3.4) to prove the following two inequalities: For any $\alpha_2 \geq 0$ and $\alpha_0 \geq 0$ there exists a positive constant C such that the following inequalities hold:

$$|||\partial_{x_2}^{\alpha_2}\partial_{x_0}^{\alpha_0}(\tau_1\tau_2)^{-1} f_1|||_{(m)} \leq C \max_{\alpha_2' \leq \alpha_2, \alpha_0' \leq \alpha_0, \beta \leq \alpha_0+\alpha_2} \|\partial_{x_2}^{\alpha_2'}\partial_{x_1}^{\beta}\partial_{x_0}^{\alpha_0'} a_\lambda\|_{(m)}, \tag{8.3.12}$$

$$|||\partial_{x_2}^{\alpha_2}\partial_{x_0}^{\alpha_0}(\tau_1\tau_2)^{-1} f_2(x_2, x_0)|||_{(m)} \leq C \max_{\alpha_2' \leq \alpha_2, \alpha_0' \leq \alpha_0, \beta \leq |\alpha_0|+|\alpha_2|+2} \|\partial_{x_2}^{\alpha_2'}\partial_{x_1}^{\beta}\partial_{x_0}^{\alpha_0'} a_\lambda\|_{(m)}. \tag{8.3.13}$$

Since

$$\partial_{x_1}^3 \Phi(x_2, x_1, x_0)|_{x_1=x_1^*} = \tau_1\tau_2\partial_{x_1}^3 \phi_{2,1}(x_2, x_1, x_0)|_{x_1=x_1^*},$$

$(\tau_1\tau_2)^{-1}\partial_{x_1}^3 \Phi(x_2, x_1, x_0)|_{x_1=x_1^*}$ is uniformly bounded, together with all its derivatives. (8.3.13) follows from this.

We shall discuss (8.3.12). Taylor's theorem says that

$$\Phi(x_2, x_1, x_0) - \Phi(x_2, x_1^*, x_0) = \frac{1}{2} H(x_2, x_1, x_0)(x_1 - x_1^*)^2,$$

where

$$H(x_2, x_1, x_0) = 1 + \tau_1\tau_2 h(x_2, x_1, x_0)$$

with

$$h(x_2, x_1, x_0) = 2\int_0^1 (1-t)\partial_{x_1}^2 \phi_{2,1}(x_2, tx_1 + (1-t)x_1^*, x_0)\, dt.$$

The assumption $T^2\kappa_2 \leq \delta_0^2\kappa_2 < 1$ implies that

$$|\tau_1\tau_2 h(x_2, x_1, x_0)| \leq \tau_1\tau_2\kappa_2 < 1,$$

because $|\partial_{x_1}^2 \phi_{2,1}(x_2, tx_1 + (1-t)x_1^*, x_0)| \leq \kappa_2$. Therefore,

$$H(x_2, x_1, x_0)^{1/2} = 1 + \sum_{n=1}^{\infty} \binom{1/2}{n} (\tau_1\tau_2 h(x_2, x_1, x_0))^n \tag{8.3.14}$$

is well-defined. Since $H_0 = H(x_2, x_1^*, x_0)$,

$$H_0^{-1/2} = 1 + \sum_{n=1}^{\infty} \binom{-1/2}{n} (\tau_1\tau_2 h(x_2, x_1^*, x_0))^n \tag{8.3.15}$$

is also well-defined. Thus we can define for all $(x_2, x_1, x_0) \in \mathbf{R}^3$

$$z = H_0^{-1/2} H(x_2, x_1, x_0)^{1/2}(x_1 - x_1^*).$$

It is clear that

$$\frac{1}{2}(H_0 z, z) = \frac{1}{2}(H(x_2, x_1, x_0)(x_1 - x_1^*), (x_1 - x_1^*)) = \Phi(x_2, x_1, x_0) - \Phi(x_2, x_1^*, x_0).$$

Thus Morse's lemma holds by the change of coordinates $x_1 \to z$.

It is sufficient for the proof of (8.3.12) to prove that $(\tau_1\tau_2)^{-1}\partial_z^2 J(x_2, x_1(z), x_0)^{-1}|_{z=0}$ is bounded, together with its derivatives of all order. By (8.3.14) and (8.3.15) we have

$$H_0^{-1/2} H(x_2, x_1, x_0)^{1/2} = (1 + \tau_1\tau_2 w(x_2, x_1, x_0))$$

with some function $w(x_2, x_1, x_0)$ depending also on parameters τ_1, τ_2. It is clear that for any $\alpha_2 \geq 0$, $\alpha_1 \geq 0$, $\alpha_0 \geq 0$, $|\partial_{x_2}^{\alpha_2}\partial_{x_1}^{\alpha_1}\partial_{x_0}^{\alpha_0} w(x_2, x_1, x_0)|$ remains bounded.

We have

$$z = H_0^{-1/2} H(x_2, x_1, x_0)^{1/2}(x_1 - x_1^*) = (1 + \tau_1\tau_2 w(x_2, x_1, x_0))(x_1 - x_1^*).$$

The Jacobian of this change of variables is

$$J(x_2, x_1, x_0) = \partial_{x_1} z = 1 + \tau_1\tau_2 w(x_2, x_1, x_0) + \tau_1\tau_2 \partial_{x_1} w(x_2, x_1, x_0)(x_1 - x_1^*).$$

Thus

$$\partial_{x_1} J(x_2, x_1, x_0)\big|_{x_1 = x_1^*} = 2\tau_1\tau_2 \partial_{x_1} w(x_2, x_1, x_0)\big|_{x_1 = x_1^*}, \tag{8.3.16}$$

$$\partial_{x_1}^2 J(x_2, x_1, x_0)\big|_{x_1 = x_1^*} = 3\tau_1\tau_2 \partial_{x_1}^2 w(x_2, x_1, x_0)\big|_{x_1 = x_1^*}. \tag{8.3.17}$$

By Remark 3.6,

$$\begin{aligned}
\partial_z^2 J(x_2, x_1(z), x_0)^{-1}|_{z=0} &= -\partial_{x_1}^2 J(x_2, x_1, x_0)|_{x_1 = x_1^*} + 2(\partial_{x_1} J(x_2, x_1, x_0)|_{x_1 = x_1^*})^2 \\
&\quad + (\partial_{x_1}^2 z(x_1)|_{x_1 = x_1^*})\,(\partial_{x_1} J(x_2, x_1, x_0)|_{x_1 = x_1^*}) \\
&= -3\tau_1\tau_2 \partial_{x_1}^2 w(x_2, x_1, x_0)\big|_{x_1 = x_1^*} + 12(\tau_1\tau_2)^2 \partial_{x_1} w(x_2, x_1, x_0)\big|_{x_1 = x_1^*}^2.
\end{aligned}$$

This proves that $(\tau_1\tau_2)^{-1}\partial_z^2 J(x_2, x_1(z), x_0)^{-1}|_{z=0}$ is bounded, together with its derivatives of all orders. Therefore, inequality (8.3.12) is proved.

Since both (8.3.12) and (8.3.13) are now proved, (8.3.4) is proved.

The theorem has been proved. □

Since (8.3.9) holds and $\nu \geq 1$, we can rewrite the previous theorem in the following form.

Corollary 8.10 *If in addition to the assumptions of Theorem 8.5 we further assume that* $\tau_1\tau_2 < 1$, *then we have*

$$
\begin{aligned}
& I(\{\tau_1, \tau_2\}; S, a_\lambda, \nu)(x_2, x_0) \\
&= \left(\frac{\nu}{2\pi i(\tau_2+\tau_1)}\right)^{1/2} e^{i\nu S^*_{2,0}(x_2,x_0)} \\
&\times \Big(D_{x_1^*}(S_2+S_1; x_2, x_0)^{-1/2} \Big(a_\lambda(x_2, x_1^*, x_0) \\
&\qquad - \frac{\tau_2\tau_1}{2\nu i(\tau_2+\tau_1)} D_{x_1^*}(S_2+S_1; x_2, x_0)^{-1} (\partial_{x_1}^2 a_\lambda)(x_2, x_1^*, x_0)\Big) \\
&\qquad + \nu^{-1}\tau_2\tau_1 r(\nu, x_2, x_0)\Big).
\end{aligned}
\tag{8.3.18}
$$

For any $K \geq 0$ *there exist positive constants* C_K *and* $M(K)$ *such that if* $|\alpha| \leq K$ *and* $|\beta| \leq K$ *we have*

$$
|||\partial_{x_2}^{\alpha}\partial_{x_0}^{\beta} r|||_{(m)} \leq C_K \max_{\alpha' \leq \alpha, \beta' \leq \beta, \gamma \leq M(K)} \|\partial_{x_2}^{\alpha'}\partial_{x_0}^{\beta'}\partial_{x_1}^{\gamma} a_\lambda\|_{(m)}. \tag{8.3.19}
$$

Proof Set

$$
\begin{aligned}
& r(\nu, x_2, x_0) \\
&= D_{x_1^*}(S_2+S_1; x_2, x_0)^{-1/2} \left(\frac{\tau_2\tau_1}{(\tau_2+\tau_1)} b_{1,\lambda}(\nu, x_2, x_0) + \frac{\tau_2\tau_1}{\nu(\tau_2+\tau_1)^2} b_{2,\lambda}(\nu, x_2, x_0)\right).
\end{aligned}
$$

Use the trivial inequality $\tau_1\tau_2 \leq 4^{-1}(\tau_1+\tau_2)^2$ and the assumption $\tau_1\tau_2 < 1$. Then inequality (8.3.19) follows from (8.3.4) and (8.3.5). □

Corollary 8.11 *We can write*

$$
\begin{aligned}
& I(\{\tau_1, \tau_2\}; S, a_\lambda, \nu)(x_2, x_0) \\
&= \left(\frac{\nu}{2\pi i(\tau_2+\tau_1)}\right)^{1/2} e^{i\nu S^*_{2,0}(x_2,x_0)} D(S_2+S_1; x_2, x_0)^{-1/2}\big(a_\lambda(x_2, x_1^*, x_0) + \tilde{R}(a_\lambda)\big).
\end{aligned}
$$

We set $R(a_\lambda)(x_2, x_0) = D(S_2+S_1; x_2, x_0)^{-1/2}\tilde{R}(a_\lambda)(x_2, x_0)$. *For any* $K \geq 0$ *there exist positive constants* C_K *and* $M(K) \geq K$ *such that if* $|\alpha| \leq K$ *and* $|\beta| \leq K$ *we have*

$$
\begin{aligned}
|||\partial_{x_2}^{\alpha_2}\partial_{x_0}^{\alpha_0}\tilde{R}(a_\lambda)|||_{(m)} &\leq C_K\nu^{-1}\left(\frac{\tau_1\tau_2}{\tau_1+\tau_2}\right) \max_{\alpha_2' \leq \alpha_2, \alpha_0' \leq \alpha_0, \beta \leq M(K)} \|\partial_{x_2}^{\alpha_2'}\partial_{x_0}^{\alpha_0'}\partial_{x_1}^{\beta} a_\lambda\|_{(m)}, \\
|||\partial_{x_2}^{\alpha_2}\partial_{x_0}^{\alpha_0} R(a_\lambda)|||_{(m)} &\leq C_K\nu^{-1}\left(\frac{\tau_1\tau_2}{\tau_1+\tau_2}\right) \max_{\alpha_2' \leq \alpha_2, \alpha_0' \leq \alpha_0, \beta \leq M(K)} \|\partial_{x_2}^{\alpha_2'}\partial_{x_0}^{\alpha_0'}\partial_{x_1}^{\beta} a_\lambda\|_{(m)}.
\end{aligned}
$$

Proof Pick

$$\tilde{R}(a_\lambda)(x_2, x_0) = -\frac{\tau_2\tau_1}{2\nu i(\tau_2+\tau_1)} D_{x_1^*}(S_2+S_1; x_2, x_0)^{-1}(\partial_{x_1}^2 a_\lambda)(x_2, x_1^*, x_0) + \frac{(\tau_2\tau_1)^2}{\nu(\tau_2+\tau_1)} b_{1,\lambda}(x_2, x_0) + \left(\frac{\tau_2\tau_1}{\nu(\tau_2+\tau_1)}\right)^2 b_{2,\lambda}(\nu, x_2, x_0).$$

Then the corollary holds. □

Corollary 8.12 *Suppose that in addition to Assumption 8.3 $a_\lambda(x_2, x_1, x_0)$ is infinitely differentiable by λ and for any $\alpha_2 \geq 0, \alpha_1 \geq 0, \alpha_0 \geq 0, \gamma \geq 0$*

$$\|\partial_{x_2}^{\alpha_2}\partial_{x_1}^{\alpha_1}\partial_{x_0}^{\alpha_0}\partial_\lambda^\gamma a_\lambda(x_2, x_1, x_0)\|_{(m)} < \infty.$$

Then for any $K \geq 0$ there exists $M(K)$ and C_K such that if $\alpha_2 \leq K, \alpha_0 \leq K, \gamma \geq 0$

$$|||\partial_{x_2}^{\alpha_2}\partial_{x_0}^{\alpha_0}\partial_\lambda^\gamma R(a_\lambda)|||_{(m)} \leq C_K \nu^{-1}\left(\frac{\tau_1\tau_2}{\tau_1+\tau_2}\right) \max_{\alpha_2' \leq \alpha_2, \alpha_0' \leq \alpha_0, \beta \leq M(K)} \|\partial_{x_2}^{\alpha_2'}\partial_{x_0}^{\alpha_0'}\partial_{x_1}^{\beta}\partial_\lambda^\gamma a_\lambda\|_{(m)}.$$

Proof The corollary is a consequence of the following formula:

$$\partial_\lambda I(\{\tau_1, \tau_2\}; S, a_\lambda, \nu)(x_2, x_0) = I(\{\tau_1, \tau_2\}; S, \partial_\lambda a_\lambda, \nu)(x_2, x_0).$$

□

8.4 Proof of Stationary Phase Method in the Case of Large Dimension

8.4.1 Hessian Determinant

By (8.1.9), the Hessian determinant of the phase at the stationary point x^* is

$$\mathrm{Hess}_{x^*} S = H_0 + W(x^*). \tag{8.4.1}$$

We set $W^* = W(x^*)$ for simplicity. We define $D(\{\tau_j\}; x_{J+1}, x_0)$ by

$$D(\{\tau_j\}; x_{J+1}, x_0) = \det(I + W^* H_0^{-1}) = \det(I + H_0^{-1} W^*).$$

Then

$$\det \mathrm{Hess}_{x^*} S = \frac{T}{\tau_1 \ldots \tau_{J+1}} D(\{\tau_j\}; x_{J+1}, x_0). \tag{8.4.2}$$

By Lemma 8.2, under the assumption $T \le \delta_0$ we have

$$(1-\kappa_2\delta_0^2)^J \le D(\{\tau_j\}; x_{J+1}, x_0) \le (1+\kappa_2\delta_0^2)^J. \tag{8.4.3}$$

Let j, l be two integers such that $1 \le j < l \le J+1$. As in Sect. 8.2, we use $T(l, j) = \tau_j + \cdots + \tau_l$ and for any fixed $(x_l, x_{j-1}) \in \mathbf{R}^2$, $(x_{l-1}^*, x_{l-2}^*, \ldots, x_j^*)$ is the stationary point of

$$S_{l,j}(x_l, \ldots, x_{j-1}) = \sum_{k=j}^{l} S_k(x_k, x_{k-1})$$

and $S_{l,j}^*(x_l, x_{j-1})$ is the stationary value of it. Since x_k^*, $j \le k < l$ is a function of (x_l, x_{j-1}), we denote it by $x_k^*(x_l, x_{j-1})$. i.e.,

$$S_{l,j}^*(x_l, x_{j-1}) = S_l(x_l, x_{l-1}^*) + \sum_{k=j+1}^{l-1} S_k(x_k^*(x_l, x_{j-1}), x_{k-1}^*(x_l, x_{j-1})) + S_j(x_j^*, x_{j-1}). \tag{8.4.4}$$

Remark 8.4 If $S_j(x_j, x_{j-1})$ is the classical action $S(T_j, T_{j-1}, x_j, x_{j-1})$ as in (4.1.4), then

$$S_{l,j}^*(x_l, x_{j-1}) = S(T_l, T_{j-1}, x_l, x_{j-1}) = S_{l.j}(x_l, x_{j-1}).$$

The following proposition follows from Proposition 8.4.

Proposition 8.5 *Suppose that* $T(l, k) = \tau_k + \tau_{k+1} + \cdots + \tau_l \le \delta_0$. *Then* $S_{l,j}^*(x_l, x_{j-1})$ *is of the form*

$$S_{l,j}^*(x_l, x_{j-1}) = \frac{|x_l - x_{j-1}|^2}{T(l, k)} + T(k, l)\phi_{k,l}(x_l, x_{j-1}). \tag{8.4.5}$$

Moreover, for any positive integer $k \ge 0$, *let* κ_k^* *be the constant in Proposition 8.4. Then*

$$|\phi_{l,j}(x_l, x_{j-1})| \le \kappa_0^*(1 + |x_{j-1}| + |x_l|)^2 \tag{8.4.6}$$

$$|\partial_{x_{j-1}}^\alpha \partial_{x_l}^\beta \phi_{l,j}(x_l, x_{j-1})| \le \kappa_1^*(1 + |x_{j-1}| + |x_l|), \quad \textit{if } |\alpha| + |\beta| = 1. \tag{8.4.7}$$

and

$$|\partial_{x_{j-1}}^\alpha \partial_{x_l}^\beta \phi_{l,j}(x_l, x_{j-1})| \le \kappa_k^*, \quad \textit{if } |\alpha| + |\beta| = k \ge 2. \tag{8.4.8}$$

In particular, in the case $k = 2$, $\kappa_2^* = 10\kappa_2(1 - \kappa_2\delta_0^2)^{-1}$ *is enough.*

We use the notation

$$D_{x^*_{l-1},\dots,x^*_j}(S_{l,j};x_l,x_{j-1}) \tag{8.4.9}$$
$$= \left(\frac{\tau_l\dots\tau_j}{\tau_l+\dots+\tau_j}\right)\det\mathrm{Hess}\Big(\sum_{k=j}^{l} S_k(x_k,x_{k-1})\Big)\Big|_{x_k=x^*_k,\,j\le k\le l-1}.$$

Here Hess means the Hessian at the critical point $(x^*_{l-1},\dots,x^*_j)$. If $j=1$ and $l=J+1$, then we have $D(\{\tau_j\};x_{J+1},x_0)=D_{x^*_J\cdots x^*_1}(S_{J+1}+\dots+S_1;x_{J+1},x_0)$.

The following lemma is a straightforward consequence of Proposition 5.1.

Lemma 8.3 *For any $k<j<l$ let $x^*_j(x_l,x_k)$ be the stationary point of $S_l(x_l,x_{l-1})+\dots+S_{k+1}(x_{k+1},x_k)$. Let $1\le j_0<j_1<j_2<\dots<j_r\le J+1$ be a sequence of integers. For any fixed $(x_{j_r},x_{j_0})\in\mathbf{R}^2$ the point $(x^*_{j_{r-1}}(x_{j_r},x_{j_0}),\dots,x^*_{j_1}(x_{j_r},x_{j_0}))$ is the only one stationary point of the function*

$$(x_{j_{r-1}-1},\dots,x_{j_1})\longrightarrow S^*_{j_r,j_{r-1}}(x_{j_r},x_{j_{r-1}-1})+\dots+S^*_{j_1,j_0}(x_{j_1},x_{j_0-1}).$$

Mimicking the proof of Lemma 5.1, we obtain

Lemma 8.4 *If $1\le j\le l$, then*

$$D_{(x^*_l,x^*_{l-1},\dots,x^*_j)}(S_{l+1,j};x_{l+1},x_{j-1}) \tag{8.4.10}$$
$$= D_{x^*_l}(S_{l+1}+S^*_{l,j};x_{l+1},x_{j-1})D_{(x^*_{l-1},\dots,x^*_j)}(S_{l,j};x_l,x_{j-1})\big|_{x_l=x^*_l}, \tag{8.4.11}$$

and

$$D_{(x^*_l,x^*_{l-1},\dots,x^*_j)}(S_{l+1,j};x_{l+1},x_{j-1})=\prod_{k=j}^{l}D_{x^*_k}(S_{k+1}+S^*_{k,j};x_{k+1},x_{j-1})\Big|_{x_k=x^*_k}. \tag{8.4.12}$$

In particular,

$$D(\{\tau_j\};x_{J+1},x_0)=\prod_{k=1}^{J}D_{x^*_k}(S_{k+1}+S^*_{k,1};x_{k+1},x_0)\Big|_{x_k=x^*_k}. \tag{8.4.13}$$

In the same way as Lemma 5.4 we can prove

Lemma 8.5 *Suppose $0<T<\delta_0$. Then for $1\le j\le l$*

$$D_{(x^*_l,x^*_{l-1},\dots,x^*_j)}(S_{l+1,j};x_{l+1},x_{j-1})=1+T(l+1,j)^2d(S_{l+1,j};x_{l+1},x_{j-1}). \tag{8.4.14}$$

There exists a positive $\rho>0$ and $C_0>0$ depending on κ_2 such that for any

$$\max_{l,j} \sup_{(x_{l+1},x_{j-1})\in\mathbf{R}^2} |d(S_{l+1,j}; x_{l+1}, x_{j-1})| < C_0. \tag{8.4.15}$$

We can choose so that $C_0 = 2\kappa_2^* e^{2\kappa_2^*\delta_0^2}$. *Moreover, for any* $K \geq 0$ *there exist constants* $\rho_K > 0$ *and* $C_K > 0$ *such that if* $\max\{\tau_j\} \leq \rho_K$

$$\max_{l,j} \sup_{\{\tau_j\}} \max_{|\alpha|\leq K, |\beta|\leq K} \sup_{(x_{l+1},x_{j-1})\in\mathbf{R}^2} |\partial_x^\alpha \partial_y^\beta d(S_{l+1,j}; x_{l+1}, x_{j-1})| \leq C_K. \tag{8.4.16}$$

8.4.2 Proof of Stationary Phase Method in Large-Dimensional Space

We prove Theorem 8.4 in the following. Recall that

$$\begin{aligned} &I(\{\tau_j\}; S, a_\lambda, \nu)(x_{J+1}, x_0) \\ &= \prod_{j=1}^{J+1} \left(\frac{-i\nu}{2\pi\tau_j}\right)^{\frac{1}{2}} \int_{\mathbf{R}^J} e^{i\nu S(x_{J+1}, x_J, \ldots, x_1, x_0)} a_\lambda(x_{J+1}, x_J, \ldots, x_1, x_0) \prod_{j=1}^{J} dx_j. \end{aligned} \tag{8.4.17}$$

We use H, G, $W(x)$, $S^*_{l,k+1}(x_l, x_k)$, $D_{x^*_{l-1},\ldots,x^*_{k+1}}(S_l + \cdots + S_{k+1}; x_l, x_k)$ in the same meaning as previous subsections.

We abbreviate $I(\{\tau_j\}; S, a_\lambda, \nu)(x_{J+1}, x_0)$ to I for simplicity. To prove the theorem we proceed in the following way. First we integrate by x_1 and we apply the stationary phase method. Using the result, we next integrate by x_2 and apply the stationary phase method, etc. In this way we perform multiple integration in succession. In doing so, we take special care in treating remainder terms to obtain estimate (8.2.5). We explain each step in detail.

Step 1. Integration by x_1.
Regarding $(\lambda, x_3, \ldots, x_{J+1})$ as a parameter, we apply Corollary 8.12. Then

$$\begin{aligned} &\left(\frac{-i\nu}{2\pi\tau_1}\right)^{\frac{1}{2}} \left(\frac{-i\nu}{2\pi\tau_2}\right)^{\frac{1}{2}} \int_{\mathbf{R}} a_\lambda(x_{J+1}, \ldots, x_0) e^{i\nu(S_2(x_2,x_1)+S_1(x_1,x_0))} dx_1 \\ &= \left(\frac{-i\nu}{2\pi T(2,1)}\right)^{\frac{1}{2}} e^{i\nu S^*_{2,1}(x_2,x_0)} \\ &\quad \times \{(P_1 a_\lambda)(x_{J+1}, \ldots, x_2, x_0) + (R_1 a_\lambda)(x_{J+1}, \ldots, x_2, x_0)\}. \end{aligned} \tag{8.4.18}$$

Here $(P_1 a_\lambda)$ is the principal term and $(R_1 a_\lambda)$ is the remainder. We know that

$$(P_1 a_\lambda)(x_{J+1}, \ldots, x_2, x_0) = D(S_1 + S_2; x_2, x_0)^{-1/2} a_\lambda(x_{J+1}, \ldots, \widehat{x_2, x_0})$$

and for any integer $k \geq 0$ there exist a constant $C_k > 0$ and an integer $M(k) > 0$ such that if $|\alpha_0| \leq k$, $\alpha_2 \leq k$ and $|\alpha_j| \leq k$, $j = 3, 4, \ldots, J+1$, then

$$(1+|\lambda|+|x_0|+|x_2|+|x_3|+\cdots+|x_{J+1}|)^{-m}\left|\partial_{x_0}^{\alpha_0}\partial_{x_2}^{\alpha_2}\prod_{j=3}^{J+1}\partial_{x_j}^{\alpha_j}(R_1a_\lambda)(x_{J+1},\ldots,x_2,x_0)\right|$$
$$\le C\nu^{-1}\frac{\tau_1\tau_2}{T(2,1)}\max_{\beta_0,\beta_1,\beta_2}\sup_{x_1\in\mathbf{R}}(1+|\lambda|+|x_0|+|x_1|+|x_2|+|x_3|+\cdots+|x_{J+1}|)^{-m}$$
$$\times\left|\prod_{j=3}^{J+1}\partial_{x_j}^{\alpha_j}\partial_{x_0}^{\beta_0}\partial_{x_2}^{\beta_2}\partial_{x_1}^{\beta_1}(a_\lambda)(x_{J+1},\ldots,x_1,x_0)\right|,$$

where $\beta_0\le\alpha_0$, $\beta_2\le\alpha_2$ and $|\beta_1|\le M(k)$. Thus

$$I=I_{P_1}+I_{R_1}, \tag{8.4.19}$$

where

$$I_{P_1}=\left(\frac{-i\nu}{2\pi T(2,1)}\right)^{\frac12}\prod_{j=3}^{J+1}\left(\frac{-i\nu}{2\pi\tau_j}\right)^{\frac12}$$
$$\times\int_{\mathbf{R}^{J-1}}e^{i\nu S(x_{J+1},x_J,\ldots,\widehat{x_2,x_0})}P_1(a_\lambda)(x_{J+1},x_J,\ldots,x_2,x_0)\prod_{j=2}^{J}dx_j,$$
$$I_{R_1}=\left(\frac{-i\nu}{2\pi T(2,1)}\right)^{\frac12}\prod_{j=3}^{J+1}\left(\frac{-i\nu}{2\pi\tau_j}\right)^{\frac12}$$
$$\times\int_{\mathbf{R}^{J-1}}e^{i\nu S(x_{J+1},x_J,\ldots,\widehat{x_2,x_0})}R_1(a_\lambda)(x_{J+1},x_J,\ldots,x_2,x_0)\prod_{j=2}^{J}dx_j.$$

Step 2. Integration by x_2.
Notice that

$$S(x_{J+1},x_J,\ldots,\widehat{x_2,x_0})=\sum_{j=3}^{J+1}S_j(x_j,x_{j-1})+S_{2,1}^*(x_2,x_0),$$

and $S_{2,1}^*(x_2,x_0)$ enjoys the property described by Proposition 8.4. The structure of (P_1a_λ) is not complicated, because $(P_1a_\lambda)(x_{J+1},\ldots,x_2,x_0)=a_\lambda(x_{J+1},\ldots,\widehat{x_2,x_0})$ is obtained from a_λ by simply replacing x_1 with $x_1^*(x_2,x_0)$. We perform integration by x_2 in I_{P_1} and apply the stationary phase method, Corollary 8.11. Then

$$\left(\frac{-i\nu}{2\pi\tau_3}\right)^{\frac12}\left(\frac{-i\nu}{2\pi T(2,1)}\right)^{\frac12}\int_{\mathbf{R}}(P_1a_\lambda)(x_{J+1},\ldots,x_2,x_0)e^{i\nu(S_3(x_3,x_2)+S_{2,1}^*(x_2,x_0))}\,dx_2$$
$$=\left(\frac{-i\nu}{2\pi T(3,1)}\right)^{\frac12}e^{i\nu S_{3,1}^*(x_3,x_0)}$$
$$\times\{(P_2P_1a_\lambda)(x_{J+1},\ldots,x_3,x_0)+(R_2P_1a_\lambda)(x_{J+1},\ldots,x_3,x_0)\}.$$

Here $(P_2P_1a_\lambda)$ is the main part and $(R_2P_1a_\lambda)$ is the remainder. We have thus

$$I_{P_1} = I_{P_2,P_1} + I_{R_2,P_1},$$

where

$$I_{P_2,P_1} = \left(\frac{-i\nu}{2\pi T(3,1)}\right)^{\frac{1}{2}} \prod_{j=3}^{J+1} \left(\frac{-i\nu}{2\pi\tau_j}\right)^{\frac{1}{2}}$$
$$\times \int_{\mathbf{R}^{J-2}} e^{i\nu S(x_{J+1},x_J,\ldots,\widehat{x_3,x_0})} P_2P_1(a_\lambda)(x_{J+1},x_J,\ldots,x_2,x_0) \prod_{j=2}^{J} dx_j,$$

$$I_{R_2,P_1} = \left(\frac{-i\nu}{2\pi T(3,1)}\right)^{\frac{1}{2}} \prod_{j=3}^{J+1} \left(\frac{-i\nu}{2\pi\tau_j}\right)^{\frac{1}{2}}$$
$$\times \int_{\mathbf{R}^{J-2}} e^{i\nu S(x_{J+1},x_J,\ldots,\widehat{x_3,x_0})} R_2P_1(a_\lambda)(x_{J+1},x_J,\ldots,x_2,x_0) \prod_{j=2}^{J} dx_j.$$

On the other hand, the amplitude (R_1a_λ) of I_{R_1} is a complicated function of x_2. We do not integrate by x_2 here and leave the operation of integration by x_2 for later use. We express this term by writing I_{id,R_1}. Therefore,

$$I = I_{P_2,P_1} + I_{R_2,P_1} + I_{id,R_1}. \tag{8.4.20}$$

Step 3. Integration by x_3.

Since the amplitude of I_{P_2,P_1} is simple as a function of x_3, we perform integration by x_3 and use the stationary phase method.

$$\left(\frac{-i\nu}{2\pi T(3,1)}\right)^{\frac{1}{2}} \int_{\mathbf{R}} (P_2P_1a_\lambda)(x_{J+1},\ldots,x_3,x_0) e^{i\nu(S_4(x_4,x_3)+S^*_{3,1}(x_3,x_0))} dx_3$$
$$= \left(\frac{-i\nu}{2\pi T(4,1)}\right)^{\frac{1}{2}} e^{i\nu S^*_{4,1}(x_4,x_0)}$$
$$\times \{(P_3P_2P_1a_\lambda)(x_{J+1},\ldots,x_4,x_0) + (R_3P_2P_1a_\lambda)(x_{J+1},\ldots,x_4,x_0)\}.$$

Thus

$$I_{P_2,P_1} = I_{P_3,P_2,P_1} + I_{R_3,P_2,P_1},$$

where

$$I_{P_3,P_2,P_1} = \left(\frac{-i\nu}{2\pi T(4,1)}\right)^{\frac{1}{2}} \prod_{j=4}^{J+1} \left(\frac{-i\nu}{2\pi\tau_j}\right)^{\frac{1}{2}}$$
$$\times \int_{\mathbf{R}^{J-3}} e^{i\nu S(x_{J+1},x_J,\ldots,\widehat{x_4,x_0})} P_3P_2P_1(a_\lambda)(x_{J+1},x_J,\ldots,x_4,x_0) \prod_{j=4}^{J} dx_j,$$

$$I_{R_3,P_2,P_1} = \left(\frac{-i\nu}{2\pi T(4,1)}\right)^{\frac{1}{2}} \prod_{j=4}^{J+1} \left(\frac{-i\nu}{2\pi \tau_j}\right)^{\frac{1}{2}}$$
$$\times \int_{\mathbf{R}^{J-3}} e^{i\nu S(x_{J+1},x_J,\dots,\widehat{x_4,x_0})} R_3 P_2 P_1(a_\lambda)(x_{J+1},x_J,\dots,x_4,x_0) \prod_{j=4}^{J} dx_j.$$

The amplitude function $R_2 P_1(a_\lambda)$ of the term I_{R_2,P_1} is a complicated function of x_3. We postpone integration by x_3. Thus we write this term by I_{id,R_2,P_1}.

The amplitude function of I_{id,R_1} is rather simple as a function of x_3, We perform integration with respect to x_3 and apply Corollary 8.11. We obtain

$$I_{id,R_1} = I_{P_3,id,R_1} + I_{R_3,id,R_1}. \tag{8.4.21}$$

Thus we have

$$I = I_{P_3,P_2,P_1} + I_{R_3,P_2,P_1} + I_{id,R_2,P_1} + I_{P_3,id,R_1} + I_{R_3,id,R_1}. \tag{8.4.22}$$

Step 4. Integration by x_4.
In the same way as above we obtain

$$\begin{aligned} I =& I_{P_4,P_3,P_2,P_1} + I_{R_4,P_3,P_2,P_1} + I_{id,R_3,P_2,P_1} + I_{P_4,id,R_2,P_1} + I_{R_4,id,R_2,P_1} \\ &+ I_{P_4,P_3,id,R_1} + I_{R_4,P_3,id,R_1} + I_{id,R_3,id,R_1}. \end{aligned}$$

Each of these terms on the right-hand side is an oscillatory integral. For instance,

$$I_{id,R_3,id,R_1} = \left(\frac{-i\nu}{2\pi T(2,1)}\right)^{\frac{1}{2}} \left(\frac{-i\nu}{2\pi T(4,3)}\right)^{\frac{1}{2}} \int_{\mathbf{R}^2} e^{i\nu(S^*_{2,1}(x_2,x_0)+S^*_{4,3}(x_4,x_2))}$$
$$\times \left(\prod_{j=5}^{J+1} \left(\frac{-i\nu}{2\pi\tau_j}\right)^{\frac{1}{2}} \int_{\mathbf{R}^{J-5}} e^{i\nu(S_{J+1,5}(x_{J+1},x_4)} \right.$$
$$\left. \times R_3 R_1(a_\lambda)(x_{J+1},\dots,x_4,x_2,x_0) \prod_{j=5}^{J} dx_j \right) dx_4 dx_2.$$

We proceed successively in this way as far as possible. Then we finally arrive at Step J, where we treat integration with respect x_J.

Step J+1.
As a result of Step J, we can write I as a sum of many terms of the form $I_{Q_J,Q_{J-1},\dots,Q_2,Q_1}$:

$$I = \sum I_{Q_J,Q_{J-1},\dots,Q_1}, \tag{8.4.23}$$

where the summation ranges over all arrangements $Q_J, Q_2, \ldots, Q_1$ that obey the following rules:

- Every Q_j is one of P_j, R_j or id.
- If $Q_j = R_j$ for some $j < J$, then Q_{j+1} must be id.
- If $Q_j = id$ for some $j \le J$, then $j > 1$ and $Q_{j-1} = R_{j-1}$.

There are only two arrangements obeying the rules for which $Q_j \neq id$ for all j. We use the following symbols for them:

$$A_0 = I_{Q_J Q_{J-1} \ldots Q_1}, \quad \text{where } Q_j = P_j \text{ for all } j = 1, 2, \ldots, J,$$
$$A_{J+1} = I_{Q_J Q_{J-1} \ldots Q_1}, \quad \text{where } Q_J = R_J \text{ and } Q_j = P_j \text{ for all } 1 \le j \le J-1.$$

Every other arrangement that obeys the rule above includes at least one $Q_j = id$. They are classified into two types, which we call type-1 and type-2.

The arrangements of type-1 are arrangements $Q_J, Q_{J-1}, \ldots, Q_2, Q_1$ obeying this rule with the following property: $Q_J = R_J$ and there exists some $1 \le j \le J-2$ with $Q_j = R_j$.

The remaining arrangements that fulfill the rule above and are not equal to either A_0 or A_{J+1} are called of type-2.

Assume that $Q_J, Q_{J-1}, \ldots, Q_2, Q_1$ is an arrangement of type-1. Then there exists a sequence of natural numbers $1 < j_1 < j_2 < \cdots < j_s < J$ such that $Q_j = id$ if and only if $j = j_k$ for some $k = 1, 2, \ldots, s$. They satisfy the relation

$$0 < j_1 - 1 < j_1 < j_2 - 1 < \cdots < j_s - 1 < j_s \le J - 1. \tag{8.4.24}$$

Conversely, if a sequence $j_1, j_2, \ldots, j_s$ satisfying this condition is given, the corresponding arrangement $Q_J, \ldots, Q_1$ of type-1 is determined uniquely.

Suppose that $Q_J, Q_{J-1}, \ldots, Q_2, Q_1$ is an arrangement of type-2. Then there exists a sequence of natural numbers $1 < j_1 < j_2 < \cdots < j_s \le J$ such that $Q_j = id$ if and only if $j = j_k$ for some $k = 1, 2, \ldots, s$. They satisfy the relation

$$0 < j_1 - 1 < j_1 < j_2 - 1 < \cdots < j_s - 1 < j_s \le J. \tag{8.4.25}$$

Conversely, if a sequence $j_1, j_2, \ldots, j_s$ satisfying this condition is given, the corresponding arrangement $Q_J, \ldots, Q_1$ of type-2 is determined uniquely.

If $Q_J, Q_{J-1}, \ldots, Q_2, Q_1$ is an arrangement of type-1 or type-2, it corresponds to the sequence $j_1, j_2, \ldots, j_s$. Thus we write $A_{j_s j_{s-1} \ldots j_1}(x_{J+1}, x_0)$ for $I_{Q_J, Q_{J-1}, \ldots, Q_2, Q_1}$. By this notation

$$\begin{aligned} &I(\{\tau_j\}, S, a, \nu)(x_{J+1}, x_0) \\ &= A_0(x_{J+1}, x_0) + A_{J+1}(x_{J+1}, x_0) + {\sum}' A_{j_s j_{s-1} \ldots j_1}(x_{J+1}, x_0) + {\sum}'' A_{j_s j_{s-1} \ldots j_1}(x_{J+1}, x_0), \end{aligned} \tag{8.4.26}$$

where $\sum'$ and $\sum''$ mean summation over all sequences $j_1, j_2, \ldots, j_s$ associated with arrangements of type-1 and type-2 respectively.

Among these terms $A_0(x_{J+1}, x_0)$ and $A_{J+1}(x_{J+1}, x_0)$ are defined as the principal part and the remainder part of the oscillatory integral:

$$\left(\frac{-i\nu}{2\pi\tau_{J+1}}\right)^{\frac{1}{2}} \left(\frac{-i\nu}{2\pi T(J,1)}\right)^{\frac{1}{2}} \times \int_{\mathbf{R}} e^{i\nu(S_{J+1}(x_{J+1},x_J)+S^*_{J,1}(x_J,x_0))}(P_{J-1}P_{J-2}\ldots P_1 a_\lambda)(x_{J+1}, x_J, x_0)\, dx_J. \tag{8.4.27}$$

Thus, we have

$$A_0(x_{J+1}, x_0) = \left(\frac{-i\nu}{2\pi T(J+1,1)}\right)^{\frac{1}{2}} e^{i\nu S^*_{J+1,1}(x_{J+1},x_0)}(P_J P_{J-1}\ldots P_2 P_1 a_\lambda)(x_{J+1}, x_0). \tag{8.4.28}$$

We have $S^*_{J+1,1}(x_{J+1}, x_0) = S(\widehat{x_{J+1}, x_0})$ and

$$(P_J P_{J-2}\ldots P_1 a_\lambda)(x_{J+1}, x_0) = \left(\prod_{k=1}^{J} D_{x^*_k}(S_{k+1} + S^*_{k,1}; x_{k+1}, x_0)^{-1/2}\right) a_\lambda(x)\Big|_{x=x^*}.$$

Apply (8.4.13) to the right-hand side of this. Then

$$A_0(x_{J+1}, x_0) = \left(\frac{-i\nu}{2\pi T(J+1,1)}\right)^{\frac{1}{2}} e^{i\nu S(\widehat{x_{J+1},x_0})} D(\{\tau_j\}; x_{J+1}, x_0)^{-1/2} a_\lambda(\widehat{x_{J+1}, x_0}).$$

This is the main term of Theorem 8.4.

We shall discuss the estimate of the remainder. The remainder term is

$$\begin{aligned}&\left(\frac{-i\nu}{2\pi T}\right)^{\frac{1}{2}} e^{i\nu S(\widehat{x_{J+1},x_0})} D(\{\tau_j\}; x_{J+1}, x_0)^{-1/2} r(\{\tau_j\}; \nu, x_{J+1}, x_0) \\ &= A_{J+1}(x_{J+1}, x_0) + \sum{}' A_{j_s j_{s-1}\ldots j_1}(x_{J+1}, x_0) + \sum{}'' A_{j_s j_{s-1}\ldots j_1}(x_{J+1}, x_0).\end{aligned} \tag{8.4.29}$$

Since $A_{J+1}(x_{J+1}, x_0)$ is the remainder part of (8.4.27), it is of the form

$$A_{J+1}(x_{J+1}, x_0) = \left(\frac{-i\nu}{2\pi T(J+1,1)}\right)^{\frac{1}{2}} e^{i\nu S^*_{J+1,1}(x_{J+1},x_0)} a_{J+1}(x_{J+1}, x_0)$$

where

$$a_{J+1}(x_{J+1}, x_0) = (R_J P_{J-1}\ldots P_2 P_1 a_\lambda)(x_{J+1}, x_0).$$

It follows from Corollary 8.11 and Assumption 8.3 that for any integer $k \geq 0$ there exists a constant $C(k) \geq 1$ and a natural number $M(k) \geq k$ such that

$$\begin{aligned}
&(1+|\lambda|+|x_{J+1}|+|x_0|)^{-m}|\partial_{x_{J+1}}^{\alpha_{J+1}}\partial_{x_0}^{\alpha_0}a_{J+1}(x_{J+1},x_0)| \\
&= (1+|\lambda|+|x_{J+1}|+|x_0|)^{-m}|\partial_{x_{J+1}}^{\alpha_{J+1}}\partial_{x_0}^{\alpha_0}(R_J P_{J-1}\dots P_2 P_1 a_\lambda)(x_{J+1},x_0)| \\
&\le C(k)\nu^{-1}\tau_{J+1}\max\sup_{x_J\in\mathbf{R}}(1+|\lambda|+|x_{J+1}|+|x_J|+|x_0|)^{-m} \\
&\quad\times|\partial_{x_{J+1}}^{\beta_{J+1}}\partial_{x_0}^{\beta_0}\partial_{x_J}^{\beta_J}(P_{J-1}\dots P_1 a_\lambda)(x_{J+1},x_J,x_0)|.
\end{aligned} \tag{8.4.30}$$

On the other hand, by (8.4.10) and Lemma 8.5, we have

$$\begin{aligned}
&(P_{J-1}\dots P_2 P_1 a_\lambda)(x_{J+1},x_J,x_0)=\prod_{j=1}^{J-1}D(S_{j+1}+S_{j,1}^*;x_{j+1},x_{j-1})^{-1/2}a_\lambda(x_{J+1},\widehat{x_J,x_0}) \\
&= D(S_{J,1};x_J,x_0)^{-1/2}a_\lambda(x_{J+1},\widehat{x_J,x_0}) \\
&= \left(1+T(J,1)^2 d(S_{J,1};x_J,x_0)\right)^{-1/2}a_\lambda(x_{J+1},\widehat{x_J,x_0}).
\end{aligned} \tag{8.4.31}$$

Since (8.4.15) and (8.4.16) hold,

$$\begin{aligned}
&|\partial_{x_{J+1}}^{\beta_{J+1}}\partial_{x_0}^{\beta_0}\partial_{x_J}^{\beta_J}(P_{J-1}\dots P_1 a_\lambda)(x_{J+1},x_J,x_0)| \\
&\le C_K\max_{\beta_J'\le\beta_J,\beta_0'\le\beta_0}|(\partial_{x_{J+1}}^{\beta_{J+1}}\partial_{x_0}^{\beta_0'}\partial_{x_J}^{\beta_J'}a_\lambda)(x_{J+1},x_J,x_0)|.
\end{aligned} \tag{8.4.32}$$

Hence by Assumption 8.3

$$\begin{aligned}
&(1+|\lambda|+|x_{J+1}|+|x_0|)^{-m}|\partial_{x_{J+1}}^{\alpha_{J+1}}\partial_{x_0}^{\alpha_0}a_{J+1}(x_{J+1},x_0)| \\
&\le C(k)\nu^{-1}\tau_{J+1}\max\sup_{x_J\in\mathbf{R}}(1+|\lambda|+|x_{J+1}|+|x_J|+|x_0|)^{-m} \\
&\quad\times|\partial_{x_{J+1}}^{\beta_{J+1}}\partial_{x_0}^{\beta_0}\partial_{x_J}^{\beta_J}a_\lambda(x_{J+1},\widehat{x_J,x_0})| \\
&\le C(k)\nu^{-1}\tau_{J+1}A_{K(k)}X_{K(k)}^3,
\end{aligned} \tag{8.4.33}$$

where max is taken over all $\beta_{J+1}\le\alpha_{J+1}$, $\beta_J\le M(k)$ and $\beta_0\le\alpha_0$ as far as $|\alpha_0|\le k$ and $|\alpha_{J+1}|\le k$.

Now we discuss other terms of (8.4.29). Let $j_1, j_2, \dots, j_s$ be the sequence associated with an arrangement $Q_J, Q_{J-1}, \dots, Q_1$ of type-1 or type-2. We set $j_0=0$ and $j_{s+1}=J+1$ as notational convention in the following discussions.

By construction $A_{j_s j_{s-1}\dots j_1}(x_{J+1},x_0)$ is an oscillatory integral over the space $\mathbf{R}^s$ of dimension $s<J$.

$$\begin{aligned}
&A_{j_s j_{s-1}\dots j_1}(x_{J+1},x_0) \\
&=\prod_{k=1}^{s+1}\left(\frac{-i\nu}{2\pi T(j_k,j_{k-1}+1)}\right)^{\frac{1}{2}}\int_{\mathbf{R}^s}e^{i\nu S_{j_s\dots j_1}^*(x_{J+1},x_{j_s},\dots,x_{j_1},x_0)} \\
&\quad\times b_{j_s\dots j_1}(x_{J+1},x_{j_s},\dots,x_{j_1},x_0)\prod_{k=1}^{s}dx_{j_k}.
\end{aligned} \tag{8.4.34}$$

Here the amplitude is

$$b_{j_s\dots j_1}(x_{J+1}, x_{j_s}, \dots, x_{j_1}, x_0) = (Q_J Q_{J-1} \dots Q_1 a_\lambda)(x_{J+1}, x_{j_s}, \dots, x_{j_1}, x_0), \tag{8.4.35}$$

and the phase function is

$$S^*_{j_s\dots j_1}(x_{J+1}, x_{j_s}, \dots, x_{j_1}, x_0) = \sum_{k=1}^{s+1} S^*_{j_k, j_{k-1}+1}(x_{j_k}, x_{j_{k-1}}). \tag{8.4.36}$$

By the assumption of Theorem 8.4 we can apply the Kumano-go–Taniguchi theorem (Theorem 7.3) to $A_{j_s j_{s-1}\dots j_1}(x_{J+1}, x_0)$. As a consequence there exists a function $a_{j_s j_{s-1}\dots j_1}(x_{J+1}, x_0)$ such that

$$A_{j_s j_{s-1}\dots j_1}(x_{J+1}, x_0) = \left(\frac{-i\nu}{2\pi T(J+1, 1)}\right)^{\frac{1}{2}} e^{i\nu S(\widehat{x_{J+1}, x_0})} a_{j_s j_{s-1}\dots j_1}(x_{J+1}, x_0). \tag{8.4.37}$$

Set

$$\begin{aligned} r(\{\tau_j\}; \nu, x_{J+1}, x_0)) &= \nu D(\{\tau_j\}; x_{J+1}, x_0)^{1/2} \\ &\times \{a_{J+1}(x_{J+1}, x_0) + \sum{}' a_{j_s j_{s-1}\dots j_1}(x_{J+1}, x_0) + \sum{}'' a_{j_s j_{s-1}\dots j_1}(x_{J+1}, x_0)\}. \end{aligned} \tag{8.4.38}$$

Then we have

$$\begin{aligned} & I(\{\tau_j\}, S, a, \nu)(x_{J+1}, x_0) \\ &= \left(\frac{\nu i}{2\pi T_{J+1}}\right)^{1/2} \exp\{i\nu S(\widehat{x_{J+1}, x_0})\} \\ &\quad \times D(\{\tau_j\}; x_{J+1}, x_0)^{-1/2}\{a_\lambda(\widehat{x_{J+1}, x_0}) + \nu^{-1} r(\{t_j\}; \nu, x_{J+1}, x_0)\}. \end{aligned} \tag{8.4.39}$$

In order to prove (8.2.5), we shall look for an estimate of $a_{j_s j_{s-1}\dots j_1}(x_{J+1}, x_0)$, which follows from bounds of derivatives of $b_{j_s\dots j_1}(x_{J+1}, x_{j_s}, \dots, x_{j_1}, x_0)$ because of the Kumano-go–Taniguchi theorem.

Lemma 8.6 *Let δ_0 be as in Theorem 8.4. Let $j_1 \dots j_s$ be a sequence associated with a type-1 arrangement. Then for any natural number K there exist a positive constant $C(K)$ and a natural number $M(K) \geq K$ such that as far as $|\alpha_{j_k}| \leq K$ for $0 \leq k \leq s+1$,*

$$\Big(1 + |\lambda| + \sum_{k=0}^{s+1} |x_{j_k}|\Big)^{-m} \Big| \Big(\prod_{k=0}^{s+1} \partial_{x_{j_k}}^{\alpha_{j_k}}\Big) b_{j_s\dots j_1}(x_{J+1}, x_{j_s}, \dots, x_{j_1}, x_0)\Big| \tag{8.4.40}$$

$$\leq C(K)^{s+2}\Big(\nu^{-1}\tau_{J+1}\prod_{k=1}^{s}\nu^{-1}\tau_{j_k}\Big)$$

$$\times \max \sup_{(x_J, x_{j_s-1},\dots,x_{j_1-1})\in\mathbf{R}^{s+1}} \Big(1+|\lambda|+|x_0|+|x_{J+1}|+|x_J|+\sum_{k=1}^{s}(|x_{j_k}|+|x_{j_k-1}|)\Big)^{-m}$$

$$\times \Big|\partial_{x_{J+1}}^{\beta_{J+1}}\partial_{x_J}^{\beta_J}\partial_{x_0}^{\beta_0}\prod_{k=1}^{s}\partial_{x_{j_k}}^{\beta_{j_k}}\partial_{x_{j_k-1}}^{\beta_{j_k-1}}a_\lambda(x_{J+1},\widehat{x_J, x_{j_s}}, x_{j_s}\widehat{-1, x_{j_{s-1}}},\dots,x_{\widehat{j_1-1}, x_0})\Big|$$

$$\leq C(K)^{s+2}\Big(\nu^{-1}\tau_{J+1}\prod_{k=1}^{s}\nu^{-1}\tau_{j_k}\Big)A_{M(K)}X_{M(K)}^{2s+1}, \tag{8.4.41}$$

where max *is taken among* β_0, β_J, β_{J+1}, β_{j_k}, β_{j_k-1}, $1\leq k\leq s+1$ *satisfying* $\beta_0\leq\alpha_0$, $\beta_{j_k}\leq\alpha_{j_k}$, $|\beta_{j_k-1}|$, $|\beta_J|\leq M(K)$. *It is sufficient to choose* $M(K)=2K+6$.

Similarly,

Lemma 8.7 *Assume that* $j_1\dots j_s$ *is associated with a type-2 arrangement. Let* δ_0 *be as in the previous lemma. Then for any natural number* K *there exist positive a constant* $C(K)$ *and a natural number* $M(K)\geq K$ *such that as far as* $|\alpha_{j_k}|\leq K$ *for* $0\leq k\leq s+1$,

$$\Big(1+|\lambda|+\sum_{k=0}^{s+1}|x_{j_k}|\Big)^{-m}\Big|\Big(\prod_{k=0}^{s+1}\partial_{x_{j_k}}^{\alpha_{j_k}}\Big)b_{j_s\dots j_1}(x_{J+1},x_{j_s},\dots,x_{j_1},x_0)\Big| \tag{8.4.42}$$

$$\leq C(K)^{s+2}\Big(\prod_{k=1}^{s}\nu^{-1}\tau_{j_k}\Big) \tag{8.4.43}$$

$$\times \max \sup_{(x_J, x_{j_s-1},\dots,x_{j_1-1})\in\mathbf{R}^{s+1}} \Big(1+|\lambda|+|x_{J+1}|+|x_0|+\sum_{k=1}^{s}(|x_{j_k}|+|x_{j_k-1}|)\Big)^{-m}$$

$$\times \Big|\partial_{x_{J+1}}^{\beta_{J+1}}\partial_{x_0}^{\beta_0}\prod_{k=1}^{s}\partial_{x_{j_k}}^{\beta_{j_k}}\partial_{x_{j_k-1}}^{\beta_{j_k-1}}a_\lambda(x_{\widehat{J+1}, x_{j_s}}, x_{j_s}\widehat{-1, x_{j_{s-1}}},\dots,x_{\widehat{j_1-1}, x_0})\Big|$$

$$\leq C(K)^{s+2}\Big(\prod_{k=1}^{s}\nu^{-1}\tau_{j_k}\Big)A_{M(K)}X_{M(K)}^{2s+2}. \tag{8.4.44}$$

Here max *is taken among* β_0, β_{j_k}, β_{j_k-1}, $1\leq k\leq s+1$ *satisfying* $\beta_0\leq\alpha_0$, $\beta_{j_k}\leq\alpha_{j_k}$, $|\beta_{j_k-1}|\leq M(K)$. *It is sufficient to choose* $M(K)=2K+6$.

We shall prove these lemmata later. Here assuming these for a moment, we shall prove Theorem 8.4 in the following way.

Suppose that $j_1 j_2,\dots,j_s$ is associated with a type-1 arrangement. Then by the Kumano-go–Taniguchi theorem (Theorem 7.3) and Lemma 8.6, for any integer $k\geq 0$ there exist a constant $C_k>0$ and a natural number $K(k)\geq k$ such that for any indices α_0, α_{J+1} with $|\alpha_0|$, $|\alpha_{J+1}|\leq k$

$$\big|(1+|\lambda|+|x_{J+1}|+|x_0|)^{-m}\partial_{x_0}^{\alpha_0}\partial_{x_{J+1}}^{\alpha_{J+1}}a_{j_s j_{s-1}\dots j_1}(x_{J+1},x_0)\big| \tag{8.4.45}$$

$$\leq C_k^{s+1}\Big(\nu^{-1}\tau_{J+1}\prod_{k=1}^{s}\nu^{-1}\tau_{j_k}\Big)A_{M(K(k))},$$

where max is taken with respect to $\beta_{j_{s+1}}, \beta_{j_{s-1}}, \ldots, \beta_{j_0}$ among those indices with $|\beta_{j_i}| \leq K(k)$ for $i = 0, 1, 2, \ldots, s+1$. (8.4.47) holds.

Taking the summation over all sequences $j_1, \ldots, j_s$ corresponding to type-1 arrangements, we have

$$\begin{aligned}
&(1+|\lambda|+|x_{J+1}|+|x_0|)^{-m}{\sum}'|\partial_{x_0}^{\alpha_0}\partial_{x_{J+1}}^{\alpha_{J+1}}a_{j_s j_{s-1}\ldots j_1}(x_{J+1},x_0)| \qquad (8.4.46)\\
&\leq A_{M(K(k))}{\sum}' C_k^{s+1}\nu^{-1}\tau_{J+1}\prod_{k=1}^{s}(\nu^{-1}\tau_{j_k})\\
&\leq A_{M(K(k))}C_k\nu^{-1}\tau_{J+1}\Big(\prod_{j=1}^{J}(1+C_k\nu^{-1}\tau_j)-1\Big).
\end{aligned}$$

Similarly, suppose $j_1 j_2, \ldots, j_s$ is associated with a type-2 arrangement. Then by the Kumano-go–Taniguchi theorem (Theorem 7.3) and Lemma 8.7, for any k there exist a constant $C_k > 0$ and a natural number $K(k) \geq k$ such that for any indices α_0, α_{J+1} with $|\alpha_0|, |\alpha_{J+1}| \leq k$

$$\begin{aligned}
&\Big|(1+|\lambda|+|x_{J+1}|+|x_0|)^{-m}\partial_{x_0}^{\alpha_0}\partial_{x_{J+1}}^{\alpha_{J+1}}a_{j_s j_{s-1}\ldots j_1}(x_{J+1},x_0)\Big| \qquad (8.4.47)\\
&\leq C_k^{s+1}\Big(\prod_{k=1}^{s}\nu^{-1}\tau_{j_k}\Big)A_{M(K(k))}.
\end{aligned}$$

Therefore, taking the summation,

$$\begin{aligned}
&(1+|\lambda|+|x_{J+1}|+|x_0|)^{-m}{\sum}''|\partial_{x_0}^{\alpha_0}\partial_{x_{J+1}}^{\alpha_{J+1}}a_{j_s j_{s-1}\cdot j_1}(x_{J+1},x_0)| \qquad (8.4.48)\\
&\leq {\sum}'' A_{M(K(k))}C_k^{s+1}\prod_{n=1}^{s}(\nu^{-1}\tau_{j_n})\\
&\leq A_{M(K(k))}C_k\Big(\prod_{j=1}^{J}(1+C_k\nu^{-1}\tau_j)-1\Big).
\end{aligned}$$

Consequently, it follows from (8.4.29), (8.4.30), (8.4.46) and (8.4.48) that

$$\begin{aligned}
&(1+|\lambda|+|x_{J+1}|+|x_0|)^{-m}\nu^{-1}\Big|\partial_{x_{J+1}}^{\alpha_{J+1}}\partial_{x_0}^{\alpha_0}\Big(r(\{\tau_j\};\nu,x_{J+1},x_0)D(\{\tau_j\};x_{J+1},x_0)^{-1/2}\Big)\Big|\\
&\leq \nu^{-1}\tau_{J+1}A_{K(k)}\\
&\quad + A_{M(K(k))}C_k\nu^{-1}\tau_{J+1}\Big(\prod_{j=1}^{J}(1+C_k\nu^{-1}\tau_j)-1\Big)\\
&\quad + A_{M(K(k))}C_k\Big(\prod_{j=1}^{J}(1+C_k\nu^{-1}\tau_j)-1\Big)
\end{aligned}$$

$$\leq A_{M(K(k))}C_k\Big(\prod_{j=1}^{J+1}(1+C_k\nu^{-1}\tau_j)-1\Big).$$

This proves estimate (8.2.5) because of (8.4.3). Theorem 8.4 has been proved up to the proof of Lemmas 8.6 and 8.7.

We now prove Lemma 8.6.

Let $0=j_0<j_1-1<j_1<j_2-1<\cdots<j_s<J$ be a sequence associated with an arrangement $Q_J, Q_{J-1}, \ldots Q_1$ of type-1 and let $j_{s+1}=J+1$. For any $r\leq s$ we define a function $\tilde{b}_{j_r j_{r-1}\cdots j_1}$ by the equation

$$\begin{aligned}&\tilde{b}_{j_r j_{r-1}\cdots j_1}(x_{J+1}, x_J, \ldots, x_{j_r+1}, x_{j_r}, x_{j_{r-1}}, \ldots, x_{j_1}, x_0)\\&=(Q_{j_r}Q_{j_r-1}\cdots Q_2Q_1a_\lambda)(x_{J+1}, x_J, \ldots, x_{j_r+1}, x_{j_r}, x_{j_{r-1}}, \ldots, x_{j_1}, x_0).\end{aligned}\tag{8.4.49}$$

It follows from (8.4.35) and (8.4.49) that $b_{j_s\ldots j_1}=R_JP_{J-1}\ldots P_{j_s+1}\tilde{b}_{j_sj_{s-1}\ldots j_1}$, in other words, $b_{j_s\ldots j_1}(x_{J+1}, x_{j_s}, \ldots, x_{j_1}, x_0)$ is the remainder term of the oscillatory integral

$$\begin{aligned}&\Big(\frac{-i\nu}{2\pi\tau_J}\Big)^{\frac{1}{2}}\int_{\mathbf{R}}e^{i\nu(S_{J+1}(x_{J+1},x_J)+S^*_{J,j_s+1}(x_J,x_{j_s}))}\\&\times D_{(x^*_{J-1},\ldots,x^*_{j_s+1})}(S_{J,j_s+1};x_J,x_{j_s})^{-1/2}\tilde{b}_{j_sj_{s-1}\ldots j_1}(x_{J+1},\widehat{x_J,x_{j_s}},x_{j_{s-1}},\ldots,x_{j_1},x_0)dx_J,\end{aligned}$$

because

$$\begin{aligned}&P_{J-1}\ldots P_{j_s+1}\tilde{b}_{j_sj_{s-1}\ldots j_1}(x_{J+1},x_J,x_{j_s},x_{j_{s-1}},\ldots,x_{j_1},x_0)\\&=D_{(x^*_{J-1},\ldots,x^*_{j_s+1})}(S_{J,j_s+1};x_J,x_{j_s})^{-1/2}\tilde{b}_{j_sj_{s-1}\ldots j_1}(x_{J+1},\widehat{x_J,x_{j_s}},x_{j_{s-1}},\ldots,x_{j_1},x_0).\end{aligned}\tag{8.4.50}$$

Variables $x_{j_{s-1}}, \ldots, x_0$ appear only in the amplitude of the above oscillatory integral. We regard $\lambda'=(\lambda, x_{j_{s-1}}, \ldots, x_0)$ as parameters. Differentiate it by x_{j_n}, $n=0, \ldots, s-1$, and set

$$\lambda'=|\lambda|+|x_0|+\sum_{n=1}^{s-1}|x_{j_n}|.$$

Then it follows from this, Lemma 8.5 and the stationary phase method (Corollary 8.11) that for any $K\geq 0$ there exists a natural number $M(K)\geq K$ and a positive constant $C_{K,0}$ such that as far as $|\alpha_{j_n}|\leq K$ for $n=1,2,\ldots,s$

$$\begin{aligned}&(1+|\lambda'|+|x_{J+1}|+|x_{j_s}|)^{-m}\sup|\partial^{\alpha_{J+1}}_{x_{J+1}}\partial^{\alpha_0}_{x_0}\prod_{n=1}^{s}\partial^{\alpha_{jn}}_{x_{jn}}b_{j_s\ldots j_1}(x_{J+1},x_{j_s},\ldots,x_{j_1},x_0)|\\&\leq C_{K,0}\nu^{-1}\frac{\tau_{J+1}T(J,j_s+1)}{T(J+1,j_s+1)}\max\sup_{x_J\in\mathbf{R}}(1+|\lambda'|+|x_{J+1}|+|x_J|+|x_{j_s}|)^{-m}\end{aligned}\tag{8.4.51}$$

$$\times \left| \partial_{x_{J+1}}^{\beta_{J+1}} \partial_{x_J}^{\beta_J} \partial_{x_{j_s}}^{\beta_{j_s}} \partial_{x_0}^{\alpha_0} \prod_{n=1}^{s-1} \partial_{x_{j_n}}^{\alpha_{j_n}} \tilde{b}_{j_s \dots j_1}(x_{J+1}, \widehat{x_J, x_{j_s}}, x_{j_{s-1}}, \dots, x_{j_1}, x_0) \right|.$$

Here max is taken over all β_{J+1}, β_J, β_{j_s} with $\beta_{J+1} \le \alpha_{J+1}$, $\beta_{j_s} \le \alpha_{j_s}$ and $|\beta_J| \le M(K)$.

We can find the bound of the right-hand side of this by virtue of the following lemma.

Lemma 8.8 *Let $a_\lambda(x_{J+1}, \dots, x_0)$ be a function that satisfies Assumption 8.3. Assume $T \le \delta_0$. Let $j_1, \dots, j_s$ be a sequence associated with the arrangement $Q_J, Q_{J-1}, \dots, Q_1$ of type-1. We set $j_0 = 0$ and $j_{s+1} = J+1$. For any $r \le s$ define $\tilde{b}_{j_r j_{r-1} \cdots j_1}$ as above in (8.4.49). For any $\alpha_{j_n}, \alpha_{j_n - 1}$, $r+1 \le n \le s+1$, let*

$$L_r = \prod_{n=r+1}^{s+1} \left(\partial_{x_{j_n}}^{\alpha_{j_n}} \partial_{x_{j_n-1}}^{\alpha_{j_n-1}} \right).$$

For any non-negative integer K there exist a positive constant $C(K)$ and a natural number $M(K) \ge K$ such that

$$\begin{aligned}
&\Big(1 + |\lambda| + \sum_{n=0}^{r} |x_{j_n}| + \sum_{n=r+1}^{s+1} (|x_{j_n}| + |x_{j_n-1}|)\Big)^{-m} \\
&\times \left| L_r \Big(\prod_{n=0}^{r} \partial_{x_{j_n}}^{\alpha_{j_n}} \Big) \tilde{b}_{j_r j_{r-1} \cdots j_1}(x_{J+1}, \widehat{x_{j_{s+1}-1}, x_{j_s}}, \dots, \widehat{x_{j_{r+1}-1}, x_{j_r}}, x_{j_{r-1}}, \dots, x_{j_1}, x_0) \right| \\
&\le C(K)^r \prod_{n=1}^{r} \left(\frac{\tau_{j_n} T(j_n - 1, j_{n-1}+1)}{\nu T(j_n, j_{n-1}+1)} \right) \\
&\times \max \sup_{(x_{j_r-1}, \dots, x_{j_1-1}) \in \mathbf{R}^r} \Big(1 + |\lambda| + |x_0| + \sum_{n=1}^{s+1} (|x_{j_n}| + |x_{j_n-1}|)\Big)^{-m} \\
&\times \left| L_r \partial_{x_0}^{\beta_0} \prod_{n=1}^{r} \partial_{x_{j_n}}^{\beta_{j_n}} \partial_{x_{j_n-1}}^{\beta_{j_n-1}} a_\lambda(x_{J+1}, \widehat{x_{j_{s+1}-1}, x_{j_s}}, \dots, \widehat{x_{j_{r+1}-1}, x_{j_r}}, , \dots, \widehat{x_{j_1-1}, x_0}) \right|,
\end{aligned} \tag{8.4.52}$$

if $|\alpha_0| \le K$ and $|\alpha_{j_n}| \le K$ for $n = 1, \dots, r$. Here max *is taken over β_0 with $\beta_0 \le \alpha_0$ and β_{j_n}, β_{j_n-1} with $\beta_{j_n} \le \alpha_{j_n}$, $|\beta_{j_n-1}| \le M(K)$, for $n = 1, \dots, r$.*

Assuming this lemma is true for the moment, we continue the proof of Lemma 8.6. It follows from Lemma 8.8 with $r = s$ and $L_s = \partial_{x_{J+1}}^{\beta_{J+1}} \partial_{x_J}^{\beta_J}$ that

$$\begin{aligned}
&\max \sup_{x_J \in \mathbf{R}} (1 + |\lambda'| + |x_{J+1}| + |x_J| + |x_{j_s}|)^{-m} \\
&\times \left| L_s \partial_{x_{j_s}}^{\beta_{j_s}} \prod_{n=0}^{s-1} \partial_{x_{j_n}}^{\alpha_{j_n}} \tilde{b}_{j_s \dots j_1}(x_{J+1}, \widehat{x_J, x_{j_s}}, x_{j_{s-1}}, \dots, x_{j_1}, x_0) \right|
\end{aligned} \tag{8.4.53}$$

$$\le C(K)^s \prod_{n=1}^{s} \left(\frac{\tau_{j_n} T(j_n - 1, j_{n-1} + 1)}{\nu T(j_n, j_{n-1} + 1)} \right)$$

$$\times \max \sup_{(x_{j_s-1},\dots,x_{j_1-1})\in \mathbf{R}^s} \left(1 + |\lambda| + |x_0| + |x_{J+1}| + |x_J| + \sum_{n=1}^{s}(|x_{j_n}| + |x_{j_n-1}|)\right)^{-m}$$

$$\times \left| \partial_{x_0}^{\beta_0} L_s \prod_{n=1}^{s} \partial_{x_{j_n}}^{\beta'_{j_n}} \partial_{x_{j_n-1}}^{\beta'_{j_n-1}} a_\lambda(x_{J+1}, \widehat{x_J, x_{j_s}}, x_{j_s}\widehat{-1, x}_{j_{s-1}}, \dots, x_{\widehat{j_1-1}, x_0}) \right|,$$

where max is taken among β_0 with $\beta_0 \le \alpha_0$, $\beta'_{j_n}, \beta'_{j_n-1}$ with $\beta'_{j_n} \le \alpha_{j_n}$, $|\beta'_{j_n-1}| \le M(K)$ for $n = 1, 2, \dots, s$. It follows from this and (8.4.51) that

$$(1 + |\lambda'| + |x_{J+1}| + |x_{j_s}|)^{-m} \left| \partial_{x_{J+1}}^{\alpha_{J+1}} \partial_{x_0}^{\alpha_0} \prod_{n=1}^{s} \partial_{x_{j_n}}^{\alpha_{j_n}} b_{j_s \dots j_1}(x_{J+1}, x_{j_s}, \dots, x_{j_1}, x_0) \right|$$

$$\le C_K C(K)^s \prod_{n=1}^{s+1} \left(\frac{\tau_{j_n} T(j_n - 1, j_{n-1} + 1)}{\nu T(j_n, j_{n-1} + 1)} \right) \tag{8.4.54}$$

$$\times \max \sup_{(x_J, x_{j_s-1},\dots,x_{j_1-1})\in \mathbf{R}^{s+1}} \left(1 + |\lambda| + |x_0| + |x_{J+1}| + |x_J| + \sum_{n=1}^{s}(|x_{j_n}| + |x_{j_n-1}|)\right)^{-m}$$

$$\times \left| \partial_{x_0}^{\beta_0} \partial_{x_{J+1}}^{\beta_{J+1}} \partial_{x_J}^{\beta_J} \prod_{n=1}^{s} \partial_{x_{j_n}}^{\beta'_{j_n}} \partial_{x_{j_n-1}}^{\beta'_{j_n-1}} a_\lambda(x_{J+1}, \widehat{x_J, x_{j_s}}, x_{j_s}\widehat{-1, x}_{j_{s-1}}, \dots, x_{\widehat{j_1-1}, x_0}) \right|,$$

where max is taken among $\beta_{J+1}, \beta_J, \beta_0, \beta'_{j_n}, \beta'_{j_n-1}, n = 1, 2, \dots s$ with $\beta_{J+1} \le \alpha_{J+1}$, $|\beta_J| \le M(K)$, $\beta_0 \le \alpha_0$, $\beta'_{j_n} \le \alpha_{j_n}$, $|\beta'_{j_n-1}| \le M(K)$, for $n = 1, \dots, s$. Inequality (8.4.54) is nothing but (8.4.40) if we choose $C(K) \ge C_K$. Lemma 8.6 has been proved up to the proof of Lemma 8.8.

We now prove Lemma 8.8.

Proof We prove the lemma by repeated use of the stationary phase method. Let r be any integer with $1 \le r \le s$. Let $L_r = \prod_{n=r+1}^{s+1} \left(\partial_{x_{j_n}}^{\alpha_{j_n}} \partial_{x_{j_n-1}}^{\alpha_{j_n-1}} \right)$. For any function $f(x_{J+1}, x_J \dots, x_{j_r+1}, x_{j_r}, x_{j_{r-1}}, \dots, x_{j_1}, x_0)$ of variables $\{x_k;\ j_r \le k \le J+1\}$ and $\{x_{j_k} : 0 \le k \le r-1\}$ we define

$$\begin{aligned} &\Theta_r f(x_{J+1}, x_{j_{s+1}-1}, x_{j_s}, \dots, x_{j_{r+1}}, x_{j_{r+1}-1}, x_{j_r}, x_{j_{r-1}}, \dots, x_{j_1}, x_0) \\ &\quad = f(x_{J+1}, x_{\widehat{j_{s+1}-1}, x_{j_s}}, \dots, x_{\widehat{j_{r+1}-1}, x}_{j_r}, x_{j_{r-1}}, \dots, x_{j_1}, x_0) \\ &\quad = f(x_{J+1}, x_J, x_{J-1}, \dots, x_{j_r+1}, x_{j_r}, x_{j_{r-1}}, \dots, x_{j_1}, x_0)\Big|_{\substack{x_k = x_k^*(x_{j_{n+1}-1}, x_{j_n}) \\ \text{for } j_n < k < j_{n+1}-1,\, r \le n \le s}} . \end{aligned}$$

We prove Lemma 8.8 by induction on r. First we discuss the case of $r = 1$. By definition

$$\tilde{b}_{j_1}(x_{J+1}, \dots, x_{j_1+1}, x_{j_1}, x_0) = \begin{cases} R_{j_1-1} P_{j_1-2}, \cdots P_1 a(x_{J+1}, \dots, x_{j_1}, x_0) & \text{if } j_1 \ge 3, \\ R_1 a(x_{J+1}, \dots, x_2, x_0) & \text{if } j_1 = 2. \end{cases}$$

In other words, if we set

$$f(x_{J+1},\ldots,x_{j_1},\,x_{j_1-1},\,x_0)=\begin{cases}P_{j_1-2}\cdots P_1a_\lambda(x_{J+1},\ldots,x_{j_1+1},\,x_{j_1},\,x_{j_1-1},\,x_0), & j_1\ge 3,\\ a(x_{J+1},\ldots,x_2,\,x_1,\,x_0), & j_1=2,\end{cases}$$

then $\tilde{b}_{j_1}(x_{J+1},\ldots,x_{j_1+1},\,x_{j_1},\,x_0)$ is given by the equality

$$\begin{aligned}&\left(\frac{-i\nu}{2\pi\tau_{j_1}}\right)^{\frac12}\left(\frac{-i\nu}{2\pi T(j_1-1,\,1)}\right)^{\frac12} \qquad (8.4.55)\\ &\quad\times\int_{\mathbf{R}} e^{i\nu(S_{j_1}(x_{j_1},\,x_{j_1-1})+S^*_{j_1-1,\,1}(x_{j_1-1},\,x_0))}f(x_{J+1},\ldots,x_{j_1+1},\,x_{j_1},\,x_{j_1-1},\,x_0)dx_{j_1-1}\\ &=\left(\frac{-i\nu}{2\pi T(j_1,\,1)}\right)^{\frac12}e^{i\nu S^*_{j_1,\,1}(x_{j_1},\,x_0)}\\ &\quad\times[(P_{j_1-1}f)(x_{J+1},\ldots,x_{j_1+1},\,x_{j_1},\,x_0)+\tilde{b}_j(x_{J+1},\ldots,x_{j_1+1},\,x_{j_1},\,x_0)].\end{aligned}$$

Here $(P_{j_1-1}f)(x_{J+1},\ldots,x_{j_1},\,x_0)$ is the principal part of the stationary phase method applied to the left-hand side. In (8.4.55) we understand that $S^*_{1,\,1}(x_1,\,x_0)=S_1(x_1,\,x_0)$ if $j_1=2$.

Notice that $L_1=\prod_{n=2}^{s+1}\partial^{\alpha_{j_n}}_{x_{j_n}}\partial^{\alpha_{j_n-1}}_{x_{j_n-1}}$. We operate $L_1\Theta_1$ on both sides of (8.4.55). Then

$$\begin{aligned}&\left(\frac{-i\nu}{2\pi\tau_{j_1}}\right)^{\frac12}\left(\frac{-i\nu}{2\pi T(j_1-1,\,1)}\right)^{\frac12} \qquad (8.4.56)\\ &\times\int_{\mathbf{R}} e^{i\nu(S_{j_1}(x_{j_1},\,x_{j_1-1})+S^*_{j_1-1,\,1}(x_{j_1-1},\,x_0))}\\ &\qquad\times(L_1\Theta_1 f)(x_{J+1},\,x_{j_{s+1}-1},\,x_{j_s},\ldots,x_{j_2-1},\,x_{j_1},\,x_{j_1-1},\,x_0)\,dx_{j_1-1}\\ &=\left(\frac{-i\nu}{2\pi T(j_1,\,1)}\right)^{\frac12}e^{i\nu S^*_{j_1,\,1}(x_{j_1},\,x_0)}\Big[(L_1\Theta_1P_{j_1-1}f)(x_{J+1},\,x_{j_{s+1}-1},\,x_{j_s},\ldots,x_{j_2-1},\,x_{j_1},\,x_0)\\ &\qquad+(L_1\Theta_1\tilde{b}_{j_1})(x_{J+1},\,x_{j_{s+1}-1},\,x_{j_s},\ldots,x_{j_2-1},\,x_{j_1},\,x_0)\Big],\end{aligned}$$

because the integral in (8.4.55) is the integral with respect to x_{j_1-1} and the other variables $x_J,\,x_{J-1},\ldots,x_{j_1+1}$ play a role of parameters.

We apply the stationary phase method to the left-hand side of (8.4.56). Then

$$\begin{aligned}&\left(\frac{-i\nu}{2\pi\tau_{j_1}}\right)^{\frac12}\left(\frac{-i\nu}{2\pi T(j_1-1,\,1)}\right)^{\frac12} \qquad (8.4.57)\\ &\times\int_{\mathbf{R}} e^{i\nu(S_{j_1}(x_{j_1},\,x_{j_1-1})+S^*_{j_1-1,\,1}(x_{j_1-1},\,x_0))}\\ &\qquad\times(L_1\Theta_1 f)(x_{J+1},\,x_{j_{s+1}-1},\,x_{j_s},\ldots,x_{j_2-1},\,x_{j_1},\,x_{j_1-1},\,x_0)\,dx_{j_1-1}\\ &=\left(\frac{-i\nu}{2\pi T(j_1,\,1)}\right)^{\frac12}e^{i\nu S^*_{j_1,\,1}(x_{j_1},\,x_0)}\Big[(P_{j_1-1}L_1\Theta_1f)(x_{J+1},\,x_{j_{s+1}-1},\,x_{j_s},\ldots,x_{j_2-1},\,x_{j_1},\,x_0)\end{aligned}$$

$$+ (R_{j_1-1}L_1\Theta_1 f)(x_{J+1},\ x_{j_{s+1}-1},\ x_{j_s},\ldots,x_{j_2-1},\ x_{j_1},\ x_0)\Big].$$

Since the stationary point $x_{j_1-1} = x^*_{j_1-1}(x_{j_1},\ x_0)$ of the phase function of (8.4.56) is independent of variables $x_{j_n},\ x_{j_n-1},\ n \geq 2$, it is clear that

$$P_{j_1-1}L_1\Theta_1 f = L_1 P_{j_1-1}\Theta_1 f = L_1\Theta_1 P_{j_1-1} f.$$

Having this equality in mind and comparing (8.4.56) with (8.4.57), we have

$$L_1\Theta_1\tilde{b}_{j_1} = R_{j_1-1}L_1\Theta_1 f. \tag{8.4.58}$$

Setting $\lambda' = (\lambda,\ x_{J+1},\ x_{j_{s+1}-1},\ldots,x_{j_2},\ x_{j_2-1})$, we apply Corollary 8.11. For any $K \geq 0$ there exist a natural number $M(K) \geq K$ and a constant $C_K \geq 1$ such that as far as $|\alpha_0| \leq K$ and $|\alpha_{j_1}| \leq K$, we have

$$\begin{aligned}
&(1+|\lambda'|+|x_{j_1}|+|x_0|)^{-m}\left|\partial_{x_{j_1}}^{\alpha_{j_1}}\partial_{x_0}^{\alpha_0}L_1\Theta_1\tilde{b}_{j_1}(x_{J+1},\ x_{j_{s+1}-1},\ x_{j_s},\ldots,x_{j_2-1},\ x_{j_1},\ x_0)\right|\\
&\leq C_K\frac{\tau_{j_1}T(j_1-1,\ 1)}{T(j_1,\ 1)\nu}\max\sup_{x_{j_1-1}\in\mathbf{R}}(1+|\lambda'|+|x_{j_1}|+|x_{j_1-1}|+|x_0|)^{-m}\\
&\times\left|\partial_{x_{j_1}}^{\beta_{j_1}}\partial_{x_{j_1-1}}^{\beta_{j_1-1}}\partial_{x_0}^{\beta_0}L_1\Theta_1 f(x_{J+1},\ x_{j_{s+1}-1},\ x_{j_s},\ldots,x_{j_2-1},\ x_{j_1},\ x_{j_1-1},\ x_0)\right|.
\end{aligned} \tag{8.4.59}$$

Here max is taken among $\beta_{j_1}\mathrm{C}\beta_{j_1-1}\mathrm{C}\beta_0$ satisfying $\beta_{j_1} \leq \alpha_{j_1}\mathrm{C}|\beta_{j_1-1}| \leq M(K)$ and $\beta_0 \leq \alpha_0$.

On the other hand, by definition of f,

$$\begin{aligned}
&\Theta_1 f(x_{J+1},\ x_{j_{s+1}-1},\ x_{j_s},\ldots,x_{j_2-1},\ x_{j_1},\ x_{j_1-1},\ x_0)\\
&=\begin{cases} D_{x^*_{j_1-2}\cdots x^*_1}(S_{j_1-1,\ 1};\ x_{j_1-1},\ x_0)^{-1/2} & \\ \quad\times a_\lambda(x_{J+1},\widehat{x_{j_{s+1}-1},\ x_{j_s}},\ldots,\widehat{x_{j_2-1},\ x_{j_1}},\widehat{x_{j_1-1},\ x_0}), & \text{if } j_1 \geq 3,\\ a(x_{J+1},\ \widehat{x_{j_{s+1}-1},\ x_{j_s}},\ldots,\widehat{x_{j_2-1},\ x_2},\ x_1,\ x_0), & \text{if } j_1 = 2.\end{cases}
\end{aligned}$$

By Leibniz's rule,

$$\begin{aligned}
&\left|\partial_{x_{j_1}}^{\beta_{j_1}}\partial_{x_{j_1-1}}^{\beta_{j_1-1}}\partial_{x_0}^{\beta_0}L_1\Theta_1 f(x_{J+1},\ x_{j_{s+1}-1},\ x_{j_s},\ldots,x_{j_2-1},\ x_{j_1},\ x_{j_1-1},\ x_0)\right|\\
&\leq C_{K,\ 1}\left|\partial_{x_{j_1}}^{\beta_{j_1}}\partial_{x_{j_1-1}}^{\beta'_{j_1-1}}\partial_{x_0}^{\beta'_0}L_1 a_\lambda(x_{J+1},\widehat{x_{j_{s+1}-1},\ x_{j_s}},\ldots,\widehat{x_{j_2-1},\ x_{j_1}},\ \widehat{x_{j_1-1},\ x_0})\right|,
\end{aligned}$$

where

$$C_{K,\ 1} = 2^{M(K)+K}\max_{\substack{|\alpha|\leq M(K),\\ |\beta|\leq K}}\ \sup_{(x_{j_1-1},x_0)\in\mathbf{R}^2}|\partial_{x_{j_1-1}}^{\alpha}\partial_{x_0}^{\beta}D_{x^*_{j_1-2}\cdots x^*_1}(S_{j_1-1,\ 1};\ x_{j_1-1},\ x_0)^{-1/2}|.$$

It follows from this and (8.4.59) that

$$(1+|\lambda'|+|x_{j_1}|+|x_0|)^{-m}\left|\partial_{x_{j_1}}^{\alpha_{j_1}}\partial_{x_0}^{\alpha_0}L_1\tilde{b}_{j_1}(x_{J+1},x_{\widehat{j_{s+1}-1},\ x_{j_s}},\ldots,x_{\widehat{j_2-1},\ x_{j_1}},x_0)\right|$$
$$\le C_K C_{K,1}\left(\frac{\tau_{j_1}T(j_1-1,\,1)}{T(j_1,\,1)\nu}\right)\max\sup_{x_{j_1-1}\in\mathbf{R}}(1+|\lambda'|+|x_{j_1}|+|x_{j_1-1}|+|x_0|)^{-m}$$
$$\times\left|\partial_{x_{j_1}}^{\beta'_{j_1}}\partial_{x_{j_1-1}}^{\beta'_{j_1-1}}\partial_{x_0}^{\beta'_0}L_1 a_\lambda(x_{J+1},\ x_{\widehat{j_{s+1}-1},\ x_{j_s}},\ldots,x_{\widehat{j_2-1},\ x_{j_1}},x_{\widehat{j_1-1},\ x_0})\right|. \tag{8.4.60}$$

Here max is taken among $\beta'_0,\beta'_{j_1},\beta'_{j_1-1}$ with $\beta'_0\le\alpha_0,\beta'_{j_1}\le\alpha_{j_1}\mathrm{C}|\beta'_{j_1-1}|\le M(K)$.
Set

$$C(K)\ge C_K C_{K,1},\quad M(K)=\max\{M(K),\ K\}.$$

Then (8.4.52) in the case $r=1$ has been proved.

In the case $s=1$ nothing further is required to prove Lemma 8.8.

In the case $s\ge 2$, we prove Lemma 8.8 by induction on r. Supposing that (8.4.52) for r is true and that $r+1\le s$, we shall prove (8.4.52) for $r+1$.

Suppose $K\ge 0$ be an arbitrary integer. Let $\alpha_{j_n},\ \alpha_{j_n-1},n=r+2,\ldots,\ s+1$ be arbitrary and

$$L_{r+1}=\prod_{n=r+2}^{s+1}\left(\partial_{x_{jn}}^{\alpha_{jn}}\partial_{x_{jn-1}}^{\alpha_{jn-1}}\right).$$

Let $\alpha_{j_n},n=0,\ldots,r+1$, be any indices with $|\alpha_{j_n}|\le K$. We set

$$\mathscr{L}=\prod_{n=r+2}^{s+1}\partial_{x_{jn}}^{\alpha_{jn}}\partial_{x_{jn-1}}^{\alpha_{jn-1}}\prod_{k=0}^{r-1}\partial_{x_{jk}}^{\alpha_{jk}}=L_{r+1}\prod_{k=0}^{r-1}\partial_{x_{jk}}^{\alpha_{jk}}. \tag{8.4.61}$$

We define

$$\begin{aligned}&f(x_{J+1},\ldots,x_{j_{r+1}+1},\ x_{j_{r+1}},\ x_{j_{r+1}-1},\ x_{j_r},\ x_{j_{r-1}},\ldots,\ x_{j_1},\ x_0) \qquad (8.4.62)\\ &=(P_{j_{r+1}-2}\cdots P_{j_r+1}\tilde{b}_{j_r\cdots j_1})(x_{J+1},\ x_J,\ldots,\ x_{j_{r+1}+1},\ x_{j_{r+1}},\ x_{j_{r+1}-1},\ x_{j_r},\ x_{j_{r-1}}\cdots,\ x_{j_1},\ x_0)\\ &=D_{x^*_{j_{r+1}-2}\cdots x^*_{j_r+1}}(S_{j_{r+1}-1,\ j_r+1};x_{j_{r+1}-1},\ x_{j_r})^{-1/2}\\ &\quad\times\tilde{b}_{j_r\cdots j_1}(x_{J+1},\ x_J,\ldots,\ x_{j_{r+1}+1},\ x_{j_{r+1}},\ x_{\widehat{j_{r+1}-1},\ x_{j_r}},\ x_{j_{r-1}},\ldots,\ x_{j_1},\ x_0).\end{aligned}$$

Then

$$\begin{aligned}&\tilde{b}_{j_{r+1}j_r\cdots j_1}(x_{J+1},\ldots,\ x_{j_{r+1}+1},\ x_{j_{r+1}},\ x_{j_r},\ x_{j_{r-1}},\ldots,\ x_{j_1},\ x_0)\\ &=R_{j_{r+1}-1}f(x_{J+1},\ldots,\ x_{j_{r+1}+1},\ x_{j_{r+1}},\ x_{j_r},\ x_{j_{r-1}},\ldots,\ x_{j_1},\ x_0).\end{aligned}$$

Thus it is defined by the equality

$$\left(\frac{-i\nu}{2\pi\tau_{j_{r+1}}}\right)^{\frac{1}{2}}\left(\frac{-i\nu}{2\pi T(j_{r+1}-1,\ j_r+1)}\right)^{\frac{1}{2}} \tag{8.4.63}$$
$$\times\int_{\mathbf{R}} e^{i\nu(S_{j_{r+1}}(x_{j_{r+1}},\ x_{j_{r+1}-1})+S^*_{j_{r+1}-1,\ j_r+1}(x_{j_{r+1}-1},\ x_{j_r}))}$$
$$\times f(x_{J+1},\ x_J,\ldots,\ x_{j_{r+1}+1},\ x_{j_{r+1}},\ x_{j_{r+1}-1},\ x_{j_r},\ x_{j_{r-1}},\ldots,\ x_{j_1},\ x_0)dx_{j_{r+1}-1}$$
$$=\left(\frac{-i\nu}{2\pi T(j_{r+1},\ j_r+1)}\right)^{\frac{1}{2}} e^{i\nu S^*_{j_{r+1},\ j_r+1}(x_{j_{r+1}},\ x_{j_r})}$$
$$\times\Big[(P_{j_{r+1}-1}f)(x_{J+1},\ x_J,\ldots,\ x_{j_{r+1}+1},\ x_{j_{r+1}},\ x_{j_r},\ x_{j_{r-1}},\ldots,\ x_{j_1},\ x_0)$$
$$+\tilde{b}_{j_{r+1}\cdots j_1}(x_{J+1},\ x_J,\ldots,\ x_{j_{r+1}+1},\ x_{j_{r+1}},\ x_{j_r},\ x_{j_{r-1}},\ \ldots,\ x_{j_1},\ x_0)\Big],$$

where $(P_{j_{r+1}-1}f)(x_{J+1},\ldots,\ x_{j_{r+1}},\ x_{j_r},\ x_{j_{r-1}},\ldots,\ x_{j_1},\ x_0)$ is the principal part of the oscillatory integral.

Operate $\mathscr{L}\Theta_{r+1}$ on both sides of (8.4.63). Then

$$\left(\frac{-i\nu}{2\pi\tau_{j_{r+1}}}\right)^{\frac{1}{2}}\left(\frac{-i\nu}{2\pi T(j_{r+1}-1,\ j_r+1)}\right)^{\frac{1}{2}} \tag{8.4.64}$$
$$\times\int_{\mathbf{R}} e^{i\nu(S_{j_{r+1}}(x_{j_{r+1}},\ x_{j_{r+1}-1})+S^*_{j_{r+1}-1,\ j_r+1}(x_{j_{r+1}-1},\ x_{j_r}))}$$
$$\times\mathscr{L}\Theta_{r+1}f(x_{J+1},\ x_J,\ldots,\ x_{j_{r+2}-1},\ x_{j_{r+1}},\ x_{j_{r+1}-1},\ x_{j_r},\ x_{j_{r-1}},\ldots,\ x_{j_1},\ x_0)dx_{j_{r+1}-1}$$
$$=\left(\frac{-i\nu}{2\pi T(j_{r+1},\ j_r+1)}\right)^{\frac{1}{2}} e^{i\nu S^*_{j_{r+1},\ j_r+1}(x_{j_{r+1}},\ x_{j_r})}$$
$$\times\Big[\mathscr{L}\Theta_{r+1}(P_{j_{r+1}-1}f)(x_{J+1},\ x_J,\ldots,x_{j_{r+2}-1},\ x_{j_{r+1}},\ x_{j_r},\ x_{j_{r-1}},\ldots,x_{j_1},\ x_0)$$
$$+\mathscr{L}\Theta_{r+1}\tilde{b}_{j_{r+1}\cdots j_1}(x_{J+1},\ x_J,\ldots,\ x_{j_{r+2}-1},\ x_{j_{r+1}},\ x_{j_r},\ x_{j_{r-1}},\ldots,\ x_{j_1},\ x_0)\Big].$$

Apply the stationary phase method to the left-hand side of (8.4.64). Then we have

$$\left(\frac{-i\nu}{2\pi\tau_{j_{r+1}}}\right)^{\frac{1}{2}}\left(\frac{-i\nu}{2\pi T(j_{r+1}-1,\ j_r+1)}\right)^{\frac{1}{2}} \tag{8.4.65}$$
$$\times\int_{\mathbf{R}} e^{i\nu(S_{j_{r+1}}(x_{j_{r+1}},\ x_{j_{r+1}-1})+S^*_{j_{r+1}-1,\ j_r+1}(x_{j_{r+1}-1},\ x_{j_r}))}$$
$$\times\mathscr{L}\Theta_{r+1}f(x_{J+1},\ x_J,\ldots,\ x_{j_{r+2}-1},\ x_{j_{r+1}},\ x_{j_{r+1}-1},\ x_{j_r},\ x_{j_{r-1}},\ldots,\ x_{j_1},\ x_0)dx_{j_{r+1}-1}$$
$$=\left(\frac{-i\nu}{2\pi T(j_{r+1},\ j_r+1)}\right)^{\frac{1}{2}} e^{i\nu S^*_{j_{r+1},\ j_r+1}(x_{j_{r+1}},\ x_{j_r})}$$
$$\times\Big[(P_{j_{r+1}-1}\mathscr{L}\Theta_{r+1}f)(x_{J+1},\ x_J,\ldots,\ x_{j_{r+2}-1},\ x_{j_{r+1}},\ x_{j_r},\ x_{j_{r-1}},\ldots,\ x_{j_1},\ x_0)$$
$$+R_{j_{r+1}-1}\mathscr{L}\Theta_{r+1}f(x_{J+1},\ x_J,\ldots,\ x_{j_{r+2}-1},\ x_{j_{r+1}},\ x_{j_r},\ x_{j_{r-1}},\ldots,\ x_{j_1},\ x_0)\Big].$$

Since the stationary point $x_{j_{r+1}-1} = x^*_{j_{r+1}-1}(x_{j_{r+1}},\ x_{j_r})$ of the phase function of the oscillatory integral (8.4.65) depends only on $x_{j_{r+1}}$ and on x_{j_r}, it is clear that

$$P_{j_{r+1}-1}\mathscr{L}\Theta_{r+1}f = \mathscr{L}\Theta_{r+1}P_{j_{r+1}-1}f.$$

Using this equality, we compare (8.4.64) and (8.4.65). Then it turns out that

$$\begin{aligned} &\mathscr{L}\Theta_{r+1}\tilde{b}_{j_{r+1}\cdots j_1}(x_{J+1},\ x_J,\ldots,\ x_{j_{r+2}-1},\ x_{j_{r+1}},\ x_{j_r},\ x_{j_{r-1}},\ldots,\ x_{j_1},\ x_0) \\ &= R_{j_{r+1}-1}\mathscr{L}\Theta_{r+1}f(x_{J+1},\ x_J,\ldots,\ x_{j_{r+2}-1},\ x_{j_{r+1}},\ x_{j_r},\ x_{j_{r-1}},\ldots,\ x_{j_1},\ x_0). \end{aligned} \tag{8.4.66}$$

Regarding

$$\lambda' = (\lambda,\ x_{j_{s+1}},\ x_{j_{s+1}-1},\ldots,\ x_{j_{r+2}},\ x_{j_{r+2}-1},\ x_{j_{r-1}},\ x_{j_{r-2}},\ldots,\ x_0) \tag{8.4.67}$$

as a parameter, we can use Corollary 8.11. Since $|\alpha_{j_r}| \le K,\ |\alpha_{j_{r+1}}| \le K$,

$$\begin{aligned} &(1 + |\lambda'| + |x_{j_{r+1}}| + |x_{j_r}|)^{-m} \\ &\times \partial_{x_{j_r}}^{\alpha_{j_r}}\partial_{x_{j_{r+1}}}^{\alpha_{j_{r+1}}}|\mathscr{L}\tilde{b}_{j_{r+1}\cdots,\ j_1}(x_{J+1}, \widehat{x_{j_{s+1}-1},\ x_{j_s}},\ldots, \widehat{x_{j_{r+2}-1},\ x_{j_{r+1}}},\ x_{j_r},\ldots,\ x_0)| \\ &\le C_K\left(\frac{\tau_{j_{r+1}}T(j_{r+1}-1,\ j_r+1)}{T(j_{r+1},\ j_r+1)\nu}\right) \\ &\quad\times \max \sup_{x_{j_{r+1}-1}\in\mathbf{R}} (1 + |\lambda'| + |x_{j_{r+1}}| + |x_{j_{r+1}-1}| + |x_{j_r}|)^{-m} \\ &\quad\times \Big|\partial_{x_{j_{r+1}}}^{\beta_{j_{r+1}}}\partial_{x_{j_{r+1}-1}}^{\beta_{j_{r+1}-1}}\partial_{x_{j_r}}^{\beta_{j_r}} \\ &\quad\mathscr{L}\Theta_{r+1}f(x_{J+1}, x_{j_{s+1}-1}, x_{j_s},\ldots, x_{j_{r+2}-1}, x_{j_{r+1}}, x_{j_{r+1}-1}, x_{j_r}, x_{j_{r-1}},\ldots,\ x_{j_1},\ x_0)\Big|. \end{aligned} \tag{8.4.68}$$

Here max is taken among $\beta_{j_{r+1}}, \beta_{j_{r+1}-1}, \beta_{j_r}$ satisfying $\beta_{j_{r+1}} \le \alpha_{j_{r+1}}, \beta_{j_r} \le \alpha_{j_r}$ and $|\beta_{j_{r+1}-1}| \le M(K)$.

On the other hand we have by definition (8.4.62) and Leibniz's rule that

$$\begin{aligned} &\Big|\mathscr{L}\partial_{x_{j_{r+1}}}^{\beta_{j_{r+1}}}\partial_{x_{j_{r+1}-1}}^{\beta_{j_{r+1}-1}}\partial_{x_{j_r}}^{\beta_{j_r}} \\ &\quad\Theta_{r+1}f(x_{J+1},\ x_{j_{s+1}-1},\ x_{j_s},\ldots,\ x_{j_{r+2}-1},\ x_{j_{r+1}},\ x_{j_{r+1}-1},\ x_{j_r},\ x_{j_{r-1}},\ldots,\ x_{j_1},\ x_0)\Big| \\ &\le C_{K,\ 1}\max\Big|\mathscr{L}\partial_{x_{j_{r+1}}}^{\beta_{j_{r+1}}}\partial_{x_{j_{r+1}-1}}^{\beta'_{j_{r+1}-1}}\partial_{x_{j_r}}^{\beta'_{j_r}} \\ &\quad\tilde{b}_{j_r\cdots j_1}(x_{J+1},\ \widehat{x_{j_{s+1}-1},\ x_{j_s}},\ldots,\ \widehat{x_{j_{r+2}-1},\ x_{j_{r+1}}},\ \widehat{x_{j_{r+1}-1},\ x_{j_r}},\ x_{j_{r-1}},\ldots,\ x_{j_1},\ x_0)\Big|, \end{aligned} \tag{8.4.69}$$

where max is taken among $\beta'_{j_r}\text{C}\beta'_{j_{r+1}-1}$ satisfying $\beta'_{j_r} \le \beta_{j_r}\text{C}\beta'_{j_{r+1}-1} \le \beta_{j_{r+1}-1}$.

We have assumed that $|\alpha_{j_n}| \le K$ for $0 \le n \le r+1$. Thus $|\beta_{j_{r+1}}| \le K, |\beta'_{j_r}| \le K$. $|\beta'_{j_{r+1}-1}| \le |\beta_{j_{r+1}-1}| \le M(K)$ if $\beta_{j_{r+1}-1}$ is as in (8.4.68). Set $L'_r = L_{r+1}\partial_{x_{j_{r+1}}}^{\beta_{j_{r+1}}}\partial_{x_{j_{r+1}-1}}^{\beta'_{j_{r+1}-1}}$. Then

$$\mathscr{L}\partial_{x_{j_{r+1}}}^{\beta_{j_{r+1}}}\partial_{x_{j_{r+1}-1}}^{\beta'_{j_{r+1}-1}}\partial_{x_{j_r}}^{\beta'_{j_r}} = L'_r\partial_{x_{j_r}}^{\beta'_{j_r}}\prod_{k=0}^{r-1}\partial_{x_{j_k}}^{\alpha_{j_k}}.$$

Now we use induction hypothesis (8.4.52) in which L_r is replaced with L'_r and α_{j_r} is replaced with β'_{j_r}. Thus we have

$$\begin{aligned}
&(1+|\lambda'|+|x_{j_{r+1}}|+|x_{j_{r+1}-1}|+|x_{j_r}|)^{-m}\Big|\mathscr{L}\partial_{x_{j_{r+1}}}^{\beta_{j_{r+1}}}\partial_{x_{j_{r+1}-1}}^{\beta'_{j_{r+1}-1}}\partial_{x_{j_r}}^{\beta'_{j_r}}\\
&\tilde{b}_{j_r,\dots,j_1}(x_{J+1},\, x_{j_s}\widehat{_{+1}-1,\, x}_{j_s},\dots,x_{j_{r+2}}\widehat{-1,\, x}_{j_{r+1}},\, x_{j_{r+1}}\widehat{-1,\, x}_{j_r},\, x_{j_r-1},\dots,\, x_{j_1},\, x_0)\Big|\\
&\le C(K)^r\prod_{n=1}^{r}\left(\frac{\tau_{j_n}T(j_n-1,\, j_{n-1}+1)}{\nu T(j_n,\, j_{n-1}+1)}\right)\\
&\times\max\sup_{(x_{j_r-1},\dots,\, x_{j_1-1})\in\mathbf{R}^r}\Big(1+|\lambda|+|x_0|+\sum_{n=1}^{s+1}(|x_{j_n}|+|x_{j_n-1}|)\Big)^{-m} \qquad (8.4.70)\\
&\Big|L'_r\partial_{x_{j_0}}^{\beta''_0}\prod_{n=1}^{r}\partial_{x_{j_n}}^{\beta''_{j_n}}\partial_{x_{j_n-1}}^{\beta''_{j_n-1}}a_\lambda(x_{J+1},\, x_{j_s}\widehat{_{+1}-1,\, x}_{j_s},\dots,\, x_{j_{r+1}}\widehat{-1,\, x}_{j_r},\, x_{j_r}\widehat{-1,\, x}_{j_r-1},\dots,\, \widehat{x_{j_1},\, x_0})\Big|
\end{aligned}$$

if $|\alpha_0|\le K,\ |\alpha_{j_n}|\le K$ for $1\le n\le r+1$. Here max is taken among $\beta''_{j_n}, \beta''_{j_n-1}$ for $1\le n\le r$, satisfying $\beta''_{j_n}\le\alpha_{j_n}$ and $|\beta''_{j_n-1}|\le M(K)$ for $1\le n\le r-1$, $\beta''_{j_r}\le\beta'_{j_r}$ and $|\beta''_{j_r-1}|\le M(K)$.

We apply (8.4.70) and (8.4.69) to the previous estimate (8.4.68). Then we have

$$\begin{aligned}
&\Big(1+|\lambda|+|x_0|+\sum_{n=r+2}^{s+1}(|x_{j_n}|+|x_{j_n-1}|)+\sum_{n=1}^{r+1}|x_{j_n}|\Big)^{-m}\\
&\quad\times\Big|L_{r+1}\Big(\prod_{n=0}^{r+1}\partial_{x_{j_n}}^{\alpha_{j_n}}\Big)\tilde{b}_{j_{r+1}\cdots j_1}(x_{J+1},\, x_{j_s}\widehat{_{+1}-1,\, x}_{j_s},\dots,\, x_{j_{r+2}}\widehat{-1,\, x}_{j_{r+1}},\, x_{j_r},\dots,\, x_0)\Big|\\
&\le C_K C_{K,\,1}C(K)^r\prod_{n=1}^{r+1}\left(\frac{\tau_{j_n}T(j_n-1,\, j_{n-1}+1)}{\nu T(j_n,\, j_{n-1}+1)}\right)\\
&\quad\times\max\sup_{(x_{j_{r+1}-1},\, x_{j_r-1},\dots,\, x_{j_1-1})\in\mathbf{R}^{r+1}}\Big(1+|\lambda|+|x_0|+\sum_{n=1}^{s+1}(|x_{j_n}|+|x_{j_n-1}|)\Big)^{-m}\\
&\quad\times\Big|L_{r+1}\partial_{x_0}^{\beta_0}\Big(\prod_{n=1}^{r+1}\partial_{x_{j_n}}^{\beta''_{j_n}}\partial_{x_{j_n-1}}^{\beta''_{j_n-1}}\Big)a_\lambda(x_{J+1},\, x_{j_s}\widehat{_{+1}-1,\, x}_{j_s},\dots,\, x_{j_r}\widehat{-1,\, x}_{j_r-1},\dots,\, \widehat{x_{j_1},\, x_0})\Big|.
\end{aligned}$$

Here max is taken among $\beta''_{j_n}\ \beta''_{j_n-1}$ for $1\le n\le r+1$ and β_0 that satisfy $\beta''_{j_n}\le\alpha_{j_n}$ $|\beta''_{j_n-1}|\le M(K)$ for $1\le n\le r+1$ and $\beta_0\le\alpha_0$. This proves (8.4.52) in the case of $r+1$ if we choose $C(K)>C_K C_{K,\,1}$.

Consequently, (8.4.52) holds for all r satisfying $1\le r\le s$. We have proved Lemma 8.8. □

Now we have completed the proof of Lemma 8.6.

Next we shall discuss the proof of Lemma 8.7. Let $0 = j_0 < j_1 - 1 < j_1 < j_2 - 1 < \cdots < j_s < J + 1$ be a sequence associated with an arrangement Q_J, Q_{J-1}, $\ldots Q_1$ of type-2. For any $r \le s$ we define a function $\tilde{b}_{j_r j_{r-1}\cdots j_1}$ by the equation

$$\begin{aligned}&\tilde{b}_{j_r j_{r-1}\cdots j_1}(x_{J+1}, x_J, \ldots, x_{j_r+1}, x_{j_r}, x_{j_{r-1}}, \ldots, x_{j_1}, x_0) \\ &\quad = (Q_{j_r} Q_{j_r-1} \cdots Q_2 Q_1 a)(x_{J+1}, \ldots, x_{j_r+1}, x_{j_r}, x_{j_{r-1}}, \ldots, x_{j_1}, x_0).\end{aligned} \tag{8.4.71}$$

Lemma 8.7 is the special case $r = s$ of the next lemma.

Lemma 8.9 *Let $a_\lambda(x_{J+1}, \ldots, x_0)$ be a function that satisfies Assumption 8.3. Assume $T \le \delta_0$. Let $j_1, \ldots, j_s$ be a sequence associated with the arrangement $Q_J, Q_{J-1},$ $\ldots, Q_1$ of type-2. Define $\tilde{b}_{j_r j_{r-1}\cdots j_1}$ as above. Let*

$$L_r = \begin{cases} \partial_{x_{J+1}}^{\alpha_{J+1}} \prod_{n=r+1}^{s} \left(\partial_{x_{j_n}}^{\alpha_{j_n}} \partial_{x_{j_n-1}}^{\alpha_{j_n-1}}\right), & \text{in the case } r < s, \\ \partial_{x_{J+1}}^{\alpha_{J+1}}, & \text{in the case } r = s. \end{cases}$$

For any non-negative integer K there exist a positive constant $C(K)$ and a natural number $M(K) \ge K$ such that we have

$$\begin{aligned}&\left(1 + |\lambda| + |x_0| + |x_{J+1}| + \sum_{n=1}^{r} |x_{j_n}| + \sum_{n=r+1}^{s} (|x_{j_n}| + |x_{j_n-1}|)\right)^{-m} \\ &\times \left| L_r \left(\prod_{n=0}^{r} \partial_{x_{j_n}}^{\alpha_{j_n}}\right) \tilde{b}_{j_r j_{r-1}\cdots j_1}(\widehat{x_{J+1}, x_{j_s}}, \cdots, \widehat{x_{j_{r+1}-1}, x_{j_r}}, x_{j_{r-1}}, \ldots, x_{j_1}, x_0) \right| \\ &\le C(K)^r \prod_{n=1}^{r} \left(\frac{\tau_{j_n} T(j_n - 1, j_{n-1} + 1)}{\nu T(j_n, j_{n-1} + 1)} \right) \\ &\quad \times \max \sup_{(x_{j_r-1}, \ldots, x_{j_1-1}) \in \mathbf{R}^r} \left(1 + |\lambda| + |x_0| + |x_{J+1}| + \sum_{n=1}^{s} (|x_{j_n}| + |x_{j_n-1}|)\right)^{-m} \\ &\quad \times \left| \partial_{x_0}^{\beta_0} L_r \prod_{n=1}^{r} \partial_{x_{j_n}}^{\beta_{j_n}} \partial_{x_{j_n-1}}^{\beta_{j_n-1}} a_\lambda(\widehat{x_{J+1}, x_{j_s}}, \widehat{x_{j_s-1}, x_{j_{s-1}}}, \ldots, \widehat{x_{j_1-1}, x_0}) \right|,\end{aligned} \tag{8.4.72}$$

if $|\alpha_0| \le K$ and $|\alpha_{j_n}| \le K$, $|\alpha_{j_n-1}| \le K$ for $n = 1, \ldots, r$. Here max *is taken over $\beta_0\, \beta_{j_n}\, \beta_{j_n-1}$, $n = 1, \ldots, r$ with $\beta_{j_n} \le \alpha_{j_n}$, $|\beta_{j_n-1}| \le M(K)$, $\beta_0 \le \alpha_0$.*

Proof of Lemma 8.9 is omitted, because it is almost the same as that of Lemma 8.8.

8.4.3 Improved Estimate

Theorem 8.4 was proved by applying Corollary 8.11 to the process of integration by x_1, x_2, ..., x_J. If we apply Theorem 8.5 instead of Corollary 8.11, then we are able to make the conclusion of Theorem 8.4 sharper. However, we require additional assumptions for that purpose.

Let

$$\Delta : 0 = T_0 < T_1 < \cdots < T_J < T_{J+1} = T$$

be the division of the interval $[0, T]$. We use notation $\tau_j = T_j - T_{j-1}, |\Delta| = \max\{\tau_j\}$ as before. Let

$$S(x_{J+1}, x_J, \ldots, x_0) = \sum_{j=1}^{J+1} S_j(x_j, x_{j-1})$$

be the phase function as before.

We call a division Δ' of the interval $[0, T]$ coarser than the division Δ if Δ is a refinement of Δ'. Let $0 = j_0 < j_1 < \cdots < j_s < j_{s+1} = J + 1$ be any subsequence of $\{0, 1, \ldots, J, J + 1\}$. Then

$$\Delta' : 0 = T_{j_0} < T_{j_1} < \cdots < T_{j_s} < T_{j_{s+1}} = T \tag{8.4.73}$$

is a division of $[0, T]$ coarser than Δ. For any function $a(x_{J+1}, x_J, \ldots, x_0)$ we define the function $\iota_{\Delta'}^{\Delta} a(x_{J+1}, x_{j_s}, \ldots, x_{j_1}, x_0)$ by

$$\iota_{\Delta'}^{\Delta} a(x_{J+1}, x_{j_s}, \ldots, x_{j_1}, x_0) = a(x_{J+1}, x_J, \ldots, x_0)\Big|_{\substack{x_k = x_k^*(x_{j_n}, x_{j_{n-1}}) \\ j_{n-1} < k < j_n, n=1,2,\ldots,s+1}} . \tag{8.4.74}$$

It is clear that

$$\iota_{\Delta'}^{\Delta} S(x_{J+1}, x_{j_s}, \ldots, x_{j_1}, x_0) = \sum_{k=1}^{s+1} S^*_{j_k, j_{k-1}+1}(x_{j_k}, x_{j_{k-1}}) \tag{8.4.75}$$

and that $\iota_{\Delta'}^{\Delta} S(x_{J+1}, x_{j_s}, \ldots, x_{j_1}, x_0)$ satisfies Assumption 8.1 for a phase function associated with Δ', because Proposition 8.4 holds.

Given any integer k with $1 \le k \le J + 1$, we write $\Delta(k)$ for the following particular division coarser than Δ:

$$\Delta(k) : T_0 = 0 < T_k < T_{k+1} < \cdots < T_J < T_{J+1} = T. \tag{8.4.76}$$

By definition $\Delta(1) = \Delta$ and $\Delta(J + 1)$ is nothing but the interval $[0, T]$ itself without any dividing point between 0 and T.

We assume in this section the following

Assumption 8.6 Let m be a non-negative constant. The amplitude function $a_\lambda(x_{J+1}, x_J, \ldots, x_0)$ with parameter λ satisfies the following properties: For any integer $K \geq 0$ there exist positive constants A_K and $X_K \geq 1$ such that for any division Δ' given by (8.4.73) coarser than Δ we have the estimate

$$\begin{aligned} &|\partial_{x_0}^{\alpha_0}\partial_{x_{j_1}}^{\alpha_{j_1}} \ldots \partial_{x_{j_s}}^{\alpha_{j_s}}\partial_{x_{J+1}}^{\alpha_{J+1}}(\iota_{\Delta'}^{\Delta} a_\lambda)(\gamma_{\Delta'})| \\ &\leq A_K X_K^{s+2}(1+|\lambda|+|x_0|+|x_{j_1}|+\cdots+|x_{j_s}|+|x_{J+1}|)^m, \end{aligned} \tag{8.4.77}$$

as far as $|\alpha_j| \leq K$, for $j = 0,\ j_1,\ \ldots,\ j_s,\ J+1$.

This assumption is stronger than Assumption 8.3. It is clear that $X_K \leq X_{K+1}$ for any K.

Now we discuss

$$\begin{aligned} &I(\Delta; S, a_\lambda, \nu)(x_{J+1}, x_0) \\ &= \prod_{j=1}^{J+1}\left(\frac{\nu}{2\pi i \tau_j}\right)^{1/2} \int_{\mathbf{R}^J} e^{i\nu S(x_{J+1}, x_J, \ldots, x_1, x_0)} a_\lambda(x_{J+1}, x_J, \ldots, x_1, x_0) \prod_{j=1}^{J} dx_j. \end{aligned} \tag{8.4.78}$$

In the next theorem, $D_{x_{j-1}^*, \ldots, x_1^*}(S_{j,1}; x_j, x_0)^{1/2}$ is regarded as a function of $(x_0, x_1, \ldots, x_J, x_{J+1})$ independent of variables x_k with $k > j$. This implies that

$$\iota_{\Delta(J+1)}^{\Delta(j)} D_{x_{j-1}^*, \ldots, x_1^*}(S_{j,1}; x_j, x_0) = D_{x_{j-1}^*, \ldots, x_1^*}(S_{j,1}; x_j, x_0)\Big|_{x_j=\gamma^*(T_j)}.$$

Similarly,

$$\iota_{\Delta(J+1)}^{\Delta(j)} \partial_{x_j}^2 D_{x_{j-1}^*, \ldots, x_1^*}(S_{j,1}; x_j, x_0) = \partial_{x_j}^2 D_{x_{j-1}^*, \ldots, x_1^*}(S_{j,1}; x_j, x_0)\Big|_{x_j=\gamma^*(T_j)}.$$

Theorem 8.7 *Suppose that $T \leq \delta_0$ and $a_\lambda(x_{J+1}, x_J, \ldots, x_1, x_0)$ satisfies Assumption 8.6. We further assume that $|\Delta| T \leq 1$. Then*

$$\begin{aligned} &I(\Delta; S, a_\lambda, \nu)(x_{J+1}, x_0) \\ &= \left(\frac{\nu}{2\pi i T}\right)^{1/2} e^{i\nu S^*(x_{J+1}, x_0)} D_{x_J^*, \ldots, x_1^*}(S; x_{J+1}, x_0)^{-1/2} \\ &\quad \times \left(\iota_{\Delta(J+1)}^{\Delta} a_\lambda(x_{J+1}, x_0) + \nu^{-1} T p(\Delta, x_{J+1}, x_0)\right) \\ &\quad + \nu^{-1} T^2 |\Delta| \left(\frac{\nu}{2\pi i T}\right)^{1/2} e^{i\nu S^*(x_{J+1}, x_0)} q(\Delta, \nu, x_{J+1}, x_0) \\ &\quad + \nu^{-2} T^2 \left(\frac{\nu}{2\pi i T}\right)^{1/2} e^{i\nu S^*(x_{J+1}, x_0)} r(\Delta, \nu, x_{J+1}, x_0). \end{aligned} \tag{8.4.79}$$

Here

$$p(\Delta, x_{J+1}, x_0) = \frac{i}{2T} \sum_{j=1}^{J} \frac{T_j \tau_{j+1}}{T_{j+1}} (\iota^{\Delta(j)}_{\Delta(J+1)} D_{x^*_{j-1},\dots,x^*_1}(S_{j,1}; x_j, x_0)^{1/2} \times \partial^2_{x_j} (D_{x^*_{j-1},\dots,x^*_1}(S_{j,1}, x_j, x_0)^{-1/2} \iota^{\Delta}_{\Delta(j)} a_\lambda))(x_{J+1}, x_0), \tag{8.4.80}$$

and functions $q(\Delta, \nu, x_{J+1}, x_0)$ *and* $r(\Delta, \nu, x_{J+1}, x_0)$ *satisfy the following estimate: For any* $K \geq 0$ *there exist an integer* $M(K) \geq 0$ *and a constant* $C_K > 0$ *independent of* Δ *such that*

$$(1 + |\lambda| + |x_{J+1}| + |x_0|)^{-m} |\partial^{\alpha_{J+1}}_{x_{J+1}} \partial^{\alpha_0}_{x_0} q(\Delta, \nu, x_{J+1}, x_0)| \leq C_K A_{M(K)+2}, \tag{8.4.81}$$

$$(1 + |\lambda| + |x_{J+1}| + |x_0|)^{-m} |\partial^{\alpha_{J+1}}_{x_{J+1}} \partial^{\alpha_0}_{x_0} r(\Delta, \nu, x_{J+1}, x_0)| \leq C_K A_{M(K)+2} \tag{8.4.82}$$

if $|\alpha_0| \leq K$ *and* $|\alpha_{J+1}| \leq K$.

Proof The proof is rather long. We divide it into several steps.

Step 1.

We perform integration of the right-hand side of (8.4.78) with respect to x_1 and apply Theorem 8.5. Let $x_1^* = x_1^*(x_2, x_0)$ denote the stationary point as before. Then

$$\begin{aligned} &\left(\frac{\nu}{2\pi i \tau_2}\right)^{1/2} \left(\frac{\nu}{2\pi i \tau_1}\right)^{1/2} \int_{\mathbf{R}} e^{i\nu(S_2(x_2,x_1)+S_1(x_1,x_0))} a_\lambda(x_{J+1}, \dots, x_1, x_0) dx_1 \\ &= \left(\frac{\nu}{2\pi i T_2}\right)^{1/2} e^{i\nu S^*_{2,1}(x_2,x_0)} D_{x_1^*}(S_{2,1}; x_2, x_0)^{-1/2} \\ &\times \Big(\iota^{\Delta}_{\Delta(2)} a_\lambda(x_{J+1}, \dots, x_2, x_0) + \frac{i\tau_2\tau_1}{2\nu T_2} b_0(a_\lambda)(x_{J+1}, x_J, \dots, x_2, x_0) \\ &+ \frac{(\tau_2\tau_1)^2}{\nu T_2} b_1(a_\lambda)(x_{J+1}, x_J, \dots, x_2, x_0) + \left(\frac{\tau_2\tau_1}{\nu i T_2}\right)^2 b_2(a_\lambda)(\nu, x_{J+1}, x_J, \dots, x_2, x_0)\Big), \end{aligned} \tag{8.4.83}$$

where

$$b_0(a_\lambda)(x_{J+1}, x_J, \dots, x_2, x_0) = D_{x_1^*}(S_2 + S_1; x_2, x_0)^{-1} (\partial^2_{x_1} a_\lambda)(x_2, x_1^*, x_0), \tag{8.4.84}$$

and there exist functions $c_{1,0}(x_2, x_0)$ and $c_{1,1}(x_2, x_0)$, which are algebraic functions of $\partial^k_{x_1}(S_2(x_2, x_1^*) + S_1(x_1^*, x_0))$, $(k = 2, 3)$, such that

$$\begin{aligned} &b_1(a_\lambda)(x_{J+1}, x_J, \dots, x_2, x_0) \\ &= \iota^{\Delta}_{\Delta(2)} a_\lambda(x_{J+1}, \dots, x_2, x_0) c_{1,0}(x_2, x_0) + \iota^{\Delta}_{\Delta(2)} \partial_{x_1} a_\lambda(x_{J+1}, \dots, x_2, x_0) c_{1,1}(x_2, x_0). \end{aligned} \tag{8.4.85}$$

And $b_2(a_\lambda)(\nu, x_{J+1}, x_J, \dots, x_2, x_0)$ is the remainder. For any $K \geq 0$ there exist a constant $C_K > 0$ and an integer $M(K) \geq 0$ such that if $|\alpha| \leq K$ and $|\beta| \leq K$, then

$$\Big(1 + |\lambda| + |x_0| + |x_2| + \sum_{j=3}^{J+1} |x_j|\Big)^{-m} |\partial_{x_0}^{\alpha} \partial_{x_2}^{\beta} b_1(a_\lambda)(x_{J+1}, x_J, \dots, x_2, x_0)|$$
$$\leq C_K \max_{\alpha' \leq \alpha, \beta' \leq \beta, |\gamma| \leq M(K)} \sup_{x_1 \in \mathbf{R}} \Big(1 + |\lambda| + |x_0| + |x_1| + |x_2| + \sum_{j=3}^{J+1} |x_j|\Big)^{-m}$$
$$\times |\partial_{x_0}^{\alpha'} \partial_{x_2}^{\beta'} \partial_{x_1}^{\gamma} a_\lambda(x_{J+1}, x_J, \dots, x_2, x_1, x_0)|, \tag{8.4.86}$$
$$\Big(1 + |\lambda| + |x_0| + |x_2| + \sum_{j=3}^{J+1} |x_j|\Big)^{-m} |\partial_{x_0}^{\alpha} \partial_{x_2}^{\beta} b_2(a_\lambda)(x_{J+1}, x_J, \dots, x_2, x_0)|$$
$$\leq C_K \max_{\alpha' \leq \alpha, \beta' \leq \beta, |\gamma| \leq M(K)} \sup_{x_1 \in \mathbf{R}} \Big(1 + |\lambda| + |x_0| + |x_1| + |x_2| + \sum_{j=3}^{J+1} |x_j|\Big)^{-m}$$
$$\times |\partial_{x_0}^{\alpha'} \partial_{x_2}^{\beta'} \partial_{x_1}^{\gamma} a_\lambda(x_{J+1}, x_J, \dots, x_2, x_1, x_0)|.$$

Now we integrate (8.4.83) with respect to $(x_J, x_{J-1}, \dots, x_2)$. Then

$$I(\Delta; S, a_\lambda, \nu)(x_{J+1}, x_0) = I_{1,0} + I_{1,1} + I_{1,2} + I_{1,3}, \tag{8.4.87}$$

where

$$I_{1,0} = \Big(\frac{\nu}{2\pi i T_2}\Big)^{1/2} \prod_{j=3}^{J+1} \Big(\frac{\nu}{2\pi i \tau_j}\Big)^{1/2} \tag{8.4.88}$$
$$\times \int_{\mathbf{R}^{J-1}} e^{i\nu(\sum_{j=3}^{J+1} S_j(x_j, x_{j-1}) + S_{2,1}^*(x_2, x_0))}$$
$$\times D_{x_1^*}(S_{2,1}; x_2, x_0)^{-1/2} (\iota_{\Delta(2)}^{\Delta} a_\lambda)(x_{J+1}, \dots, x_2, x_0) \prod_{j=2}^{J} dx_j,$$

$$I_{1,1} = \frac{i\tau_2\tau_1}{2\nu T_2} \Big(\frac{\nu}{2\pi i T_2}\Big)^{1/2} \prod_{j=3}^{J+1} \Big(\frac{\nu}{2\pi i \tau_j}\Big)^{1/2}$$
$$\times \int_{\mathbf{R}^{J-1}} e^{i\nu(\sum_{j=3}^{J+1} S_j(x_j, x_{j-1}) + S_{2,1}^*(x_2, x_0))}$$
$$\times D_{x_1^*}(S_{2,1}; x_2, x_0)^{-1/2} b_0(a_\lambda)(x_{J+1}, x_J, \dots, x_2, x_0) \prod_{j=2}^{J} dx_j, \tag{8.4.89}$$

$$I_{1,2} = \frac{(\tau_2\tau_1)^2}{\nu T_2}\left(\frac{\nu}{2\pi i T_2}\right)^{1/2}\prod_{j=3}^{J+1}\left(\frac{\nu}{2\pi i \tau_j}\right)^{1/2} \times \int_{\mathbf{R}^{J-1}} e^{i\nu(\sum_{j=3}^{J+1} S_j(x_j,x_{j-1})+S_{2,1}^*(x_2,x_0))} \times D_{x_1^*}(S_{2,1}; x_2, x_0)^{-1/2} b_1(a_\lambda)(x_{J+1}, x_J, \dots, x_2, x_0)\prod_{j=2}^{J} dx_j, \tag{8.4.90}$$

$$I_{1,3} = \left(\frac{\tau_2\tau_1}{\nu i T_2}\right)^2\left(\frac{\nu}{2\pi i T_2}\right)^{1/2}\prod_{j=3}^{J+1}\left(\frac{\nu}{2\pi i \tau_j}\right)^{1/2} \times \int_{\mathbf{R}^{J-1}} e^{i\nu(\sum_{j=3}^{J+1} S_j(x_j,x_{j-1})+S_{2,1}^*(x_2,x_0))} \times D_{x_1^*}(S_{2,1}; x_2, x_0)^{-1/2} b_2(a_\lambda)(x_{J+1}, x_J, \dots, x_2, x_0)\prod_{j=2}^{J} dx_j. \tag{8.4.91}$$

We no longer discuss $I_{1,0}$ in Step 1 and leave it to Step 2. We now discuss $I_{1,1}$. □

Proposition 8.6 *Suppose that Assumption 8.6 holds. Let $k_0, k_1, \dots, k_s$ be a sequence of integers with $1 < k_0 < k_1 < \cdots < k_s < k_{s+1} = J+1$. Let*

$$\Delta' : T_0 < T_{k_0} < T_{k_1} < \cdots < T_{k_s} < T_{k_{s+1}} = T_{J+1} \tag{8.4.92}$$

be a division coarser than $\Delta(k_0)$. Let Δ'' be the division defined by

$$\Delta'' : T_0 < T_{k_0-1} < T_{k_0} < T_{k_1} < \cdots < T_{k_s} < T_{k_{s+1}} = T_{J+1}. \tag{8.4.93}$$

Then for any $K \geq 0$ there exists a constant $C_K > 0$ such that if $|\alpha_{k_n}| \leq K$, $n = 0, 1, \dots, s+1$, $|\alpha_0| \leq K$ and $1 \leq |\alpha_{k_0-1}| \leq 2$, then

$$\left(1 + |\lambda| + |x_0| + \sum_{n=0}^{s+1} |x_{k_n}|\right)^{-m} \left|\partial_{x_0}^{\alpha_0} \prod_{n=0}^{s+1} \partial_{x_{k_n}}^{\alpha_{k_n}} \iota_{\Delta'}^{\Delta''} \partial_{x_{k_0-1}}^{\alpha_{k_0-1}} (\iota_{\Delta''}^{\Delta} a_\lambda)(x_{J+1}, x_{k_s}, \dots, x_{k_0}, x_0)\right| \leq C_K A_{K+2} X_{K+2}^{s+3}. \tag{8.4.94}$$

Proof We prove the proposition in the case $\alpha_{k_0-1} = 2$. The other case is simpler. Suppose $|\alpha_{k_n}| \leq K$, $n = 1, \dots, s+1$ and $|\alpha_0| \leq K$. If $\beta_0 \leq \alpha_0$, $\beta_{\alpha_{k_0-1}}$, $\beta_{k_0} \leq \alpha_{k_0}$, then by Assumption 8.6,

$$\left(1 + |\lambda| + |x_0| + |x_{k_0-1}| + \sum_{n=1}^{s+1} |x_{k_n}|\right)^{-m}$$

$$\times \left| \partial_{x_0}^{\beta_0} \partial_{x_{k_0-1}}^{\beta_{k_0-1}} \partial_{x_{k_0}}^{\beta_{k_0}} \prod_{n=1}^{s+1} \partial_{x_{k_n}}^{\alpha_{k_n}} \partial_{x_{k_0-1}}^2 (\iota_{\Delta''}^{\Delta} a_\lambda)(x_{J+1}, x_{k_s}, \ldots, x_{k_0}, x_{k_0-1}, x_0) \right|$$
$$\leq A_{K+2} X_{K+2}^{s+3}.$$

By definition

$$\iota_{\Delta'}^{\Delta''} \partial_{x_{k_0-1}}^2 (\iota_{\Delta''}^{\Delta} a_\lambda)(x_{J+1}, x_{k_s}, \ldots, x_{k_0}, x_0) \tag{8.4.95}$$
$$= \partial_{x_{k_0-1}}^2 (\iota_{\Delta''}^{\Delta} a_\lambda)(x_{J+1}, x_{k_s}, \ldots, x_{k_0}, x_{k_0-1}, x_0)\Big|_{x_{k_0-1}=x_{k_0-1}^*(x_{k_0}, x_0)}.$$

On the other hand, as a consequence of Corollary 8.4 and Proposition 8.3 there exists a constant $B_K > 0$ such that if $|\beta_0| \leq K$ and $|\beta_{k_0-1}| \leq K$

$$|\partial_{x_0}^{\beta_0} \partial_{x_{k_0}}^{\beta_{k_0}} x_{k_0-1}^*(x_{k_0}, x_0)| \leq B_K. \tag{8.4.96}$$

It follows from these and the chain rule that there exists a positive constant C_K such that if $|\alpha_0| \leq K$, $|\alpha_{k_n}| \leq K$

$$\left(1 + |\lambda| + |x_0| + \sum_{n=0}^{s+1} |x_{k_n}|\right)^{-m} \left|\partial_{x_0}^{\alpha_0} \prod_{n=0}^{s+1} \partial_{x_{k_n}}^{\alpha_{k_n}} \iota_{\Delta'}^{\Delta''} \partial_{x_{k_0-1}}^2 (\iota_{\Delta''}^{\Delta} a_\lambda)(x_{J+1}, x_{k_s}, \ldots, x_{k_0}, x_0)\right|$$
$$\leq C_K A_{K+2} X_{K+2}^{s+4}.$$

This proves (8.4.94). □

Proposition 8.7 *$I_{1,1}$ is of the following form:*

$$I_{1,1} = \left(\frac{i\tau_1\tau_2}{2\nu T_2}\right)\left(\frac{\nu}{2\pi i T}\right)^{1/2} e^{i\nu S^*(x_{J+1},x_0)} D_{x_J^*\ldots x_1^*}(S; x_{J+1}, x_0)^{-1/2} p_1(x_{J+1}, x_0)$$
$$+ \frac{\tau_1\tau_2 T}{\nu^2 T_2}\left(\frac{\nu}{2\pi i T}\right)^{1/2} e^{i\nu S^*(x_{J+1},x_0)} r_{1,1}(\nu, x_{J+1}, x_0), \tag{8.4.97}$$

where $p_1(x_{J+1}, x_0) = (\iota_{\Delta(J+1)}^{\Delta} \partial_{x_1}^2 a_\lambda)(x_{J+1}, x_0)$. For any non-negative integer K there exists C_K and $M(K)$ such that if $|\alpha_0| \leq K$ and $|\alpha_0| \leq K$, then

$$|\partial_{x_0}^{\alpha_0} \partial_{x_{J+1}}^{\alpha_{J+1}} r_{1,1}(\nu, x_{J+1}, x_0)| \leq C_K A_{M(K)+2}(1 + |\lambda| + |x_{J+1}| + |x_0|)^m. \tag{8.4.98}$$

Proof In the process of integration of (8.4.89), we perform integration with respect to variables $x_3, x_4, \ldots, x_J$ prior to integration with respect to x_2:

$$\prod_{j=3}^{J+1}\left(\frac{\nu}{2\pi i \tau_j}\right)^{1/2} \int_{\mathbf{R}^{J-2}} e^{i\nu(\sum_{j=3}^{J+1} S_j(x_j,x_{j-1}))} (\iota_{\Delta(2)}^{\Delta} \partial_{x_1}^2 a_\lambda)(x_{J+1}, \ldots, x_2, x_0)) \prod_{j=3}^{J} dx_j.$$

Consider division $T_2 < T_3 < \cdots < T_J < T_{J+1}$ of the interval $[T_2, T_{J+1}]$. Then Proposition 8.6 implies that $(\iota^{\Delta}_{\Delta(2)}\partial^2_{x_1}a_\lambda)(x_{J+1}, \ldots, x_2, x_0)$ satisfies Assumption 8.6 with X_K, A_K replaced with X_{K+2}, A_{K+2}, respectively, if we regard λ and x_0 as parameters. Therefore, it follows from Theorem 8.4 that

$$\begin{aligned}
&\prod_{j=3}^{J+1}\left(\frac{\nu}{2\pi i\tau_j}\right)^{1/2}\int_{\mathbf{R}^{J-2}} e^{i\nu\sum_{j=3}^{J+1}S_j(x_j,x_{j-1})}\iota^{\Delta}_{\Delta(2)}(\partial^2_{x_1}a_\lambda)(x_{J+1},\ldots,x_2,x_0)\prod_{j=3}^{J}dx_j\\
&=\left(\frac{\nu}{2\pi iT(J+1,3)}\right)^{1/2} e^{i\nu S^*_{J+1,3}(x_{J+1},x_2)}D_{x^*_J\ldots x^*_3}(S_{J+1,3};x_{J+1},x_2)^{-1/2}\\
&\times(\iota^{\Delta(2)}_{\Delta^\flat}\iota^{\Delta}_{\Delta(2)}\partial^2_{x_1}a_\lambda)(x_{J+1},x_2,x_0)+\nu^{-1}T(J+1,3)u_1(x_{J+1},x_2,x_0),
\end{aligned}\tag{8.4.99}$$

where $\Delta^\flat$ means the division

$$\Delta^\flat : T_0 = 0 < T_2 < T_{J+1} = T$$

and $u_1(x_{J+1}, x_2, x_0)$ is the remainder term. For any integer $K \geq 0$ there exists a positive constant C_K and an integer $M(K)$ such that we have the estimate:

$$\begin{aligned}
&(1+|\lambda|+|x_0|+|x_2|+|x_{J+1}|)^{-m}|\partial^\alpha_{x_{J+1}}\partial^\beta_{x_2}\partial^\gamma_{x_0}u_1(\nu,x_{J+1},x_2,x_0)|\\
&\leq C_K A_{M(K)+2}X^3_{M(K)+2},
\end{aligned}\tag{8.4.100}$$

as far as $|\alpha|, |\beta|, |\gamma| \leq K$.

Substitute (8.4.99) into (8.4.89) and integrate by x_2. Then

$$I_{1,1} = I_{1,1,0} + I_{1,1,1},\tag{8.4.101}$$

where

$$\begin{aligned}
&I_{1,1,0}\\
&=\left(\frac{i\tau_1\tau_2}{2\nu T_2}\right)\left(\frac{\nu}{2\pi iT_2}\right)^{1/2}\left(\frac{\nu}{2\pi iT(J+1,3)}\right)^{1/2}\int_{\mathbf{R}} e^{i\nu(S^*_{J+1,3}(x_{J+1},x_2)+S^*_{2,1}(x_2,x_0))}\\
&\times\left(D_{x^*_1}(S_{2,1};x_2,x_0)D_{x^*_J\ldots x^*_3}(S_{J+1,3};x_{J+1},x_2)\right)^{-1/2}(\iota^{\Delta}_{\Delta^\flat}\partial^2_{x_1}a_\lambda)(x_{J+1},x_2,x_0)dx_2,
\end{aligned}\tag{8.4.102}$$

$$\begin{aligned}
&I_{1,1,1}\\
&=\left(\frac{\tau_1\tau_2}{2\nu iT_2}\right)\left(\frac{\nu}{2\pi iT_2}\right)^{1/2}\left(\frac{\nu}{2\pi iT(J+1,3)}\right)^{1/2}\int_{\mathbf{R}} e^{i\nu(S^*_{J+1,3}(x_{J+1},x_2)+S^*_{2,1}(x_2,x_0))}\\
&\times\left(D_{x^*_1}(S_{2,1};x_2,x_0)D_{x^*_J\ldots x^*_3}(S_{J+1,3};x_{J+1},x_2)\right)^{-1/2}\nu^{-1}T(J+1,3)u_1(x_{J+1},x_2,x_0)dx_2.
\end{aligned}\tag{8.4.103}$$

Apply Corollary 8.11 to $I_{1,1,0}$. Then it turns out that

$$I_{1,1,0} = I_{1,1,0,0} + I_{1,1,0,1}.$$

Here $I_{1,1,0,0}$ is the principal part of $I_{1,1,0}$ and $I_{1,1,0,1}$ is the remainder. By Proposition 5.1 the stationary point of $x_2 \to S^*_{J+1,2}(x_{J+1}, x_2) + S^*_{2,0}(x_2, x_0)$ coincides with $x_2^* = x_2^*(x_{J+1}, x_0)$, which is the x_2 component of the stationary point of the function $(x_J, x_{J-1}, \ldots, x_2, x_1) \to S(x_{J+1}, x_J, \ldots, x_2, x_1, x_0)$, and $S^*_{J+1,3}(x_{J+1}, x_2^*) + S^*_{2,1}(x_2^*, x_0) = S^*(x_{J+1}, x_0) = S(x_{J+1}, x_J^*, \ldots, x_1^*, x_0)$. Moreover, by Lemma 5.2

$$D_{x_1^*}(S_{2,1}; x_2, x_0) D_{x_J^* \ldots x_3^*}(S^*_{J+1,3}; x_{J+1}, x_2) D_{x_2^*}(S^*_{J+1,3} + S^*_{2,1}; x_{J+1}, x_0)\Big|_{x_2 = x_2^*}$$
$$= D_{x_J^* \ldots x_1^*}(S; x_{J+1}, x_0).$$

Thus

$$I_{1,1,0,0} = \left(\frac{i\tau_1\tau_2}{2\nu T_2}\right)\left(\frac{\nu}{2\pi i T}\right)^{1/2} e^{i\nu S^*(x_{J+1},x_0)}$$
$$\times D_{x_J^* \ldots x_1^*}(S; x_{J+1}, x_0)^{-1/2} (\iota^{\Delta}_{\Delta(J+1)} \partial^2_{x_1} a_\lambda)(x_{J+1}, x_0),$$
$$I_{1,1,0,1} = \left(\frac{\tau_1\tau_2}{2\nu i T_2}\right)\left(\frac{T_2 T(J+1,3)}{\nu T}\right)\left(\frac{\nu}{2\pi i T}\right)^{1/2} e^{i\nu S^*(x_{J+1},x_0)}$$
$$\times D_{x_J^* \ldots x_1^*}(S; x_{J+1}, x_0)^{-1/2} v_1(\nu, x_{J+1}, x_0).$$

It follows from Corollary 8.11 and Proposition 8.6 that for any K there exist positive constants C_K and $M(K)$ such that as far as $|\alpha_{J+1}| \le K$, $|\alpha_0| \le K$

$$(1 + |\lambda| + |x_0| + |x_{J+1}|)^{-m} |\partial^{\alpha_{J+1}}_{x_{J+1}} \partial^{\alpha_0}_{x_0} v_1(x_{J+1}, x_0)|$$
$$\le C_K \max_{|\gamma| \le M(K)} \sup_{x_2} (1 + |\lambda| + |x_0| + |x_2| + |x_{J+1}|)^{-m}$$
$$\times |\partial^{\alpha_{J+1}}_{x_{J+1}} \partial^{\alpha_0}_{x_0} \partial^{\gamma}_{x_2} \left(D_{x_1^*}(S_{2,1}; x_2, x_0) D_{x_J^* \ldots x_3^*}(S_{J+1,3}; x_{J+1}, x_2)\right)^{-1/2} \iota^{\Delta}_{\Delta^\flat} \partial^2_{x_1} a_\lambda(x_{J+1}, x_2, x_0)|$$
$$\le C_K A_{M(K)+2} X^3_{M(K)+2}.$$

Similarly, applying Corollary 8.9, we have

$$I_{1,1,1} = \nu^{-1} T(J+1,3) \left(\frac{\tau_1\tau_2}{2\nu i T_2}\right)\left(\frac{\nu}{2\pi i T}\right)^{1/2} e^{i\nu S^*(x_{J+1},x_0)} w_1(\nu, x_{J+1}, x_0).$$

It follows from Corollary 8.9 and (8.4.100) that for any K there exist C_K and $M(K)$ such that as far as $|\alpha_{J+1}| \le K$, $|\alpha_0| \le K$

$$(1 + |\lambda| + |x_{J+1}| + |x_0|)^{-m} |\partial^{\alpha_{J+1}}_{x_{J+1}} \partial^{\alpha_0}_{x_0} w_1(\nu, x_{J+1}, x_0)| \tag{8.4.104}$$
$$\le C_K \max_{|\gamma| \le M(K)} \sup_{x_2} (1 + |\lambda| + |x_{J+1}| + |x_2| + |x_0|)^{-m} |\partial^{\alpha_{J+1}}_{x_{J+1}} \partial^{\gamma}_{x_2} \partial^{\alpha_0}_{x_0} u_1(\nu, x_{J+1}, x_2, x_0)|$$
$$\le C_K A_{M(K)+2} X^3_{M(K)+2}.$$

Now set

$$r_{1,1}(\nu, x_{J+1}, x_0) = T^{-1}T(J+1,3)w_1(\nu, x_{J+1}x_0) + \left(\frac{T_2 T(J+1,3)}{iT^2}\right) D_{x^*_J \dots x^*_1}(S; x_{J+1}, x_0)^{-1/2} v_1(\nu, x_{J+1}, x_0).$$

Then we have the following equality:

$$\begin{aligned} I_{1,1} &\qquad (8.4.105)\\ &= \frac{i\tau_1\tau_2}{2\nu T_2}\left(\frac{\nu}{2\pi i T}\right)^{1/2} e^{i\nu S^*(x_{J+1},x_0)} D_{x^*_J \dots x^*_1}(S; x_{J+1}, x_0)^{-1/2} \iota^{\Delta}_{\Delta(J+1)} \partial^2_{x_1} a_\lambda)(x_{J+1}, x_0)\\ &+ \nu^{-1} T \frac{\tau_1\tau_2}{2\nu i T_2}\left(\frac{\nu}{2\pi i T}\right)^{1/2} e^{i\nu S^*(x_{J+1},x_0)} r_{1,1}(\nu, x_{J+1}, x_0). \end{aligned}$$

Moreover, for any $K \geq 0$ there exist constants $C_K > 0$ and $M(K) > 0$ such that

$$(1 + |\lambda| + |x_{J+1}| + |x_0|)^{-m} |\partial^{\alpha_{J+1}}_{x_{J+1}} \partial^{\alpha_0}_{x_0} r_{1,1}(\nu, x_{J+1}, x_0)| \leq C_K A_{M(K)+2}, \quad (8.4.106)$$

if $|\alpha_{J+1}| \leq K$ and $|\alpha_0| \leq K$. Proposition 8.7 has been proved. □

Now we discuss $I_{1,2}$.

Proposition 8.8 *Let $k_0, k_1, \dots, k_s$ be a sequence of integers with $2 = k_0 < k_1 < \dots < k_s < k_{s+1} = J+1$. Let*

$$\Delta' : T_0 < T_{k_0} < T_{k_1} < \dots < T_{k_s} < T_{k_{s+1}} = T_{J+1} \quad (8.4.107)$$

be a division coarser than $\Delta(k_0)$. Then for any $K \geq 0$ there exists a constant $C_K > 0$ such that if $|\alpha_{k_n}| \leq K$, then

$$\begin{aligned} &\left(1 + |\lambda| + |x_0| + \sum_{n=0}^{s+1} |x_{k_n}|\right)^{-m} \left|\partial^{\alpha_0}_{x_0} \prod_{n=0}^{s+1} \partial^{\alpha_{k_n}}_{x_{k_n}} \iota^{\Delta''}_{\Delta'} \partial^2_{x_{k_0-1}} (\iota^{\Delta}_{\Delta''} b_1(a_\lambda)(x_{J+1}, x_{k_s}, \dots, x_{k_0}, x_0)\right| \\ &\leq C_K A_{K+2} X^{s+3}_{K+2}. \end{aligned} \quad (8.4.108)$$

Proof The functions $c_0(x_2, x_1)$ and $c_1(x_2, x_1)$ in (8.4.85) are independent of x_j with $j \geq 3$. Thus the proposition follows from Assumption 8.6 and Proposition 8.6. □

It follows from (8.4.85) and Proposition 8.6 that $b_1(a_\lambda)$ satisfies Assumption 8.6 regarded as a function of $(x_{J+1}, x_J, \dots, x_2)$ with parameter λ, x_0. The next proposition follows from Proposition 8.8 and Corollary 8.9:

Proposition 8.9 *We can write*

$$I_{1,2} = \frac{(\tau_1\tau_2)^2}{\nu T_2}\left(\frac{\nu}{2\pi i T}\right)^{1/2} e^{i\nu S^*(x_{J+1},x_0)} q_1(\nu, x_{J+1}, x_0). \quad (8.4.109)$$

For any $K \geq 0$ there exist $C_K > 0$ and $M(K)$ such that if $|\alpha_{J+1}| \leq K$ and $|\alpha_0| \leq K$, then

$$(1+|\lambda|+|x_0|+|x_{J+1}|)^{-m}|\partial_{x_0}^{\alpha_0}\partial_{x_{J+1}}^{\alpha_{J+1}}q_1(\nu,x_{J+1},x_0)| \leq C_K A_{M(K)+2}X_{M(K)+2}^3. \tag{8.4.110}$$

Proof We perform integration with respect to $x_3,\ldots,x_J$ of (8.4.90). Then by Corollary 8.9,

$$\prod_{j=3}^{J+1}\left(\frac{\nu}{2\pi i\tau_j}\right)^{1/2}\int_{\mathbf{R}^{J-2}} e^{i\nu(\sum_{j=3}^{J+1}S_j(x_j,x_{j-1}))}b_1(a_\lambda)(x_{J+1},x_J,\ldots,x_3,x_2,x_0)\prod_{j=3}^{J}dx_j$$
$$=\left(\frac{\nu}{2\pi iT(J+1,3)}\right)^{1/2}e^{i\nu S^*_{J+1,3}(x_{J+1},x_2))}\tilde{u_{\lambda}}_1(\nu,x_{J+1},x_2,x_0). \tag{8.4.111}$$

For any $K \geq 0$ there exist $C_K > 0$ and an integer $M(K) > 0$ such that

$$(1+|\lambda|+|x_0|+|x_2|+|x_{J+1}|)^{-m}|\partial_{x_0}^{\alpha_0}\partial_{x_2}^{\alpha_2}\partial_{x_{J+1}}^{\alpha_{J+1}}\tilde{u_{\lambda}}_1(\nu,x_{J+1},x_2,x_0)| \tag{8.4.112}$$
$$\leq C_K A_{M(K)+2}X_{M(K)+2}^3.$$

Replace (8.4.111) in (8.4.90). Then

$$I_{1,2}=\frac{(\tau_1\tau_2)^2}{\nu T_2}\left(\frac{\nu}{2\pi iT_2}\right)^{1/2}\left(\frac{\nu}{2\pi iT(J+1,3)}\right)^{1/2} \tag{8.4.113}$$
$$\times\int_{\mathbf{R}} e^{i\nu(S^*_{J+1,3}(x_{J+1},x_2)+S^*_{2,1}(x_2,x_0))}\tilde{u_{\lambda}}_1(x_{J+1},x_2,x_0)dx_2.$$

Using (8.4.112), we apply Corollary 3.6 to the above equality. Then we have Proposition 8.9. □

We discuss $I_{1,3}$.

Consider division $T_2 < T_3 < \cdots < T_J < T_{J+1} = T$ of the interval $[T_2, T]$. Though $b_2(a_\lambda)$ is much more complicated than $b_1(a_\lambda)$, it also satisfies Assumption 8.3 corresponding to the interval $[T_2, T]$ instead of $[T_0, T]$ if x_0 is regarded as a parameter and constants A_K, X_K are suitably changed. The following proposition similar to Proposition 8.6 holds.

Proposition 8.10 *Let $k_0, k_1, \ldots, k_s$ be a sequence of integers with $2 = k_0 < k_1 < \cdots < k_s < k_{s+1} = J+1$. Let*

$$\Delta' : T_0 < T_2 = T_{k_0} < T_{k_1} < \cdots < T_{k_s} < T_{k_{s+1}} = T_{J+1} \tag{8.4.114}$$

be a division coarser than $\Delta(k_0)$. Then for any $K \geq 0$ there exists a constant $C_K > 0$ such that if $|\alpha_{k_n}| \leq K$, then

$$\left(1+|\lambda|+|x_0|+\sum_{n=0}^{s+1}|x_{k_n}|\right)^{-m}|\partial_{x_0}^{\alpha_0}\prod_{n=0}^{s+1}\partial_{x_{k_n}}^{\alpha_{k_n}}\iota_{\Delta'}^{\Delta(2)}b_2(a_\lambda)(x_{J+1},x_{k_s},\ldots,x_{k_0},x_0)|$$
$$\leq C_K A_{K+2} X_{K+2}^{s+3}. \tag{8.4.115}$$

Proof Let $k_0, k_1, \ldots, k_s$ be a sequence of integers with $2 = k_0 < k_1 < \cdots < k_s < k_{s+1} = J+1$. Let

$$\Delta' : T_0 < T_{k_0} = T_2 < T_{k_1} < \cdots < T_{k_s} < T_{k_{s+1}} = T_{J+1} \tag{8.4.116}$$

be a division coarser than $\Delta(2)$. For any sequence of multi-indices α_{k_j}, $j=1, \ldots, s+1$, we define

$$P(\partial) = \prod_{j=1}^{s+1}\partial_{x_{k_j}}^{\alpha_{k_j}}. \tag{8.4.117}$$

Let Δ'' be the division defined by

$$\Delta'' : T_0 < T_1 < T_{k_0} = T_2 < T_{k_1} < \cdots < T_{k_s} < T_{k_{s+1}} = T_{J+1}. \tag{8.4.118}$$

Notice that

$$\iota_{\Delta'}^{\Delta''}P(\partial)\iota_{\Delta''}^{\Delta} = P(\partial)\iota_{\Delta'}^{\Delta''}\iota_{\Delta''}^{\Delta} = P(\partial)\iota_{\Delta'}^{\Delta}, \tag{8.4.119}$$

because $x_1^*(x_2, x_0)$ is independent of x_j with $j \geq 3$. Operate $P(\partial)\iota_{\Delta'}^{\Delta(2)}$ on both sides of (8.4.83) and use the equality $\iota_{\Delta'}^{\Delta(2)}\iota_{\Delta(2)}^{\Delta} = \iota_{\Delta'}^{\Delta}$. Then

$$\left(\frac{\nu}{2\pi i\tau_2}\right)^{1/2}\left(\frac{\nu}{2\pi i\tau_1}\right)^{1/2}P(\partial)\iota_{\Delta'}^{\Delta(2)}\int_{\mathbf{R}}e^{i\nu(S_2(x_2,x_1)+S_1(x_1,x_0))}a_\lambda(x_{J+1},\ldots,x_1,x_0)dx_1$$
$$=\left(\frac{\nu}{2\pi i T_2}\right)^{1/2}e^{i\nu S_{2,1}^*(x_2,x_0)}D_{x_1^*}(S_{2,1};x_2,x_0)^{-1/2} \tag{8.4.120}$$
$$\times\Big(P(\partial)\iota_{\Delta'}^{\Delta(2)}\iota_{\Delta(2)}^{\Delta}a_\lambda(x_{J+1},\ldots,x_2,x_0)+\frac{\tau_2\tau_1}{2\nu i T_2}P(\partial)\iota_{\Delta'}^{\Delta(2)}b_0(a_\lambda)(x_{J+1},x_J,\ldots,x_2,x_0)$$
$$+\frac{(\tau_2\tau_1)^2}{\nu T_2}P(\partial)\iota_{\Delta'}^{\Delta(2)}b_1(a_\lambda)(x_{J+1},x_J,\ldots,x_2,x_0)$$
$$+\left(\frac{\tau_2\tau_1}{\nu i T_2}\right)^2P(\partial)\iota_{\Delta'}^{\Delta(2)}b_2(a_\lambda)(x_{J+1},x_J,\ldots,x_2,x_0)\Big).$$

Since the stationary point $x_j^*(x_{k_n}, x_{k_{n-1}})$, $k_{n-1} < j < k_n$, of $S_{k_n,k_{n-1}+1}$ for $n = 1, 2, \ldots, s+1$ are independent of x_1, we can replace x_j with $x_j^*(x_{k_n}, x_{k_{n-1}})$ under the integral sign. We can also operate $P(\partial)$ under the integral sign. Thus

$$\prod_{j=1}^{2}\left(\frac{\nu}{2\pi i\tau_j}\right)^{1/2}P(\partial)\iota_{\Delta'}^{\Delta(2)}\int_{\mathbf{R}}e^{i\nu(S_2(x_2,x_1)+S_1(x_1,x_0))}a_\lambda(x_{J+1},\ldots,x_1,x_0)dx_1$$

$$= \prod_{j=1}^{2} \left(\frac{\nu}{2\pi i \tau_j} \right)^{1/2} \int_{\mathbf{R}} e^{i\nu(S_2(x_2,x_1)+S_1(x_1,x_0))} P(\partial)\iota^{\Delta}_{\Delta''} a_\lambda(x_{k_{s+1}}, \dots, x_{k_1}, x_2, x_1, x_0) dx_1. \tag{8.4.121}$$

We can apply Corollary 8.11 to the right-hand side of (8.4.121). Then

$$\begin{aligned} &\prod_{j=1}^{2} \left(\frac{\nu}{2\pi i \tau_j} \right)^{1/2} \int_{\mathbf{R}} e^{i\nu(S_2(x_2,x_1)+S_1(x_1,x_0))} P(\partial)\iota^{\Delta}_{\Delta''} a_\lambda(x_{J+1}, x_{k_s}, \dots, x_{k_1}, x_2, x_1, x_0) dx_1 \\ &= \left(\frac{\nu}{2\pi i T_2} \right)^{1/2} e^{i\nu S^*_{2,1}(x_2,x_0)} D_{x_1^*}(S_{2,1}; x_2, x_0)^{-1/2} \qquad (8.4.122) \\ &\quad \times \Big[\iota^{\Delta''}_{\Delta'} P(\partial)\iota^{\Delta}_{\Delta''} a_\lambda(x_{J+1}, \dots, x_2, x_0) + \frac{\tau_2\tau_1}{2\nu i T_2} b_0(P(\partial)\iota^{\Delta}_{\Delta''} a_\lambda)(x_{J+1}, x_J, \dots, x_2, x_0) \\ &\quad + \frac{(\tau_2\tau_1)^2}{\nu T_2} b_1(P(\partial)\iota^{\Delta}_{\Delta''} a_\lambda)(x_{J+1}, x_J, \dots, x_2, x_0) \\ &\quad + \left(\frac{\tau_2\tau_1}{\nu i T_2} \right)^2 b_2(P(\partial)\iota^{\Delta}_{\Delta''} a_\lambda)(x_{J+1}, x_J, \dots, x_2, x_0) \Big]. \end{aligned}$$

Since (8.4.119) holds and $x_1^*(x_2, x_0)$ is independent of x_j with $j \geq 3$,

$$\begin{aligned} &b_0(P(\partial)\iota^{\Delta}_{\Delta''} a_\lambda)(x_{J+1}, x_{k_s}, \dots, x_{k_1}, x_2, x_0) \qquad (8.4.123) \\ &= \partial^2_{x_1} P(\partial)\iota^{\Delta}_{\Delta''} a_\lambda(x_{J+1}, x_{k_s}, \dots, x_{k_1}, x_2, x_1, x_0) \Big|_{x_1 = x_1^*(x_2,x_0)} \\ &= \iota^{\Delta''}_{\Delta'} P(\partial) \partial^2_{x_1} \iota^{\Delta}_{\Delta''} a_\lambda(x_{J+1}, x_{k_s}, \dots, x_{k_1}, x_2, x_0) \\ &= P(\partial)\iota^{\Delta''}_{\Delta'} \iota^{\Delta}_{\Delta''} \partial^2_{x_1} a_\lambda(x_{J+1}, x_{k_s}, \dots, x_{k_1}, x_2, x_0) \\ &= P(\partial)\iota^{\Delta}_{\Delta'} \partial^2_{x_1} a_\lambda(x_{J+1}, x_{k_s}, \dots, x_{k_1}, x_2, x_0) \\ &= P(\partial)\iota^{\Delta(2)}_{\Delta'} b_0(a_\lambda)(x_{J+1}, x_{k_s}, \dots, x_{k_1}, x_2, x_0). \end{aligned}$$

Similarly,

$$\begin{aligned} &b_1(P(\partial)\iota^{\Delta}_{\Delta''} a_\lambda)(x_{J+1}, x_{k_s}, \dots, x_{k_1}, x_2, x_0) \qquad (8.4.124) \\ &= P(\partial)\iota^{\Delta(2)}_{\Delta'} b_1(a_\lambda)(x_{J+1}, x_{k_s}, \dots, x_{k_1}, x_2, x_0), \end{aligned}$$

because $c_0(x_2, x_1)$ and $c_1(x_2, x_0)$ are independent of x_j, $j \geq 3$.

Using (8.4.119), (8.4.121), (8.4.123) and (8.4.124), we compare (8.4.120) with (8.4.122). Then we obtain

$$\begin{aligned} &P(\partial)\iota^{\Delta(2)}_{\Delta'} b_2(a_\lambda)(x_{J+1}, x_{k_s}, \dots, x_{k_1}, x_2, x_0) \qquad (8.4.125) \\ &= b_2(P(\partial)\iota^{\Delta}_{\Delta''} a_\lambda)(x_{J+1}, x_{k_s}, \dots, x_{k_1}, x_2, x_0). \end{aligned}$$

Apply (8.3.5) in Theorem 8.5 to the right-hand side of (8.4.125). Then for any non-negative integer K there exist a positive constant C_K and a positive integer $M(K)$

such that if $|\alpha_0| \le K$, $|\alpha_2| \le K$ and $|\alpha_{j_n}| \le K$ $(1 \le n \le s+1)$, then

$$\left(1 + |\lambda| + |x_0| + |x_2| + \sum_{n=1}^{s+1} |x_{j_n}|\right)^{-m} \tag{8.4.126}$$
$$\times |\partial_{x_0}^{\alpha_0} \partial_{x_2}^{\alpha_2} P(\partial) \iota_{\Delta'}^{\Delta(2)} b_2(a_\lambda)(\nu, x_{j_{s+1}}, x_{j_s}, \dots, x_{j_1}, x_2, x_0)|$$
$$= \left(1 + |\lambda| + |x_0| + |x_2| + \sum_{n=1}^{s+1} |x_{j_n}|\right)^{-m} \tag{8.4.127}$$
$$\times |\partial_{x_0}^{\alpha_0} \partial_{x_2}^{\alpha_2} b_2(P(\partial) \iota_{\Delta''}^{\Delta} a_\lambda)(\nu, x_{j_{s+1}}, x_{j_{s+1}-1}, \dots, x_{j_1-1}, x_2, x_0)|$$
$$\le C_K \max_{\substack{\beta_0 \le \alpha_0, \beta_2 \le \alpha_2 \\ |\beta_1| \le M(K)}} \sup_{x_1 \in \mathbf{R}} \left(1 + |\lambda| + |x_0| + |x_1| + |x_2| + \sum_{n=1}^{s+1} |x_{j_n}|\right)^{-m}$$
$$\times |\partial_{x_1}^{\beta_1} \partial_{x_0}^{\beta_0} \partial_{x_2}^{\beta_2} P(\partial) \iota_{\Delta''}^{\Delta} a_\lambda(x_{j_{s+1}}, x_{j_s}, \dots, x_{j_1}, x_2, x_1, x_0)|$$
$$\le C_K A_{M(K)} X_{M(K)}^{2s+5}.$$

Proposition 8.8 is proved. □

This implies that the function $b_2(a_\lambda)(\nu, x_{J+1}, x_J, \dots, x_3, x_2, x_0)$ satisfies Assumption 8.6 if it is regarded as a function of $(x_{J+1}, x_J, \dots, x_2)$ with parameter λ, x_0. In just the same way as we obtained Proposition 8.9 from Proposition 8.8 we obtain:

Proposition 8.11 *We can write*

$$I_{1,3} = \left(\frac{\tau_1 \tau_2}{\nu T_2}\right)^2 \left(\frac{\nu}{2\pi i T}\right)^{1/2} e^{i\nu S^*(x_{J+1}, x_0)} r_1(\nu, x_{J+1}, x_0). \tag{8.4.128}$$

For any $K \ge 0$ there exist $C_K > 0$ and $M(K)$ such that if $|\alpha_{J+1}| \le K$ and $|\alpha_0| \le K$,

$$(1 + |\lambda| + |x_0| + |x_{J+1}|)^{-m} |\partial_{x_0}^{\alpha_0} \partial_{x_{J+1}}^{\alpha_{J+1}} r_1(\nu, x_{J+1}, x_0)| \le C_K A_{M(K)} X_{M(K)}^3. \tag{8.4.129}$$

Step 1 ends here.

Step 2.

Here we discuss $I_{1,0}$ given by (8.4.88). The oscillatory integral $I_{1,0}$ is of the same type as integral $I(\Delta; S, a_\lambda, \nu)$ of (8.4.78). More precisely, define

$$a_{\lambda,1}(x_{J+1}, x_J, \dots, x_2, x_0) = D_{x_1^*}(S_{2,1}; x_2, x_0)^{-1/2} \iota_{\Delta(2)}^{\Delta} a_\lambda(x_{J+1}, x_J, \dots, x_2, x_0). \tag{8.4.130}$$

Then $I_{1,0}$ is the oscillatory integral corresponding to the division $\Delta(2)$ of the interval $[0, T]$ with the phase function $S_{J+1,3} + S_{2,1}^*$ and the amplitude function $a_{\lambda,1}$, i.e.,

$$I_{1,0}(x_{J+1}, x_0) = I(\Delta(2); S_{J+1,3} + S^*_{2,1}, a_{\lambda,1})(\nu, x_{J+1}, x_0).$$

For any $K \geq 0$ let

$$B_{2,K} = \max_{|\alpha_0| \leq K, |\alpha_2| \leq K,} \sup_{\substack{(x_2, x_0) \in \mathbf{R}^2 \\ 0 < \tau_1 \leq \delta, 0 < \tau_2 \leq \delta}} |\partial^{\alpha_2}_{x_2} \partial^{\alpha_0}_{x_0} D(S_{2,1}; x_2, x_0)^{-1/2}|.$$

Then the amplitude function $a_{\lambda,1} = D_{x_1^*}(S_{2,1}, x_2, x_0)^{-1/2} \iota^{\Delta}_{\Delta(2)} a_\lambda$ of $I_{1,0}$ satisfies Assumptions 8.6 with respect to variables $(x_{J+1}, \ldots, x_2, x_0)$ if we replace A_K with $2^{2K} B_K A_K$ because of Leibniz's formula. Therefore, we can follow discussions in Step 1 almost word for word.

Perform integration with respect x_2 in (8.4.88). Then

$$\begin{aligned}
&\left(\frac{\nu}{2\pi i T_2}\right)^{1/2} \left(\frac{\nu}{2\pi i \tau_3}\right)^{1/2} \int_{\mathbf{R}} e^{i\nu(S_3(x_3, x_2) + S^*_{2,1}(x_2, x_0))} a_{\lambda,1}(x_{J+1}, \ldots, x_2, x_0) dx_2 \\
&= \left(\frac{\nu}{2\pi i T_3}\right)^{1/2} e^{i\nu S^*_{3,1}(x_3, x_0)} D_{x_2^*}(S_3 + S^*_{2,1}; x_3, x_0)^{-1/2} \qquad (8.4.131) \\
&\times \Big[\iota^{\Delta(2)}_{\Delta(3)} a_{\lambda,1}(x_{J+1}, \ldots, x_3, x_0) + \frac{i\tau_3 T_2}{2\nu T_3} b_0(a_{\lambda,1})(x_{J+1}, x_J, \ldots, x_3, x_0) \\
&+ \frac{(\tau_3 T_2)^2}{\nu T_3} b_1(a_{\lambda,1})(x_{J+1}, x_J, \ldots, x_3, x_0) + \left(\frac{\tau_3 T_2}{\nu i T_3}\right)^2 b_2(a_{\lambda,1})(\nu, x_{J+1}, x_J, \ldots, x_3, x_0)\Big],
\end{aligned}$$

where

$$b_0(a_{\lambda,1})(x_{J+1}, x_J, \ldots, x_3, x_0) = \iota^{\Delta(2)}_{\Delta(3)} (\partial^2_{x_2} a_{\lambda,1})(x_{J+1}, x_J, \ldots, x_3, x_0), \qquad (8.4.132)$$

and there exist functions $c_{2,0}(x_3, x_0)$ and $c_{2,1}(x_3, x_0)$, which are algebraic functions of $\partial^k_{x_2}(S_3(x_3, x_2^*) + S^*_{2,1}(x_2^*, x_0))$, $(k = 2, 3)$, such that

$$\begin{aligned}
b_1(a_{\lambda,1})(x_{J+1}, x_J, \ldots, x_3, x_0) =& \iota^{\Delta(2)}_{\Delta(3)} a_{\lambda,1}(x_{J+1}, \ldots, x_3, x_0) c_{2,0}(x_3, x_0) \qquad (8.4.133) \\
&+ \iota^{\Delta(2)}_{\Delta(3)} \partial_{x_2} a_\lambda(x_{J+1}, \ldots, x_3, x_0) c_{2,1}(x_3, x_0),
\end{aligned}$$

and $b_2(a_{\lambda,1})(\nu, x_{J+1}, x_J, \ldots, x_3, x_0)$ is the remainder. Corresponding to (8.4.87) we have

$$I_{1,0} = I_{2,0} + I_{2,1} + I_{2,2} + I_{2,3},$$

where

$$\begin{aligned}
I_{2,0} =& \left(\frac{\nu}{2\pi i T_3}\right)^{1/2} \prod_{j=4}^{J+1} \left(\frac{\nu}{2\pi i \tau_j}\right)^{1/2} \\
&\times \int_{\mathbf{R}^{J-2}} e^{i\nu(\sum_{j=4}^{J+1} S_j(x_j, x_{j-1}) + S^*_{3,1}(x_3, x_0))} D_{x_2^*}(S_3 + S^*_{2,1}; x_3, x_0)^{-1/2}
\end{aligned}$$

$$\times (\iota^{\Delta(2)}_{\Delta(3)} a_{\lambda,1})(x_{J+1},\dots,x_3,x_0)\prod_{j=3}^{J} dx_j,$$

$$I_{2,1} = \frac{i\tau_3 T_2}{2\nu T_3}\left(\frac{\nu}{2\pi i T_3}\right)^{1/2}\prod_{j=4}^{J+1}\left(\frac{\nu}{2\pi i \tau_j}\right)^{1/2} \times \int_{\mathbf{R}^{J-2}} e^{i\nu(\sum_{j=4}^{J+1} S_j(x_j,x_{j-1})+S^*_{3,1}(x_3,x_0))} D_{x_2^*}(S_3+S^*_{2,1};x_3,x_0)^{-1/2} \times b_0(a_{\lambda,1})(x_{J+1},x_J,\dots,x_3,x_0)\prod_{j=3}^{J} dx_j, \tag{8.4.134}$$

$$I_{2,2} = \frac{(\tau_3 T_2)^2}{\nu T_3}\left(\frac{\nu}{2\pi i T_3}\right)^{1/2}\prod_{j=4}^{J+1}\left(\frac{\nu}{2\pi i \tau_j}\right)^{1/2} \times \int_{\mathbf{R}^{J-2}} e^{i\nu(\sum_{j=4}^{J+1} S_j(x_j,x_{j-1})+S^*_{3,0}(x_3,x_0))} D_{x_2^*}(S_3+S^*_{2,1};x_3,x_0)^{-1/2} \times b_1(a_{\lambda,1})(x_{J+1},x_J,\dots,x_3,x_0)\prod_{j=3}^{J} dx_j, \tag{8.4.135}$$

and

$$I_{2,3} = \left(\frac{\tau_3 T_2}{\nu i T_3}\right)^2\left(\frac{\nu}{2\pi i T_3}\right)^{1/2}\prod_{j=4}^{J+1}\left(\frac{\nu}{2\pi i \tau_j}\right)^{1/2} \times \int_{\mathbf{R}^{J-2}} e^{i\nu(\sum_{j=4}^{J+1} S_j(x_j,x_{j-1})+S^*_{3,0}(x_3,x_0))} D_{x_2^*}(S_3+S^*_{2,1};x_3,x_0)^{-1/2} \times b_2(a_{\lambda,1})(\nu,x_{J+1},x_J,\dots,x_3,x_0)\prod_{j=3}^{J} dx_j. \tag{8.4.136}$$

Proposition 8.12

$$I_{2,0} = \left(\frac{\nu}{2\pi i T_3}\right)^{1/2}\prod_{j=4}^{J+1}\left(\frac{\nu}{2\pi i \tau_j}\right)^{1/2}\int_{\mathbf{R}^{J-2}} e^{i\nu(\sum_{j=4}^{J+1} S_j(x_j,x_{j-1})+S^*_{3,1}(x_3,x_0))} \times D_{x_2^*,x_1^*}(S_{3,1};x_3,x_0)^{-1/2}(\iota^{\Delta}_{\Delta(3)} a_\lambda)(x_{J+1},\dots,x_3,x_0)\prod_{j=3}^{J} dx_j.$$

Proof Since we know by (8.4.13) that

$$D_{x_2^*}(S_3+S_{2,1}^*;x_3,x_0)\iota_{\Delta(3)}^{\Delta(2)}(D_{x_1^*}(S_{2,0};x_2,x_0))=D_{x_2^*,x_1^*}(S_{3,1};x_3,x_0),$$

we see that

$$\begin{aligned}&D_{x_2^*}(S_3+S_{2,1}^*;x_3,x_0)^{-1/2}(\iota_{\Delta(3)}^{\Delta(2)}a_{\lambda,1})(x_{J+1},\dots,x_3,x_0)\\&=D_{x_2^*,x_1^*}(S_{3,1};x_3,x_0)^{-1/2}(\iota_{\Delta(3)}^{\Delta}a_\lambda)(x_{J+1},\dots,x_3,x_0).\end{aligned} \tag{8.4.137}$$

□

We do not discuss $I_{2,0}$ any further in Step 2 and we leave it to Step 3.

Mimicking discussions in Step 1, we have the following proposition:

Proposition 8.13 *We can write*

$$\begin{aligned}I_{2,1}&=\left(\frac{iT_2\tau_3}{2\nu T_3}\right)\left(\frac{\nu}{2\pi iT}\right)^{1/2}e^{i\nu S^*(x_{J+1},x_0)}\\&\quad\times\left(D_{x_J^*\dots x_1^*}(S;x_{J+1},x_0)^{-1/2}p_2(x_{J+1},x_0)+\nu^{-1}Tr_{2,1}(\nu,x_{J+1},x_0)\right),\\I_{2,2}&=\frac{(\tau_3T_2)^2}{\nu T_3}\left(\frac{\nu}{2\pi iT}\right)^{1/2}e^{i\nu S^*(x_{J+1},x_0)}q_2(\nu,x_{J+1},x_0),\\I_{2,3}&=\left(\frac{T_2\tau_3}{\nu T_3}\right)^2\left(\frac{\nu}{2\pi iT}\right)^{1/2}e^{i\nu S^*(x_{J+1},x_0)}r_2(\nu,x_{J+1},x_0).\end{aligned} \tag{8.4.138}$$

Here

$$\begin{aligned}&p_2(x_{J+1},x_0)\\&=\iota_{\Delta(J+1)}^{\Delta(2)}\left(D_{x_1^*}(S_{2,1};x_2,x_0)^{1/2}\partial_{x_2}^2(D_{x_1^*}(S_{2,1};x_2,x_0)^{-1/2}\iota_{\Delta(2)}^{\Delta}a_\lambda)\right)(x_{J+1},x_0),\end{aligned} \tag{8.4.139}$$

and for any $K\ge 0$ there exist a constant $C_K>0$ and an integer $M(K)\ge 0$ such that

$$\begin{aligned}&(1+\lambda+|x_0|+|x_{J+1}|)^{-m}\left(|\partial_{x_{J+1}}^{\alpha_{J+1}}\partial_{x_0}^{\alpha_0}r_{2,1}(\nu,x_{J+1},x_0)|+|\partial_{x_{J+1}}^{\alpha_{J+1}}\partial_{x_0}^{\alpha_0}q_2(x_{J+1},x_0)|\right.\\&\left.+|\partial_{x_{J+1}}^{\alpha_{J+1}}\partial_{x_0}^{\alpha_0}r_2(\nu,x_{J+1},x_0)|\right)\le C_KA_{M(K)+2}X_{M(K)+2}^4,\end{aligned} \tag{8.4.140}$$

as far as $|\alpha_{J+1}|\le K$ and $|\alpha_0|\le K$. The number $M(K)$ is the same as the one in Step 1.

Proof We have only to prove (8.4.139). In the same way as we proved Proposition 8.7, we obtain

$$\begin{aligned}&p_2(x_{J+1},x_0)\\&=D_{x_J^*x_{J-1}^*\dots x_2^*}\left(S_{J+1,3}+S_{2,1}^*;x_{J+1},x_0\right)^{-1/2}(\iota_{\Delta(J+1)}^{\Delta(2)}\partial_{x_2}^2a_{\lambda,1})(x_{J+1},x_0).\end{aligned} \tag{8.4.141}$$

On the other hand, by Lemma 5.2

$$\begin{aligned} &D_{x_J^* x_{J-1}^* \dots x_1^*}\left(S_{J+1,1}; x_{J+1}, x_0\right) \\ &= D_{x_J^* x_{J-1}^* \dots x_2^*}\left(S_{J+1,3} + S_{2,1}^*; x_{J+1}, x_0\right) \iota_{\Delta(J+1)}^{\Delta(2)} D_{x_1^*}(S_{2,1}; x_2, x_0). \end{aligned}$$

Substituting this in (8.4.141) and using (8.4.130), we can prove (8.4.139). □

Step 2 ends here.

Continuing this process k times, $(2 \le k \le J-1)$, we discuss Step k.

Step k.

We discuss $I_{k-1,0}$ in the same way as in Step 2. Define

$$a_{\lambda,k-1}(x_{J+1}, x_J, \dots, x_k, x_0) = D_{x_{k-1}^* \dots x_1^*}(S_{k,1}; x_k, x_0)^{-1/2} \iota_{\Delta(k)}^{\Delta} a_\lambda(x_{J+1}, x_J, \dots, x_k, x_0).$$

Then

$$\begin{aligned} I_{k-1,0} &= \left(\frac{\nu}{2\pi i T_k}\right)^{1/2} \prod_{j=k+1}^{J+1} \left(\frac{\nu}{2\pi i \tau_j}\right)^{1/2} \\ &\times \int_{\mathbf{R}^{J-k+1}} e^{i\nu(\sum_{j=k+1}^{J+1} S_j(x_j, x_{j-1}) + S_{k,1}^*(x_k, x_0))} a_{\lambda,k-1}(x_{J+1}, x_J, \dots, x_k, x_0) \prod_{j=k}^{J} dx_j \\ &= I(\Delta(k); S_{J+1,k+1} + S_{k,1}^*, a_{\lambda,k-1})(\nu, x_{J+1}, x_0). \end{aligned}$$

Consider this as an oscillatory integral corresponding to division $\Delta(k)$ of interval $[0, T]$ with the amplitude function $a_{\lambda,\, k-1}$.

It follows from Lemma 5.4 that there exists a positive constant B_K such that

$$B_K = \max_{\substack{|\alpha_0| \le K, |\alpha_k| \le K, \\ k=2,3,\dots J}} \sup_{(x_k, x_0) \in \mathbf{R}^2} |\partial_{x_k}^{\alpha_k} \partial_{x_0}^{\alpha_0} D_{x_{k-1}^*, \dots, x_1}(S_{k,1}; x_k, x_0)^{-1/2}| < \infty.$$

Therefore, we can follow the discussions in Step 2 and prove:

Proposition 8.14 *In the case* $2 \le k \le J-1$

$$I_{k-1,0} = I_{k,0} + I_{k,1} + I_{k,2} + I_{k,3}.$$

Here

$$\begin{aligned} I_{k,0} &= \left(\frac{\nu}{2\pi i T_{k+1}}\right)^{1/2} \prod_{j=k+2}^{J+1} \left(\frac{\nu}{2\pi i \tau_j}\right)^{1/2} \\ &\times \int_{\mathbf{R}^{J-k}} e^{i\nu(\sum_{j=k+2}^{J+1} S_j(x_j, x_{j-1}) + S_{k+1,1}^*(x_{k+1}, x_0))} D_{x_k^* \dots x_1^*}(S_{k+1,1}; x_{k+1}, x_0)^{-1/2} \end{aligned}$$

$$\times\ (\iota^{\Delta}_{\Delta(k+1)}a_\lambda)(x_{J+1},\ldots,x_{k+1},x_0)\prod_{j=k+1}^{J}dx_j,$$

$$\begin{aligned} I_{k,1} &= \frac{iT_k\tau_{k+1}}{2\nu T_{k+1}}\left(\frac{\nu}{2\pi iT}\right)^{1/2}e^{i\nu S^*(x_{J+1},x_0)} \\ &\quad\times\left(D_{x_J^*x_{J-1}^*\ldots x_1^*}(S;x_{J+1},x_0)^{-1/2}p_k(x_{J+1},x_0)+\nu^{-1}Tr_{k,1}(\nu,x_{J+1},x_0)\right), \\ I_{k,2} &= \frac{(T_k\tau_{k+1})^2}{\nu T_{k+1}}\left(\frac{\nu}{2\pi iT}\right)^{1/2}e^{i\nu S^*(x_{J+1},x_0)}q_k(\nu,x_{J+1},x_0), \\ I_{k,3} &= \left(\frac{\tau_{k+1}T_k}{\nu T_{k+1}}\right)^2\left(\frac{\nu}{2\pi iT}\right)^{1/2}e^{i\nu S^*(x_{J+1},x_0)}r_k(\nu,x_{J+1},x_0). \end{aligned} \tag{8.4.142}$$

Here

$$\begin{aligned} p_k(x_{J+1},x_0) &= \iota^{\Delta(k)}_{\Delta(J+1)}\Big[D_{x_{k-1}^*x_{k-1}^{\cdot}\ldots,x_1^*}(S_{k,1};x_k,x_0)^{1/2} \\ &\quad\times\partial^2_{x_k}(D_{x_{k-1}^*x_{k-1}^{\cdot}\ldots,x_1^*}(S_{k,1};x_k,x_0)^{-1/2}\iota^{\Delta}_{\Delta(k)}a_\lambda)\Big](x_{J+1},x_0), \end{aligned} \tag{8.4.143}$$

and for any $K\geq 0$ there exist a constant C_K and an integer $M(K)$ such that

$$\begin{aligned} &(1+\lambda+|x_0|+|x_{J+1}|)^{-m}\left(|\partial^{\alpha_{J+1}}_{x_{J+1}}\partial^{\alpha_0}_{x_0}q_k(x_{J+1},x_0)|+|\partial^{\alpha_{J+1}}_{x_{J+1}}\partial^{\alpha_0}_{x_0}r_k(\nu,x_{J+1},x_0)|\right. \\ &\qquad\left.+|\partial^{\alpha_{J+1}}_{x_{J+1}}\partial^{\alpha_0}_{x_0}r_{k,1}(\nu,x_{J+1},x_0)|\right) \\ &\leq C_K A_{M(K)+2}X^4_{M(K)+2}, \end{aligned} \tag{8.4.144}$$

as far as $|\alpha_{J+1}|\leq K$ and $|\alpha_0|\leq K$. We can choose so that C_K and $M(K)$ are independent of k.

Step k ends here.

Step J.

We have

$$I_{J-1,0}=\left(\frac{\nu}{2\pi iT_J}\right)^{1/2}\left(\frac{\nu}{2\pi i\tau_{J+1}}\right)^{1/2}$$
$$\int_{\mathbf{R}^1}e^{i\nu(S_{J+1}(x_{J+1},x_J)+S^*_{J,1}(x_J,x_0))}a_{\lambda,J-1}(x_{J+1},x_J,x_0)dx_J.$$

By the stationary phase method, Theorem 8.5,

Proposition 8.15

$$I_{J-1,0}=I_{J,0}+I_{J,1}+I_{J,2}+I_{J,3}, \tag{8.4.145}$$

where

$$I_{J,0} = \left(\frac{\nu}{2\pi i T}\right)^{1/2} e^{\nu S^*(x_{J+1},x_0)} D_{x_J^* \dots x_1^*}(S; x_{J+1}, x_0)^{-1/2} (\iota_{\Delta(J+1)}^{\Delta} a_\lambda)(x_{J+1}.x_0)$$

and

$$\begin{aligned} I_{J,1} &= \frac{iT_J \tau_{J+1}}{2\nu T} \left(\frac{\nu}{2\pi i T}\right)^{1/2} e^{i\nu S^*(x_{J+1},x_0)} D_{x_J^* \dots x_1^*}(S; x_{J+1}, x_0)^{-1/2} p_J(x_{J+1}, x_0), \\ I_{J,2} &= \frac{(T_J \tau_{J+1})^2}{\nu T} \left(\frac{\nu}{2\pi i T}\right)^{1/2} e^{i\nu S^*(x_{J+1},x_0)} q_J(\nu, x_{J+1}, x_0), \\ I_{J,3} &= \left(\frac{\tau_{J+1} T_J}{\nu T_{J+1}}\right)^2 \left(\frac{\nu}{2\pi i T}\right)^{1/2} e^{i\nu S^*(x_{J+1},x_0)} r_J(\nu, x_{J+1}, x_0). \end{aligned} \tag{8.4.146}$$

Here

$$\begin{aligned} p_J(x_{J+1}, x_0) = \iota_{\Delta(J+1)}^{\Delta(J)} \Big(& D_{x_{J-1}^*, \dots, x_1^*}(S_{J,1}; x_J, x_0)^{1/2} \\ & \times \partial_{x_J}^2 (D_{x_{J-1}^*, \dots, x_1^*}(S_{J,1}; x_J, x_0)^{-1/2} \iota_{\Delta(J)}^{\Delta} a_\lambda) \Big)(x_{J+1}, x_0), \end{aligned} \tag{8.4.147}$$

and for any $K \geq 0$ there exist a constant $C_K > 0$ and an integer $M(K) \geq 0$ such that

$$\begin{aligned} &(1 + \lambda + |x_0| + |x_{J+1}|)^{-m} \left(|\partial_{x_{J+1}}^{\alpha_{J+1}} \partial_{x_0}^{\alpha_0} q_J(x_{J+1}, x_0)| + |\partial_{x_{J+1}}^{\alpha_{J+1}} \partial_{x_0}^{\alpha_0} r_J(\nu, x_{J+1}, x_0)| \right) \\ &\leq C_K A_{M(K)+2} X_{M(K)+2}^4 \end{aligned} \tag{8.4.148}$$

as far as $|\alpha_{J+1}| \leq K$ and $|\alpha_0| \leq K$. □

Step J ends here.
The final step of proof of Theorem 8.7.

As a consequence of Step 1 to Step J, it turns out that

$$I(\Delta; S, a, \nu)(x_{J+1}, x_0) = I_{J,0} + \sum_{k=1}^{J} \left(I_{k,1} + I_{k,2} + I_{k,3} \right).$$

Define

$$p(\Delta, x_{J+1}, x_0) = T^{-1} \sum_{k=1}^{J} \frac{iT_k \tau_{k+1}}{2T_{k+1}} p_k(x_{J+1}, x_0). \tag{8.4.149}$$

Then

$$\sum_{k=1}^{J} I_{k,1} = \left(\frac{\nu}{2\pi i T}\right)^{1/2} e^{i\nu S^*(x_{J+1},x_0)} \tag{8.4.150}$$

$$\times \Big(D_{x_J^* x_{J-1}^* \dots x_1^*}(S; x_{J+1}, x_0)^{-1/2} \nu^{-1} T p(\Delta, x_{J+1}, x_0) + \nu^{-1} T \sum_{k=1}^{J} \frac{i T_k \tau_{k+1}}{2\nu T_{k+1}} r_{k,1}(\nu, x_{J+1}, x_0)\Big).$$

Define

$$q(\Delta, \nu, x_{J+1}, x_0) = T^{-2} |\Delta|^{-1} \sum_{k=1}^{J} \frac{(\tau_{k+1} T_k)^2}{T_{k+1}} q_k(\nu, x_{J+1}, x_0). \tag{8.4.151}$$

Then

$$\sum_{k=1}^{J} I_{k,2} = \nu^{-1} T^2 |\Delta| \left(\frac{\nu}{2\pi i T}\right)^{1/2} e^{i\nu S^*(x_{J+1}, x_0)} q(\Delta, \nu, x_{J+1}, x_0). \tag{8.4.152}$$

Define

$$r(\Delta, \nu, x_{J+1}, x_0) = T^{-2} \sum_{k=1}^{J} \left(\frac{\tau_{k+1} T_k}{T_{k+1}}\right)^2 r_k(\nu, x_{J+1}, x_0) + T^{-1} \sum_{k=1}^{J-1} \frac{T_k \tau_{k+1}}{2i T_{k+1}} r_{k,1}(\nu, x_{J+1}, x_0).$$

Then

$$\sum_{k=1}^{J} (I_{k,1} + I_{k,3}) = \left(\frac{\nu}{2\pi i T}\right)^{1/2} e^{i\nu S^*(x_{J+1}, x_0)} \Big(p(\Delta, x_{J+1}, x_0) + \nu^{-2} T^2 r(\Delta, \nu, x_{J+1}, x_0)\Big).$$

Therefore, equality (8.4.79) in Theorem 8.7 holds.

Notice that

$$\sum_{k=1}^{J} \frac{(\tau_{k+1} T_k)^2}{T_{k+1}} \le |\Delta| T^2, \quad \sum_{k=1}^{J} \frac{\tau_{k+1} T_k}{T_{k+1}} \le T, \quad \text{and} \quad \sum_{k=1}^{J} \left(\frac{\tau_{k+1} T_k}{T_{k+1}}\right)^2 \le |\Delta| T \le T^2.$$

It follows from these together with (8.4.98), (8.4.110), (8.4.129), (8.4.144) and (8.4.148) that

$$(1 + \lambda + |x_0| + |x_{J+1}|)^{-m} \Big(|\partial_{x_{J+1}}^{\alpha} \partial_{x_0}^{\beta} q(\Delta, \nu, x_{J+1}, x_0)| + |\partial_{x_{J+1}}^{\alpha} \partial_{x_0}^{\beta} r(\Delta, \nu, x_{J+1}, x_0)|\Big) \le C_K A_{M(K)+2} X_{M(K)+2}^4.$$

Thus (8.4.81) and (8.4.82) are proved.

Theorem 8.7 has been proved. □

Chapter 9
L^2-boundedness of Oscillatory Integral Operators

Abstract The proof of L^2-boundedness of oscillatory integral operators is given in this chapter in a little more general form than cited in Theorem 6.3.

9.1 L^2-boundedness of Oscillatory Integral Operators

We shall discuss the following integral transformations with a parameter $\nu \geq 1$:

$$A(\nu)\ f(x) = \iint_{\mathbf{R}^n \times \mathbf{R}^m} a(x,\theta,y) e^{i\nu\phi(x,\theta,y)} f(y)\, dy d\theta. \tag{9.1.1}$$

The phase function ϕ and the amplitude a satisfy the following conditions:

(A-I) $\phi(x,\theta,y)$ is a real-valued C^∞-function of $(x,\theta,y) \in \mathbf{R}^n \times \mathbf{R}^m \times \mathbf{R}^n$.

(A-II) There exists a constant $\kappa_0 > 0$ such that $|\det D(\phi)(x,\theta,y)| \geq \kappa_0$, where $D(\phi)(x,\theta,y)$ is the $m+n$ square matrix

$$D(\phi)(x,\theta,y) = \begin{pmatrix} \partial_x \partial_y \phi(x,\theta,y) & \partial_x \partial_\theta \phi(x,\theta,y) \\ \partial_\theta \partial_y \phi(x,\theta,y) & \partial_\theta \partial_\theta \phi(x,\theta,y) \end{pmatrix}. \tag{9.1.2}$$

(A-III) Every entry of the matrix $D(\phi)(x,\theta,y)$ is uniformly bounded, together with its all derivatives, i.e., for any multi-indices α, β, γ there exists a constant $C_{\alpha,\beta,\gamma} > 0$ such that

$$|\partial_x^\alpha \partial_y^\beta \partial_\theta^\gamma d(\phi)(x,\theta,y)| \leq C_{\alpha,\beta,\gamma}, \tag{9.1.3}$$

where $d(\phi)(x,\theta,y)$ denotes each of entries of $D(\phi)(x,\theta,y)$.

(A-IV) $a(x,\theta,y)$ is uniformly bounded, together with its all derivatives, i.e., for any multi-indices α, β, γ there exists a constant $C_{\alpha,\beta,\gamma} > 0$ such that

$$|\partial_x^\alpha \partial_y^\beta \partial_\theta^\gamma a(x,\theta,y)| \leq C_{\alpha,\beta,\gamma}. \tag{9.1.4}$$

D. Fujiwara, *Rigorous Time Slicing Approach to Feynman Path Integrals*,
Mathematical Physics Studies, DOI 10.1007/978-4-431-56553-6_9

The following symbols are convenient: Let $k \geq 0$ be an integer. For a function $a(x, \theta, y)$

$$\|a\|_{\mathscr{B}^k} = \sum_{|\alpha|+|\beta|+|\gamma|\leq k} \sup_{(x,\theta,y)\in\mathbf{R}^{n+m+n}} |\partial_x^\alpha \partial_y^\beta \partial_\theta^\gamma a(x, \theta, y)|, \tag{9.1.5}$$

and

$$\mathscr{B}^k(\mathbf{R}^{n+m+n}) = \{a(x, \theta, y); \|a\|_{\mathscr{B}^k} < \infty\}. \tag{9.1.6}$$

Since (A-II) implies that the matrix $D(\phi)(x, \theta, y)$ is non-singular, we write its inverse $D(\phi)^{-1}(x, \theta, y)$. Let $\kappa(2, \ell + 2)$ be a constant such that

$$\max \sum_{|\alpha|+|\beta|+|\gamma|\leq \ell} a \sup_{x,\theta,y} \left|\partial_x^\alpha \partial_y^\beta \partial_\theta^\gamma d(\phi)(x, \theta, y)\right| \leq \kappa(2, \ell + 2), \tag{9.1.7}$$

where max is taken among all entries of the matrix $D(\phi)$. We abbreviate $\kappa(2, 2)$ to κ_2. It follows from Cramére's formula that there exists a constant $C(m, n) > 0$ depending only on m, n such that every entry of $|D(\phi)^{-1}(x, \theta, y)| \leq C(m, n)\kappa_0^{-1}\kappa_2^{m+n}$.

From now on we write $C(m, n)$ for various positive constants depending only on m, n.

Let $\mathrm{dist}((x, \theta), (x', \theta'))$ denote the distance between two points (x, θ) and (x', θ') in $\mathbf{R}^{n+m}$. i.e.,

$$\mathrm{dist}((x, \theta), (x', \theta')) = (|x - x'|^2 + |\theta - \theta|^2)^{1/2}. \tag{9.1.8}$$

Lemma 9.1 *Suppose that the phase function ϕ fulfills (A-I), (A-II) and (A-III). Then there exists a constant $\kappa_3 > 0$ depending only on m, n, κ_0, κ_2 such that*

$$\begin{aligned} &\left|\partial_y\phi(x, \theta, y) - \partial_y\phi(x', \theta', y)\right|^2 + \left|\partial_\theta\phi(x, \theta, y) - \partial_\theta\phi(x', \theta', y)\right|^2 \\ &\quad \geq \kappa_3 \mathrm{dist}((x, \theta), (x', \theta'))^2, \\ &\left|\partial_x\phi(x, \theta, y) - \partial_x\phi(x, \theta', y')\right|^2 + \left|\partial_\theta\phi(x, \theta, y) - \partial_\theta\phi(x, \theta', y')\right|^2 \\ &\quad \geq \kappa_3 \mathrm{dist}((y, \theta), (y', \theta'))^2. \end{aligned} \tag{9.1.9}$$

Proof Let $z = \partial_y\phi(x, \theta, y)$ and $\eta = \partial_\theta\phi(x, \theta, y)$. If y is fixed, then it turns out from the global implicit function theorem of Hadamard [58] that

$$T_y : \mathbf{R}^{n+m} \ni (x, \theta) \to (z, \eta) \in \mathbf{R}^{n+m}$$

is a global diffeomorphism. We write $(x(z, \eta, y), \theta(z, \eta, y))$ for $T_y^{-1}(z, \eta)$. Since every entry of the Jacobian matrix $\partial(x, \theta)/\partial(z, \eta) = D(\phi)^{-1}(x, \theta, y)$ is bounded, there exists a constant $C(m, n) > 0$ depending only on m, n, κ_0, κ_2 such that

$$\begin{aligned} &\left|x(y, \eta, z) - x(y, \eta', z')\right|^2 + \left|\theta(y, \eta, z) - \theta(y, \eta', z')\right|^2 \\ &\quad \leq C(m, n)(|z - z'|^2 + |\eta - \eta'|^2). \end{aligned} \tag{9.1.10}$$

Let $x' = x(y, \eta', z')$ and $\theta' = \theta(y, \eta', z')$. Then

$$z' = \partial_y \phi(x', \theta', y) \qquad \eta' = \partial_\theta \phi(x', \theta', y).$$

Putting these in (9.1.10), we have

$$\begin{aligned}
&|x - x'|^2 + |\theta - \theta'|^2 \\
&\quad \le C(m, n) \Big(\big|\partial_y \phi(x, \theta, y) - \partial_y \phi(x', \theta', y)\big|^2 \\
&\qquad\qquad + \big|\partial_\theta \phi(x, \theta, y) - \partial_\theta \phi(x', \theta', y)\big|^2 \Big) .
\end{aligned}$$

This is equivalent to the first inequality of the lemma. Similarly, we can prove the inequality equivalent to the second inequality. If we choose $\kappa_3 > 0$ suitably, then the lemma holds. □

The main aim in the chapter is to give the proof of the following theorem, cf. [10].

Theorem 9.1 (L^2-boundedness theorem) *Let $\nu \ge 1$. Suppose that $\phi(x, \theta, y)$ fulfills (A-I), (A-II) and (A-III) above and that $a(x, \theta, y)$ satisfies (A-IV). Then there exists a positive C depending on m, n, κ_0, κ_2 and $\kappa(2, 8(m + n))$ such that*

$$\|A(\nu) f\| \le C\nu^{-(m+n)/2} \|a\|_{\mathscr{B}^{8(m+n)+2}} \|f\| \tag{9.1.11}$$

for any $f \in C_0^\infty(\mathbf{R}^n)$. Here $\|a\|$ means the usual norm in $L^2(\mathbf{R}^n)$ with respect to Lebesgue's measure dx in $\mathbf{R}^n$.

Remark 9.1 In the case $m = 0$, (9.1.1) then reduces to

$$A(\nu) f(x) = \iint_{\mathbf{R}^n} a(x, y) e^{i\nu\phi(x,y)} f(y)\, dy, \tag{9.1.12}$$

and the matrix $D(\phi)$ reduces to the $n \times n$ matrix

$$D(\phi)(x, y) = \left(\partial_x \partial_y \phi(x, \theta, y)\right) . \tag{9.1.13}$$

The conditions above reduce to Assumptions 6.1 and 6.2. Theorem 6.3 follows from Theorem 9.1, cf. also [23].

Proof We begin with:

Proposition 9.1 *Inequality (9.1.11) is reduced to the case $\nu = 1$.*

Proof Suppose (9.1.11) is true in the case $\nu = 1$. Let $\nu \ge 1$. Then we shall prove (9.1.11) in the case $\nu \ge 1$. Set $(X, \Theta, Y) = \nu^{1/2}(x, \theta, y)$, $F(Y) = f(y) = f(\nu^{-1/2} Y)$ and $G(X) = A(\nu) f(x) = (A(\nu) f)(\nu^{-1/2} X)$. Then

$$G(X) = \nu^{-(m+n)/2} \iint_{\mathbf{R}^n \times \mathbf{R}^m} b(X, \Theta, Y) e^{i\psi(X,\Theta,Y)} F(Y)\, dY d\Theta, \tag{9.1.14}$$

where

$$b(X, \Theta, Y) = a(\nu^{-1/2}X, \nu^{-1/2}\Theta, \nu^{-1/2}Y) \tag{9.1.15}$$

and

$$\psi(X, \Theta, Y) = \nu\phi(\nu^{-1/2}X, \nu^{-1/2}\Theta, \nu^{-1/2}Y). \tag{9.1.16}$$

By definition

$$\begin{aligned}
&\partial_X^\alpha \partial_\Theta^\beta \partial_Y^\gamma \psi(X, \Theta, Y) \\
&= \nu^{1-(|\alpha|+|\beta|+|\gamma|)/2} \partial_x^\alpha \partial_\theta^\beta \partial_y^\gamma \phi\Big|_{(x,\theta,y)=\nu^{-1/2}(X,\Theta,Y)},
\end{aligned} \tag{9.1.17}$$

$$\begin{aligned}
&\partial_X^\alpha \partial_\Theta^\beta \partial_Y^\gamma b(X, \Theta, Y) \\
&= \nu^{-(|\alpha|+|\beta|+|\gamma|)/2} \partial_x^\alpha \partial_\theta^\beta \partial_y^\gamma a\Big|_{(x,\theta,y)=\nu^{-1/2}(X,\Theta,Y)}.
\end{aligned} \tag{9.1.18}$$

Thus

$$D(\psi)(X, \Theta, Y) = D(\phi)(x, \theta, y), \quad \text{for } (x, \theta, y) = \nu^{-1/2}(X, \Theta, Y).$$

In particular, ϕ and Ψ have common κ_0 and κ_2 in assumptions (A-II) and (A-III). We can also choose common $\kappa(2, \ell + 2)$ for ϕ and ψ, because $\nu \geq 1$ and (9.1.17) with $2 + \ell \geq 2$ hold. Similarly, for any $k \geq 0$

$$\|b\|_{\mathscr{B}^k} \leq \|a\|_{\mathscr{B}^k}. \tag{9.1.19}$$

As we assumed that (9.1.11) is true in the case $\nu = 1$, we apply it to (9.1.14). Then

$$\|G\| \leq C(m, n, \kappa_0, \kappa_2, \kappa(2, 8(m+n))\nu^{-(m+n/2)} \|b\|_{\mathscr{B}^{8(m+n)+2}} \|F\|. \tag{9.1.20}$$

This together with equalities

$$\|F\| = \nu^{n/2}\|f\|, \quad \|G\| = \nu^{n/2}\|A(\nu) f\|$$

prove (9.1.11). □

Now we prove Theorem 9.1 in the case $\nu = 1$. We abbreviate $A(1)$ to A. We use the following partition of unity on $\mathbf{R}^n$ and $\mathbf{R}^m$: Let $\varphi(x)$ be a non-negative function of class C^∞ such that $\int_{\mathbf{R}^n} \varphi(x)dx = 1$ and $\varphi(x) = 0$ outside the unit ball $\{x \in \mathbf{R}^n; |x| \leq 1\}$. Let $\psi(\theta)$ be a similar function on $\mathbf{R}^m$.

Then we define

$$a_{s,\sigma,t}(x,\theta,y) = a(x,\theta,y)\varphi(x-s)\psi(\theta-\sigma)\varphi(y-t) \tag{9.1.21}$$

and

$$A_{s,\sigma,t}\, f(x) = \iint_{\mathbf{R}^m\times\mathbf{R}^n} a_{s,\sigma,t}(x,\theta,y)e^{i\phi(x,\theta,y)} f(y)\, d\theta dy. \tag{9.1.22}$$

We can write

$$Af(x) = \lim_{k\to\infty} \iiint_{|s|^2+|\sigma|^2+|t|^2\le k^2} A_{s,\sigma,t} f(x)\, ds d\sigma dt. \tag{9.1.23}$$

Since (9.1.23) converges strongly in $L^2(\mathbf{R}^n)$ if $f \in C_0^\infty(\mathbf{R}^n)$, Theorem 9.1 follows from the next lemma.

Lemma 9.2 *For any $f \in C_0^\infty(\mathbf{R}^n)$ and for any compact susbset $K \subset \mathbf{R}^{n+m+n}$ we have the estimate*

$$\left\| \iiint_K A_{s,\sigma,t} f\, ds d\sigma dt \right\| \le C\|a\|_{\mathscr{B}^{8(m+n)+2}} \|f\|, \tag{9.1.24}$$

where C is a positive constant depending only on $m, n, \kappa_0, \kappa_2, \kappa(2, 2+8(m+n))$.

Proof Let $p = (s,\sigma,t)$ and $p' = (s',\sigma',t')$. For the proof of the lemma we use Cotlar's lemma [12] on almost orthogonal operators.

Lemma 9.3 (Cotlar's lemma) *Let H be a Hilbert space and $(Z, d\mu)$ be a σ-finite measure space. Let $\{T_z; z \in Z\}$ be a family of bounded linear operators in H such that for any $x, y \in H$ correspondence $Z \ni z \to (T_z x, y)$ is a measurable. Moreover there exists a positive constant M_0 such that*

$$\|T_z\| \le M_0.$$

Suppose there exists a function $k(z,z')$ measurable in $Z \times Z$ with the following two properties:

- *For any $(z,z') \in Z \times Z$*

$$\|T_z T_{z'}^*\| \le k(z,z')^2, \qquad \|T_z^* T_{z'}\| \le k(z,z')^2, \tag{9.1.25}$$

where T_z^* is the adjoint operator of T_z.
- *Define the integral transform on $L^2(Z, d\mu)$ by*

$$Kf(z) = \int_Z k(z,z') f(z') d\mu(z') \quad \textit{for } f(z) \in L^2(Z, d\mu).$$

Then K is a bounded linear operator in $L^2(Z, d\mu)$ with norm M.

Suppose $\{T_z : z \in Z\}$ satisfies the above assumption. Then for any measurable set $W \subset Z$ with finite measure the following estimate holds:

$$\left\| \int_W T_z d\mu(z) \right\| \le M. \tag{9.1.26}$$

Cotlar's lemma will be proved in the next section. Here we continue the proof of Lemma 9.2 using Cotlar's lemma, i.e., we have only to prove the following estimates:

1. There exists a constant C independent of p such that

$$\|A_p\| \le C. \tag{9.1.27}$$

2. Let A_p^* be the adjoint of A_p. Then there exist a function $h(p, p')$ and a constant κ_4 depending only on $m, n, \kappa_0, \kappa_2, \kappa(2, 2 + 8(m + n))$ such that

$$\|A_p A_{p'}^*\| \le h(p, p')^2, \tag{9.1.28}$$

$$\|A_p^* A_{p'}\| \le h(p, p')^2, \tag{9.1.29}$$

$$\sup_p \int_{\mathbf{R}^{(n+m+n)}} h(p, p')\, dp' \le \kappa_4, \tag{9.1.30}$$

$$\sup_{p'} \int_{\mathbf{R}^{(n+m+n)}} h(p, p')\, dp \le \kappa_4. \tag{9.1.31}$$

We first prove (9.1.27). By definition (9.1.22) of A_p

$$|A_p f(x)| \le \|a\|_{\mathscr{B}^0} \varphi(x - s) \int_{\mathbf{R}^n} \varphi(y - t)|f(y)|\, dy. \tag{9.1.32}$$

It follows from Young's inequality that

$$\|A_p f\| \le \|a\|_{\mathscr{B}^0} \|\varphi\|_{\mathscr{B}^0(\mathbf{R}^n)} \|f\|. \tag{9.1.33}$$

(9.1.27) is proved.

We now discuss (9.1.28).

$$A_p A_{p'}^* g(x) = \int_{\mathbf{R}^n} k_{pp'}(x, x') g(x')\, dx', \tag{9.1.34}$$

where

$$k_{pp'}(x, x') = \iiint_{\mathbf{R}^{(n+m+n)}} a_p(x, \theta, y) \overline{a_{p'}(x', \theta', y)} e^{i\Phi(x,\theta,y,\theta',x')}\, dy d\theta d\theta',$$

$$\Phi(x, \theta, y, \theta', x') = \phi(x, \theta, y) - \phi(x', \theta', y).$$

and

$$A_p^* A_{p'} g(x) = \int_{\mathbf{R}^n} \tilde{k}_{pp'}(x, x') g(x')\, dx', \tag{9.1.35}$$

where

$$\begin{aligned}\tilde{k}_{pp'}(x, x') &= \iiint_{\mathbf{R}^{(n+m+n)}} \overline{a_p(y, \theta, x)} a_{p'}(y, \theta', x') e^{i\tilde{\Phi}(x,\theta,y,\theta',x')}\, dy d\theta d\theta', \\ \tilde{\Phi}(x, \theta, y, \theta', x') &= -\phi(y, \theta, x) + \phi(y, \theta', x').\end{aligned}$$

Now we discuss $A_p A_{p'}^*$. Note that

$$\begin{aligned}& a_p(x, \theta, y)\overline{a_{p'}(x', \theta', y)} \\ &= \varphi(x - s)\varphi(x' - s')\psi(\theta - \sigma)\psi(\theta' - \sigma')\varphi(y - t)\varphi(y - t')a(x, \theta, y)\overline{a(x', \theta', y)}.\end{aligned}$$

It is clear that function $k_{pp'}(x, x') = 0$ if $|t - t'| \geq 4$. We shall discuss the case $|t - t'| \leq 4$. Let $\nabla\Phi$ denote the $(m + n + m)$-vector

$$\begin{aligned}\nabla\Phi &= (\partial_\theta \Phi, \partial_y \Phi, \partial_{\theta'} \Phi) \\ &= (\partial_\theta \phi(x, \theta, y), (\partial_y \phi(x, \theta, y) - \partial_y \phi(x', \theta', y)), -\partial_{\theta'} \phi(x', \theta', y)).\end{aligned}$$

It follows from Lemma 9.1 that

$$\begin{aligned}|\nabla\Phi|^2 &\geq 2^{-1}\left(|\partial_\theta \varphi(x, \theta, y) - \partial_{\theta'} \varphi(x', \theta', y)|^2 + |(\partial_y \varphi(x, \theta, y) - \partial_y \varphi(x', \theta', y)|^2\right) \\ &\geq 2^{-1}\kappa_3(|x - x'|^2 + |\theta - \theta'|^2).\end{aligned} \tag{9.1.36}$$

Let $\rho = (1 + |\nabla\Phi|^2)^{1/2}$ and L be the following differential operator of order 1:

$$L = \rho^{-2}\Big(1 + \sum_{j=1}^m \partial_{\theta_j}\Phi \cdot \partial_{\theta_j} + \sum_{j=1}^m \partial_{\theta'_j}\Phi \cdot \partial_{\theta'_j} + \sum_{j=1}^m \partial_{y_j}\Phi \cdot \partial_{y_j}\Big).$$

Then the adjoint L^* of L is

$$L^* = -\rho^{-2}\Big(\sum_{j=1}^m \partial_{\theta_j}\Phi \cdot \partial_{\theta_j} + \sum_{j=1}^m \partial_{\theta'_j}\Phi \cdot \partial_{\theta'_j} + \sum_{j=1}^m \partial_{y_j}\Phi \cdot \partial_{y_j}\Big) + B$$

with

$$B = \rho^{-2} - \sum_{j=1}^m \partial_{\theta_j}(\rho^{-2}\partial_{\theta_j}\Phi) - \sum_{j=1}^m \partial_{\theta'_j}(\rho^{-2}\partial_{\theta'_j}\Phi) - \sum_{j=1}^m \partial_{y_j}(\rho^{-2}\partial_{y_j}\Phi). \tag{9.1.37}$$

Since $Le^{i\Phi} = ie^{i\Phi}$, for any integer $\ell \geq 0$

$$k_{pp'}(x, x') = (-i)^\ell \iiint_{\mathbf{R}^{(n+m+n)}} a_p(x, \theta, y)\overline{a_{p'}(x', \theta', y)} L^\ell e^{i\Phi(x,\theta,y,\theta',x')}\, dyd\theta d\theta'.$$

Integration by parts leads us to

$$\begin{aligned} &k_{pp'}(x, x') \\ &= (-i)^\ell \iiint_{\mathbf{R}^{(n+m+n)}} L^{*\ell} \left(a_p(x, \theta, y)\overline{a_{p'}(x', \theta', y)}\right) e^{i\Phi(x,\theta,y,\theta',x')}\, dyd\theta d\theta'. \end{aligned} \tag{9.1.38}$$

Similarly, in order to treat $A_p^* A_{p'}$, we define

$$\begin{aligned} \nabla\tilde{\Phi} &= (\partial_\theta \tilde{\Phi}, \partial_y \tilde{\Phi}, \partial_{\theta'} \tilde{\Phi}) \\ &= (-\partial_\theta \phi(y, \theta, x),\ (\partial_y \phi(y, \theta', x') - \partial_y \phi(y, \theta, x)),\ \partial_{\theta'} \phi(y, \theta', x')). \end{aligned}$$

Then

$$\begin{aligned} |\nabla\tilde{\Phi}|^2 &= |\partial_\theta \phi(y, \theta, x)|^2 + |\partial_y \phi(y, \theta', x') - \partial_y \phi(y, \theta, x)|^2 + |\partial_{\theta'} \phi(y, \theta', x')|^2 \\ &\geq 2^{-1}(|\partial_\theta \phi(y, \theta, x) - \partial_{\theta'} \phi(y, \theta', x')|^2 + |\partial_y \phi(y, \theta', x') - \partial_y \phi(y, \theta, x)|^2) \\ &\geq 2^{-1}\kappa_3(|x - x'|^2 + |\theta - \theta|^2). \end{aligned} \tag{9.1.39}$$

Let $\tilde{\rho} = (1 + |\nabla\tilde{\Phi}|^2)^{1/2}$ and

$$\tilde{L} = \tilde{\rho}^{-2}\left(1 + \sum_{j=1}^{m} \partial_{\theta_j} \tilde{\Phi} \cdot \partial_{\theta_j} + \sum_{j=1}^{n} \partial_{y_j} \tilde{\Phi} \cdot \partial_{y_j} + \sum_{j=1}^{m} \partial_{\theta'_j} \tilde{\Phi} \cdot \partial_{\theta'_j}\right).$$

Let $\tilde{L}^*$ be its adjoint operator. Then for any integer $\ell \geq 0$

$$\begin{aligned} &\tilde{k}_{pp'}(x, x') \\ &= (-i)^\ell \iiint_{\mathbf{R}^{(n+m+n)}} \tilde{L}^{*\ell} \left(\overline{a_p(y, \theta, x)} a_{p'}(y, \theta', x')\right) e^{i\tilde{\Phi}(x,\theta,y,\theta',x')}\, dyd\theta d\theta'. \end{aligned} \tag{9.1.40}$$

Lemma 9.4 *For any integer $\ell \geq 0$, there exists a constant $\kappa_5(\ell)$ such that*

$$\left| L^{*\ell} \left(a_p(x, \theta, y)\overline{a_{p'}(x', \theta', y)}\right) \right| \leq \kappa_5(\ell)\varphi(x - s)\varphi(x' - s')\rho^{-\ell} \|a\|^2_{\mathscr{B}^\ell} \tag{9.1.41}$$

if

$$|x - s| \leq 1, |\theta - \sigma| \leq 1, |y - t| \leq 1, |y - t'| \leq 1, |\theta' - \sigma'| \leq 1, |x' - s'| \leq 1. \tag{9.1.42}$$

Otherwise,

$$\left|L^{*\ell}\left(a_p(x,\theta,y)\overline{a_{p'}(x',\theta',y)}\right)\right| = 0.$$

Similarly,

$$\left|\tilde{L}^{*\ell}\left(\overline{a_p(y,\theta,x)}a_{p'}(y,\theta',x')\right)\right| \leq \kappa_5(\ell)\varphi(x-t)\varphi(x'-t')\tilde{\rho}^{-\ell}\|a\|^2_{\mathscr{B}^\ell} \quad (9.1.43)$$

if

$$|x-t| \leq 1, |\theta-\sigma| \leq 1, |y-s| \leq 1, |y-s'| \leq 1, |\theta'-\sigma'| \leq 1, |x'-t'| \leq 1. \quad (9.1.44)$$

Otherwise,

$$\left|\tilde{L}^{*\ell}\left(\overline{a_p(y,\theta,x)}a_{p'}(y,\theta',x')\right)\right| = 0.$$

We can choose so that

$$\kappa_5(\ell) = C(\ell,m,n)(1+\kappa(2,4\ell))^{3\ell}, \quad (9.1.45)$$

where $C(\ell,m,n)$ is a positive constant depending only on ℓ, m and n.

Proof Let D denote any one of differential operators $\partial_{\theta_j}, \partial_{\theta'_j}$ for $j = 1,2,\ldots,m$ and ∂_{y_j} for $j = 1,2,\ldots,n$. Let $D^2 = D \cdot D$. That is, D^2 represents one of the differential operators $\partial_{\theta_j}\partial_{\theta_k}$, $\partial_{\theta_j}\partial_{\theta'_k}$, $\partial_{\theta'_j}\partial_{\theta'_k}$, $\partial_{\theta_j}\partial_{y_q}$, $\partial_{\theta'_j}\partial_{y_q}$ and $\partial_{y_q}\partial_{y_r}$ for $j,k = 1,2,\ldots,m$, $q,r = 1,2,\ldots,n$. Similarly, we shall use symbols D^3, D^4, ...

We claim that

$$\begin{aligned} &L^{*\ell}\left(a_p(x,\theta,y)\overline{a_{p'}(x',\theta',y)}\right) \\ &= \sum \rho^{-k}\cdot D^{r_1}\Phi\cdot D^{r_2}\Phi\cdots D^{r_q}\Phi\cdot D^u(a_p\overline{a_{p'}}), \end{aligned} \quad (9.1.46)$$

where the summation ranges over

$$2\ell \leq k \leq 4\ell, \quad -2\ell \leq q-k \leq -\ell, \quad r_1+r_2+\cdots+r_q \leq k, \quad u \leq \ell. \quad (9.1.47)$$

Assuming for the moment that (9.1.46) and (9.1.47) are true, we prove Lemma 9.4. Since $|D\Phi(x,\theta,y,\theta',x')| \leq \rho$ and $|D^r\Phi(x,\theta,y,\theta',x')| \leq \kappa(2,r)\rho$,

$$\begin{aligned} &|\rho^{-k}D^{r_1}\Phi\cdot D^{r_2}\Phi\cdots D^{r_q}\Phi\cdot D^u(a_p\overline{a_{p'}})| \\ &\leq \rho^{-k}(1+\kappa(2,4\ell))^q\rho^q\|a_p\overline{a_{p'}}\|_{\mathscr{B}^\ell} \\ &\leq C(\ell,m,n)\rho^{-\ell}(1+\kappa(2,4\ell))^{3\ell}\varphi(x-s)\varphi(x'-s')\|a\|^2_{\mathscr{B}^\ell}. \end{aligned}$$

(9.1.41) and (9.1.45) follow from this and (9.1.46).

We prove (9.1.46) and (9.1.47). We call the term $\rho^{-k} D^{r_1}\Phi \cdot D^{r_2}\Phi \cdots D^{r_q}\Phi \cdot D^u(a_p\overline{a_{p'}})$ of type $(-k, q-k, r_1+r_2+\cdots+r_q-k, u)$, which is an element of $\mathbf{Z}^4$. For instance $D\Phi$ is of type (0,1,1,0) and $D\rho$ is a sum of terms of type $(-1, 1, 2, 0)$, because $D\rho = (2\rho)^{-1}\sum D^2\Phi \cdot D\Phi$.

If two terms one of which is of type $(-k, q-k, r_1+r_2+\cdots+r_q-k, u)$ and the other is of type $(-k', q'-k', r_1'+r_2'+\cdots+r_{q'}'-k', 0)$ are given, then their product is of type $(-k-k', q+q'-(k+k'), r_1+r_2+\cdots+r_q+r_1'+r_2'+\cdots+r_{q'}'-(k+k'), u)$.

Since $D(\rho^{-2}D\Phi) = -\rho^{-4}\sum D\Phi D\Phi D^2\Phi + \rho^{-2}D^2\Phi$, this is a sum of the terms of types $(-4, -1, 0, 0)$ and $(-2, -1, 0, 0)$. Therefore, B of (9.1.37) is a sum of terms of the types $(-2, -2, -2, 0)$, $(-4, -1, 0, 0)$ and $(-2, -1, 0, 0)$. Computation shows that

$$
\begin{aligned}
&D(\rho^{-k} \cdot D^{r_1}\Phi \cdot D^{r_2}\Phi \cdots D^{r_q}\Phi \cdot D^u(a_p\overline{a_{p'}})) \\
&= -k\rho^{-k-1} D\rho \cdot D^{r_1}\Phi \cdot D^{r_2}\Phi \cdots D^{r_q}\Phi \cdot D^u(a_p\overline{a_{p'}}) \\
&+ \rho^{-k} D(D^{r_1}\Phi \cdot D^{r_2}\Phi \cdots D^{r_q}\Phi) \cdot D^u(a_p\overline{a_{p'}}) \\
&+ \rho^{-k} D^{r_1}\Phi \cdot D^{r_2}\Phi \cdots D^{r_q}\Phi \cdot D^{u+1}(a_p\overline{a_{p'}}).
\end{aligned}
$$

This is a sum of terms of types

$$
\begin{gathered}
(-k-2, q-k, r_1+r_2+\cdots+r_q-k+1, u), \\
(-k, q-k, r_1+r_2+\cdots+r_q-k+1, u)
\end{gathered}
$$

and

$$
(-k, q-k, r_1+r_2+\cdots+r_q-k, u+1).
$$

Since $\rho^{-2}D\Phi$ is of type $(-2, -1, -1, 0)$,

$$
L^*(\rho^{-k} \cdot D^{r_1}\Phi \cdot D^{r_2}\Phi \cdots D^{r_q}\Phi \cdot D^u(a_p\overline{a_{p'}})
$$

is a sum of terms of types

$$
\begin{gathered}
(-k-4, q-k-1, r_1+r_2+\cdots+r_q-k, u), \\
(-k-2, q-k-1, r_1+r_2+\cdots+r_q-k, u), \\
(-k-2, q-k-1, r_1+r_2+\cdots+r_q-k-1, u+1), \\
(-k-2, q-k-2, r_1+r_2+\cdots+r_q-k-2, u).
\end{gathered}
$$

Starting from $a_p\overline{a_{p'}}$ which is of the type (0, 0, 0, 0), we apply this rule ℓ times and obtain (9.1.46) and (9.1.47). We have proved the claim and Lemma 9.4. □

We continue the proof of Lemma 9.2 by using Lemma 9.4 with $\ell_0 = 2(m+2n+1)$. Condition (9.1.42) implies that $|t - t'| \le |y - t| + |y - t'| \le 2$. Let $\omega(y)$ be the characteristic function of the ball $\{y \in \mathbf{R}^n; |y| \le 4\}$. Then

$$|k_{pp'}(x,x')| \tag{9.1.48}$$
$$\le \kappa_5(\ell_0)\|a\|^2_{\mathscr{B}^{\ell_0}}\omega(t-t')\varphi(x-s)\varphi(x'-s')\iiint_{\substack{|\theta-\sigma|\le 1,|\theta'-\sigma'|\le 1,\\ |y-t|\le 1,|y-t'|\le 1,}} \rho^{-\ell_0}\,d\theta d\theta' dy.$$

Since the volume of unit-balls $\mathbf{R}^n$ or $\mathbf{R}^m$ is finite, it follows from Lemma 9.1 that

$$|k_{pp'}(x,x')| \tag{9.1.49}$$
$$\le C(m,n)\kappa_5(\ell_0)\|a\|^2_{\mathscr{B}^\ell}\omega(t-t')\varphi(x-s)\varphi(x'-s')P(s,s',\theta,\theta')^{-\ell_0},$$

where

$$P(s,s',\sigma,\sigma') = \inf_{\substack{|\theta-\sigma|\le 1,|\theta'-\sigma'|\le 1\\ |x-s|\le 1,|x'-s'|\le 1}} (1+2^{-1}\kappa_3(|x-x'|^2+|\theta-\theta'|^2))^{1/2}. \tag{9.1.50}$$

This leads us to

$$sup_{x'\in\mathbf{R}^n}\int_{\mathbf{R}^n}|k_{pp'}(x,x')|\,dx \le C(m,n)\kappa_5(\ell_0)\|a\|^2_{\mathscr{B}^\ell}\omega(t-t')P(s,s',\sigma,\sigma')^{-\ell_0},$$
$$sup_{x\in\mathbf{R}^n}\int_{\mathbf{R}^n}|k_{pp'}(x,x')|\,dx' \le C(m,n)\kappa_5(\ell_0)\|a\|^2_{\mathscr{B}^\ell}\omega(t-t')P(s,s',\sigma,\sigma')^{-\ell_0}.$$

It follows from these and Young's inequality that

$$\|A_pA^*_{p'}\| \le C(m,n)\kappa_5(\ell_0)\|a\|^2_{\mathscr{B}^\ell}\omega(t-t')P(s,s',\sigma,\sigma')^{-\ell_0}. \tag{9.1.51}$$

We shall look for the bound of $P(s,s',\sigma,\sigma')^{-\ell_0}$. We write $\mathrm{dist}((x,\theta),(s,\sigma)) = (|x-s|^2+|\theta-\sigma|^2)^{1/2}$. It is the distance between two points (x,θ) and (s,σ) in $\mathbf{R}^n\times\mathbf{R}^m$. Since the conditions $|\theta-\sigma|\le 1$, $|\theta'-\sigma'|\le 1$, $|x-s|\le 1$, $|x'-s'|\le 1$ imply that $\mathrm{dist}((x,\theta),(s,\sigma))\le\sqrt{2}$ and $\mathrm{dist}((x',\theta'),(s',\sigma'))\le\sqrt{2}$,

$$P(s,s',\theta,\theta')$$
$$\ge \inf_{\substack{\mathrm{dist}((x,\theta),(s,\sigma))\le\sqrt{2},\\ \mathrm{dist}((x',\theta'),(s',\sigma'))\le\sqrt{2}}} (1+2^{-1}\kappa_3\mathrm{dist}((x,\theta),(x',\theta'))^2)^{1/2}.$$

Suppose that $\mathrm{dist}((s,\sigma),(s',\sigma'))\ge 4$. Then the conditions $\mathrm{dist}((x,\theta),(s,\sigma))\le\sqrt{2}$ and $\mathrm{dist}((x',\theta'),(s',\sigma'))\le\sqrt{2}$ imply that

$$2^{-3/2}\mathrm{dist}((s,\sigma),(s',\sigma')) > \mathrm{dist}((x,\theta),(s,\sigma)),$$
$$2^{-3/2}\mathrm{dist}((s,\sigma),(s',\sigma')) > \mathrm{dist}((x,\theta),(s,\sigma)).$$

Therefore, we have

$$\begin{aligned}&\mathrm{dist}((x,\theta),(x',\theta'))\\&\ge \mathrm{dist}((s,\sigma),(s',\sigma'))-\mathrm{dist}((x,\theta),(s,\sigma))-\mathrm{dist}((x',\theta'),(s',\sigma'))\\&\ge (1-2^{-1/2})\mathrm{dist}((s,\sigma),(s',\sigma')).\end{aligned}$$

We set $c_0 = 2^{-1}(1-2^{-1/2})^2$. Then

$$\begin{aligned}&(1+2^{-1}\kappa_3\mathrm{dist}((x,\theta),(x',\theta'))^2)^{1/2}\\&\qquad\ge (1+c_0\kappa_3\mathrm{dist}((s,\sigma),(s',\sigma'))^2)^{1/2}.\end{aligned}\tag{9.1.52}$$

On the other hand in the case $\mathrm{dist}((s,\sigma),(s',\sigma'))<4$,

$$\begin{aligned}&(1+2^{-1}\kappa_3\mathrm{dist}((x,\theta),(x',\theta'))^2)^{1/2}\\&\ge (1+c_0\kappa_3 4^2)^{-1/2}(1+c_0\kappa_3\mathrm{dist}((s,\sigma),(s',\sigma'))^2)^{1/2}.\end{aligned}\tag{9.1.53}$$

It follows from this and (9.1.52) that for all (s,θ) and (s',σ')

$$P(s,s',\theta,\theta')\ge (1+\kappa_3 c_0 4^2)^{-1/2}(1+c_0\kappa_3\mathrm{dist}((s,\sigma),(s',\sigma'))^2)^{1/2}.\tag{9.1.54}$$

On the other hand

$$\omega(t-t')\le (1+c_0\kappa_3 4^2)^{\ell_0/2}(1+c_0\kappa_3|t-t'|^2)^{-\ell_0/2}.\tag{9.1.55}$$

Therefore,

$$\begin{aligned}&\omega(t-t')P(s,s',\sigma,\sigma')^{-\ell_0}\\&\le (1+c_0\kappa_3 4^2)^{\ell_0}(1+c_0\kappa_3|t-t'|^2)^{-\ell_0/2}(1+c_0\kappa_3\mathrm{dist}((s,\sigma),(s',\sigma'))^2)^{-\ell_0/2}\\&\le (1+c_0\kappa_3 4^2)^{\ell_0}(1+c_0\kappa_3(|t-t'|^2+\mathrm{dist}((s,\sigma),(s',\sigma'))^2))^{-\ell_0/2}.\end{aligned}\tag{9.1.56}$$

It follows from this and (9.1.62) that

$$\|A_pA_{p'}^*\|\le h(p,p')^2,\tag{9.1.57}$$

where

$$\begin{aligned}&h(p,p')\\&=(C(m,n)\kappa_5(\ell_0))^{1/2}\,\|a\|_{\mathscr{B}^\ell}(1+c_0\kappa_3 4^2)^{\ell_0/2}\\&\qquad\times\ (1+c_0\kappa_3(|t-t'|^2+\mathrm{dist}((s,\sigma),(s',\sigma'))^2)^{-\ell_0/4}.\end{aligned}\tag{9.1.58}$$

Next we discuss $\|A_p^*A_{p'}\|$. It follows from (9.1.40) that

$$|\tilde{k}_{pp'}(x,x')| \tag{9.1.59}$$
$$\le C(m,n)\kappa_5(\ell_0)\|a\|^2_{\mathscr{B}^\ell}\varphi(x-t)\varphi(x'-t')\iiint_{\substack{|\theta-\sigma|\le 1,|\theta'-\sigma'|\le 1,\\ |y-s|\le 1,|y-s'|\le 1,}} \tilde{\rho}^{-\ell_0}\,d\theta d\theta' dy.$$

Since by Lemma 9.1

$$\tilde{\rho}^{-\ell_0} \le (1+2^{-1}\kappa_3 \mathrm{dist}((x,\theta),(x',\theta'))^2)^{-\ell_0/2},$$

$$|\tilde{k}_{pp'}(x,x')| \tag{9.1.60}$$
$$\le C(m,n)\kappa_5(\ell_0)\|a\|^2_{\mathscr{B}^\ell}\omega(s-s')\varphi(x-t)\varphi(x'-t')P(t,t',\theta,\theta')^{-\ell_0},$$

where

$$P(t,t',\sigma,\sigma') = \inf_{\substack{|\theta-\sigma|\le 1,|\theta'-\sigma'|\le 1\\ |x-t|\le 1,|x'-t'|\le 1}} (1+2^{-1}\kappa_3(|x-x'|^2+|\theta-\theta'|^2)). \tag{9.1.61}$$

It turns out that the right-hand side of (9.1.60) is obtained by interchanging (s,s') and (t,t') on the right-hand side of (9.1.49). Since the function $h(p,p')$ is invariant under this interchange, we have

$$\|A_p^* A_{p'}\| \le h(p,p')^2. \tag{9.1.62}$$

Consequently, both (9.1.28) and (9.1.29) are proved.

Since $\ell_0/2 = m+2n+1$,

$$\iint_{\mathbf{R}^{n+m+n}} (1+c_0\kappa_3(|x|^2+|\theta|^2+|y|^2))^{-\ell_0/4}\,dxd\theta dy < \infty.$$

Therefore, there exists a constant $C>0$ depending on m, n, κ_0, κ_2, $\kappa(2,2m+4n+1)$ such that

$$\sup_p \iiint_{\mathbf{R}^{m+2n}} h(p,p')dp' < C\|a\|_{\mathscr{B}^{2m+4n+1}}$$
$$\sup_{p'} \iiint_{\mathbf{R}^{m+2n}} h(p,p')dp < C\|a\|_{\mathscr{B}^{2m+4n+1}}.$$

It follows from these and Young's inequality that (9.1.30) and (9.1.31) hold.

We have proved Lemma 9.2. Consequently, Theorem 9.1 has been proved up to the proof of Cotlar's lemma, which will be proved in the next section. □

9.2 Cotlar's Lemma

Here we present the proof of Cotlar's lemma, Lemma 9.3, following Calderon–Vaillancourt [12].

Proof Let

$$B_m \equiv \|T_{z_1}T^*_{z_2}\cdots T_{z_{2m-1}}T^*_{z_{2m}}\|.$$

Then the following two inequalities hold:

$$B_m \le \|T_{z_1}T^*_{z_2}\|\cdots\|T_{z_{2m-1}}T^*_{z_{2m}}\|$$
$$B_m \le \|T_{z_1}\|\|T^*_{z_2}T_{z_3}\|\cdots\|T^*_{z_{2m-2}}T_{z_{2m-1}}\|\|T^*_{z_{2m}}\|.$$

By taking the product of these, we have

$$B_m^2 \le \|T_{z_1}\|\|T_{z_1}T^*_{z_2}\|\|T^*_{z_2}T_{z_3}\|\cdots\|T_{z_{2m-1}}T^*_{z_{2m}}\|\|T^*_{z_{2m}}\|.$$

Taking the square root and using the assumption of the lemma, we obtain

$$B_m \le M_0 k(z_1, z_2)k(z_2, z_3)\cdots k(z_{2m-1}, z_{2m}).$$

Let N be a subset of Z with finite measure S and let $\chi_N(z)$ be its characteristic function. Then

$$S = \int_N \chi_N(z)d\mu(z).$$

Moreover, we have

$$\begin{aligned}
&\left\|\left[\left(\int_N T_z d\mu(z)\right)\left(\int_N T_z d\mu(z)\right)^*\right]^m\right\|^{1/m} \qquad (9.2.1)\\
&\le \left[\int\cdots\int_{z_i\in N}\|T_{z_1}T^*_{z_2}\cdots T^*_{z_{2m}}\|d\mu(z_1)d\mu(z_2)\cdots d\mu(z_{2m})\right]^{1/m}\\
&\le \Bigg|M_0\int\int_{z_1,z_{2m}\in N} d\mu(z_1)d\mu(z_{2m})\\
&\quad\times\int\cdots\int_{Z^{2m-2}} k(z_1,z_2)\cdots k(z_{2m-1},z_{2m})d\mu(z_2)\cdots, d\mu(z_{2m-1})\Bigg|^{1/m}\\
&= \left|M_0\int\int_{z_1,z_{2m}\in N} k^{(2m-1)}(z_1,z_{2m})d\mu(z_1)d\mu(z_{2m})\right|^{1/m}.
\end{aligned}$$

Here

$$k^{(2m-1)}(z, z_{2m})$$
$$= \int_{Z^{2m-1}} k(z_1, z_2)k(z_2, z_3)\cdots k(z_{2m-1}, z_{2m})d\mu(z_2)\cdots d\mu(z_{2m-1}).$$

This is the integral kernel of the bounded operator K^{2m-1} in $L^2(Z, d\mu)$. $\chi_N(z)$ is an element of $L^2(Z, d\mu)$, because $N \subset Z$ is a set of finite measure. Let (,) be the inner product in $L^2(Z, d\mu)$. Then the right-hand side of (9.2.1) is bounded by

$$\begin{aligned} \left|M_0(\chi_N, K^{2m-1}\chi_N)\right|^{\frac{1}{m}} &\leq (M_0\|\chi_N\|^2\|K^{2m-1}\|)^{\frac{1}{m}} \\ &\leq (M_0 S^2 M^{2m-1})^{\frac{1}{m}}. \end{aligned}$$

Tending $m \to \infty$, we obtain

$$\left\| \int_N T_z d\mu(z) \right\|^2 \leq M^2.$$

Since M is independent of N. Cotlar's lemma is proved. □

References

1. Albeverio, S., Boutet de Monvel-Berthier, A., Brezeźniak, Z.: Stationary phase method in infinite dimensions by finite dimensional approximations: applications to the Schrödinger equation. Potential Anal. **4**, 469–502 (1995)
2. Albeverio, S., Brzeźniak, Z., Haba, Z.: On the Schrödinger equation with potentials which are Laplace transform of measures. Potential Anal. **9**, 65–82 (1998)
3. Albeverio, S., Guatteri, S., Mazzucchi, S.: Phase space Feynman path integrals. J. Math. Phys. **43**, 2847–2857 (2002)
4. Albeverio, S., Hoegh-Krohn, R.J.: Oscillatory integrals and the method of stationary phase in infinitely many dimensions, with application to the classical limit of quantum mechanics. Invent. Math. **40**, 59–106 (1977)
5. Albeverio, S., Mazzucchi, S.: Feynman path integrals for polynomially growing potentials. J. Funct. Anal. **221**, 83–121 (2005)
6. Albeverio, S., Mazzucchi, S.: The time-dependent quartic oscillator—a Feynman path integral approach. J. Funct. Anal. **238**, 471–488 (2006)
7. Albeverio, S., Mazzucchi, S.: Infinite dimensional oscillatory integrals as projective systems of functionals. J. Math. Soc. Japan **67**, 1295–1316 (2015)
8. Albeverio, S.A., Hoegh-Krohn, R.J.: Mathematical theory of Feynman path integrals, 1st edn. Lecture Notes in Mathematics, vol. 523. Springer (1976)
9. Albeverio, S.A., Hoegh-Krohn, R.J., Mazzucchi, S.: Mathematical theory of Feynman path integrals, 2nd edn. Lecture Notes in Mathematics, vol. 523. Springer (2005)
10. Asada, K., Fujiwara, D.: On some oscillatory integral transformations in $L^2(\mathbf{R}^d)$. Jpn. J. Math. **4**, 299–361 (1978)
11. Birkhoff, G.D.: Quantum mechanics and asymptotic series. Bull. Am. Math. Soc. **39**, 681–700 (1933)
12. Caldelón, A.P., Vaillencourt, R.: On the boundedness of Pseudo-differential operators. J. Math. Soc. Jpn. **23**, 374–378 (1970)
13. Cameron, R.H.: A family of integrals serving to connect the Wiener and Feynman integrals. J. Math. Phys. (MIT) **39**, 126–140 (1960)
14. Cameron, R.H.: The Ilstow and Feynman integrals. J. d'Anal. Math. **10**, 287–361 (1962)
15. Cameron, R.H., Storvick, D.: A simple definition of the Feynman integral with application. Mem. AMS **288**. American Mathematical Society (1983)
16. Dirac, P.: The Principle of Quantum Mechanics. 3rd edn, Oxford (1947)
17. Duistermaat, J.J.: Oscillatory integrals, Lagrange immersions and unfolding of singularities. Comm. Pure Appl. Math. **27**, 207–281 (1974)
18. Duistermaat, J.J., Hörmander, L.: Fourier integral operators II. Acta Math. **128**, 183–269 (1972)
19. Dunford, N., Schwartz, J.T.: Linear operators, part 1: general theory, pure and applied mathematics, vol. 7. Interscience (1964)

D. Fujiwara, *Rigorous Time Slicing Approach to Feynman Path Integrals*, Mathematical Physics Studies, DOI 10.1007/978-4-431-56553-6

20. Elworthy, K.D., Truman, A.: Classical mechanics, the diffusion (heat) equations. Schrödinger equation on Riemannian manifolds. Preprint. Univ, Warwick (1979)
21. Feynman, R.P.: Space time approach to non relativistic quantum mechanics. Rev. Mod. Phys. **20**, 367–387 (1948)
22. Feynman, R.P., Hibbs, A.R.: Quantum Mechanics and Path Integrals. McGraw-Hill, New York (1965)
23. Fujiwara, D.: On the boundedness of integral transformations with highly oscillatory kernels. Proc. Jpn. Acad. **51**, 96–99 (1975)
24. Fujiwara, D.: A construction of the fundamental solution for the Schrödinger equations. J. d'Anal. Math. **35**, 41–96 (1979)
25. Fujiwara, D.: Remarks on convergence of the Feynman path integrals. Duke Math. Jour. **47**, 559–600 (1980)
26. Fujiwara, D.: The stationary phase method with an estimate of the remainder term on a space of large dimension. Nagoya Math. J. **124**, 61–97 (1991)
27. Fujiwara, D.: Some Feynman path integrals as oscillatory integrals over a Sobolev manifolds. In: Proceedings of the International Conference on Functional Analysis in memory of Prof. Kôsaku Yosida. Lecture Notes in Mathematics, vol. 1540, pp. 39–53. Springer (1993)
28. Fujiwara, D.: Mathematical Method for Feynman Path Integrals. Springer, Tokyo (1999)
29. Fujiwara, D.: An integration by parts formula for Feynman path integrals. J. Math. Soc. Jpn. **65**, 1273–1318 (2013)
30. Fujiwara, D., Kumano-go, N.: Smooth functional derivatives in Feynman path integrals by time slicing approximation. Bull. Sci. Math. **129**, 57–79 (2005)
31. Fujiwara, D., Kumano-go, N.: The second term of semi-classical asymptotic expansion for Feynman path integrals with integrand of polynomial growth. J. Math. Soc. Jpn. **58**, 837–867 (2006)
32. Fujiwara, D., Kumano-go, N.: Phase space Feynman path integrals via piecewise bicharacterisic paths and their semiclassical approximations. Bull. Sci. Math. **132**, 313–357 (2008)
33. Fujiwara, D., Kumano-go, N., Taniguchi, K.: A proof of Estimate of Kumano-go-Taniguchi Type for multiproduct of Fourier integral operators. Funk. Ekv. **40**, 459–470 (1997)
34. Fujiwara, D., Tsuchida, T.: The time slicing approximation of the fundamental solution for the Schrödinger equation with electromaganetic fields. J. Math. Soc. Jpn. **49**, 299–327 (1997)
35. Gelfand, I.M., Yaglom, A.M.: Integrals in functional spaces and its applications in quantum physics(Translated from Uspekhi Matematicheskikh Nauk, vol. 11 (1956)). J. Math. Phys. **1**, 48–69 (1960). Translated from Uspekhi Matematicheskikh Nauk, vol. 11 (1956)
36. Hardy, G.H., Littlewood, J.E., Pólya, G.: Inequalities, 2nd edn. Cambridge University Press, Cambridge (1967)
37. Hida, T.: Analysis of Brownian functionals. No. 13 in Carleton Math. Notes (1975)
38. Hida, T.: White noise approach to Feynman integrals. J. Korean Math. Soc. **38**, 275–281 (2001)
39. Hörmander, L.: Fourier integral operators I. Acta Math. **127**, 79–183 (1971)
40. Ichinose, T.: Path integral for a hyperbolic system of the first order. Duke Math. J. **51**, 1–36 (1984)
41. Ichinose, W.: A mathematical theory of the phase space Feynman path integral of the functional. Comm. Math. Phys., 739–779 (2006)
42. Ito, K.: Generalized uniform complex measure in Hilbert space and its application to the Feynman path integrals. In: Proceedings of the 5th Berkeley symposium on Mathematical Statistics and Probability part 1, vol. 2, pp. 145–161. University of California Press, Berkeley (1967)
43. Kato, T.: Perturbation theory for linear operators: Die Grundlehlen der mahemtischen Wissenschaften in Einzeldarstellungen, vol. 132. Springer (1966)
44. Kitada, H., Kumano-go, H.: A family of Fourier integral operators and the fundamental solution for a Schrödinger equation. Osaka J. Math. **18**, 291–360 (1981)
45. Kumano-go, H., Taniguchi, K.: Fourier integral operators of multiphase and the fundamental solution for a hyperbolic system. Funk. Ekv. **22**, 161–196 (1979)

46. Kumano-go, N.: Feynman math integrals as analysis on path space by time slicing approximation. Bull. Sci. Math. **128**, 197–251 (2004)
47. Kumano-go, N.: Phase space Feynman path integrals with smooth functional derivatives by time slicing approximation. Bull. Sci. Math. **135**, 936–987 (2011)
48. Kumano-go, N., Chiba, Y. (eds.): Introductory Workshop on Path Integrals and Pseudo-Differential Operators, *RIMS kôkyûroku*, vol. 1958. Research Institute for Mathematical Sciences, Kyoto Univ., Kyoto, Japan (2015)
49. Maslov, V.P.: Theórie des perturbations et méthodes asymptotiques (French translation). Dunod (1970)
50. Mazzucchi, S.: Mathematical Feynman Path Integrals and Their Application. World Scientific (2009)
51. Morette, C.: On the definition and approximation of Feynman path integrals. Phys. Rev. **81**, 848–852 (1951)
52. Morse, M.: The calculus of variations in the large, AMS Colloquium Publications, vol. 18. American Math, Soc. AMS Colloquium Publications (1934)
53. Nelson, E.: Feynman integrals and Schrödinger equation. J. Math. Phys. **5**, 332–343 (1964)
54. Pauli, W.: Selected topic in field quantization. The Pauli Lectures on Physics. MIT press (1977)
55. Reed, M., Simon, B.: Methods of Modern Mathematical Physics: 1 Functional Analysis. Academic Press, New York (1972)
56. Riesz, F., Nagy, B.S.: Functional Analysis. Dover (1990)
57. Schulman, L.S.: Techniques and Applications of Path Integration. Wiley (1981)
58. Schwartz, J.T.: Nonlinear Functional Analysis. Gordon and Breach Science Publishers (1969)
59. Simon, B.: Trace ideals and their applications. London Mathematical Society Lecture Note Series, vol. 35. Cambridge University Press (1979)
60. Streit, L., Hida, T.: Generalized Brownian functionals and the Feynman integrals. Stoch. Process. Appl. **16**, 55–69 (1983)
61. Truman, A.: Classical mechanics, the diffusion (heat) equation and Schrödinger equation. J. Math. Phys. **18**, 2308–2315 (1977)
62. Tsuchida, T.: Remarks on Fujiwara's stationary phase method on a space of large dimension with a phase function involving electromagnetic fields. Nagoya Math. J. **136**, 157–189 (1994)
63. Van Vleck, J.H.: The correspondence principle in the statistical interpretation of quantum mechanics. Proc. Nat. Acad. Sci. USA **14**, 178–188 (1928)
64. Weyl, H.: The Theory of Groups and Quantum Mechanics, (English translation). Dover (1950)
65. Whittaker, E.T.: A treatise on the analytical dynamics of particles and rigid bodies; with an introduction to the problem of three bodies, paperback edition of 4th edn (1937). Cambridge University Press (1959)
66. Yajima, K.: The quasi-classical limit of quantum scattering theory. Comm. Math. Phys. **69**, 101–129 (1979)
67. Yajima, K.: Schrödinger evolution equations with magnetic fields. J. d'Anal. Math. **56**, 29–76 (1991)
68. Yajima, K.: Smoothness and non-smoothness of the fundamental solution of time dependent Schrödinger equations. Comm. Math. Phys. **181**, 605–629 (1996)

Index

D. Fujiwara, *Rigorous Time Slicing Approach to Feynman Path Integrals*,
Mathematical Physics Studies, DOI 10.1007/978-4-431-56553-6

MIX
Papier aus verantwortungsvollen Quellen
Paper from responsible sources
FSC® C105338

If you have any concerns about our products,
you can contact us on
ProductSafety@springernature.com

In case Publisher is established outside the EU,
the EU authorized representative is:
Springer Nature Customer Service Center GmbH
Europaplatz 3, 69115 Heidelberg, Germany

Printed by Libri Plureos GmbH
in Hamburg, Germany